Sulfur-Centered Reactive Intermediates in Chemistry and Biology

NATO ASI Series

Advanced Science Institutes Series

A series presenting the results of activities sponsored by the NATO Science Committee, which aims at the dissemination of advanced scientific and technological knowledge, with a view to strengthening links between scientific communities.

The series is published by an international board of publishers in conjunction with the NATO Scientific Affairs Division

A	Life Sciences	Plenum Publishing Corporation
B	Physics	New York and London
C	Mathematical and Physical Sciences	Kluwer Academic Publishers Dordrecht, Boston, and London
D	Behavioral and Social Sciences	
E	Applied Sciences	
F	Computer and Systems Sciences	Springer-Verlag
G	Ecological Sciences	Berlin, Heidelberg, New York, London,
H	Cell Biology	Paris, and Tokyo

Recent Volumes in this Series

Series A: Life Sciences

Sulfur-Centered Reactive Intermediates in Chemistry and Biology

Edited by

Chryssostomos Chatgilialoglu

Consiglio Nazionale delle Richerche
Ozzano Emilia (Bologna), Italy

and

Klaus-Dieter Asmus

Hahn-Meitner-Institut
Berlin, Federal Republic of Germany

Plenum Press
New York and London
Published in cooperation with NATO Scientific Affairs Division

Proceedings of a NATO Advanced Study Institute on
Sulfur-Centered Reactive Intermediates in Chemistry and Biology,
held June 18–30, 1989,
in Maratea, Italy

Library of Congress Cataloging-in-Publication Data

NATO Advanced Study Institute on Sulfur-Centered Reactive
    Intermediates in Chemistry and Biology (1989 : Maratea, Italy)
      Sulfur-centered reactive intermediates in chemistry and biology
    edited by Chryssostomos Chatgilialoglu and Klaus-Dieter Asmus.
         p.   cm. -- (NATO ASI series. Series A, Life sciences ; v.
    197)
      "Proceedings of a NATO Advanced Study Institute on Sulfur-Centered
    Reactive Intermediates in Chemistry and Biology, held June 18-30,
    1989, in Maratea, Italy"--Verso t.p.
      "Published in cooperation with NATO Scientific Affairs Division."
      Includes bibliographical references and index.
      ISBN 0-306-43723-6
      1. Organosulphur compounds--Congresses.   I. Chatgilialoglu,
    Chryssostomos.   II. Asmus, Klaus-Dieter.   III. North Atlantic Treaty
    Organization.   Scientific Affairs Division.   IV. Title.   V. Series.
    QD412.S1N37   1989
    547'.06--dc20   90-14295
                                                                        CIP

© 1990 Plenum Press, New York
A Division of Plenum Publishing Corporation
233 Spring Street, New York, N.Y. 10013

Printed in the United States of America

FOREWORD

A wonderfully successful NATO Advanced Study Institute on "Sulfur-Centered Reactive Intermediates in Chemistry and Biology" was held 18-30 June, 1989, at the Hotel Villa del Mare in Maratea, Italy. Despite the beautiful setting with mountains behind us and over-looking the clear blue Mediterranean Sea under a cloudless sky (and with a private beach available), the lectures were extremely well attended. While some credit can go to the seriousness of the students, more must go to the calibre of speakers and the high quality of their presentations. The Director, Dr. C. Chatgilialoglu, and Co-Director, Professor K.-D. Asmus, are to be congratulated for putting together such an outstanding scientific program. Dr. Chatgilialoglu is also to be commended for arranging an equally stimulating social pro-gram which included bus, train and boat trips to many local sites of interest.

It was particularly fitting that a meeting on the chemistry and biochemistry of sulfur should be held in Italy since Italian chemists have made major contributions to our under-standing of the organic chemistry of sulfur, including the chemistry of its reactive inter-mediates. The early Italian interest in sulfur chemistry arose from the fact that Italy, or more specifically, Sicily, was a major world producer of sulfur prior to the development and exploitation of the Frasch process in Texas and Louisiana. More recently, under stimulating guidance of the late, great Italian chemist, Professor Angelo Mangini of Bologna, studies on the organic chemistry of sulfur and, particularly, on sulfur-centered reactive intermediates have received new impetus and have flourished in Italy. Several of Professor Mangini's scientific "children" attended the meeting and so did some of their scientific children (Mangini's "grandchildren"). I know that he would have approved of this meeting and believe that he was with us in spirit throughout the two weeks.

The speakers and the students were mainly chemists. However, their scientific back-grounds and research interests varied enormously - theoretical chemistry, gas phase kinetics, thermodynamics and so on, all the way to the mechanisms of enzyme catalyzed reactions. This interdisciplinary approach, which covered the entire field of sulfur chemistry, is a credit to the organizers of the meeting and their judgement in choosing speakers. Certainly, it made the meeting especially valuable to all attendees. After each talk, one had learned something new, interesting and, quite frequently, of direct relevance to one's own research program. Because of the interdisciplinary nature of the meeting audience participation was exceptionally strong. Long and vigorous discussion followed each presentation and, indeed, often continued long into the night!

For all scientists interested in the chemistry and biochemistry of sulfur this book provides a valuable and permanent record of NATO's 1989 ASI on sulfur. Please read and enjoy the 39 papers presented by 26 scientists from 7 NATO countries and Japan.

5 July 1989

K. U. Ingold
Ottawa, Canada

Sulfur-Centered Reactive Intermediates in Chemistry and Biology, Edited by
C. Chatgilialoglu and K.-D. Asmus, Plenum Press, New York, 1990

PREFACE

This book contains the main lectures given at the NATO Advanced Studies Institute on "Sulfur-Centered Reactive Intermediates in Chemistry and Biology" held in Acquafredda di Maratea (Italy) June 18-30, 1989. The first chapters consider theoretical aspects and give a survey of general thermodynamic properties of sulfur functionalities in molecules. A second larger group of articles covers the variety of experimental techniques which have most successfully been applied for the investigation of sulfur-centered reactive intermediates. The generation and the properties of these species, particularly radical and non-radical cations, are extensively described in the following papers. Many interesting and important studies in this field could not have been performed without especially tailored molecules. Consequently, their synthesis is a subject covered in this book. Equally interesting will be the other synthesis papers describing the application of sulfur-organic compounds and sulfur-centered radicals for the preparation of new compounds and the understanding of reaction mechanisms. The final, large section highlights the role of the sulfur compounds and sulfur-centered radical species in biochemistry and biology. The reader will realize how many of the interpretations and conclusions in this "life science" oriented subject have benefited from the fundamental knowledge described in the first part of the book.

While we could not cover all possible topics in the still expanding subject of sulfur-centered reactive intermediates, an effort has been made to provide the state-of-the-art on, at least, some key aspects in this field. We believe this book will be useful for the scientific community as a reference work as well as an introduction to the various fields presented in the individual chapters by authors who are top experts in their respective disciplines.

The meeting received a great impetus from the fact that scientists got together who would not necessarily attend the same conferences, but could now discover the full value of broad and interdisciplinary discussions. This, and also common interests in the many cultural and social activities offered by our Italian hosts (including the appreciation of surprising talents of young as well as established scientists in the game of football/soccer), were the basis for the high spirit of this, as we feel, most successful summer school.

Our special thanks are due to NATO, Consiglio Nazionale delle Ricerche, Progetto Finalizzato "Chimica Fine II", Università di Basilicata, Hahn-Meitner-Institut Berlin, Regione di Basilicata, Azienda Autonoma Soggiorno e Turismo di Maratea, Glaxo S.p.A. and Farmitalia Carlo Erba who kindly provided the financial support.

Finally, a word on the preparation of this book. Despite all the work which had to be invested into collecting, reading, editing, typing, and correcting the contributions, it has been an interesting and scientifically rewarding task. If completion took a bit longer than originally anticipated our only excuse is that we, a small editing crew, (and all official guide- and deadlines) totally underestimated the real amount of work coming up. Fortunately we could count on the professional and always interested help of Mrs. Kim Kube for the typing and Dr. Wolfgang Hoyer for the computer drawn chemical structures. They and all those who in one way or the other contributed to the finishing of this book deserve our sincere thanks.

K.-D. Asmus	C. Chatgilialoglu
Berlin, F. R. Germany	Bologna, Italy

Sulfur-Centered Reactive Intermediates in Chemistry and Biology, Edited by
C. Chatgilialoglu and K.-D. Asmus, Plenum Press, New York, 1990

vii

CONTENTS

Sulfur-Centered Reactive Intermediates in Chemistry and Biology, Edited by
C. Chatgilialoglu and K.-D. Asmus, Plenum Press, New York, 1990

(

FORCE-FIELD AND MOLECULAR ORBITAL CALCULATIONS

IN ORGANOSULFUR CHEMISTRY

Timothy Clark

Institut für Organische Chemie der
Friedrich-Alexander-Universität Erlangen-Nürnberg
Henkestraße 42, 8520 Erlangen, Fed. Rep. Germany

Chemical structure and energy calculations[1] are unique among the available research tools in that they do not require the molecule in question to have been made or isolated - or even that it is capable of existence. Calculations are capable of delivering structures, energies and electronic properties such as dipole moments, charge distributions or ionization potentials with reasonable accuracy at a fraction of the cost (both financial and in time and effort) of comparable experimental studies. Two main types of structure and energy calculation are available: force-field (molecular-mechanics) and molecular orbital (MO) calculations. The latter can be subdivided into the semiempirical (MINDO/3, MNDO, AM1, PM3 etc.) and *ab initio* techniques. The object of this article is to provide an overview of the cost (in computer time), applicability and accuracy of the various methods.

COMPUTER REQUIREMENTS

Computer requirements for the different methods differ enormously. Simple molecular mechanics optimizations on reasonably sized molecules can be done easily on a PC. However, large molecular dynamics simulations, which use the same basic model as molecular mechanics, may require a supercomputer. Semiempirical MO calculations can be done on PC's for small molecules, but require a workstation or supermini for large (ca. 150 atom) molecules. *Ab initio* can be performed very effectively on the better workstations, but may require supermini or supercomputer performance for very large or very high level calculations. Scheme 1 provides a rough overview of the situation.

Reference 1 gives a table (p. 3) of the computer times needed to optimize the geometry of propane using the various theoretical techniques.[1] These vary from 0.8 seconds for molecular mechanics to almost 5,000 seconds for an *ab initio* 3-31G* calculation. This is by no means the highest practicable *ab initio* calculation for a molecule of this size, so that a factor of 10,000 between computer time requirements for molecular mechanics and high level *ab initio* is not unreasonable. As a rough guideline, the computer time ratio *force-field : semiempirical MO : ab initio* is about 1 : 10 - 100 : 500 - 10,000. The last number could equally well be infinity because the level at which *ab initio* calculations can be performed is open-ended. These large differences mean that an understanding of the characteristics of each method is essential for the effective use of computer facilities.

Sulfur-Centered Reactive Intermediates in Chemistry and Biology, Edited by
C. Chatgilialoglu and K.-D. Asmus, Plenum Press, New York, 1990

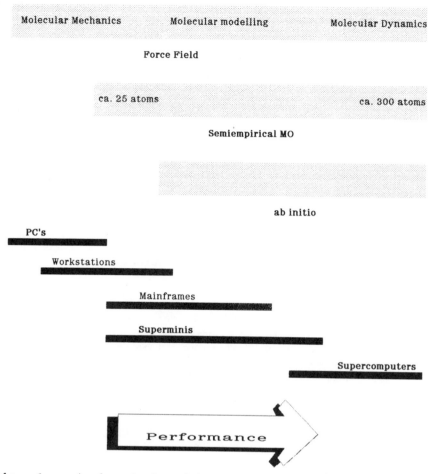

| Molecular Mechanics | Molecular modelling | Molecular Dynamics |

Force Field

ca. 25 atoms ca. 300 atoms

Semiempirical MO

ab initio

PC's

Workstations

Mainframes

Superminis

Supercomputers

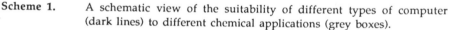

Performance

Scheme 1. A schematic view of the suitability of different types of computer (dark lines) to different chemical applications (grey boxes).

FORCE-FIELD METHODS

Force-field, or molecular mechanics, methods[1,2] are based on a simple mechanical model of molecules. Bonds between atoms are considered to behave like springs that obey a modified version of Hooke's law. Similarly, bond angles have preferred values from which they can be deformed according to a potential that depends on the square of the deformation plus higher terms. Torsional potentials (1-, 2- and 3-fold) are used for each pair of vicinal bonds and compound terms that depend both on the bond lengths to a given atom and to the angles between the two bonds (stretch-bend terms) are also used. Steric interactions are taken into account by considering van der Waals interactions between individual atoms. Van der Waals interactions between geminal atoms are not usually considered because they are accounted for by the angle-bending and stretch-bend terms. The sum of these potentials is known as the force-field. Force-fields are parametrized by varying the preferred lengths and angles and the force constants governing the individual potentials until as good a fit as possible to experimental values (usually heats of formation and geometries) is obtained. Because there are any number of different force-fields that can give a good fit to a given set of experimental data, the individual energies (stretching, torsional, van der Waals etc.) do not have any physical meaning and should not be used to interpret experimental findings.

Because force-field methods are parametrized to fit experiment, the quality of the results depends heavily on the quality and quantity of the parametrization data-set. Thus, the most accurate force-fields are available for the alkanes, for which a large body of experimental data exists. Molecular mechanics for alkane heats of formation and structures can be of experimental quality or better. Structural elements for which less data are available, such as sulfur-containing groups, do not give results of such high accuracy, but are nevertheless well treated. The two great weaknesses of force-field calculations are their inability to deal with unusual electronic effects, which are not included in the theoretical model, and the fact that they are limited to stable equilibrium geometries because the parametrization data-sets can only include such species. Clearly, they cannot deal with reactions in which bonds are made and broken without an extra parametrization for this process, but they may also give poor results for the activation energies of processes such as conformational interconversions, in which no bonds are made or broken.

Force-field programs for simple molecular mechanics optimizations range from Allinger's MM287 (available from the Quantum Chemistry Program Exchange, QCPE) to the commercially available "molecular modelling" packages that provide color graphic input and output facilities. Care should be taken with such packages that the force-field being used really is what it is supposed to be. This can be checked by comparing results for some test molecules with the original literature.

SEMIEMPIRICAL MOLECULAR ORBITAL METHODS

Although there are many different semiempirical MO-methods, the best known and most widely used are those developed by Michael Dewar and his school. MINDO/3, the first such method to gain general acceptance, is now outdated and has been replaced by MNDO,[3] AM1[4] and PM3.[5] These are all NDDO-based methods and the latter two (which differ only in the parameter sets used) were developed to eliminate some of the known weaknesses of MNDO. Two programs that incorporate these methods, MOPAC and AMPAC, are available from QCPE. Vectorized versions that provide very high performance on superminis and supercomputers are available for some machines. Scheme 2 shows the elements for which MNDO, AM1 and PM3 have been parametrized.

There are several important points to consider when calculating sulfur compounds – especially with MNDO. The first is that there are two sets of MNDO parameters for sulfur and silicon. The original parametrizations for these elements were not as successful as had been hoped and were later repeated and new parameter sets introduced. Care should be taken when using older programs that the correct parameter set is installed. MNDO-calculations should not be used for sulfones and sulfoxides or other "hypervalent" sulfur compounds, for which they give very large errors. This was originally thought to be due to the fact that MNDO uses only s- and p-orbitals and that d-orbitals are necessary in order to describe sulfones and sulfoxides correctly. However, Stewart included hypervalent compounds in the parametrization set for PM3 and was able to obtain satisfactory results for sulfones and sulfoxides. Dewar's AM1-parameters for sulfur are also able to treat sulfones and sulfoxides adequately.

Care should be taken when using MNDO because it does not reproduce hydrogen bonds. For applications where this is important, AM1 or PM3 should be used. These two methods are comparable in performance, although PM3 is probably more widely applicable and AM1 has the edge if rotation barriers in conjugated systems, such as peptides, are important.

Generally, semiempirical MO-methods offer a good compromise between efficiency and accuracy for questions that cannot be answered by force-field calculations. This includes all problems in which electronic effects are important as well as reactivity problems. Molecular mechanics is probably more useful if purely conformational problems are of interest.

3

MNDO-Parameters:

H						
Li	Be	B	C	N	O	F
	Mg	Al	Si	P	S	Cl
	Zn		Ge			Br
			Sn			I
	Hg		Pb			

AM1-Parameters:

H					
		B	C	N	O	F
	Al	Si	P	S	Cl
	Zn		Ge		Br
					I
	Hg				

PM3-Parameters:

H				
	C	N	O	F
Si	P	S	Cl
			Br
			I

Scheme 2.   Elements for which MNDO, AM1 and PM3 parameters are available (Nov. 1989). Lithium and magnesium were parametrized for MNDO by groups other than that of Prof. Dewar.

Semiempirical calculations can now be done for many experimental compounds and are being used for increasingly larger systems, such as enzyme models, as computer hardware becomes more sophisticated. A RISC-workstation dedicated to semiempirical calculations is a very powerful research tool indeed.

AB INITIO CALCULATIONS

Ab initio molecular orbital theory[6] is the most computationally demanding of the three techniques covered here. That does not always mean that it is the most accurate. *Ab initio* calculations that are too large to allow the use of a reasonable basis set or that cannot be fully optimized are likely to be less reliable than (far cheaper) AM1 or PM3 calculations on the same system. If, however, the molecules being studied are small enough for a full *ab initio* investigation at an adequate level of theory, the best results can almost always be expected from the *ab initio* calculations.

One major difference between *ab initio* and semiempirical techniques is that the level at which the calculation is performed can be varied using *ab initio* theory. There are two major ways to vary the level of the calculation *via* the basis set and the level at which electron correlation is treated. The smallest basis sets (the atomic orbitals that are combined to give the molecular orbitals) are known as minimal bases because they contain only enough orbitals per atom to accomodate the electrons of the neutral atom and to maintain spherical symmetry. Thus, a minimal basis for carbon consists of $1s$, $2s$, $2p_x$ $2p_y$ and $2p_z$ orbitals, and for hydrogen only $1s$. STO-3G is probably the best known minimal basis set. The acronym means Slater Type Orbitals simulated by 3 Gaussian functions. Gaussian functions are used rather than Slater orbitals because integrals between Gaussian functions are easier to calculate than those between Slater functions. Minimal basis sets are, however, poorly suited for sulfur calculations. Split-valence bases are more sopisticated. Their valence orbitals are split into compact (inner) and diffuse (outer) components. Linear combinations of inner and outer components give a range of orbital size in the MO's and thus provide more flexibility than a minimal basis. The most widely used split-valence basis is probably 3-21G (3 Gaussians for the core orbitals, 2 for the inner and 1 for the outer valence orbitals), which was developed for fast geometry optimizations. A variation of 3-21G is 3-21G(*) (sometimes written as 3-21G*), which has an extra set of d-functions for second row elements. A more satisfactory, although more expensive, solution is to use d-functions on all non-hydrogen atoms in a polarization basis set such as 6-31G*. In most molecules, the role of the d-functions is to polarize the p-orbitals, rather than to act as valence orbitals. This effect is, however, very important. Geometry optimizations at 6-31G* reliably predict structures and relative energies of sulfur compounds.

Normally, geometry optimizations are performed at the self-consistent field (SCF), or Hartree-Fock (HF) level. In this approximation, the electron-electron repulsion is calculated by considering the repulsion between a given electron and the mean field of all the others. Because, however, the individual electrons tend to avoid each other, this mean field approximation calculates a repulsion energy that is too high. Additional corrections must be applied to take the correlated motions of the electrons into account. This electron correlation is often not important for geometry optimizations, which are usually carried out at the SCF-level, but has important energetic consequences. Therefore, energy calculations are often done on the SCF-geometries with some sort of correlation correction. This may be by means of configuration interaction (CI), in which many electronic configurations are allowed to mix, or by perturbation theory. The best known perturbational techniques are the Møller-Plesset methods. Møller-Plesset corrections can be calculated to second, third or fourth order (MP2, MP3 and MP4, respectively). Fourth order corrections may include only single and double excitations (MP4sd), singles, doubles and quadruples (MP4sdq) or singles, doubles, triples and quadruples (MP4sdtq). Although energies are usually quoted at the last level, MP4sdq is the default level for the Gaussian programs, so that the exact MP4-level of a given calculation should be checked carefully. Hehre, Radom, Schleyer and Pople[6] give a useful overview of *ab initio* techniques.

Ab initio programs are more complex than their semiempirical counterparts, and therefore often more expensive. The best known are the Gaussian series of programs from Gaussian Inc. The most recent version is Gaussian 88, but Gaussian 86 is still the most modern version available for some types of machines. A very useful (and much cheaper) alternative to the Gaussian programs is Monstergauss from the University of Toronto. Although not as versatile as Gaussian 88, Monstergauss is far less machine-dependent and can easily be adapted to most computers. For vector computers (superminis and supercomputers), Cadpac is usually much faster than Gaussian 88, especially for MP2 optimizations and frequency-calculations and for molecules of high symmetry. Cadpac is available from the University of Cambridge. Older *ab initio* programs, such as HONDO or Gamess, are still in use for specialized applications.

SUMMARY

The three types of calculations described here provide a wide range of useful applications in sulfur chemistry. Force-field methods are cheap and well suited to conformational problems or those in which the relative stabilities of a series of stable molecules are of interest. Semiempirical MO-calculations are less accurate but far more flexible and can now be applied routinely to experimentally accessible molecules including sulfoxides and sulfones. *Ab initio* calculations at adequate levels are well suited to accurate studies on small molecules and for basic research into the electron properties of sulfur-containing moieties.

REFERENCES

1. T. Clark, "A Handbook of Computational Chemistry", Wiley, New York (1985).
2. U Burkert and N.L. Allinger, "Molecular Mechanics", ACS Monograph 177, American Chemical Society, Washington, DC (1982).
3. M.J.S. Dewar and W. Thiel, *J. Am. Chem. Soc.* 99:4899, 4907 (1977).
4. M.J.S. Dewar, E.G. Zoebisch, E.F. Healy, and J.J.P. Stewart, *J. Am. Chem. Soc.* 107:3902 (1985).
5. J.J.P. Stewart, *J. Comput. Chem.* 10:209, 221 (1989).
6. W.J. Hehre, L. Radom, P.v.R. Schleyer, and J.A. Pople, "Ab Initio Molecular Orbital Theory", Wiley, New York (1986).

ELECTRONIC TRANSITIONS IN SULFUR-CENTERED RADICALS

BY MEANS OF MSXα CALCULATIONS

Maurizio Guerra

I. Co. C. E. A.
Consiglio Nazionale delle Ricerche
40064 Ozzano Emilia (Bologna), Italy

Highly reactive sulfur-centered radicals, which play an important role in air pollution and in biological systems, are usually idientified by their UV/visible absorption and/or ESR spectra. Spectral information are sometimes insufficient for an unequivocal characterization of the transient species. Their identification could be achieved by comparing the optical absorption spectra with the energy and intensity of electronic transitions computed using quantum mechanical methods. The MSXα method[1] has proved to be a powerful tool for assigning optical transitions in radicals,[2] and is used to assign the spectral bands of transient sulfur-centered radical species to specific electronic transitions.[3]

CALCULATIONS OF THE ELECTRONIC TRANSITIONS IN RADICALS BY MEANS OF THE MSXα METHOD

The MSXα Method

The MSXα method is a "first principle" and basis set independent method based upon an approximate potential. The statistical treatment of the exchange interaction reduces the N-electron Schrödinger equation to a set of one-electron differential equations. (in atomic units)

$$[-1/2 \, \nabla^2 + V_C(1) + V_{xc}(1)] \, \Psi(1) \;=\; \varepsilon \Psi(1) \tag{1}$$

The first two terms represent the kinetic and Coulombic potentials, respectively, and V_{xc} is a local potential which takes into account both of the exchange potential and electron correlation. The simplest and most widely utilized exchange-correlation potential is related to the local electronic charge density, ρ, and a scaling factor α:

$$V_{xc} \;=\; -6\alpha \, [3\rho(r)/8\pi]^{1/3} \tag{2}$$

The multiple scattering procedure allows an effective solution of the monoelectronic equations. The coordinate space of the molecule is partitioned into three regions, as shown for the methyl radical in Figure 1. In the atomic regions (I) which are inside the spheres centered on the atoms and in the extra-molecular region (III), which is outside the outer-sphere surrounding the entire molecule, the potential is assumed to be spherically symmetric and the electronic wavefunction is expanded in real spherical harmonics. In the intersphere

Sulfur-Centered Reactive Intermediates in Chemistry and Biology, Edited by
C. Chatgilialoglu and K.-D. Asmus, Plenum Press, New York, 1990

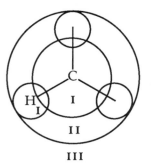

Fig. 1. Division of space in the MXα method into atomic (I), intersphere (II), and extramolecular (III) regions for the methyl radical.

region (II), which is the region between the atomic spheres and the outer-sphere surrounding the entire molecule, the potential is taken to be a constant, determined by the volume average of the potential, and a multicenter expansion of the wavefunction is used. The expansion coefficients are determined by solving a set of homogeneous linear equations under the condition that the wavefunction and its first derivative are continuous at the sphere boundaries.

The Transition State Concept for Computing Electronic Transitions

In the Xα theory the orbital eigenvalues ε_i are related to the Xα statistical total energy, $<E_{X\alpha}>$, as the first derivatives of the total energy with respect to the orbital occupation number n_i:

$$\varepsilon_i = \partial<E_{X\alpha}>/\partial n_i \tag{3}$$

The eigenvalues ε_i are then equal to the slope of the total energy function rather than the difference between two values of the total energy as in Hartree-Fock theory (Koopmans' theorem).[4] Consequently the energy variation occurring in an electronic transition can be obtained by performing calculations on a state (transition state) where the occupation number of the orbitals are halfway between those of the initial and final electronic states.[5] For an excitation from the i-*th* to the j-*th* orbital

$$\Delta E_{i \rightarrow j} = (\varepsilon_j - \varepsilon_i)_{ts} + \text{third order terms} \tag{4}$$

Efficiency

The MSXα approach has a number of advantages over the more traditional LCAO-SCF method for assigning optical absorption spectra in radicals. The radial flexibility of the wavefunction which can efficiently account for both single center and multicenter charge distribution, allows both the valence and Rydberg states to be treated at the same level of accuracy. Some of the pitfalls of using an atom-centered LCAO expansion for describing Rydberg states are thus avoided. The existence of an excited state depends on the potential and not on the basis set. Imaginary excited states which are computed in LCAO approaches with split-valence basis sets do not occur. The transition state method, used to evaluate the transition energies, takes into account the electron relaxation occurring during the excitation as in the ΔSCF approach. The transition state procedure has, however, two advantages over the more traditional ΔSCF approach. Transition energies are evaluated as the difference of eigenvalues and not as the difference of the total energy of separate calculations. Errors in numerical procedure are thus avoided. The variational collaps of higher excited states on the lowest excited state of the same symmetry which complicates calculations of the corresponding transition energies does not occur. Obviously the best theoretical tool for assigning the absorption spectra is to perform large scale CI calculations, however, the MSXα method has the advantage of requiring limited computer resources so that large polyatomic systems can be easily investigated. Furthermore, higher atomic number elements can be treated by including scalar relativistic corrections. For example, electronic transition energies were computed for systems as large as triphenyl-silyl, -germyl and -stannyl radicals. For the latter radical scalar relativistic calculations were performed.[6]

Reliability

The best procedure for computing optical transitions in radicals with the MSXα method was established by performing different types of MSXα calculations on alkyl and $H_3M^\bullet$ (M = Si, Ge) radicals. In highly symmetric radicals $R_3X^\bullet$, vertical transition energies to Rydberg orbitals were reproduced with an accuracy comparable to that obtained with CI calculations employing large basis sets. The deviation from experiment is about 2000 cm^{-1} both for CI and the MSXα method. The valence transitions have not been experimentally determined for these radicals, but the difference between MSXα and CI valence transition energies is less than 1000 cm^{-1}. In asymmetric radicals valence transitions are described with the same accuracy as found with highly symmetric radicals, whereas the Rydberg transitions are described

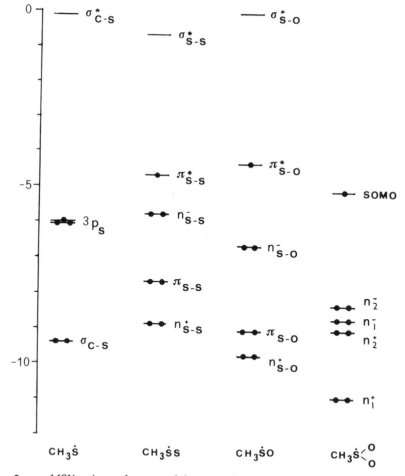

Fig. 2. MSXα eigenvalues, ε_i, of the ground state for sulfur-centered radicals.

with less accuracy. For exmaple, in the ethyl radical Rydberg transition energies are underestimated by about 7000 cm^{-1}. The approximation of the spherical potential in the outer-sphere region and the great increase of the intersphere volume with respect to the atomic volume in asymmetric radicals could be responsible for the larger deviation from experiment for Rydberg transitions. MSXα calculations of partitioned clusters[7] should improve the agreement with experiment in branched radicals.

ELECTRONIC TRANSITIONS IN SULFUR-CENTERED RADICALS

In this section we shall briefly survey the assignment of absorption spectra of sulfur-centered radicals. Comparison with oxy-analogues will also be reported. The eigenvalues ε_i computed with MSXα calculations for sulfur-centered radicals are displayed in Figure 2 as a guide for comparing electronic transition energies.

Thiyl

The electronic ground state of $CH_3S^\bullet$ has an orbitally degnerate ^2E symmetry, where the three outermost electrons occupy the degenerate 3p orbitals on the sulfur atom. The low-lying valence transition occurs from the σ_{CS} MO (a_1 symmetry) to the degenerate 3p sulfur orbitals (e symmetry). The frequency (28800 cm^{-1}) computed with the MSXα method is close to that of the weak band observed in the absorption spectra of $RS^\bullet$ radicals[8] occurring at about 30000 cm^{-1} and is, as expected, blue-shifted with respect to that of the band (26500 cm^{-1}) observed in gas phase for the laser induced fluorescence spectrum obtained from photolysis of CH_3SCH_3 and attributed to the thiyl radical.[9] Calculations also reproduced the blue shift (6000 cm^{-1}) caused by substitution of sulfur by the more electronegative oxygen atom. The most significant difference on going from the thiyl to the oxyl radical is the absence of the valence transition to the σ^*_{CO} MO in the far UV region. The corresponding transition for the thiyl radical was computed to be at lower frequency (50000 cm^{-1}) than the first member of the ns Rydberg series. This agrees with the observation that vacant carbon-heteroatom antibonding MOs were characterized, at low energy, by means of Electron Transmission Spectroscpy (ETS measures the energy at which slow electrons are temporarely trapped by a molecular target) and MSXα calculations[10] for sulfur but not oxygen derivatives. The participation of sulfur d orbitals is significant and contributes, together with other factors, towards the stabilization of this MO.

Perthiyl

The outermost occupied MOs of $CH_3SS^\bullet$ (C_s symmetry) derive from the out-of-phase interaction of the 3p sulfur AOs (π^*_{SS}, SOMO) and of the sulfur lone pairs lying in the molecular plane (n^-_{SS}). Owing to the close similarity of these interactions, the transition between these MOs ($n^-_{SS} \rightarrow \pi^*_{SS}$) was expected to occur at very low energy. In fact, the excitation frequency was computed in the near infrared region at about 10000 cm^{-1}. The corresponding transition was observed in the spectrum of the oxy analogue ($CH_3OO^\bullet$)[11] at 7400 cm^{-1} and computed at 9000 cm^{-1}. The transitions from the in-phase MOs, (π_{SS}, $n^+_{SS} \rightarrow \pi^*_{SS}$), were estimated to occur in the visible region at about 23000 cm^{-1} and in the UV region at 32000 cm^{-1}, respectively. The transition to the lowest unoccupied σ^*_{SS} MO was estimated to occur at a frequency near the latter one, while, as for thiyl, the transition to σ^*_{CS} lies in the far UV region. The experimental characterization of the perthiyl radical has been the subject of controversy. In fact, the same spectrum observed in the flash photolysis of tert-butyl disulfide,[12] was interpreted according to equation 5, and the laser flash photolysis of tert-butyl tetrasulfide and t-BuSSCl[13], according to equations 6 and 7.

$$t\text{-BuSSBu-}t \quad \rightarrow \quad 2\ t\text{-BuS}^\bullet \tag{5}$$

$$t\text{-BuS}_4\text{Bu-}t \quad \rightarrow \quad 2\ t\text{-BuSS}^\bullet \tag{6}$$

$$t\text{-BuSSCl} \quad \rightarrow \quad t\text{-BuSS}^{\bullet} + Cl^{\bullet} \tag{7}$$

The spectrum shows a strong band at 370 nm (27000 cm^{-1}) and a weak band at 550 nm (18200 cm^{-1}). Experimental evidence supported, in a convincing way, the formation of the perthiyl radical. Furthermore, MSXα calculations on perthiyl suggested that the strong band could be attributed to the superposition of the $\pi^*_{SS} \rightarrow \sigma^*_{SS}$ and $n^+_{SS} \rightarrow \pi^*_{SS}$ transitions and the weak band to the $\pi_{SS} \rightarrow \pi^*_{SS}$ transition. However, the agreement between theory and experiment is slightly worse than that obtained for other sulfur and oxygen-centered radicals the frequencies being overestimated by about 5000 cm^{-1}. This deviation could be partially due to the use of the methyl instead of tert-butyl group in performing the calculations.

Sulfinyl

Regarding the valence electrons, the sulfinyl radical is isoelectronic with $CH_3SS^{\bullet}$. The large difference of electronegativity between S and O atoms causes a strong polarization in the π MOs. The SOMO becomes mainly localized at sulfur, consequently the valence transitions are strongly blue-shifted. The $n^- \rightarrow \pi^*$ transition shifts from the near infrared to the visible region. The computed frequency (19000 cm^{-1}) is, as expected, blue-shifted with respect to that of the O–O band observed in the chemiluminescence spectrum of HSO$^{\bullet}$ [14] at 14500 cm^{-1}. The transitions from the corresponding in-phase MOs ($\pi_{SO}, n^+_{SO} \rightarrow \pi^*_{SO}$) are shifted in the UV region at frequencies higher than those computed for the first member of the ns Rydberg series and for the transition to the lowest vacant MO ($\pi^*_{SO} \rightarrow \sigma^*_{SO}$) at about 35000 cm^{-1}.

Sulfonyl

Sulfonyl is a σ radical with a pyramidal configuration at sulfur. The five outermost MOs result from interaction of the four non-bonding oxygen 2p orbitals and the sulfur 3p orbital, which lies at higher energy. The 3p orbital can interact for reason of symmetry only with the in-phase combination of the 2p orbitals (n^+_1, n^+_2). The SOMO is localized to a large extent also on the oxygen atoms owing to the large overlap occurring between the n^+_1 and 3p orbitals. The transitions from the non-interacting (n^-_1, n^-_2) and weakly interacting (n^+_2) MOs to the SOMO were computed in the UV region over a short range of frequencies (27000 - 31000 cm^{-1}). The excitation frequency (30300 cm^{-1}) of the transition with the higher oscillator strength ($n^-_1 \rightarrow$ SOMO) is in good agreement with that of the band observed at about 29000 cm^{-1} in the spectra of the alkyl sulfonyl radical. The red shift observed with aryl substitution was also reproduced by calculations.[15]

REFERENCES

1. K.H. Johnson, *Adv. Quantum Chem.* 7:143 (1973).
2. C. Chatgilialoglu and M. Guerra, *J. Am. Chem. Soc.* 112:0000 (1990).
3. M. Guerra and C. Chatgilialoglu, unpublished results.
4. T.A. Koopmans, *Physica* 1:104 (1933).
5. J.C. Slater, "Computational Methods in Band Theory", Plenum Press, New York (1971).
6. C. Chatgilialoglu and M. Guerra, Electronic spectra of group-IVB centered radicals using MSXα method, presented at EUCHEM Conf. on Organic Free Radicals, Assisi, Italy, Sept. 22-26, 1986.
7. R. Kjellander, *Chem. Phys.* 12:469 (1976).
8. C. Chatgilialoglu, "Handbook of Organic Photochemistry", J.C. Scaiano, ed., CRC Press, Boca Raton, Part IV, Section 1 (1989).
9. M. Suzuki, G. Inoue, and H.J. Akimoto, *Chem. Phys.* 81:5405 (1984).
10. M. Guerra, G. Distefano, D. Jones, F.P. Colonna, and A. Modelli, *Chem. Phys.* 91:383 (1984).

11. H.E. Hunziker and H.R. Wendt, *J. Chem. Phys.* 64:3488 (1976).
12. O. Ito and M. Matsuda, *Bull. Chem. Soc. Jpn.* 51:427 (1978).
13. T.J. Burkey, J.A. Hawari, F.P. Lossing, J. Lusztyk, R. Sutcliffe, and D. Griller, *J. Org. Chem.* 50:4966 (1985).
14. O. Schurath, M. Weber, and H.K. Becker, *J. Chem. Phys.* 67:110 (1977).
15. C. Chatgilialoglu, D. Griller, and M. Guerra, *J. Phys. Chem.* 81:3747 (1987).

THE ELECTRONIC PROPERTIES OF SULFUR-CONTAINING

SUBSTITUENTS AND MOLECULES: AN AB INITIO STUDY

Timothy Clark

Institut für Organische Chemie der
Friedrich-Alexander-Universität Erlangen-Nürnberg
8520 Erlangen, Fed. Rep. Germany

This article presents a general overview of a series of *ab initio* molecular orbital calculations[1] designed to elucidate the electronic interactions between carbon-centered cationic, radical and anionic reaction sites and a directly bonded sulfide, sulfoxide or sulfone substituent. These three substituents have been modelled by the $-SH$, $-(SO)$ and $-(SO_2)H$ groups, respectively. Hartree-Fock (RHF for closed-shell and UHF for radical systems) optimizations for XCH_3, XCH_2^+, $XCH_2^{\bullet}$ and XCH_2^- [X = SH, (SO)H and (SO$_2$)H] were used to obtain molecular structures. The calculated energies were refined by carrying out calculations using a fourth order Møller-Plesset correction for electron correlation on the Hartree-Fock geometries. MP4 calculations included single, double, triple and quadruple excitations (MP4sdtq) and did not include the non-valence orbitals. The energies quoted in the text and tables refer to this level of calculation. A second topic covered in this article is three-electron bonding to oxidized sulfur centers.

ELECTRONIC EFFECTS OF SULFUR SUBSTITUENTS

Substituent stabilization energies are usually calculated as methyl stabilization energies according to equation (1):

$$CH_3^{+/\bullet/-} \;\; + \;\; XCH_3 \;\; \rightarrow \;\; CH_4 \;\; + \;\; XCH_2^{+/\bullet/-} \tag{1}$$

Equation (1) is a so-called isodesmic equation. This means that there are equal numbers of CH(6) and CX(1) bonds on both sides of the equation. Because the correlation energy for a given type of bond is essentially constant, isodesmic equations help to reduce errors that arise from the incomplete calculation of the correlation energy, only about 70% of which is contained in the MP4sdtq correction. The calculated energies for the nine systems considered are shown in Scheme 1.

Several points arise from the data exhibited in Scheme 1. Firstly, the stabilization energies found for $XCH_2^{\bullet}$ are far smaller than those for XCH_2^+ and XCH_2^-. This is expected from the theory of one- and three-electron bonds,[2] which are only strong for charged radical systems, not for neutral radicals. Secondly, there is a steady change in electronic properties in going from X = SH, which stabilizes cations very effectively and anions less so, to X = (SO$_2$)H, which is most effective in stabilizing anions. A closer analysis of the results with the help of the natural bond orbital (NBO)[3] approach reveals the various electronic factors

Sulfur-Centered Reactive Intermediates in Chemistry and Biology, Edited by
C. Chatgilialoglu and K.-D. Asmus, Plenum Press, New York, 1990

X	+	•	−
–SH	−64.9	−7.9	−22.9
–(SO)H	−45.2	−5.2	−62.5
–(SO$_2$)H	−29.4	+0.5	−70.9

Scheme 1. Calculated (MP4sdtq/6-31G*//6-31G*) methyl stabilization (kcal mol^{-1}) energies for the sulfur-substituted methyl cations, radicals and anions, CH$_3$X.

responsible for the effects obtained. It is appropriate to consider the effect of d-orbitals on sulfur here. The following analysis will not use d-orbital participation as a major cause of stabilization effects, but will rather concentrate on electronegativity differences. The NBO-analysis does not reveal significant d-orbital participation other than the normal role as polarization functions (both on carbon and on sulfur). In molecules such as sulfones, this polarization effect is important for a proper description of the molecule. It is, however, equally important for CH$_3$X as for XCH$_2^-$ and therefore does not contribute significantly to stabilization effects.[4]

CATION-STABILIZATION BY SULFUR SUBSTITUENTS

The 6-31G* optimized geometries of the CH$_3$X molecules are shown in Scheme 2 and those of the XCH$_2^+$ cations in Scheme 3.

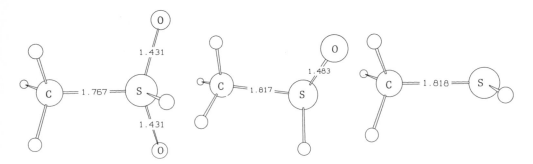

Scheme 2

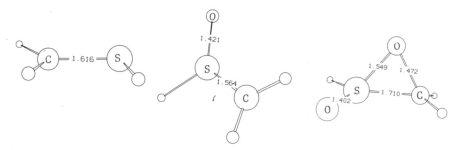

Scheme 3

Sulfide and sulfoxide substituents provide their very high stabilizations of the cationic center by classical π-conjugation. In the sulfide case, the π-system is essentially a C=S double bond and for the sulfoxide a 4π-electron C–S–O allylic-type π-system:

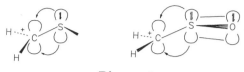

Diagram 1

The planar structures of $HSCH_2^+$ and $H(SO)CH_2^+$ are consistent with this type of interaction (see Scheme 3). Classical π-stabilization of an adjacent cationic center by a sulfone group is, however, not possible. In this case, neighboring group participation by one of the sulfone oxygens is the only alternative:

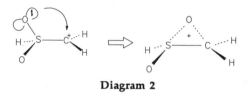

Diagram 2

This leads to the observed bridged structure of $H(SO_2)CH_2^+$ (Scheme 3). Note, however, that this bridging is relatively ineffective in stabilizing the molecule and that $-(SO_2)H$ is the worst of the three sulfur substituents in this respect.

SULFUR SUBSTITUENT EFFECTS ON RADICAL CENTERS

As outlined above, radical stabilization energies in neutral systems are generally far smaller than those of cations or anions.[2] The upper limit for 1- or 3-electron π-stabilization in neutral systems seems to be in the range 10 - 12 kcal mol^{-1}. Nevertheless, stabilization energies of this magnitude are ample for controlling the course of a reaction – even if they are small in comparison with the closed-shell systems. The UHF/6-31G* optimized geometries of the three $XCH_2^{\bullet}$ radicals are shown in Scheme 4.

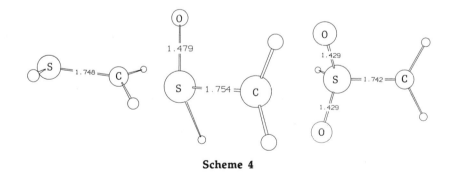

Scheme 4

$HSCH_2^{\bullet}$ and $H(SO)CH_2^{\bullet}$ show remnants of the π-conjugation effects found in $HSCH_2^+$ and $H(SO)CH_2^+$, but now as 3- and 5-electron systems, respectively. The structures (Scheme 4) are no longer exactly planar and the CS-bonds are longer than in the cationic systems.

H(SO$_2$)CH$_2$$^\bullet$ is remarkable for its lack of stabilizing effects. Both its geometry and the slight calculated destabilization indicate that there is no stabilization of radical centers α-to sulfone groups.

ANION-STABILIZATION BY SULFUR-SUBSTITUENTS

The structures of the three XCH$_2$$^-$ anions are shown in Scheme 5.

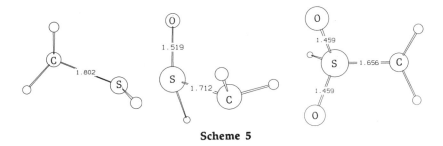

Scheme 5

The structure of the least strongly stabilized of the three anions, HSCH$_2$$^-$, shows a pyramidal anionic center with a long CS-bond. This sort of structure is typical for carbon anions stabilized by σ-acceptors and can best be understood as a resonance hybrid between the carbon-centered anion structure and a complex between methylene and a thiyl anion. In MO-terms, this interaction involves donation from the carbon lone-pair to the σ^*_{CS} orbital:

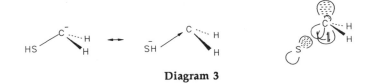

Diagram 3

In contrast, H(SO)CH$_2$$^-$ and H(SO$_2$)CH$_2$$^-$ both show shortened CS-bonds (Scheme 5) relative to H(SO)CH$_3$ and H(SO$_2$)CH$_3$ (Scheme 2). The SO-bonds are, however, lengthened in both cases. These geometry changes reveal the cause of the very high stabilization energies for these two molecules - negative hyperconjugation. One of the most important differences between first- and second-row elements of the same group is that the second-row elements are more electropositive. Thus, the electronegativities of sulfur and oxygen are 3.50 and 2.44, respectively. This means that SO-bonds are strongly polarized towards oxygen and, more importantly in this case, that σ^*_{SO} orbitals are concentrated on sulfur:

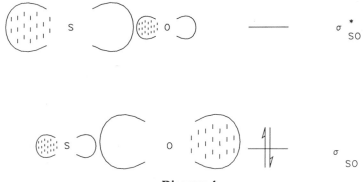

Diagram 4

This and the low energy of the σ^*_{SO} orbitals makes them excellent acceptors:

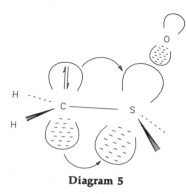

Diagram 5

A partial C=S double bond and a lengthened SO-bond result. This type of negative hyperconjugation is well known for β-halogeno carbanions,[5] but is remarkably effective in the SO-systems, as shown by the large geometry changes, conformations and very high stabilization energies.

THREE-ELECTRON BONDS TO SULFIDE RADICAL CATIONS

One-electron oxidation of sulfides under normal conditions usually leads to sulfide dimer radical cations in which two sulfur atoms are joined by a three-electron bond with a dissociation energy around 30 kcal mol^{-1}.[1b] The MP2/6-31G*-calculated dissociation energy of $H_2S \ldots SH_2^{+\bullet}$ agrees well with the experimental value for $Me_2S \ldots SMe_2^{+\bullet}$.[6] (For experimental examples of three-electron bonded species see also article by K.-D. Asmus in this book, who introduced the symbol "•∴•" as notation for this type of bond).

The three-electron bond is formed when the singly occupied orbital of a sulfide radical cation interacts with the corresponding doubly occupied σ^*_{SS} orbital. The energy of these bonds has been shown[2] to depend on $e^{-(\delta IP)}$, where δIP is the difference in ionization potentials between the two interacting partners. The practical question of a competition between complexation to another sulfide and to water (to give $R_2S \ldots OH_2^{+\bullet}$) arises in aqueous solution. $H_2S \ldots OH_2^{+\bullet}$ was found[2] to have an SO-bond energy of 21 kcal mol^{-1}, but is not necessarily a good model for $R_2S \ldots OH_2^{+\bullet}$. We have therefore optimized $Me_2S \ldots OH_2^{+\bullet}$ at UHF/6-31G*. The geometry is shown in Scheme 6.

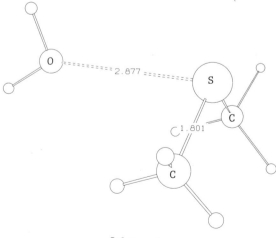

Scheme 6

Note that the orientation of the water ligand is different to that in $H_2O \ldots SH_2^{+\bullet}$ as a consequence of the change in ionization potential between H_2S and Me_2S. The complex is found to have a dissociation energy of 16.8 kcal mol^{-1}, so that the weakest observable $R_2S \ldots SR_2^{+\bullet}$ three-electron bonded complex in aqueous solution must have an SS-bond energy at least this high.

CONCLUSIONS

The electronic properties of sulfur-containing substituents can be described satisfactorily using the π-donor characteristics of the S-atom and the σ-acceptor characteristics of SO-bonds. The latter effect is caused by negative hyperconjugation that is a result of the electro-negativitiy difference between oxygen and sulfur. d-Orbitals are not necessary in order to describe the electronic properties of sulfur substituents. Oxidized sulfur centers readily form three-electron bonds to donor molecules. The observable range of S–S three-electron bonds in aqueous solution lies roughly between 16 and 21 kcal mol^{-1}.

REFERENCES

1 a. See T. Clark, "A Handbook of Computational Chemistry", Wiley, N.Y. (1985).
   b. W.J. Hehre, L. Radom, P.v.R. Schleyer, and J.A. Pople, "Ab Initio Molecular Orbital Theory", Wiley, N.Y. (1986).
2. T. Clark, *J. Am. Chem. Soc.* 110:1672 (1988) and references therein.
3. See A.E. Reed, F. Weinhold, and LA. Curtiss, *Chem. Rev.* 88:899 (1988) and references therein.
4. F. Bernardi, I.G. Czismadia, A. Mangini, H.B. Schlegel, M.-H. Whangbo, and S. Wolfe, *J. Am. Chem. Soc.* 97:2209 (1975).
5. A.J. Kos and P.v.R. Schleyer, *Tetrahedron* 39:1141 (1983).
6. A.J: Illies, P. Livant, and M.L. McKee, *J. Am. Chem. Soc.* 110:7980 (1988).

REACTIVITY OF SULFUR-CENTERED NUCLEOPHILES TOWARDS DIA- AND PARAMAGNETIC REAGENTS. THE FRONTIER ORBITALS APPROACH

Michel Arbelot, André Samat, Michel Rajzmann, Monique Meyer, André Gastaud, and Michel Chanon

LCIM case 561 U.A. CNRS no. 126
Faculté des Sciences de St Jérome
13397 Marseille Cedex 13, France

Considering the large polarisability of sulfur, any nucleophiles centered on sulfur have often been classified as "soft" nucleophiles and their general reactivity has been rationalized along this line. It is the purpose of the present report to examine the reactivity of sulfur-centered nucleophiles in the following reactions:

$$\underset{R'}{\overset{R}{\diagdown}}\underset{Y}{\overset{Z-X}{\diagup}}C=S + CH_3I \xrightarrow{Me_2CO} \underset{R'}{\overset{R}{\diagdown}}\underset{Y}{\overset{Z-X}{\diagup}}C=\overset{+}{S}CH_3 \ I^- \tag{1}$$

X, Y, Z = NR, sp ^{2}N, sp^{2}C, S, O
R, R' = H, Me, Ph
The C–Z bond in the cycle may be also unsaturated

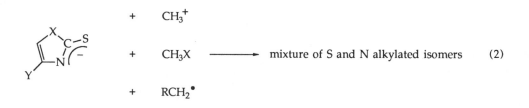

$$+ \quad CH_3^+$$
$$+ \quad CH_3X \longrightarrow \text{mixture of S and N alkylated isomers} \tag{2}$$
$$+ \quad RCH_2^{\bullet}$$

For equation (1), the reactivity of a population of about 40 different compounds, selected for changing the electronic characteristics of the thiocarbonyl group, was kinetically measured through conductimetric measurements.[1,2] Scheme 1 displays some of the trends in reactivity.

The main qualitative features are:

1. the activating effect of the atom β to the nucleophilic center decreases in the order sp^{2}C > NR > S > O
2. annelation decreases the nucleophilicity of S
3. ring opening deactivates the nucleophilicity of S.

Sulfur-Centered Reactive Intermediates in Chemistry and Biology, Edited by
C. Chatgilialoglu and K.-D. Asmus, Plenum Press, New York, 1990

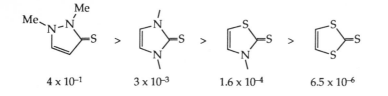

| 4 x 10⁻¹ | 3 x 10⁻³ | 1.6 x 10⁻⁴ | 6.5 x 10⁻⁶ |

Annelation deactivates (≈ 1.1 to 1.9 kcal Mol⁻¹)

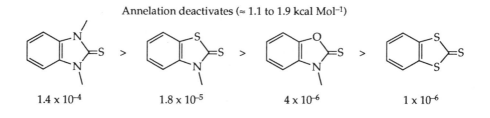

| 1.4 x 10⁻⁴ | 1.8 x 10⁻⁵ | 4 x 10⁻⁶ | 1 x 10⁻⁶ |

Ring opening deactivates

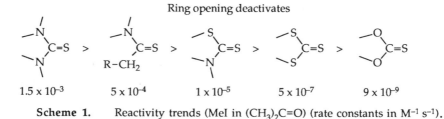

| 1.5 x 10⁻³ | 5 x 10⁻⁴ | 1 x 10⁻⁵ | 5 x 10⁻⁷ | 9 x 10⁻⁹ |

Scheme 1.   Reactivity trends (MeI in $(CH_3)_2C=O$) (rate constants in $M^{-1} s^{-1}$).

The wide range of reactivity spanning over 7 orders of magnitude calls attention to the risk of quick classification of an atom as "soft" without looking at its structural environment. The following two rate constant ratios (referring to MeI) illustrate this point:

The thiocarbonyl group has been classified as a *soft* nucleophile.

$$\frac{k \; \langle \rangle - S^-}{k \; \langle \rangle - O^-} = 10^6$$

$$= 10^7$$

Therefore, for a given polarisable nucleophilic center, structural effects may be more important than mere classification into "polarisable" vs. "non polarisable".

This converges with previous remarks on a too loose use of the terms "hard" and "soft" in reactivity.[3] The range of nucleophilic reactivity ratios of the thiocarbonyl group would even exceed this 10^7 value if one were to add compounds such as those shown in Scheme 2.

Best nucleophiles   *Poorest nucleophiles*

Scheme 2

Energy (eV) of MO's

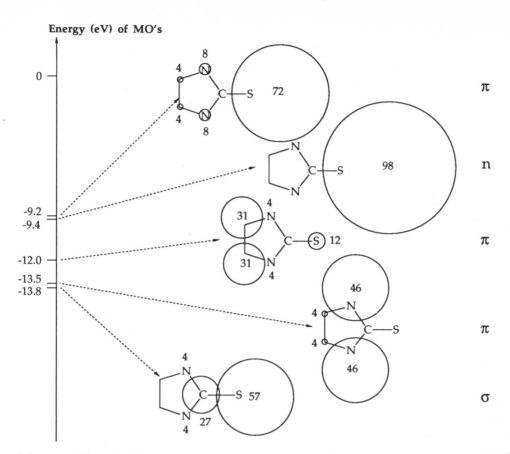

Scheme 3. Theoretical treatment of a typical thiocarbonyl.[4d] The numbers in or near circles refer to atomic orbital participation for the considered MO.

The thiocarbonyl group allows even more interesting considerations besides these qualitative trends. The reason why this is so may be found in the PES spectra: for thiocarbonyl compounds one may experimentally and selectively reach the energy of frontier orbitals of π and n symmetries.[4] This contrasts with carbonyl compounds where the PES spectra usually do not allow an easy distinction of π and n orbitals. The MO interpretation [4d] of a typical thiocarbonyl compound is shown in Scheme 3. For this compound (imidazoline thione-2), the HOMO is of π symmetry close to a MO of n symmetry. These two orbitals are ca. 2.5 eV higher in energy than the next lower MO and, as a consequence, these thiocarbonyl derivatives appear to be particularly suited for an experimental verification of Klopman approach.[5] In this approach, it is proposed that the importance of the interaction between a nucleophile and an electrophile is measured as in equation 3:

$$\Delta E \;=\; \underbrace{q_s q_c \, \Gamma_{sc}}_{Q} \;+\; \underbrace{\Delta E_{solv}}_{S} \;+\; \underbrace{2 \sum_{i}^{occ.} \sum_{j}^{vac.} \frac{c_{ri}^2 \, c_{sj}^2 \, \beta_{rs}^2}{E_i - E_j}}_{F} \tag{3}$$

q_s = charge on sulfur

q_c = charge on C in CH_3I

Γ_{sc} = coulombic interaction between S and C

r = nucleophilic atom

s = electrophilic atom

β_{rs} = overlap between r and s for the considered couple of orbitals

HOMO C=S ———— $\uparrow \; E_i - E_j$ ———— LUMO CH_3X

The parameters c_{ri} and c_{sj} are the coefficients of the atomic orbitals centered on sulfur and carbon (CH$_3$X), respectively, for the considered interacting molecular orbitals. For example, one will successively compare the LUMO of CH$_3$X with the HOMO of the thiocarbonyl, then the same LUMO of CH$_3$X with the next HOMO of the thiocarbonyl, and so on. It is clear though that the terms for which $E_i - E_j$ is too large should have comparatively little weight in the overall perturbation approach.

Under favorable conditions one may hope that, in a series of compounds, ΔE of eq. 3 correlates with the relative values of rate constants. In our case we have verified that neither Q nor S terms in equ. (3) are the leading parameters for the series of considered thiocarbonyl compounds. We are now going to see that F (frontier term) is the leading reactivity parameter and, to be more specific, *that the β term plays a determining role in the approach*. Indeed, the usual qualitative application of the pertubation method for frontier controlled reactions concentrates only on the interactions between the HOMO of the nucleophile and the LUMO of the electrophile.[6] Within the series of thiocarbonyl compounds considered here, a couple of frontier orbitals may play a role. We expected, within the frontier orbital approach, either a correlation of reactivity with the HOMO of the thiocarbonyl, or a correlation with a linear combination of the HOMO and next HOMO.

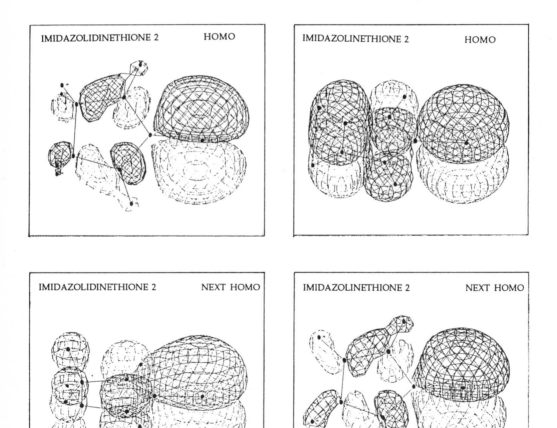

Scheme 4

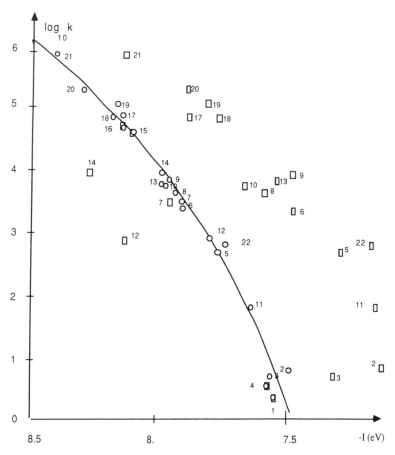

Fig. 1. Correlation PES nucleophilic reactivity.

1 = 1,2-Dimethylpyrazoline-3-thione;
3 = 1,3-Dimethyl-1,2,3-triazolo-4-sulfur;
5 = 1,3-Dimethylimidazoline-2-thione;
7 = 1,3-Dimethylimidazolidine-2-thione;
9 = 1,3-Dimethylbenzimidazolidine-2-thione;
11 = 1,3,4,5-Tetramethylimidazoline-2-thione;
13 = 3,4,5-Trimethyloxazoline-2-thione;
15 = 4,5-Dimethyl-1,2-dithiol-3-thione;
17 = 3,5-Dimethyl-1,3,4-thiadiazoline-2-thione;
19 = 3-Dimethyl-5-phenyl-1,3,4-thiadiazoline-2-thione;
21 = Benzo-1,3-dithiol-2-thione;

2 = 1,3,5-Trimethyl-1,2,3-triazolo-4-sulfur;
4 = 1,2,4-Trimethylpyrazoline-3-thione;
6 = 3,4,5-Trimethylthiazoline-2-thione;
8 = 3,5-Dimethylthiazoline-2-thione;
10 = 3-Methylthiazoline-2-thione;
12 = NN-Tetramethylthiourea
14 = 3-Methylthiazolidine-2-thione;
16 = 5-Phenyl-1,2-dithiol-3-thione
18 = 3-Methylbenzothiazoline-2-thione
20 = 3-Methylbenzoxazoline-2-thione;
22 = 1,3-Dimethyl-4-phenylimidazoline-2-thione

Neither of these correlations were found to exist, but something even more interesting and specific emerged.

To understand this specificity one first has to consider a unique property of the set of studied thiocarbonyls. This property is displayed in Scheme 4. One sentence summarizes it: The HOMO of most of the studied thiocarbonyl compounds is of π symmetry (Fig. 1) but an appropriate structural modification (saturation) in the heterocyclic structural moiety leads to an inversion in the depth of the considered couple of frontier orbitals. For saturated compounds, the HOMO is of n symmetry whereas for their unsaturated counterparts this HOMO is of π symmetry.

The graph between the *log* of rate constants and the depth of orbitals[7] (Fig. 1) shows that there is considerable scatter in the correlation between the nucleophilic reactivity for the set of thiocarbonyl compounds and the depth of their frontier orbitals of π symmetry (squares). A far better correlation holds between the depth of the frontier orbitals of n symmetry (circles) and reactivity. Furthermore, even for the compounds where the orbital of n symmetry is not the HOMO, the energy of this orbital correlates better with reactivity than does the orbital of π symmetry.

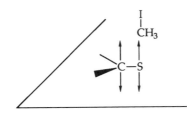

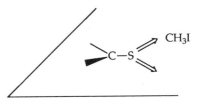

a) Discarded by PES-reactivity correlation and by rate studies with sterically hindered RI

b) Very probable angular attack

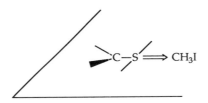

c) Discarded by rate studies with sterically hindered RI

Scheme 5. Directionality of the interaction

This series of thiocarbonyl compounds therefore provides us with an unprecedented experimental illustration of the principle of *symmetry active frontier orbital*. It is concluded that the β term in equ.(3) which measures the overlap between the interacting frontier orbitals is dominant. This strongly suggests that there is definitely a directionality of attack of the thiocarbonyl group on the electrophilic center (Scheme 5b); this has been, in fact, confirmed by the use of steric structural effects[8a,b] and converges with theoretical calculations.[8c] It holds irrespective of the leaving group attached to the carbon electrophilic center, as shown by the linear relationship between the activation energies for the reactions between a set of thiocarbonyl compounds and methyl tosylate [equ. (4)]:

$$\Delta G_{MeTos} = 0.92\ \Delta G_{MeI} + 4.19 \tag{4}$$

The structure of the transition state may be evaluated even more precisely by the quantity of the steric strain release in the transition state of the reaction between 3-alkyl Δ-4 thiazoline thiones-2 and methyl halides. Indeed, in the final product the "size" of the thiomethyl group is less important the the "size" of the thiocarbonyl group in the starting material. This may be quantitatively measured by using the N-iPr group conformation as a monitor of these sizes.[8b] As the transition state displays a strain release of about 0.4 kcal/mole, and the strain release in going from substrate to products amounts to 0.7 kcal/mole, it follows that the C=S group has already considerably changed its shape (60%) in the transition state. This consideration is important because it is *often postulated that the pertubation approach [equ. (3)] should only apply for reactions with an early transition state (transition state like reagent); our results show that this is not necessarily so*. From this viewpoint they converge with S. Wold's warnings about using extrathermodynamic relationships to gain insight into the structure of transition states.[18]

The ratio of rate constants between a given nucleophile and the MeI/MeTos couple has been proposed as an indicator of the electron transfer ability of the nucleophile.[9] Within this approach the thiocarbonyl group (k_I/k_{Tos} 10^2) is still very far from the situation in highly polarisable organometallics, for example, the k_I/k_{Tos} for various organometallics are:

$Pd(PEt_3)_3$	$Ni(PEt_3)_4$	$CpW(CO)_3^-$
3×10^5	8×10^5	1.5×10^5

We have checked, by the absence of photostimulation effects in the reaction between thiocarbonyl and CH_3X and the reactivity of thiocarbonyl toward efficient free radical clocks, that no outer sphere electron transfer component is present in these reactions. Finally, it is noted that the difference of 0.8 in ionisation potential corresponds to 7 orders in magnitude in rate constants (Figure 1). This suggests that in the sulfur compounds, described in Glass's chapter in this book, very definite differences in nucleophilic reactivity should appear concomitantly with the splitting interactions of lone pair orbitals associated with specific spatial interactions.

The ambient reactivity of various heterocyclic thioamides further illustrates the role of frontier orbitals in controlling the reactivity of nucleophiles centered on sulfur. Again, structural variations in the heterocyclic moiety allow a smoooth tuning of the sulfur electronic properties [eq. (5) and Scheme 6].

Alkylation of heterocycles by diazomethane in THF

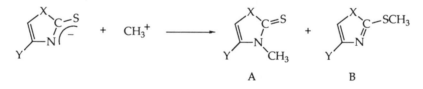

A   B

Alkylation predominantly yields B (except for rhodanines) with   (5)
a preference varying between

and

67 %   92 %

Scheme 6

The reactions of the neutral heterocyclic thioamid (0.05 M in THF) with diazomethane may be represented by equation (5). Except for rhodanine for which mainly N-alkylation occurs, the dominant site of alkylation is sulfur. The percentages of S-alkylation on sulfur smoothly vary between 67% and 92% (Figure 2). CNDO calculations performed on the population of experimentally studied ambient anions yield the π charge density, $\sigma + \pi$ charges, and HOMO coefficients on S and N in the various ambient anions (Table 1).

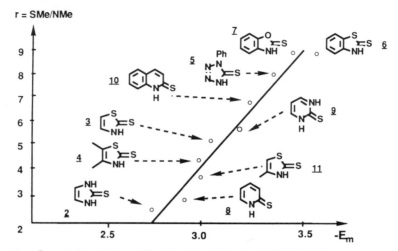

Figure 2. Correlation (N *vs.* S) selectivity/depth of HOMO. (Does not apply to saturated substrates.)

Table 1. Ambient Anion Parameters Provided by CNDO Calculations for Compounds *1* to *11* (code of numbering see Fig. 2). *1* is Rhodanine

	Atom	π charge density	$\pi + \sigma$ charge	c in HOMO	E_m (eV)
1	N	1.35	−0.31	0.47	
	S	1.60	−0.51	1.23	−4.12
2	N	1.25	−0.16	0.34	
	S	1.81	−0.63	1.3	−2.54
3	N	1.22	−0.22	0.18	
	S	1.85	−0.64	1.47	−3.02
4	N	1.30	−0.26	0.25	
	S	1.76	−0.61	1.33	−2.98
5	N	1.36	−0.28	0.51	
	S	1.74	−0.60	1.40	−3.3
6	N	1.41	−0.30	0.5	
	S	1.63	−0.53	1.15	−3.53
7	N	1.35	−0.32	0.34	
	S	1.83	−0.67	1.46	−3.4
8	N	1.20	−0.24	0.235	
	S	1.81	−0.66	1.46	−2.93
9	N	1.18	−0.25	1.21	
	S	1.81	−0.64	1.47	−3.13
10	N	1.21	−0.26	0.28	
	S	1.78	−0.64	1.40	−3.13
11	N	1.35	−0.27	0.20	
	S	1.88	−0.69	1.6	−3.0

None of these parameters is suitable for direct correlation with the experimentally observed values r = {SMe}/{NMe}. A good correlation is, however, obtained when r is plotted against the energy of the HOMO in the ambident anions obtained by removal of a proton from the heterocyclic structures displayed in Figure 2. From eq. (3) one would have expected a correlation where the difference of E_n (energy of the LUMO in CH_3^+) – E_m (energy of the HOMO in the ambident anion) was to play a role. This equation was also hinting at a possible role of the atomic coefficients of N and S in the HOMO of the ambident anion.

A closer look at the CNDO parameters associated with the various ambident anions reveals a correlation between E_m and the atomic coefficients of sulfur and nitrogen in the HOMO of ambident anions [eq. (6)].

$$E_m = 2a'(C_S^m)^2 + 2b'(C_N^m)^2 \tag{6}$$

E_m : energy HOMO
C_S^m : coefficient of atomic orbital of S in HOMO
C_N^m : coefficient of atomic orbital of N in HOMO
a' and b': constants

This yields the overall relation (7) for r (ratio SMe/NMe in the alkylation reaction)

$$r = \alpha(C_S^m)^2 + \beta(C_N^m)^2 + \gamma \tag{7}$$

where $\alpha = 16 \pm 3$, $\beta = 38.5 \pm 4$, $\gamma = -28 \pm 7$.

Therefore, the reactivity of the ambident anions toward a carbenium like reagent seems essentially frontier controlled.[10] This result is somewhat unexpected if one considers eq. (5): indeed, in this equation, at least formally, a negatively charged nucleophile reacts with a positively charged carbocation. One could, therefore, have expected a stronger control by the charges [Q term in eq. (3)].

In terms of the HSAB (Hard and Soft Acids and Bases)[3] approach, a CH_3^+ would be located nearer to the hard electrophile centers than would a $^\bullet CH_2R$ radical. In other words, this means that the LUMO of CH_3^+ is deeper (–11.2 eV) than the SOMO of an alkyl radical (–10.3 eV for $CH_3^\bullet$).[11] We, therefore, have studied the reactivity of some of the preceding ambident anions towards α-p-dinitrocumene [eq. (8)].[12] In this reaction, and depending upon the studied heterocyclic compound, one obtains the different products shown in eq. (8) (Table 2). The effects of scavengers, analogies with literature data, and photostimulation suggest that one is dealing with a short chain SR_N1 process.

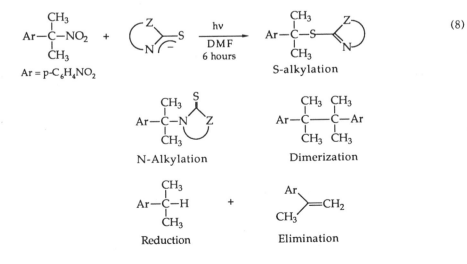

(8)

Table 2. Photostimulated reaction of sodium salts of various ambient anions towards α-p-dinitrocumene. The small percentage of N-isomer reported for imidazolinethione-2 is the result of a photochemical rearrangement of the primary S-isomer.[12] r = S-alkylation/N-alkylation in the reaction of the neutral thioamide with diazomethane (see Fig. 2)

	S-alkyl	N-Alkyl	Dimer	Reduction	Elimination	r
(benzothiazoline-2-thione structure)	36	0	18	20	26	9
(2-methyl thiazoline-thione structure)	86	0	0	traces	13	3.7
(imidazoline-2-thione structure)	92	0	3	5	0	3.7
(N-methyl imidazoline-2-thione structure)	47	10	1	0	23	2.3

In such an SR$_N$1 mechanism (Scheme 7), the N versus S selectivity (shown as r in Table 2) is probably settled in step 3. In this step, the p-nitrocumyl radical reacts with the ambident anion. There has been some controversy[13] in this type of reaction whether the overall selectivity is the result of the relative stabilities of anion radicals N$^{\bullet-}$ and S$^{\bullet-}$ or if the kinetic selectivity of the ambident anion is the dominant factor. In an independant study,[14] we have shown that the N-substituted isomers are definitely more stable than their S-counterparts. Therefore, in Table 2, one expects a greater thermodynamic stability for the N alkyl products than for their S isomers. In the precursor radical anions of these products

Origin of selectivity (Ar = p-NO$_2$C$_6$H$_4$)

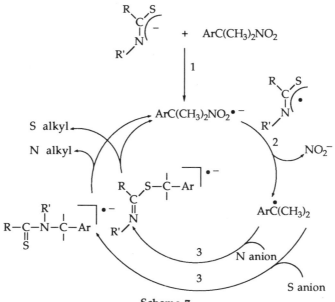

Scheme 7

28

(Scheme 7) the LUMO is, in both isomers, mainly located in the p-nitrophenyl moiety of the structure. This localisation indicates that there should not be an inversion of stability of the isomeric radical anions with respect to the final products: the N$^{\bullet-}$ radical anion (Scheme 7) thermodynamically should be more stable than its S$^{\bullet-}$ isomer. In such a line of reasoning the large predominance of the S-alkylated product in our studies seems to be the result of kinetic selectivity in step 3 (Scheme 7). This kinetic selectivity would be consistent with a frontier controlled type of reactivity for these ambident anions. One must keep in mind, however, that at this point the treatment of relative thermodynamic stabilities of isomeric radical anions is still something of an art and any explanation in this field must still be regarded as tentative only: indeed there are reports[17] where the position of a protomeric equilibrium is reverted when passing from neutral species to their radical anion counterparts.

In the selectivity of ambident anions toward electrophilic centers, the following rule (Kornblum's rule)[15] is often verified: "as the character of a given reaction changes from an S_N1-like to an S_N2-like, an ambident nucleophile becomes more likely to attack with its less electronegative atom." If one wants to extend this rule to rationalize the reactivity of ambident anions toward radical centers by forming 3e bonds one will probably have to consider that, in the spectrum of S_N1-like to S_N2-like situations, the case of carbon-centered radicals will definitely be on the S_N2 side. This similarity with S_N2 is expected to vary with the depth of the SOMO in the considered radical (Table 3).

The radicals with the least deep SOMO lie beyond the S_N2 extreme in the foregoing spectrum. In such an extension of Kornblum's rule, the words of caution already operating in ambident reactivity (absence of equilibration between the products, role of ion pairing and counter ions) have to be kept in mind. Table 3 gives some typical values of ionization potential (I.P.) for radical species to facilitate a systematic exploration of selectivity in the ambident reactivity of anions leading to 3e bonds. For such an extension, one should keep in mind Clark's relation (see T. Clark's article on 3e bonds in this book) which indicates that if a too large difference exists in electronegativities between two atoms, the 3e bond that they may form between them should be highly unstable from a thermodynamic point of view.

Table 3. Representative values of radical I.P. (eV) to provide an approximate idea of their SOMO's depths. (For more details see ref. 16)

Cyclopropenyl$^{\bullet}$	5.8, 7.6	F$_3$C$^{\bullet}$	9.2
p-MeOC$_6$H$_4$CH$_2$$^{\bullet}$	6.9	H$_3$C$^{\bullet}$	9.9
I-adamantyl$^{\bullet}$	6.2	HS$^{\bullet}$	10.5
(CH$_3$)$_3$C$^{\bullet}$	6.7	NCCH$_2$$^{\bullet}$	10.9
H$_3$Si$^{\bullet}$	7.6	HOO$^{\bullet}$	11.5
(CH$_3$)$_3$CCH$_2$$^{\bullet}$	7.9	Br$^{\bullet}$	11.8
H$_2$NCH$_2$$^{\bullet}$	7.6	HO$^{\bullet}$	12.9
CH$_3$S$^{\bullet}$	8.1	Cl$^{\bullet}$	14.5
PhO$^{\bullet}$	8.7	NH$_3$$^{\bullet+}$	23.5
Cl$_3$C$^{\bullet}$	8.8		

REFERENCES

1. M. Arbelot, R. Gallo, M. Chanon, and J. Metzger, *Int. J. Sulfur Chem.* 9:201 (1976).
2. M. Arbelot, *Thesis*, Marseille (1980).
3. M. Arbelot and M. Chanon, *Nouv. J. Chem.* 7:499 (1983).
4 a. M. Arbelot, C. Guimon, D. Gombeau, and G. Pfister-Guillouzu, *J. Mol. Struct.* 20:487 (1974).
   b. C. Guimon, G. Pfister-Guillouzo, M. Arbelot, and M. Chanon, *Tetrahedron* 30:3831 (1974).
   c. C. Guimon, M. Arbelot, and G. Pfister-Guillouzo, *Spectrochim. Acta* 31A:985 (1975).
   d. C. Guimon, G. Pfister-Guillouzo, and M. Arbelot, *Tetrahedron* 31:2769 (1975).
   e. C. Guimon, G. Pfister-Guillouzo, and M. Arbelot, *J. Mol. Struct.* 30:339 (1976).
5. G. Klopman, *J. Am. Chem. Soc.* 90:223 (1968).
6. I. Fleming, "Frontier Orbitals and Chemical Reactions", Wiley, Chichester (1976).
7. M. Arbelot, J. Metzger, M. Chanon, C. Guimon, and G. Pfister-Guillouzo, *J. Am. Chem. Soc.* 96:6217 (1974).
8 a. M. Arbelot, M. Chanon, R. Gallo, C. Roussel, J. Metzger, and M. Begtrub, *J. Chem. Soc. Perkin 2* 1169 (1977).
   b. C. Roussel, R. Gallo, M. Chanon, and J. Metzger, *J. Chem. Soc. Perkin 2* 1304 (1974).
   c. G. Gombeau, *Thesis*, Pau (1978).
9 a. R.G. Pearson and P. E. Figdore, *J. Am. Chem. Soc.* 102:1541 (1980).
   b. M. Chanon, *Bull. Soc. Chim. Fr.* II 197 (1982).
10. K. Fukui, *J. Chem. Phys.* 20:722 (1952); 22:1433 (1954); *Acc. Chem Res.* 4:57 (1971).
11. D.J. Pasto, *J. Am. Chem. Soc.* 110:8164 (1988).
12. M. Meyer, A. Samat, M. Chanon, *Heterocycles* 24:1013 (1986).
13a. N. Kornblum, P. Ackerman, and R.T. Swiger, *J. Org. Chem.* 45:5294 (1980).
   b. L.M. Tolbert and A. Siddiqui, *Tetrahedron* 38:1079 (1982); *J. Org. Chem.* 49:1744 (1984).
   c. G.A. Russell, B. Mudryk, F. Ros, and M. Jawdowsiuk, *Tetrahedron* 38:1059 (1982).
   d. R.K. Norris and D. Randles, *J. Org. Chem.* 47:1047 (1982).
   e. W.R. Bowman, D. Rackshit, and M.D. Valmas, *J. Chem. Soc. Perkin Trans. 1* 2327 (1984).
14a. C. Roussel, M. Chanon, and R. Barone, "Thiazole and Derivatives", J. Metzger, ed., Wiley, New York, Part 2, p. 377 (1979).
   b. M. Chanon, M. Conte, J. Micozzi, and J. Metzger, *Int. J. Sulfur Chem. C* 6:85 (1971).
15a. N. Kornblum, R.A. Smiley, R.K. Blackwood, and D.C. Iffland, *J. Am. Chem. Soc.* 77:6269 (1955).
   b. R. Gompper and H.U. Wagner, *Angew. Chem. Int. Ed.* 15:321 (1976).
16. M. Chanon, M. Razjmann, and F. Chanon, *Tetrahedron*, in press.
17. G. Frenking and H. Schwarz, *Z. Naturforsch.* 37b:1602 (1982).
18a. M. Sjöström and S. Wold, *Acta Chem. Scand. Ser B* 35:537 (1981).
   b. P.R. Young and W.P. Jencks, *J. Am. Chem. Soc.* 101:3288 (1979)

ALKANETHIYLPEROXYL RADICALS

Chryssostomos Chatgilialoglu and Maurizio Guerra

I. Co. C. E. A.
Consiglio Nazionale delle Ricerche
40064 Ozzano Emilia (Bologna), Italy

Alkanethiylperoxyl radicals are the adducts of alkanethiyl radicals with molecular oxygen, viz.,

$$RS^{\bullet} + O_2 \quad \rightarrow \quad RSOO^{\bullet} \tag{1}$$

Most of the work on the structural characteristics of these radicals, their formation and reactions is very recent and, to a certain extent, this class of radicals represent the expansion and importance of organosulfur reactive intermediates in various branches of chemistry. In fact, it is believed that these reactive species play important roles in the atmospheric sulfur cycle[1] and in biological systems.[2] Some authors have reported alkanethiylperoxyl radicals as $RSO_2^{\bullet}$ rather than $RSOO^{\bullet}$ and this fact has already caused some confusion because $RSO_2^{\bullet}$ represents the well-known class of alkanesulfonyl radicals in which the sulfur atom is bound to two oxygen atoms as well as to R.[3]

In the gas phase, several studies on the photo-oxidation of thiols, sulfides and disulfides have been carried out.[4] Based on product studies it has been suggested that the major pathway should be the formation of sulfonyl radicals via the alkanethiylperoxyl adduct formed by reaction 1, although the precise mechanistic scheme is still under debate. Following the discovery by Suzuki et al.[5] of the laser-induced fluorescence spectrum of the $CH_3S^{\bullet}$ radical, the rate constant for reaction 1 is found to be less than 1.5×10^3 M^{-1} s^{-1}.[6] These results are in contrast with earlier conclusions which indicated a rate constant several orders of magnitude faster. One explanation for this discrepancy is that $CH_3S^{\bullet}$ and O_2 form an adduct in a reversible manner. Although a mechanistic scheme which attempts to accomodate all the data obtained in the gas phase experiments has recently been suggested,[7] neither spectroscopic evidence nor theoretical calculations on alkanethiylperoxyl radicals has yet appeared.

In the condensed phase,[2] pulse radiolysis studies amongst others indicate that alkanethiyl radicals react with molecular oxygen to form a radical adduct which absorbs at ca. 550 nm. Rate constants for reaction 1, obtained either by direct methods or by using a competitive reaction, were found to be in the range of $10^7 - 10^{10}$ M^{-1} s^{-1}.[8] Such a large range of measured rate constants clearly indicates a complex nature for reaction 1. On the other hand, the visible absorption at 550 nm, attributed to $RSOO^{\bullet}$, is unknown for the alkylperoxyl radicals,[9] indicating two possibilities; that is, either the $RSOO^{\bullet}$ radicals have different structural characteristics from $ROO^{\bullet}$ radicals or the 550 nm absorption is due to another species.

Sulfur-Centered Reactive Intermediates in Chemistry and Biology, Edited by
C. Chatgilialoglu and K.-D. Asmus, Plenum Press, New York, 1990

Electron spin resonance investigation of the irradiated thiols in a number of aqueous and organic matrices in the presence of oxygen has clearly shown the production of sulfinyl radicals,[10] viz.,

$$RS^\bullet + O_2 \rightarrow \rightarrow RSO^\bullet \qquad (2)$$

Two distinct mechanistic schemes have been proposed for the formation of $RSO^\bullet$ radicals, whose identities have been confirmed using ^{17}O-labelled molecular oxygen.

To summarize, although the $RSOO^\bullet$ radicals have been invoked many times as reactive intermediates in the reactions of thiyl radicals with molecular oxygen, these species remain elusive.

THEORETICAL STUDIES

Molecular orbital calculations can provide an insight to possible reaction paths for the reaction of thiyl radicals with molecular oxygen, the structure of the reactive intermediates and the assignment of their optical absorption spectra. Indeed, the structural parameters of a molecule in its ground state can be accurately determined by performing *ab initio* SCF calculations employing a double zeta basis set plus polarization functions (3-21G*, 6-31G*), reaction paths can be investigated with confidence by also taking into account the correlation energy either using configuration interaction (CI) methods or with Moeller-Plesset perturbation theory at second (MP2), third (MP3), and fourth (MP4) orders,[11] and optical absorption spectra can be assigned using the Multiple Scattering Xα (MSXα) method.[12]

Reaction of Methanethiyl with Molecular Oxygen

Two reaction paths were explored for the reaction of the methanethiyl radical with molecular oxygen:[13] an asymmetric reaction path which leads to the methanethiylperoxyl radical (eq. 3) and a symmetric reaction path which leads to a cyclic dioxathiirane type structure (eq. 4).

$$CH_3S^\bullet + {}^3O_2 \rightarrow CH_3SOO^\bullet \qquad (3)$$

$$CH_3S^\bullet + {}^3O_2 \rightarrow CH_3-\overset{\bullet\bullet}{S}\overset{O}{\underset{O}{\diagup}} \qquad (4)$$

However, the closure of the methanethiylperoxyl radical to the cyclic dioxathiirane (eq. 5) and its further rearrangement to methanesulfonyl radicals (eq. 6) were also investigated in detail.[13] The energy correlation diagram for these reactions is reported in Figure 1.

$$CH_3SOO^\bullet \rightarrow CH_3-\overset{\bullet\bullet}{S}\overset{O}{\underset{O}{\diagup}} \qquad (5)$$

$$CH_3-\overset{\bullet\bullet}{S}\overset{O}{\underset{O}{\diagup}} \rightarrow CH_3-\overset{\bullet\bullet}{S}\overset{O}{\underset{O}{\diagdown}} \qquad (6)$$

The ground state of molecular oxygen is a triplet state. Two unpaired electrons occupy the degenerate π^* antibonding MOs derived from the out-of-phase interaction of the 2p orbitals which are perpendicular to the molecular axis. These orbitals are mainly involved in the reactivity of molecular oxygen by mixing with the degenerate (e symmetry) high-lying occupied 3p orbitals on sulfur and the low-lying unoccupied σ_{CS}^* MOs of the methanethiyl radical.

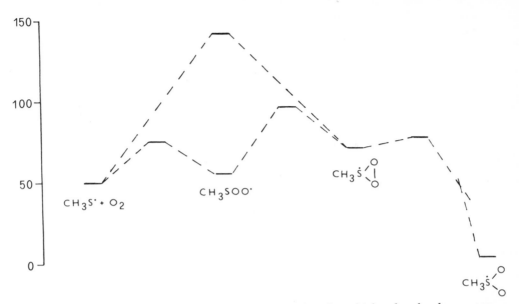

150 —

100 —

50 —

0 —

CH₃S˙ + O₂

CH₃SOO˙

CH₃Ṡ⟨O/O (with double bond to O)

CH₃Ṡ⟨O/O

Fig. 1. MP2/3-21G* energy profile for the reaction of methanethiyl and molecular oxygen.

In the asymmetric reaction pathway, the bond between the sulfur and the oxygen atom is formed by interaction of the 3p singly occupied MO on sulfur with the π^* MO on molecular oxygen. For this type of reaction a relatively low energy barrier is expected. In fact, the energy barrier, from MP2/3-21G* calculations, was estimated to be 25 kcal mol⁻¹ and the methanethiylperoxyl radical is about 5 kcal mol⁻¹ less stable than the reagents. The two reagents approach each other along a perpendicular pathway keeping the alkyl group trans to the terminal oxygen. The methyl group lies slightly out ($\approx 20°$) of the plane defined by the sulfur and oxygen atoms. The σ/π mixing is small and this radical can be considered as a π-type. The unpaired electron occupies the antibonding π^* MO resulting from the interaction of the singly occupied π^* MO of $^3O_2$ with the doubly occupied 3p sulfur orbital and is mainly localized on the terminal oxygen.

Along the symmetric pathway the thiyl radical approaches the molecular oxygen keeping the C–S bond perpendicular to the SOO plane and in the cyclic structure the alkyl group adopts an apical position, the value of the angle between the C–S bond and the ring plane being about 80°. The energy barrier for formation of the cyclic structure is much higher ($\Delta E = 91$ kcal mol⁻¹) than that computed for the asymmetric pathway ($\Delta E = 25$ kcal mol⁻¹). The dioxathiirane type radical is found to be 20 kcal mol⁻¹ less stable than the methanethiylperoxyl radical. In fact, in the cyclic structure the three outermost electrons occupy two sigma antibonding MOs, σ_{SO}^* (singly occupied) and pseudo-π_{OO}^* (doubly occupied) making the radical highly unstable. Formation of the sulfonyl radical occurs through the increase of the O–S–O angle. The vacant σ_{OO}^* MO is stabilized due to a reduction of overlap between the 2p orbital of the oxygen atoms which are antibonding and becomes more stable than the σ_{SO}^* SOMO. Since both orbitals have the same symmetry (a"), owing to avoided crossing, a barrier ($\Delta E = 14$ kcal mol⁻¹) occurs between the cyclic structure and the excited sulfonyl radical having the unpaired electron localized mainly at oxygen atoms. For large values of the O–S–O angle the energy of the σ_{OO}^* SOMO becomes comparable to that of an oxygen lone pair due to the small overlap between the 2p oxygen atomic orbitals, and is found to be more stable than the MO deriving from the antibonding interaction between the sulfur 3p orbital and the in-phase 2p oxygen atomic orbital which are strongly destabilized owing to the reduction of overlap as shown in Figure 2. At an O–S–O angle of about 95° the potential energy curve of the sulfonyl radical ($^2A'$) crosses that of the excited sulfonyl radical ($^2A''$), the two states having different symmetry, and the more stable sulfonyl radical can be formed.

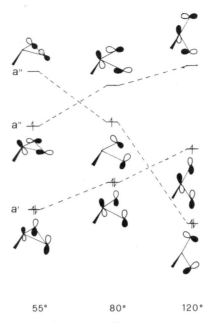

55°   80°   120°

Fig. 2. Correlation energy diagram of the MOs mainly involved in reaction 6. MOs for cyclic dioxathiirane (left), excited state (center) and ground state (right) of sulfonyl.

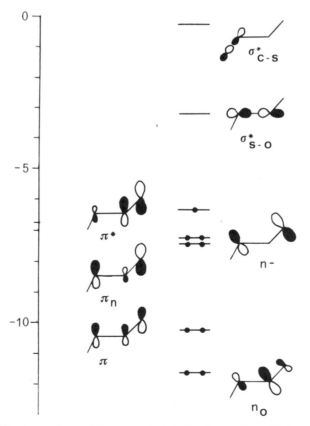

Fig. 3. MSXα eigenvalues of the ground state for the methanethiylperoxyl radical.

The radical center becomes pyramidal, the angle between the C–S bond and the SOO plane decreasing from 80° to 50°. The energy diagram in Figure 1 shows that the sulfonyl radical is the most stable species, the other transient species being more than 50 kcal/mol higher in energy.

Electron Transitions in Alkanethiylperoxyl Radicals[13]

The eigenvalues of the alkanethiylperoxyl computed with the MSXα method are reported in Figure 3 and may serve as a guide for comparing electron transition energies. (As discussed in the article by M. Guerra in this book they do not correspond to ionization potentials or electron affinities as in Hartree-Fock theory.) The excitation from non-bonded pseudo-π and n orbitals to the pseudo-π* SOMO were computed at frequencies in the near infrared region at 8300 cm^{-1} and 9700 cm^{-1} (λ = 1205 nm, 1030 nm) as in peroxyl and perthiyl radicals.[12] The transitions from the bonding pseudo-π MO and the n MO mainly localized at central oxygen to the SOMO were found in the UV region at 35100 cm^{-1} (λ = 285 nm), and 44400 cm^{-1} (λ = 225 nm), respectively. In the near UV region the transition to the vacent σ_{SO}^{*} was computed at 26700 cm^{-1} (λ = 375 nm) but no transition is calculated for the visible. The transition to σ_{CS}^{*} was evaluated in the far UV region at 50500 cm^{-1} just above the first member of the ns Rydberg series.

CONCLUSIONS

Our calculations indicate that the reaction of methanethiyl radicals with oxygen either via an asymmetric path (reaction 3) or via a symmetric path (reaction 4) is endothermic (5 and 20 kcal mol^{-1}, respectively). However, there can be little doubt that the first step in the reaction of the methanethiyl radical with oxygen is:

$$CH_3S^{\bullet} + O_2 \rightarrow CH_3SOO^{\bullet} \tag{3}$$

as the computed activation energy for this path is 66 kcal mol^{-1} less than the corresponding activation energy for the symmetric path.

Although the formation of the cyclic dioxathiirane type structure via reactions 4 and 5 requires relatively high activation energies, such a cylcic intermediate would lead to the more stable sulfonyl radicals as shown in Figure 1.

The *ab initio* calculations indicate that alkanethiylperoxyl radicals, structurally similar to alkylperoxyl radicals, should show electron transitions around 240 nm and 1200 nm. Indeed, MSXα calculations predict both these transitions for the methanethiylperoxyl radical. Consequently, the weak absorption band at 550 nm assigned to alkanethiylperoxyl radicals (see introduction) should be due to another species. Based on the available data on the chemiluminescence spectrum[14] of HSO$^{\bullet}$ and on the MSXα calculations of the sulfur-centered species,[12] candidates for the 550 nm absorption could be the RSO$^{\bullet}$ radicals. Therefore, for the reaction of thiyl radicals with oxygen in the presence of high thiol concentration, we tentatively suggest the following mechanism:

$$RS^{\bullet} + O_2 \rightarrow RSOO^{\bullet} \tag{1}$$

$$RSOO^{\bullet} + RSH \rightarrow RSOO\overset{\bullet}{S}(H)R \rightarrow RSO^{\bullet} + RSOH \tag{7}$$

Moreover, this suggestion finds further support from ESR experiments with irradiated thiols in a number of aqueous and organic matrices in the presence of molecular oxygen.[10]

REFERENCES

1. See for example: H. MacLeod, S.M. Aschmann, R. Atkinson, E.C. Tuazon, J.A. Sweetmann, A.M. Winer, and T.N. Pitts Jr., *J. Geophys. Res.* 91:5338 (1986).

2. C. von Sonntag, "The Chemical Basis of Radiation Biology", Taylor & Francis, New York, pp. 353 (1987).

3. C. Chatgilialoglu, *in:* "The Chemistry of Sulphones and Sulphoxides", S. Patai, Z. Rappoport, and C.J.M. Stirling, eds., Wiley, Chichester, Chapter 25 (1988).

4. S. Hatakeyama and H. Akimoto, *J. Phys. Chem.* 87:2387 (1983) and references cited therein for the earlier work.

5. M. Suzuki, G. Inoue and H. Akimoto, *J. Chem. Phys.* 81:5405 (1984).

6. G.S. Tyndall and A.R. Ravishankara, *J. Phys. Chem.* 93:2426 (1989).

7. C. Chatgilialoglu, *in:* "The Chemistry of Sulphenic Acids and Their Derivatives", S. Patai, ed., Wiley, Chichester , (1990).

8. J. Mönig, K.-D. Asmus, L.G. Forni, and R.L. Willson, *Int. J. Radiat. Biol.* 52:589 (1987).

9. C. Chatgilialoglu, *in:* "Handbook of Organic Photochemistry", Vol. II, J.C. Scaiano, ed., CRC Press, Boca Raton, Chapter 1 (1989).

10. S.G. Swarts, D. Becker, S. DeBolt, and M.D. Sevilla, *J. Phys. Chem.* 93:115 (1989).

11. see: T. Clark in this volume.

12. see: M. Guerra in this volume.

13. C. Chatgilialoglu and M. Guerra, manuscript in preparation.

14a. K.H. Becker, M.A. Inocencio, and U. Schurath, *Int. J. Chem. Kinet. Symp.* 1:205 (1979).

  b. U. Schurath, M. Weber, and K.H. Becker, *J. Chem. Phys.* 67:110 (1977).

THERMOCHEMISTRY OF SULFUR-CENTERED INTERMEDIATES

David Griller, José A. Martinho Simões*, and Daniel D.M. Wayner

Division of Chemistry
National Research Council of Canada
Ottawa, Ontario, Canada K1A 0R6

*Centro de Química Estrutural, Complexo I
Instituto Superior Técnico
1096 Lisboa Codex, Portugal

Sulfur-centered intermediates play important roles in the chemistry of coal, oil, atmospheric pollution and biological systems. If we are to fully understand this chemistry and to arm ourselves with some predictive power about the behavior of transient sulfur-centered species, we need to have reliable thermochemical data to guide us.

Sidney Benson recognized the importance of having a reliable database of this kind and wrote an excellent review of the subject.[1] In this work, he drew upon the solid thermochemical information that was available on sulfur containing molecules. While only limited data was available on transient intermediates, Benson was able to expand the applicability of these results by the use of group additivity contributions.

Much of the data in that review[1] remains current. However, in addition to new experiments on sulfur-containing species that have been carried out in the last decade, significant changes have occurred in data on the heats of formation of alkyl radicals These affect calculations of the strengths of carbon-sulfur bonds. In this review, we have taken account of these changes and have recalculated much of the thermochemistry that relates to sulfur-centered radicals. We have also used mass spectrometric and electrochemical data to illustrate how the thermochemical properties of sulfur-containing molecules, radicals, ions, and radical ions can be tied together.

THERMOCHEMICAL DATA

The overwhelming majority of values selected for the standard enthalpies of formation of organic sulfur compounds[2,3] have been derived from rotating-bomb combustion calorimetry experiments and are thus believed to be quite reliable.[4,5] From these data, unknown enthalpies of formation can often be easily estimated by using additivity schemes.[2,6] This relative wealth of accurate data for stable molecules contrasts with the values available for the enthalpies of formation of sulfur-centered organic radicals, where errors are typically ± 8 kJ mol^{-1}.

Sulfur-Centered Reactive Intermediates in Chemistry and Biology, Edited by
C. Chatgilialoglu and K.-D. Asmus, Plenum Press, New York, 1990

Sulfur-Hydrogen Bond Dissociation Enthalpies

The values reported for the first bond dissociation enthalpy of H_2S, D(HS–H), are shown in Table 1 in chronological order, starting with the one given in Benson's review.[1] The highest value, 389 ± 8 kJ mol^{-1},[8] was derived by combining the measured electron affinity of

Table 1. Sulfur-Hydrogen Bond Dissociation Enthalpies (kJ mol^{-1})

LH	$\Delta H_f^0(LH,g)^a$	L	$\Delta H_f^0(L,g)^b$	D(L–H)	Ref.
H_2S	-20.6 ± 0.5	HS	146 ± 4	385 ± 4	1
			140 ± 5	379 ± 5	7
			150 ± 8^c	389 ± 8^c	8
			142.7	381.3	9
			142 ± 4	381 ± 4	10
			138.6 ± 0.4	377.2 ± 0.6	11
			139 ± 5	378 ± 5	12
		Selected	*139 ± 4*	*378 ± 4*	
HS	139 ± 4^d	S	*278.81*	*358 ± 4*	
MeSH	-22.9 ± 0.7	MeS	143 ± 6	384 ± 6	1
			128 ± 8^c	369 ± 8^c	8
			139 ± 6	380 ± 8	10
			130 ± 4	371 ± 4	14
		Selected	*129 ± 8*	*370 ± 8*	
RSH^e		RS		*370 ± 8*	
PhSH	112.4 ± 0.9	PhS	243 ± 8	*349 ± 8*	10
		Selected	*233 ± 8*	*338 ± 8*	
H_2S_2	15.52^f	HS_2	*92 ± 10*	*295 ± 10*	
HS_2	92 ± 10^d	S_2	*128.4*	*254 ± 10*	
H_2S_n		HS_n		*295 ± 10*	
HS_3	108^d	S_3	*133*	*243*	
HS_4	121^d	S_4	*137*	*234*	
HS_5	135^d	S_5	*124*	*207*	
$RSSH^e$		RS_2		*295*	
RS_nH^e		RS_n		*295*	

[a] Data from ref. 3 unless indicated otherwise.
[b] Enthalpies of formation of S, S_n ($n \geq 2$), and H quoted from ref. 9,
[c] See text.
[d] Selected value.
[e] R = alkyl.
[f] Ref. 9.

HS, 2.31 ± 0.01 eV,[8] with the gas-phase acidity of H_2S, 1478.6 ± 8.4 kJ mol^{-1}, and the ionization energy of hydrogen atom, 13.598 eV,[9] eq. 1.

$$D(HS-H) = EA(HS) + \Delta H_{acid}(H_2S) - IE(H) \qquad (1)$$

The ΔH_{acid} values presented in a recent compilation[13], are, however, 6 - 10 kJ mol^{-1} lower than the one used in ref. 8, implying a similar decrease in D(HS–H), i.e. D(HS–H) = 379 - 383 kJ mol^{-1}. On the other hand, the "high" value recommended in Benson's review, 385 ± 4 kJ mol^{-1}, was quoted from an early edition of JANAF tables which now give D(HS–H) = 378 ± 5 kJ mol^{-1},[12] a result that relies on photoionization data. The value obtained by Hwang and Benson from gas-phase iodination studies,[7] 379 ± 5 kJ mol^{-1}, is in good agreement with the ones recommended in JANAF[12] and NBS[9] tables. Finally, the number quoted from McMillan and Golden's review,[10] 381 ± 4 kJ mol^{-1}, is an average between Hwang and Benson's value and an early JANAF recommendation.

The value selected in Table 1 for D(HS–H) reflects the above comments and also the most recent photoionization result, 377.2 ± 0.6 kJ mol^{-1}.[11] The second sulfur-hydrogen bond dissociation enthalpy in H_2S, D(S–H), Table 1, was derived from the value selected for $\Delta H_f^0(HS,g)$.

The first value for D(MeS–H) in Table 1, quoted from Benson's review,[1] was later reassessed by McMillan and Golden[10] as 380 ± 8 kJ mol^{-1}. More recent iodination studies by Shum and Benson[14] led to a lower result, 371 ± 4 kJ mol^{-1}, which is in good agreement with the one derived from the electron affinity to MeS together with $\Delta H_{acid}(MeSH) = 1502.1 \pm 8.4$ kJ mol^{-1} and IE(H), i.e. D(MeS–H) = 369 ± 8 kJ mol^{-1}.[8] However, this value decreases by 9 kJ mol^{-1} when the recently recommended value for $\Delta H_{acid}(MeSH)$ is used.[13] The relatively large uncertainty assigned to the selection in Table 1 is the outcome of the above discrepancies.

It has long been recognized that sulfur-hydrogen bond dissociation enthalpies D(RS–H) are independent of the length and configuration of the alkyl chain.[1,15] This has been confirmed in an electron photodetachment study by Janousek et al.[8] of a series of RSH molecules (R = Me, Et, Pr, i-Pr, and t-Bu). For these compounds, D(RS–H) were found to be constant to within ca. ± 1 kJ mol^{-1}.

According to the selected values in Table 1, replacing a hydrogen by an alkyl group in H_2S leads to a decrease of 8 kJ mol^{-1} in D(S–H). A larger effect (29 kJ mol^{-1}) is of course induced by a phenyl group, due to the resonance stabilization of the PhS radical. The value selected in Table 1 for D(PhS–H) is lower than the one given by Benson,[1] 343 ± 6 kJ mol^{-1} (later reassessed in ref. 10 as 349 ± 8 kJ mol^{-1}). It relies on more recent data for $\Delta H_f^0(PhSMe,g)$,[3] on $\Delta H_f^0(Me,g)$ (Appendix), and on the enthalpy of decomposition of methylphenylsulfide, 282 kJ mol^{-1},[16] which produces PhS and Me radicals.

The enthalpy of formation of HSS$^\bullet$ radical is not known but an approximate value can be estimated by using experimental data[3,9] and an additivity method:

$$\Delta H_f^0(H_2S_2,g) = 15.52 \text{ kJ mol}^{-1} = 2[S-(H)(S)]$$

$$\Delta H_f^0(S_2Me_2,g) = -24.2 \text{ kJ mol}^{-1} = 2[C-(H)_3(S)] + 2[S-(C)(S)]$$

$$\begin{aligned} \Delta H_f^0(HSS,g) - \Delta H_f^0(MeSS,g) &= [S-(H)(S^\bullet)] - [S-(C)(S^\bullet)] - [C-(H)_3(S)] \\ &= (15.52 + 24.2)/2 = 19.9 \text{ kJ mol}^{-1} \end{aligned}$$

It is of course assumed that $[S-(H)(S^\bullet)] - [S-(C)(S^\bullet)] \approx [S-(H)(S)] - [S-(C)(S)]$. The enthalpy of formation of MeSS radical is available (72 ± 8 kJ mol^{-1}; see chapter "Sulfur-carbon bond dissociation enthalpies") yielding $\Delta H_f^0(HS_2,g)$ and D(HSS–H) in Table 1. An analogous exercise led Benson to similar results.[1]

The first-sulfur-hydrogen bond dissociation enthalpy in polysulfides H_2S_n (n>2) is expected to be nearly constant and similar to D(HSS–H), if group additivity methods are applicable. Consider, for example, HS_3. Its enthalpy of formation can be estimated from $\Delta H_f^0(HS_2,g)$ by adding the group value for [S–(S)(S$^\bullet$)], assumed equal to [S–(S)$_2$].[1] The result obtained, 105 kJ mol^{-1}, together with $\Delta H_f^0[H_2S_3,g] = 30.5$ kJ mol^{-1},[9] leads to D(HS$_3$–H) = 293 kJ mol^{-1}. Identical values are obtained for D(HS$_4$–H) and D(HS$_5$–H).

The constancy of sulfur-hydrogen first bond dissociation enthalpies in di- and polysulfides is not observed for the second bond dissociation enthalpies, D(S$_n$–H), n ≥ 2. These were derived from the estimated enthalpies of formation of HS$_n$ and from tabulated data for $\Delta H_f^0(S_n,g)$.[9] The large decrease in D(S$_n$–H) with increasing n is due to stabilization of S$_n$ molecules which undergo cyclization when S–H bond is cleaved. The smaller n, the higher the ring strain, implying a smaller stabilization and a larger bond dissociation enthalpy.

Reliable estimates of D(RSS–H) are easy to make. It has been shown[17] that D(RSS–Bu-t) does not change with R (see chapter "Sulfur-carbon bond dissociation enthalpies"). This constancy, together with the group additivity scheme used to obtain $\Delta H_f^0(RSSH,g)$ and $\Delta H_f^0(RSSBu-t,g)$,[17] implies that D(RSS–H) is also independent of R. Consider, for example, R=Me. In this case, $\Delta H_f^0(MeSSMe,g)$ is taken as the average between $\Delta H_f^0(H_2S_2,g)$[9] and $\Delta H_f^0(MeSSMe,g)$.[3] The enthalpy of formation of MeSS$^\bullet$ radical, 72 ± 8 kJ mol^{-1} (see chapter "Sulfur-carbon bond dissociation enthalpies"), leads to the value shown in Table 1 for D(RSS–H). The additivity method implies also that this bond dissociation enthalpy is equal to D(RS$_n$–H), n>2 and R=H, alkyl.

Sulfur-Sulfur Bond Dissociation Enthalpies

Sulfur-sulfur bond dissociation enthalpies in di- and polysulfides, H_2S_n, shown in Table 2, were calculated from the enthalpies of formation of the radicals in Table 1. It is observed

Table 2.   Sulfur-Sulfur Bond Dissociation Enthalpies (kJ mol^{-1})

LL'	$\Delta H_f^0(LL',g)^a$	L	$\Delta H_f^0(L,g)^b$	D(S–S)
S_2	128.4	S	278.81	429.2
H_2S_2	15.52	HS	139 ± 4	262
H_2S_3	30.50	HS$_2$	92 ± 10	201
H_2S_4	44.22	HS$_3$	108	203
		HS$_2$	92 ± 10	140
H_2S_5	57.91	HS$_4$	121	202
		HS$_3$	108	142
RSSR'^{c}		RS		285 ± 16
PhSSPh	243.5 ± 4.1^d	PhS	233 ± 8	223 ± 16
RS$_4$R'^{c}		RS$_2$		142 ± 16

[a] Data from ref. 9 unless indicated otherwise.
[b] Enthalpy of formation of S from ref. 9.
[c] R,R' = alkyl.
[d] Ref. 3.

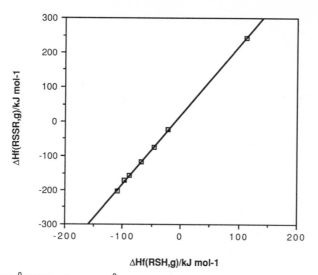

Fig. 1. ΔH_f^0(RSSR, g) vs. ΔH_f^0(RSH, g). R = Me, Et, Pr, Bu, i-Bu, t-Bu, and Ph.
Data from ref. 3.

that D(HS–SH) is about 60 kJ mol⁻¹ higher than D(HS$_n$–SH), which has a constant value, *ca.* 202 kJ mol⁻¹. This large decrease is attributed to higher stabilization energies of HS$_n$ fragments, which are nearly constant for $n \geq 2$. Indeed, when the group SH in HS$_n$–SH is replaced by HS$_{n'}$, the bond dissociation enthalpies also decrease by 60 kJ mol⁻¹: D(HS$_n$–S$_m$H) $\approx$ 140 kJ mol⁻¹.

Sulfur-sulfur bond dissociation enthalpies in organic disulfides RSSR' do not depend on the nature of alkyl groups R and R'. It is easily demonstrated that this constancy in D(RS–SR') stems from the fact that D(RS–H) are also constant. A plot of ΔH_f^0(RSSR,g) *vs.* ΔH_f^0(RSH,g) leads to an excellent linear correlation (Figure 1):

$$\Delta H_f^0(RSSR,g) = (1.999 \pm 0.014)\,\Delta H_f^0(RSH,g) \; + \; (19.3 \pm 1.2) \tag{2}$$

The meaning of this correlation can be understood by considering the reaction

$$RSSR(g) \; + \; H_2(g) \;\; \rightarrow \;\; 2RSH(g) \tag{3}$$

from which

$$\Delta H_f^0(RSSR,g) \;\; = \;\; 2\Delta H_f^0(RSH,g) \; - \; \Delta H_r^0 \tag{4}$$

A comparison between equations 2 and 4 enables one to conclude that the enthalpy of reaction 3, $\Delta H_r^0 = -19.3$ kJ mol⁻¹ for any R. Finally, this enthalpy can be expressed in terms of bond dissociation enthalpies:

$$\Delta H_r^0 \;\; = \;\; D(RS–SR) \; - \; 2D(RS–H) \; + \; D(H–H) \tag{5}$$

Using D(H–H) = 436.0 kJ mol⁻¹ [9] and D(RS–H) = 370 ± 8 kJ mol⁻¹ (Table 1), D(RS–SR) = 285 ± 16 kJ mol⁻¹ is obtained.

Another interesting feature of the above correlation is that it includes the data for R=Ph. Therefore, the nearly perfect constancy of ΔH_r^0 reflects the fact that D(RS–SR) and

D(RS–H) follow parallel trends. Equation 5 and D(PhS–H) = 338 ± 8 kJ mol^{-1} (Table 1) lead to D(PhS–SPh) = 221 ± 16 kJ mol^{-1}, very close to the value in Table 2, which was directly obtained from ΔH_f^0(PhSSPh,g).

When R = alkyl and R' = aryl, the additivity method implies that D(RS–SR') is simply the average of D(RS–SR) and D(R'S–SR'). The same applies to D(RS–SH) and D(R'S–SH).

A constant sulfur-sulfur bond dissociation enthalpy D(RSS–SSR') is also observed for the tetrasulfides RS$_4$R' (R, R' = alkyl). This is not unexpected since it follows from the additivity method and from the constancy of D(RSS–Bu–t) for different R (see chapter "Sulfur-carbon bond dissociation enthalpies"). Consider R=R'=Me. In this case, ΔH_f^0(MeS$_4$Me,g) is estimated as 2.6 ± 3.0 kJ mol^{-1} by adding two [S–(S)$_2$] groups[1] to ΔH_f^0(MeS$_2$Me,g).[3] Using ΔH_f^0(MeSS,g) = 72 ± 8 kJ mol^{-1} (see chapter "Sulfur-carbon bond dissociation enthalpies"), D(MeSS–SSMe) = D(RSS–SSR') = 142 ± 16 kJ mol^{-1}. Note also that this value matches the one obtained for D(HSS–SSH), which indicates similar relaxation energies for fragments HS$_2$ and RS$_2$. It is therefore most likely that D(RS$_n$–S$_m$R') values (n,m ≥ 2) have a constant value of ca. 140 kJ mol^{-1}.

A similar procedure could be used to calculate D(PhSS–SSPh) but ΔH_f^0(PhSS,g) is not available. It is, however, expected that the influence of the phenyl group on the stabilization energy of PhSS is substantially smaller than in the case of PhS. Therefore, D(PhSS–SSPh) may not be far from the upper limit of 140 kJ mol^{-1}.

Other sulfur-sulfur bond dissociation enthalpies, such as D(HSS–SR) (R = alkyl, Ph), are easily calculated from the data in Table 2. Note, for example, that D(HSS–SH) = [D(HSS–SSH) + D(HS–SH)]/2. Similarly, D(HSS–SR) = [D(HSS–SSH) + D(RS–SR)]/2 ≈ 213 or 182 kJ mol^{-1} for R = alkyl or phenyl, respectively.

Sulfur-Carbon Bond Dissociation Enthalpies

Sulfur-carbon bond dissociation enthalpies D(HS–R) presented in Table 3 rely on the value selected for ΔH_f^0(HS,g), Table 1, and those for ΔH_f^0(R,g) which are given in the Appendix. As expected, D(HS–R) are fairly constant (ca. 303 kJ mol^{-1}) (except for R = Me). They decrease by only 4 kJ mol^{-1} in the cases of i-Pr and s-Bu, and by 10 kJ mol^{-1} for R = t-Bu. The noteworthy feature of the set of data is that they fall in a narrow range, 16 kJ mol^{-1}, as compared with the spread of 43 kJ mol^{-1} observed for D(R–H) in the corresponding hydrocarbons. The approximate constancy of D(HS–R), R = alkyl, is illustrated by the plot in Figure 2 and by eq. 6 (r = 0.995).

$$\Delta H_f^0(RSH,g) = (0.894 \pm 0.035)\ \Delta H_f^0(R,g) - (153.7 \pm 3.1) \tag{6}$$

The intercept for eq. 6, when added to ΔH_f^0(SH,g) gives an average value for D(HS–R) of 293 kJ mol^{-1}, which is close to the average obtained directly from D(HS–R) in Table 3. It is also noted that similar plots involving alcohols ROH, amines RNH$_2$, and alkyl chlorides RCl, lead to a similar conclusion: the length and the configuration of alkyl radicals have little effect on D(R–X), X = OH, NH$_2$, Cl.

The influence of Pauling's electronegativity of X on variations of D(R–X) has been discussed before,[18] the conclusion being that those variations "increase almost exponentially as the electronegativity of the leading atom in X decreases". This statement has been quantified in a recent paper by Luo and Benson.[19] Here, Pauling electronegativity has been replaced by the so-called unshielded core potential, V_x defined by the ratio between the number of valence electrons of X, n_x and the covalent radius of X, r_x. Plots of ΔH_f^0(RX,g) - ΔH_f^0(MeX,g) versus V_x for several R yielded good linear correlations from which D(R–X) – D(Me–X) can be

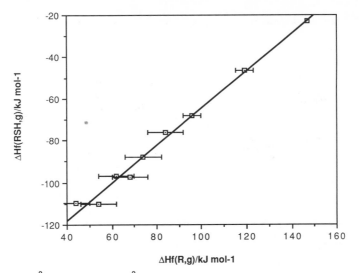

Fig. 2. ΔH_f^0(RSH, g) vs. ΔH_f^0(R, g). R = Me, Et, Pr, i-Pr, Bu, s-Bu, i-Bu, t-Bu, and Pe.

Table 3. Sulfur-Carbon Bond Dissociation Enthalpies (kJ mol^{-1})

LL'	ΔH_f^0(LL',g)[a]	L'	ΔH_f^0(L',g)	D(L–L')
MeSH	–22.9 ± 0.7	SH	139 ± 4	309 ± 4
EtSH	–46.3 ± 0.6			304 ± 6
PrSH	–67.9 ± 0.7			303 ± 6
i-PrSH	–76.2 ± 0.7			299 ± 9
BuSH	–88.1 ± 1.2			301 ± 9
s-BuSH	–96.9 ± 0.9			298 ± 9
i-BuSH	–97.3 ± 0.9			304 ± 9
t-BuSH	–109.6 ± 0.9			293 ± 7
PeSH	–110.1 ± 1.1			303 ± 9
PhSH	112.4 ± 0.9			357 ± 9
BzSH	92.8 ± 1.1			246 ± 7
MeSR[b]		SR		313 ± 8
MeSBu-t	–121.3 ± 0.8	SBu-t	42 ± 8	310 ± 8
MeSPh	97.3 ± 0.8	SPh	233 ± 8	283 ± 8
EtSR[b]		SR		308 ± 9
PrSMe	–82.3 ± 1.0	SMe	129 ± 8	307 ± 9
i-PrSMe	–90.5 ± 0.8	SMe		304 ± 11
t-BuSMe	–121.3 ± 0.8	SMe		294 ± 10

43

Table 3 (continued)

LL'	$\Delta H_f^0(LL',g)^a$	L'	$\Delta H_f^0(L',g)$	D(L–L')
PrSEt	−104.7 ± 0.8	SEt	106 ± 8	307 ± 9
i-PrSEt	−117.2 ± 2.4	SEt	•	307 ± 12
t-BuSEt	−148.0 ± 2.5	SEt		298 ± 10
PrSPr	−125.3 ± 0.9	SPr	84 ± 8	305 ± 9
i-PrSPr-i	−142.0 ± 1.5	SPr-i	76 ± 8	302 ± 11
t-BuSBu-t	−188.9 ± 1.0	SBu-t	42 ± 4	275 ± 7
PhSMe	97.3 ± 0.8	SMe	129 ± 8	362 ± 11
PhSEt	77.0 ± 2.6	SEt	106 ± 8	359 ± 12
		SPh	233 ± 8	275 ± 9
PhSPh	231.0 ± 2.9	SPh		332 ± 12
MeSSR^c		S_2R		243 ± 10
EtSSR^c		S_2R		241 ± 9
i-PrSSR^c		S_2R		235 ± 9
t-BuSSR^c		S_2R		230 ± 5
PhSSR^c		S_2R		292 ± 12
MeC(O)SH	−175.1 ± 8.5	SH	139 ± 4	290 ± 10
MeC(O)SEt	−228.2 ± 0.9	SEt	106 ± 4	310 ± 4
MeC(O)SPr	−250.4 ± 0.9	SPr	84 ± 8	310 ± 8
MeC(O)SPr-i	−256.3 ± 1.0	SPr-i	76 ± 8	308 ± 8
MeC(O)SBu	−270.6 ± 1.3	SBu	64 ± 8	310 ± 8
MeC(O)SBu-t	−283.6 ± 1.1	SBu-t	42 ± 4	301 ± 4
R_2SO^d		SO	5.0 ± 1.3	225 ± 4^e
Ph_2SO	106.8 ± 3.1	SO		279 ± 8^e
Me_2SO_2	−373.1 ± 3.0	SO_2	−296.8 ± 0.2	185 ± 2^e
$Me(Et)SO_2$	−408.6 ± 3.0	SO_2		189 ± 3^e
Et_2SO_2	−429.3 ± 2.6	SO_2		185 ± 4^e
$Me(i-Pr)SO_2$	−434.0 ± 2.8	SO_2		184 ± 4^e
$Me(t-Bu)SO_2$	−473.5 ± 3.8	SO_2		184 ± 4^e
$(t-Bu)_2SO_2$	−546.3 ± 3.7	SO_2		169 ± 6^e
Ph_2SO_2	−118.7 ± 3..3	SO_2		241 ± 8^e
Bz_2SO_2	−157.1 ± 3.2	SO_2		130 ± 6^e

[a] Data from ref. 3 unless indicated otherwise.
[b] R = alkyl.
[c] R = H, alkyl.
[d] R = n-alkyl (but probably valid for any alkyl).
[e] Mean bond dissociation enthalpy.

derived as a function of V_x. Take, for example, $R = t\text{-Bu}$. In this case $\Delta H_f{}^0(t\text{-BuX,g})$ – $\Delta H_f{}^0(\text{MeX,g}) = -33.9 - 9.62\,V_x$.[19] Using the values for $\Delta H_f{}^0(t\text{-Bu,g})$ and $\Delta H_f{}^0(\text{Me,g})$ given in the Appendix one obtains

$$D(\text{Me–X}) \;-\; D(t\text{-Bu–X}) \;\; = \;\; 69.0 - 9.62\,V_x \tag{7}$$

When $X = SH$, $V_x = 5.77$[19] and $D(\text{Me–SH}) - D(t\text{-Bu–SH}) = 13.5 \text{ kJ mol}^{-1}$, quite close to the difference in Table 3. For low electronegativity elements, such as hydrogen, the corresponding difference will be large and positive, whereas for high electronegativity elements, such as oxygen (in OH), $D(t\text{-Bu–X})$ exceeds $D(\text{Me–X})$.

The constancy of $D(\text{RS–Me}) = 313 \pm 8 \text{ kJ mol}^{-1}$ and $D(\text{RS–Et}) = 308 \pm 9 \text{ kJ mol}^{-1}$ (R =alkyl) could be expected from a constant $D(\text{RS–H})$, Table 1. Likewise, $D(\text{RS–Me}) - D(\text{PhS–Me}) = 30 \text{ kJ mol}^{-1}$ is very close to $D(\text{RS–Et}) - D(\text{PhS–Et}) = 33 \text{ kJ mol}^{-1}$ and to $D(\text{RS–H}) - D(\text{PhS–H}) = 32 \text{ kJ mol}^{-1}$ (Table 1). The same trends must occur for other $D(\text{RS–R'})$.

The small but probably significant decrease (5 kJ mol^{-1}) when going from $R' = \text{Me}$ to $R' = \text{Et}$ in $D(\text{RS–R'})$ is not likely to continue for higher n-alkyls. For example, in the case of $R' = \text{Pr}$, $D(\text{RS–Pr}) \approx 306 \text{ kJ mol}^{-1}$ (R = Me, Et, Pr). It is therefore possible to assign a constant value to $D(\text{RS–R'})$ of $307 \pm 8 \text{ kJ mol}^{-1}$, R' being a n-alkyl group other than methyl.

$D(\text{RS–R'})$ is not, in fact, sensitive to the nature of R, since the two alkyl groups are kept far apart by the large sulfur atom. For example, $D(t\text{-BuS–Me}) = 310 \pm 8 \text{ kJ mol}^{-1}$ is close to $D(\text{RS–Me}) = 313 \pm 8 \text{ kJ mol}^{-1}$ and $D(t\text{-BuS–Et}) = 309 \pm 6 \text{ kJ mol}^{-1}$ (not included in Table 3) is similar to $D(\text{RS–Et}) = 308 \pm 9 \text{ kJ mol}^{-1}$. However, when R' is a branched alkyl group, a small decrease in $D(\text{RS–R'})$ is apparent: $D(\text{MeS–Pr}) > D(\text{MeS–Pr-}i) > D(\text{MeS–Bu-}t)$ and $D(\text{EtS–Pr}) > D(\text{EtS–Bu-}t)$. These trends are not in contradiction with the statement about the negligible interactions between R and R' since they are only due to different stabilization energies of R' radicals, i.e. the stabilization energy for $t\text{-Bu}^\bullet$ radical will be more negative than for $i\text{-Pr}^\bullet$ and $\text{Pr}^\bullet$. The lack of interactions is not only indicated by the above equalities, e.g. $D(t\text{-BuS–Et}) = D(\text{RS–Et})$ (R = alkyl), but also by a plot of $\Delta H_f{}^0(\text{RSR,g})$ versus $\Delta H_f{}^0(\text{RSH,g})$, Figure 3.

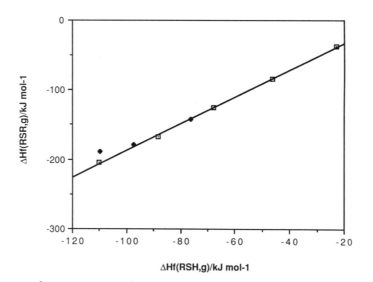

Fig. 3. $\Delta H_f{}^0(\text{RSR, g})$ vs. $\Delta H_f{}^0(\text{RSH, g})$. The line is the least square fitting to the data for R = Me, Et, Pr, Bu, and Pe (open squares). The remaining three points are for $i\text{-Pr}$, $i\text{-Bu}$, and $t\text{-Bu}$.

The least squares for the n-alkyl data (r = 0.9996),

$$\Delta H_f^0(RSR,g) \;=\; (1.936 \pm 0.031)\,\Delta H_f^0(RSH,g) \;+\; (6.1 \pm 2.3) \tag{8}$$

makes it possible to reproduce the experimental data[3] for other sulfides, RSR', within ca. ± 1 kJ mol⁻¹, including i-PrSPr-i, MeSPr-i, and MeSBu-t (in this case $\Delta H_f^0(RSH,g)$ used in equation 8 is simply the average of $\Delta H_f^0(MeSH,g)$ and $\Delta H_f^0(t$-BuSH,g)). Therefore, those sulfides are all free from steric effects caused by interactions between R and R' and D(t-BuS–Me) – D(MeS–Bu-t) = 16 kJ mol⁻¹ and D(i-PrS–Me) – D(MeS–Pr-i) = 9 kJ mol⁻¹ must reflect the negative values of the relaxation energies of t-Bu and i-Pr, respectively, relative to the relaxation energy of Me.

Additional steric effects (relative to t-BuSH) are noticeable in t-BuSBu-t and can be quantified by comparing its experimental enthalpy of formation with the value calculated from equation 8 (see also Figure 3). The result, 17 kJ mol⁻¹, can be compared with the one obtained from D(MeS–Bu-t) – D(t-BuS–Bu-t), 19 kJ mol⁻¹.

It is important to stress that the previous discussion relies on values which span a range of 19 kJ mol⁻¹ (excluding the di-$tert$-butyl compound). The general conclusion is therefore that sulfur-carbon bond dissociation enthalpies are rather constant (ca. 303 kJ mol⁻¹) in alkyl thiols and dialkyl sulfides.

Sulfur-phenyl bond dissociation enthalpies, D(RS–Ph) R being an alkyl group, are in the expected range, 360 kJ mol⁻¹, and are very close to D(HS–Ph). When R = Ph, the bond enthalpy is smaller due to the resonance stabilization of PhS: D(PhS–Ph) = 332 ± 12 kJ mol⁻¹.

The discussion of sulfur-carbon bond dissociation enthalpies in disulfides (Table 3) begins by recalling the comment made in the preceding chapter "Sulfur-sulfur bond dissociation enthalpies" that D(HS–SH) – D(HS₂–SH) ≈ D(HS₂–SH) – D(HS₂–S₂H) ≈ 60 kJ mol⁻¹ and D(HS₂–S₂H) ≈ D(RS₂–S₂R'), indicating similar relaxation energies for fragments HS₂ and RS₂. Those trends suggest that D(HS–R') will exceed D(HS₂–R') and D(RS₂–R') by a similar amount. This is, in fact, observed. For example, D(HS–Me) – D(HS₂–Me) = 66 kJ mol⁻¹, and D(HS–Bu-t) – D(t-BuS₂–Bu-t) = 63 kJ mol⁻¹.

The bond dissociation enthalpies D(RSS–R') in the disulfides show almost the same trends as those found for D(RS–R') in the sulfides. The only difference is that now the interaction between groups R and R' are negligible due to the extra sulfur atom, even in the case of t-BuSSBu-t. The enthalpy of formation of this molecule merely reflects the enthalpy of formation of the two separate t-BuS moieties (see chapter "Sulfur-sulfur bond dissociation enthalpies" and Figure 1), no additional correction due to repulsive interactions being required. In other words, the point for t-BuSSBu-t is fitted by equation 2. It should be remembered that a similar plot involving sulfides RSR', Figure 3, shows that the point for R=R'=t-Bu lies above the least squares fit to the n-alkyl data, equation 8.

Recent experimental measurements of the appearance energies of t-Bu⁺ ions from a series of disulfides RSSBu-t (R = Me, Et, i-Pr, t-Bu) yielded a constant value for that quantity, AE(t-Bu⁺) = 9.08 ± 0.04 eV,[17] which in turn implies a constant sulfur-t-butyl bond dissociation enthalpy. D(RSS–Bu-t) = 230 ± 5 kJ mol⁻¹, calculated by subtracting the adiabatic ionization energy of t-Bu radical, 6.70 ± 0.03 eV,[13] from the measured AE(t-Bu⁺), can then be used to derive other bond dissociation enthalpies, D(RSS–R'). Consider, for example, R = Me and R' = Et. $\Delta H_f^0(MeSSEt,g)$ is not tabulated but can be estimated as –50.0 ± 2.0 kJ mol⁻¹ by using equation 2 or simply by taking the average between $\Delta H_f^0(MeSSMe,g)$ and $\Delta H_f^0(EtSSEt,g)$.[3] As $\Delta H_f^0(MeSSBu-t,g)$ can also be estimated by the same method, giving –113.3 ± 2.0 kJ mol⁻¹, the above D(RSS–Bu-t) value yields $\Delta H_f^0(MeSS,g)$ = 72 ± 8 kJ mol⁻¹ and D(MeSS–Et) = 241 ± 9 kJ mol⁻¹. Additivity methods also indicate that sulfur-carbon bond dissociation enthalpies in polysulfides, D(RS$_n$–R'), will be identical to D(RSS–R').

The thermochemical data for ethanethioic acid and some of its esters, MeC(O)SR (R=H, alkyl),[3] together with the derived enthalpies of formation of SR radicals and $\Delta H_f^0(MeCO,g)$ (Appendix), led to the values of D[MeC(O)–SR] shown in Table 3. They are similar to other sulfur-carbon bond dissociation enthalpies, e.g. D(R–SH) and D(RS–R'), except in the case of ethanethioic acid, for which D[MeC(O)–SH] is significantly lower than D(R–SH). There is, however, a relatively large uncertainty in the enthalpy of formation of the acid in the gas phase. A plot of $\Delta H_f^0[MeC(O)SR,g]$ vs. $\Delta H_f^0(RSH,g)$ as in Figure 3, reveals that the i-Pr and t-Bu esters are destabilized by 2 and 9 kJ mol^{-1} respectively, i.e. these points lie above the line defined by n-alkyl data. This destabilization is in keeping with the bond enthalpy values in Table 3.

Sulfur-carbon bond dissociation enthalpies in sulfoxides, R_2SO, and sulfones, R_2SO_2, are not available. Mean $\overline{D}(S–R)$ values can, however, be derived by using the enthalpies of formation of those molecules and $\Delta H_f^0(SO,g)$ or $\Delta H_f^0(SO_2,g)$ (see Table 4), and $\Delta H_f^0(R,g)$ (Appendix). It is seen that $\overline{D}(OS–R) = [D(OS(R)–R) + D(OS–R)]/2 = 225$ kJ mol^{-1} are identical for R = Me, Et, and Pr. Also, the difference $\overline{D}(OS–Ph) – \overline{D}(OS–R) = 54$ kJ mol^{-1} is in the range observed in other families of compounds. For sulfones, $\overline{D}(O_2S–R)$ are not as constant, but their spread is still small, ca. 185 ± 4 kJ mol^{-1}, for R = n-alkyl. Interactions between the two alkyl groups are negligible, except perhaps when R = t-butyl. It is worth noting that the internal consistency of the thermochemical data for the sulfoxides is not as good as in previous cases, as judged by a plot between the enthalpies of formation and $\Delta H_f^0(RSR',g)$. The decrease of sulfur-carbon mean bond dissociation enthalpies from the sulfoxides to the sulfones was to be expected as a result of the increase of oxidation state of the sulfur atom.

Sulfur-Oxygen Bond Dissociation Enthalpies

The chemical stability of SO_2 relative to SO is evidenced by its strong sulfur-oxygen bond dissociation enthalpy, D(OS–O), which is almost 30 kJ mol^{-1} higher than D(S–O). In other words, the reaction

$$SO(g) + {}^1/_2O_2(g) \rightarrow SO_2(g) \tag{9}$$

is very exothermic, $\Delta H^0 = -301.8$ kJ mol^{-1} and $\Delta G^0 = -279.1$ kJ mol^{-1}. On the other hand, the oxidation of SO_2,

$$SO_2(g) + {}^1/_2O_2(g) \rightarrow SO_3(g) \tag{10}$$

has rather less favorable thermodynamics ($\Delta H^0 = -99.0$ kJ mol^{-1}, $\Delta G^0 = -71.0$ kJ mol^{-1}), reflecting a much weaker $O_2S–O$ bond (Table 4).

The trend $D_1(S–O) > D_2(S–O)$ in SO_2 is also observed for the sulfur-oxygen bonds in sulfoxides and sulfones, Table 4: $D[Me_2S(O)–O]$ are about 100 kJ mol^{-1} higher than $D(Me_2S–O)$.

There seems to be an increase in $D(R_2S–O)$ with the length of n-alkyl chain. This trend is not apparent for $D[R_2S(O)–O]$ but, as stated before, data for sulfones are probably less reliable than for sulfoxides and sulfides. For example, a comparison of $D[(Pr)_2S(O)–O]$ with the data for the remaining alkyls suggests that the value tabulated for $\Delta H_f^0[(Pr)_2SO_2,g]$ is ca. 10 kJ mol^{-1} too high.

Mean sulfur-oxygen σ-bond dissociation enthalpies in sulfites and sulfates are shown in Table 4. A decrease of ca. 50 kJ mol^{-1} from $\overline{D}(OS–OR)$ to $\overline{D}(O_2S–OR)$ is apparent (recall that $\overline{D}(OS–R) – \overline{D}(O_2S–R) \approx 40$ kJ mol^{-1}), but the trends are nearly parallel, i.e. there seems to be a slight decrease in the bond dissociation enthalpy with increasing size of alkyl chain. $D[(RO)_2S(O)–O]$ are seen to be constant and ca. 10 - 20 kJ mol^{-1} weaker than $D[R_2S(O)–O]$.

Table 4. Sulfur-Oxygen Bond Dissociation Enthalpies (kJ mol^{-1})

LL'	ΔH_f^0(LL',g)[a]	L'	ΔH_f^0(L',g)	D(L–L')
SO	5.0 ± 1.3[c]	O	249.2 ± 0.1	523.0 ± 1.3
SO$_2$	−296.8 ± 0.2	SO	5.0 ± 1.3	551.0 ± 1.3
SO$_3$	−395.8 ± 0.7[c]	SO$_2$	−296.8 ± 0.2	348.2 ± 0.7
Me$_2$SO	−151.3 ± 0.8	Me$_2$S	−37.5 ± 0.6	363.0 ± 1.0
Et$_2$SO	−205.6 ± 1.5	Et$_2$S	−83.6 ± 0.8	371.2 ± 1.7
Pr$_2$SO	−254.9 ± 1.5	Pr$_2$S	−125.3 ± 0.9	378.8 ± 1.7
Ph$_2$SO	106.8 ± 3.1	Ph$_2$S	231.2 ± 2.9	374 ± 4
Me$_2$SO$_2$	−373.1 ± 3.0	Me$_2$SO	−151.3 ± 0.8	471 ± 3
Et$_2$SO$_2$	−429.3 ± 2.6	Et$_2$SO	−205.6 ± 1.5	473 ± 3
Pr$_2$SO$_2$	−463.3 ± 2.6	Pr$_2$SO	−254.9 ± 1.5	463 ± 3
Ph$_2$SO$_2$	−118.7 ± 3.3	Ph$_2$SO	106.8 ± 3.1	475 ± 5
(MeO)$_2$SO	−483.4 ± 2.0	SO	5.0 ± 1.3	262 ± 4[d]
MeO(EtO)SO	−524.0 ± 2.1	SO		265 ± 4[d]
(EtO)$_2$SO	−552.2 ± 1.9	SO		262 ± 4[d]
(PrO)$_2$SO	−588.3 ± 2.1	SO		256 ± 4[d]
(BuO)$_2$SO	−625.3 ± 4.4	SO		252 ± 4[d]
(MeO)$_2$SO$_2$	−687.0 ± 1.9	SO$_2$	−296.8 ± 0.2	213 ± 4[d]
		(MeO)$_2$SO	−483.4 ± 2.0	453 ± 3
(EtO)$_2$SO$_2$	−756.3 ± 2.0	SO$_2$	−296.8 ± 0.2	213 ± 4[d]
		(EtO)$_2$SO	−552.2 ± 1.9	453 ± 3
(PrO)$_2$SO$_2$	−792.0 ± 2.1	SO$_2$	−296.8 ± 0.2	207 ± 4[d]
		(PrO)$_2$SO	−588.3 ± 2.1	453 ± 3
(BuO)$_2$SO$_2$	−828.9 ± 3.1	SO$_2$	−296.8 ± 0.2	203 ± 4[d]
		(BuO)$_2$SO	−625.3 ± 4.4	453 ± 5

[a] Data from ref. 3, unless indicated otherwise.
[b] Enthalpies of formation of O and S quoted from ref. 9.
[c] Ref. 12.
[d] Mean bond dissociation enthalpy.

FINAL COMMENTS

The bond dissociation enthalpies derived in the previous section rely on gas-phase data. It is important to ask if those values can be used to predict the thermodynamics of chemical processes in solution.

The similarity between gas-phase and solution-determined bond enthalpy data has been evidenced by a wealth of recent results obtained by photoacoustic calorimetry.[20] Even in polar solvents like water, is has been shown that the solvation enthalpies of a variety of organic substates, RH, are nearly identical to the solvation enthalpies of radicals R$^{\bullet}$, so that the solvent effect on the enthalpy of the reaction

$$RH \quad \rightarrow \quad R^{\bullet} + H^{\bullet} \tag{11}$$

is almost negligible (the solvation enthalpy of the hydrogen atom in water is small, about -4 kJ mol^{-1}). In conclusion, the answer to the above question is a cautious "yes": gas-phase bond dissociation enthalpies can be used for calculations involving species in solution, provided that a cancellation of solvation enthalpies occurs.

In addition to the large body of gas phase enthalpy data for neutral radicals, there exists a significant amount of photoionization and mass spectrometric data which describe the thermochemistry of the related cations and anions. Most of the reliable data for the ionization energies and electron affinities of sulfur-containing species are compiled in a very recent reference.[13]

It is possible to derive gas phase bond enthalpies for the ions if the appropriate ionization energies (IE) and/or appearance energies (AE) are available. For example, the cleavage of a radical cation, $R–R^{\bullet+}$, can be defined using either of the two following thermochemical cycles:

$$R–R \quad \rightarrow \quad R–R^{\bullet+} \qquad\qquad IE(RR)$$

$$R–R \quad \rightarrow \quad R^+ + R^{\bullet} \qquad\qquad AE(R^+)$$

$$\overline{\phantom{R-R \quad \rightarrow \quad R^+ + R^{\bullet} \qquad\qquad}}$$

$$R–R^{\bullet+} \rightarrow \quad R^+ + R^{\bullet} \qquad\qquad D(R–R^{\bullet+}) = AE(R^+) - IE(RR)$$

or,

$$R–R \quad \rightarrow \quad 2R^{\bullet} \qquad\qquad D(R–R)$$

$$R–R \quad \rightarrow \quad R–R^{\bullet+} \qquad\qquad IE(RR)$$

$$R^{\bullet} \quad \rightarrow \quad R^+ \qquad\qquad IE(R^{\bullet})$$

$$\overline{\phantom{R-R \quad \rightarrow \quad R^+ + R^{\bullet} \qquad\qquad}}$$

$$R–R^{\bullet+} \rightarrow \quad R^+ + R^{\bullet} \qquad\qquad D(R–R^{\bullet+}) = D(R–R) - IE(RR) + IE(R^{\bullet})$$

In this case the cycles are simplified since the equations describe the cleavage of a symmetric ion. Obviously, for the cleavage of a cation such as $R–R'^{\bullet+}$ both modes of cleavage (i.e. the formation of either R^+ or R'^+) should be considered.

There are enough data[13] to apply this approach to a simple disulfide such as dimethyldisulfide. The enthalpies for the various cleavage processes of dimethyldisulfide are given below:

ΔH/kJ mol^{-1}

MeSSMe	$\rightarrow$	$2MeS^{\bullet}$	285
MeSSMe	$\rightarrow$	$MeSS^{\bullet} + Me^{\bullet}$	243
MeSSMe$^{\bullet+}$	$\rightarrow$	$MeS^+ + MeS^{\bullet}$	349
MeSSMe$^{\bullet+}$	$\rightarrow$	$MeSS^{\bullet} + Me^+$	478
MeSSMe$^{\bullet+}$	$\rightarrow$	$MeSS^+ + Me^{\bullet}$	301

The removal of an electron from the disulfide leads to the formation of a three electron bond and this interaction results in an overall increase in the bond dissociation enthalpy. Of the two possible S–C cleavages, the process forming the methyl radical is predicted to be favored. However, both of these cleavage processes are predicted to be less favorable than the S–C cleavage of the neutral disulfide.

Although the reduction of disulfides is an important biochemical reaction, very little reliable data related to the enthalpies of formation of sulfur-containing anions in the gas phase are available. Pulse radiolysis data[21,22] suggest that the sulfur-sulfur bond dissociation Gibbs energy, in aqueous solution, for simple disulfide radical anions, eq. 12, is as low as 20 kJ mol^{-1}. If the standard entropy change for this reaction is the same as that for the neutral disulfide, a bond enthalpy of ca. 65 kJ mol^{-1} is derived; a dramatic decrease compared to the neutral disulfide!

$$RSSR^{\bullet -} \quad \rightarrow \quad RS^- + RS^{\bullet} \tag{12}$$

If it is assumed that the gas phase bond dissociation enthalpy of the disulfide radical anion is similar to that in solution (this is found to be the case for several radical cations[23]), then it is possible to calculate the gas phase enthalpy of formation of MeSSMe$^{\bullet -}$ as ca. 16 kJ mol^{-1}. Using this number and a thermochemical cycle similar to that outlined above, it is predicted the C–S bond is also substantially weakened in the radical anion compared to the neutral disulfide, i.e. D(MeSS–Me$^-$) $\approx$ 195 kJ mol^{-1}, eq. 13. The decrease in the bond dissociation enthalpy upon the one electron reduction of the disulfide must be due to the fact that the electron is placed into an antibonding σ^* orbital (see article by K.-D. Asmus in this book).

$$MeSSMe^{\bullet -} \quad \rightarrow \quad MeSS^{\bullet} + Me^- \tag{13}$$

The electron affinity of MeSS$^{\bullet}$ is not available, so it is not possible to determine the thermochemistry for the cleavage of the disulfide radical anion to form MeSS$^-$ and the methyl radical.

In principle, it is possible to obtain the solution equivalent of ionization energies and electron affinities by making electrochemical measurements. Recently, we have described how these electrochemical measurements can be combined with the homolytic Gibbs bond energies in order to calculate the thermochemistry for all of the possible cleavage reactions of neutral compounds and their related radical cations and radical anions.[23,24] We have found that for simple hydrocarbon radical cations, the bond energies in the gas phase and in solution do not differ significantly. We suggested that the reason for the lack of solvent effect was because the radical cations and the product cations were solvated to a similar extent.[23]

In practice, obtaining meaningful electrochemical data is not always straightforward. The polarographic reduction of disulfides and many other organosulfur compound[25] is complicated by adsorption of the starting materials and the products causing shifts of the half-wave potentials of varying magnitude and direction. In some cases, products containing mercury have been isolated. In aqueous solution, the waves are generally pH dependent since the proton and electron transfer equilibria are both established on the electrochemical time scale.

Recently, Surdhar and Armstrong[26,27] have used a thermochemical cycle similar to that described above, to estimate E^0 (the thermodynamically significant potentials) for RS$^-$/RS$^{\bullet}$ (0.84 V vs. NHE), 2RS$^-$/RSSR$^{\bullet -}$ (0.65 V vs. NHE) and RSSR$^{\bullet -}$/RSSR (–1.64 V vs. NHE). These data were used to rationalize the observation that cysteine was more susceptible to oxidation in alkaline aqueous solution than it was at low pH.

In summary, it is possible to derive thermodynamic properties of organosulfur ions in the gas phase and in solution. The difficulty is in determining the reliability of the relevant data (i.e. ionization energies, electron affinities, and electrochemical oxidation and reduction potentials). In contrast to the bond enthalpies of the neutral species, bond enthalpies for the ions determined in the gas phase will not always be applicable to solution. The reason for this is that the solvation enthalpies of small ions (e.g. the proton) will often be much larger than those of the larger organic fragments. In cases where the fragments are of similar size (as in the case of the symmetric disulfides), the differences between the gas phase and solution bond enthalpies are expected to be small.[23]

REFERENCES

1. S.W. Benson, *Chem. Reviews* 78:23 (1978).
2. J.D. Cox and G. Pilcher, "Thermochemistry of Organic and Organometallic Compounds", Academic Press, London (1970).
3. J.B. Pedly, R.D. Naylor, and S.P. Kirby, "Thermochemical Data of Organic Compounds", Chapman and Hall, London (1986).
4. G. Waddington, S. Sunner, and W.N. Hubbard, *in*: "Experimental Thermochemistry", Vol. 1, F.D. Rossini, ed., Interscience, New York, Chapter 7 (1956).
5. A.J. Head and W.D. Good, *in*: "Experimental Chemical Thermodynamics", Vol. 1, Combustion Calorimetry, S.Sunner and M. Mansson, eds., Pergamon, Oxford, Chapter 9 (1979).
6. S.W. Benson, "Thermochemical Kinetics", Wiley, New York (1976).
7. R.J. Hwang and S.W. Benson, *Int. J. Chem. Kinet.* 11:579 (1979).
8. B.K. Janousek, K.J. Reed, and J.I. Brauman, *J. Am. Chem. Soc.* 102:3125 (1980).
9. D.D. Wagman, W.H: Evans, V.B. Parker, R.H. Schumm, I. Halow, S.M. Bailey, K.L. Churney, and R.L. Nuttall, *J. Phys. Chem. Ref. Data.* 11, Suppl. No. 2 (1982).
10. D.F. McMillen and D.M. Golden, *Ann. Rev. Phys. Chem.* 33:493 (1982).
11. J.C. Traeger, *Org. Mass. Spectrom.* 19:514 (1984).
12. M.W. Chase, Jr., C.A. Davies, J.R. Downey, Jr., D.J. Frurip, R.A. McDonald, and A.N. Syverud, *J. Phys. CHem. Ref. Data* 14, Suppl. No. 1 (1985).
13. S.G. Lias, J.E. Bartmess, J.F. Liebman, J.L. Holmes, R.D. Levin, and W.G. Mallard, *J. Phys. Chem. Ref. Data* 17, Suppl. No. 1 (1988).
14. L.G.S. Shum, S.W. Benson, *Int. J. Chem. Kinet.* 15:433 (1983).
15. D.H. Fine and J.B. Westmore, *Can. J. Chem.* 48:395 (1970) and refs. cited therein.
16. A.J. Colussi and S.W. Benson, *Int. J. Chem. Kinet.* 9:295 (1977).
17. J.A. Hawari, D. Griller, and F.P. Lossing, *J. Am. Chem. Soc.* 108:3273 (1986).
18. D.Griller, J.M. Kanabus-Kaminska, and A. Maccoll, *J. Mol. Structure (Theochem.)* 163:125 (1988).
19. Y.-R. Luo and S.W. Benson, *J. Phys. Chem.* 92:5255 (1988).
20. J.M. Kanabus-Kaminska, B.C. Gilbert, and D. Griller, *J. Am. Chem. Soc.* 111:3311 (1989) and refs. cited therein.
21. W. Karmann, G. Meissner, and A. Henglein, *Z. Naturforsch.* 22:274 (1967).
22. J.E. Packer, *in*: "The Chemistry of The Thiol Group", Part 2, S. Patai, ed., Wiley, London (1974).
23. D.D.M. Wayner, J.J. Dannenberg, and D. Griller, *Chem. Phys. Lett.* 131:189 (1986).
24. D. Griller, J.A. Martinho Simões, P. Mulder, B.A. Sim, and D.D.M. Wayner, *J. Am. Chem. Soc.* 11:7872 (1989).
25. J.Q. Chambers, *in*: "Encyclopedia of Electrochemistry of the Elements", Vol. 12, A.J. Bard and H. Lund, eds., Marcel Dekker, New York (1978).
26. P.S. Surdhar and D.A. Armstrong, *J. Phys. Chem.* 90:5915 (1986).
27. P.S. Surdhar and D.A. Armstrong, *J. Phys. Chem.* 91:6532 (1987).
28. P.D. Pacey and J.H. Wimalasena, *J. Phys. Chem.* 88:5657 (1984).
29. M. Brouard, P.D. Lightfoot, and M.J. Pilling, *J. Phys. Chem.* 90:445 (1986).
30. A.L. Castelhano, P.R. Marriot, and D. Griller, *J. Am. Chem. Soc.* 103:4262 (1981).
31. A.L. Castelhano and D. Griller, *J. Am. Chem. Soc.* 104:3655 (1982).
32. J.J. Russell, J.A. Seetula, and D. Gutman, *J. Am. Chem. Soc.* 110:3092 (1988).
33. J.L. Holmes, F.P. Lossing, and A. Maccoll, *J. Am. Chem. Soc.* 110:7339 (1988).
34. S.S. Parmar and S.W. Benson, *J. Am. Chem. Soc.* 111:57 (1989).

35. W. Tsang, *J. Am. Chem. Soc.* 107:2872 (1985).
36. C.E. Canosa and R.M. Marshall, *Int. J. Chem. Kinet.* 13:303 (1981).
37. G.H. Kruppa and J.L. Beauchamp, *J. Am. Chem. Soc.* 108:2162 (1986).
38. K. Hayashibara, G.H. Kruppa, and J.L. Beauchamp, *J. Am. Chem. Soc.* 108:5441 (1986).
39. J.J. Russell, J.A. Seetula, R.S. Timonen, D. Gutman, and D.F. Nava, *J. Am. Chem. Soc.* 110:3084 (1988).
40. J.H. Kiefer, L.J. Mizerka, M.R. Patel, and H.-C. Wei, *J. Phys. Chem.* 89:2013 (1985).
41. Y. Malinovich and C. Lifshitz, *J. Phys. Chem.* 90:2200 (1986).

Appendix   Auxiliary Thermochemical Data

Radical	$\Delta H_f^0(g)$, kJ mol^{-1} [a]	Ref.
Me	146.9 ± 0.6	10
Et	119 ± 4	18,28-34
Pr	96 ± 4	18,30,31,33,35
i-Pr	84 ± 8	18,30-33,35
Bu	(74 ± 8)[b]	
s-Bu	62 ± 8	18,30,31,33,35
i-Bu	(68 ± 8)[c]	
t-Bu	44 ± 6	18,30,31,33,35-39
Pe	(54 ± 8)	
Ph	330 ± 8	18,40,41
Bz	200 ± 6	18
MeCO	−24.3 ± 1.7	10

[a] Estimated data in parentheses.
[b] The only available experimental value[3] is 76.7 ± 2.5 kJ mol^{-1}.
[c] A recent experimental value[3] is 66.1 ± 1.3 kJ mol^{-1}.

EXPERIMENTAL PROCEDURES IN ELECTROLYSIS

Henning Lund

Department of Organic Chemistry
University of Århus
8000 Århus C, Denmark

Electrolysis of organic compounds has certain inherent advantages and disadvantages. The electrolytic method presents the possibility of controlling the activity of the reagent, the electron, over a wide range by proper choice of the electrode potential. Another advantage is that the transfer of electrons can occur at low temperatures, and at a chosen pH, or under aprotic conditions, so that sensitive compounds may be oxidized or reduced under mild and well-defined conditions. Especially for large-scale preparations it is also important that the electron is a non-polluting reagent and leaves no waste products.

An obvious disadvantage of the method is that the reaction of one mole of a substance requires n x 96500 coulombs (n is the number of electrons in the electrode reaction); this requires well-designed cells for industrial applications. Another limitation is the necessity of employing a medium capable of conducting the electrical current; this requires a salt which is soluble in the medium.

Reviews on practical problems in organic electrochemistry[1a,b] and on electroanalytical methods[2a-c] useful for obtaining relevant parameters for the optimization of electrochemical reactions have been published.

In an electrochemical reduction an electron is transferred from the electrode to the lowest unoccupied (LUMO) orbital of the molecule and in an oxidation an electron is removed from the highest occupied orbital (HOMO) to the electrode. This is illustrated in Fig. 1. In Fig. 1a (left) the energy of the electrons in the electrode is not high enough for them to be transferred to the LUMO of the substrate, but by raising the energy level of the electrons (making the electrode potential more negative, Fig. 1a, right) a transfer can take place. To keep a measurable current running an overpotential has to be applied. Conversely, by lowering the energy level of the electrons in the electrode (making the electrode potential more positive, Fig. 1b) an abstraction of an electron from HOMO may take place.

The important feature of controlling the reaction is a control of the potential of the electrode, that is the potential difference between the electrode and a point just outside the electrical double layer; the voltage between anode and cathode includes the potentials of both electrodes and the voltage due to the ohmic resistance, $E = iR$. The potential of the

Sulfur-Centered Reactive Intermediates in Chemistry and Biology, Edited by
C. Chatgilialoglu and K.-D. Asmus, Plenum Press, New York, 1990

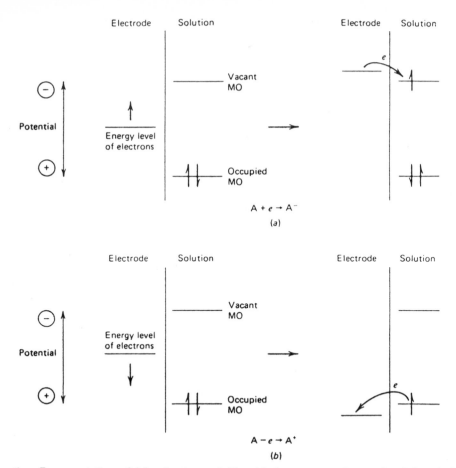

Fig. 1. Representation of (*a*) reduction and (*b*) oxidation process of a species A in solution. The molecular orbitals (MO) of species A shown are the highest occupied MO. and the lowest vacant MO. As shown, these correspond in an approximate way to the E^0's of the A/A$^-$ and A$^+$/A couples, respectively (from ref. 2a).

electrode of interest ("the working electrode") may be controlled by controlling either the current or the potential.

The difference in the two ways of controlling the electrolytic reaction is illustrated in Fig. 2.[1a] In this Figure curve I depicts schematically the connection between the current through the cell and the potential of the working electrode. The initial solution contains two reducible compounds or one compound with two groups reducible at different potentials. When the potential at the cathode is between 0 and E_A no electron transfer across the electrical double layer can take place and thus no current runs through the cell. If the cathode potential is made more negative, the electron transfer becomes possible; that is, the reduction of the most easily reducible compound or group starts. Between E_A and E_B the current rises dependent on the potential, but when the value E_B has been reached, all the molecules that arrive at the electrode and which can undergo the first reduction are reduced as soon as they reach the electrode. In the potential interval E_B to E_C the current is limited by the transport of the reducible compound to the cathode. This current is called the

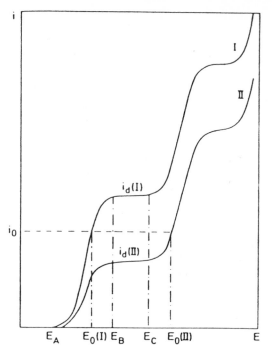

Fig. 2. Schematic representation of the connection between the current and the potential of the working electrode in a solution containing a compound with two groups reducible at different potentials. Curve I, before electrolysis; curve II, after the passage of some current; I_0 is an applied current, $E_0(I)$ and $E_0(II)$ the potentials corresponding to i_0; i_d is the limiting current, and E_A, E_B, and E_C are applied potentials (from ref. 1a).

limiting current i_d, and under fixed conditions it is proportional to the concentration of the electroactive compound.

A further change of the electrode potential in the negative direction results in the occurrence of the second electrode reaction and a rising current; a similar S-shaped curve results from this reduction. At more negative potentials a third reaction or a reduction of the medium takes place.

If a suitable current $i_0[i_0 < i_d((I)]$ is sent through the cell, the cathode potential assumes the value $E_0(I)$ and for $i_0 < i_d(I)$ this is well below the potential (E_C) at which the second electrode reaction starts; a selective reduction thus occurs at the beginning of the electrolysis. During the electrolysis the concentration of the reducible compound, and thus its limiting current, diminishes and after a while (curve II) the limiting current becomes smaller than the applied current $[i_0 > i_d(II)]$. The cathode potential has then, by necessity, reached the value $E_0(II)$ and at this potential the second electrode reaction also takes place; the electrolysis is no longer selective.

If the second electrode process does not interfere with the first one (e.g., it may be due to the background), the method may still be acceptable in the laboratory. For industrial processes, however, a high current yield is generally desirable and waste of current on the background will be less tolerable. The current yield is the theoretical amount of electricity divided by the amount actually employed for the production of a particular substance (usually expressed in percent).

On the other hand, when the electrode potential is the factor controlled and kept at a suitable value, e.g. at E_B, the second electrode process cannot take place, and the reduction remains selective to the end. The current through the cell is never higher than the limiting current corresponding to the first electrode reaction. This means that the current decreases during the reduction and becomes very small towards the end of the reaction, as the limiting current is proportional to the concentration of the electroactive material.

Another way of keeping the potential constant is to maintain a certain concentration of the reducible compound, e.g., corresponding to the limiting current i_d in Fig. 2. By continuously adding substrate and removing product, and by keeping the current at a constant value smaller than $i_d(I)$, e.g., i_y the potential will stay at $E_o(I)$ and the reaction will remain selective. This is the preferred way in large-scale electrolysis.

If one wants to make electrolysis of organic compounds the simplest way is to place two electrodes into the solution, send a dc-current through the solution and see what happens. If one is not satisfied with that approach, different electroanalytical methods give you the possibility to obtain information on, e.g., redox potentials, intermediates, and rate constants. Classical polarography and cyclic voltammetry (CV) are probably the most widely used methods for that purpose.

The polarographic data may be used as a guide for electrolysis, as illustrated by the following example.[3a] The half-wave potentials and limiting currents at different pH of 2,3-dihydro-2,3-dimethyl-1,4-phthalazinedione (1) are plotted in Fig. 3. In acid solution up to about pH 6 the half-wave potentials vary linearly with pH; about pH 7 the first wave disappears and another one appears at a more negative potential which does not vary with pH. This indicates that for the species reduced in acid solution one more proton is involved in the reaction before the potential-determining step than that reduced in alkaline, and in this case the monoprotonated compound is reduced at low pH. The limiting current in strongly acid solution is approximately three times as high as that in alkaline medium for the same concentration which means that three times as many electrons are involved in the electrode reaction responsible for the first wave at pH 0 as for the second wave. Between pH 1 and 5 the height of the first wave diminishes gradually and reaches the height of a two-electron reduction at pH 5.

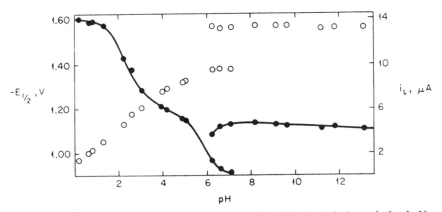

Fig. 3. Dependence on pH of the limiting current [μA] (full circles), and the half-wave potentials [SCE] (open circles) of 2,3-dihydro-2,3-dimethyl-1,4-phthalazinedione (1). Concentration 2.0 x 10^{-4} M (from ref. 3a).

The electrode reactions were investigated by reducing the compound at pH 0 and 9; the reduction in acid solution produced 2-methylphthalimidine (2) by a six-electron reaction, whereas that at pH 9 yielded 3,4-dihydro-2,3-dimethyl-4-hydroxy-1-phthalazinone (3). This compound was then investigated polarographically (Fig. 4); it is reduced in acid solution in two two-electron waves, at pH 4 in two one-electron waves, and is not reducible in alkaline solution. It is of interest that the compound is, when reducible, more easily reduced than the starting material.

Compound 3 was then reduced in hydrochloric acid at the potential of the first wave (−0.7 V vs. SCE); the product obtained is 3,4-dihydro-2,3-dimethyl-1-phthalazinone (4), this product is also obtained from 1 in a four-electron reaction at pH 5.

The fact that the polarographic data for 1 at pH 5 suggests a two-electron reaction, whereas the preparative results prove a four-electron reduction, means that a slow chemical step occurs after the uptake of the first two electrons. The partly reduced molecule diffuses from the microelectrode before the chemical follow-up reaction has occurred, but this does not matter in an exhaustive, preparative reaction. The rate of the slow step is pH-dependent and this step is not apparent at low pH where it is sufficiently fast. The slow step is suggested to be the acid-catalyzed dehydration of 2 to the quaternary phthalazinone (5).

Finally, 4 was polarographed (Fig. 4); the wave of 4 appears at the same potential as that of the second wave of 3. Reduction of 4 yields a mixture of 2-methylphthalimidine and its precursor, N-methyl-2-(methylaminomethyl)benzamide (6).

The results are presented in Scheme 1, and it is postulated that the reduction of 1 to 2-methylphthalimidine (2) proceeds through 3, 5, 4 , and 6.[3a]

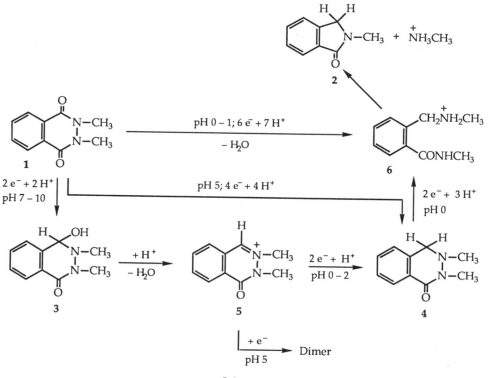

Scheme 1

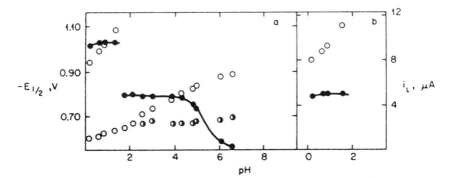

Fig. 4. Dependence on pH of the limiting current [μA] (full circles), and the half-wave potentials [SCE] of (a) 3,4-Dihydro-2,3-dimethyl-4-hydroxy-1-phthalazinone (**3**) (open circles) and (b) 3,4-dihydro-2,3-dimethyl-1-phthalazinone (**4**) (half-filled circles). Concentration 2.0×10^{-4} M (from ref. 3a).

The information thus obtained is very valuable for the choice of experimental conditions; it must, however, be kept in mind that there are cases where differences between the results obtained in micro- and macroelectrolysis occur. This is primarily caused by differences between micro- and macroscale experiments with respect to the concentrations ordinarily employed and to the time scale of the experiments.

The concentration influences a competition between a first-order and a second-order reaction. In order to predict the outcome it is important to identify first-order/second-order reactions and their rate constants. Cyclic voltammetry is a valuable tool for that purpose.

In a cyclic voltammogram (Fig. 5) the current is measured as a function of the potential. For a reduction the potential is changed to more negative values at a constant rate, "sweep rate", v (V s⁻¹), and then, at a certain potential, the sweep is reversed, usually employing the same (numeric) sweep rate as in the forward direction. The shape of the voltammogram is dependent on the value of the heterogeneous electron-transfer rate constant k_h; for $k_h \geq$ 0.1 cm s⁻¹ the voltammogram is reversible, at lower values the voltammogram becomes quasi-reversible or irreversible.

In a reversible cyclic voltammogram the reoxidation of the anion radical $A^{\bullet-}$, formed during the forward sweep, is seen as an anodic peak; if, however, $A^{\bullet-}$ reacts chemically in some way, it is not available for reoxidation on the reverse scan, so only a small or no anodic peak is seen. In the usual electrochemical nomenclature, an electron transfer reaction is called E and a chemical follow-up reaction C. The process in question would thus be an EC reaction; the chemical step after a reduction in most cases would be a reaction with an electrophile, including protons, a cleavage reaction where a nucleophile is expelled, or a dimerization; for oxidation reactions with a nucleophile, loss of a proton or dimerization would be the most common follow-up reaction.

The chemical follow-up reaction causes a displacement of the peak potential for reductions to less negative potentials. At the potential where the current begins to flow, the equilibrium required by the Nernst equation is disturbed by the chemical reaction. The Nernst equation requires a certain proportion between the oxidized and reduced forms at a given potential, but if the reduced form $A^{\bullet-}$ is removed from the equilibrium, then the electrode tries to re-establish the required proportion $A/A^{\bullet-}$ by reducing more A, which means that the current at a given potential becomes higher than in the simple, reversible case. In a reductive EC reaction, the peak potential is thus shifted towards positive values (and towards negative values for oxidations); for a reaction with first-order kinetics, the $E_p(\text{red})$ shifts 30 mV into the positive direction when k_c is increased ten-fold.

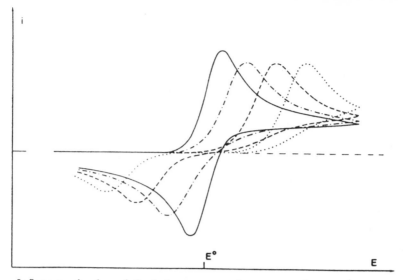

Fig. 5. Influence of value of the heterogeneous electron-transfer rate constant k_h on the shape of the cyclic voltammograms. ——, k_h = 0.1 cm sec^{-1}; –·–·–, k_h = 0.01 cm sec^{-1}; -----, k_h = 10^{-3} cm sec^{-1}, ·····, k_h = 10^{-4} cm sec^{-1}. When $k_h \geq 0.1$ cm sec^{-1} the cyclic voltammogram is "reversible" (from ref. 2c).

The influence of the chemical follow-up reaction depends on the ratio of the rate constant k_c of the C step and the sweep rate v. The higher v, the less influence has the follow-up reaction; for chemical reactions with first-order rate constants of $k_c \leq 10^3$ s^{-1} it is possible to "outrun" the chemical reaction and obtain a reversible cyclic voltammogram at high v. The E_p(red) for a given system with first-order kinetics is shifted by 30 mV into the negative direction when v is increased ten-fold.[2] By plotting E_p vs. log v, one gets curves from which the value of k_c can be obtained. This is illustrated in Fig. 6 for a reaction where the chemical step is a cleavage.

Fig. 6 shows a plot of log v vs. E for 5-chloro-8-methoxyquinoline (QCl), alone and in the presence of carbon dioxide.[3b] At slow sweep rates only the cathodic peak is seen; the lifetime of the anion radical is too short to influence the reverse scan. The follow-up reaction is a cleavage of the primarily formed anion radical according to eqs. (1) and (2).

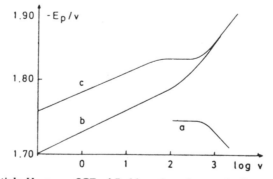

Fig. 6. Peak potentials V versus SCE of 5-chloro-8-methoxyquinoline in dependence of the sweep rate v without and in the presence of CO_2. Curve a, anodic peak potential without CO_2; b, cathodic peak potential in the presence of CO_2; c, cathodic peak potential without CO_2; (from ref. 3b).

$$QCl + e^- \underset{\rightleftarrows}{\overset{k_h}{}} QCl^{\bullet-} \tag{1}$$

$$QCl^{\bullet-} \xrightarrow{k_c} Q^{\bullet} + Cl^- \tag{2}$$

The purpose of the experiments, whose results are shown in Fig. 6, was to investigate whether the cleavage of the anion radical was faster than the reaction of $QCl^{\bullet-}$ with carbon dioxide, or vice versa. If the cleavage turned out to be slower than the carboxylation, it should be possible to carboxylate QCl without loss of the Cl substituent.

Curve c represents the E_p(red) of QCl $vs.$ v; at $v < 10^2$ V s^{-1}, $dE_p/d \log v$ is -30 mV; at $v \sim 10^2$ V s^{-1}, the chemical reaction is being outrun, the anodic peak appears (curve a) and the voltammogram assumes the shape of a single, reversible reaction (1) without a complicating follow-up reaction (2). From the sweep rate, where the line $dE_p/d \log v$ = -30 mV intersects the horizontal line k_c, the first-order rate constant, can be obtained. At still higher v, the peak separation increases, which means that the heterogenous rate constant k_h is becoming small compared to v, and the voltammogram assumes the shape of a quasi-reversible system. If the follow-up reaction had been a second order reaction, $dE_p/d \log v$ had been -19 mV and $dE_p/d \log c$ = 19 mV.

Curve b shows E_p(red) of QCl $vs.$ v in the presence of CO_2. It is seen that E_p is shifted towards positive values compared to curve c and that it is not possible to outrun the chemical follow-up reaction [eq. (3)] within the available range of sweep rates ($v \leq 10^3$ V s^{-1}).

$$QCl^{\bullet-} + CO_2 \rightarrow Cl-Q-CO_2^{\bullet-} \tag{3}$$

The CV investigation thus indicates that it should be possible to carboxylate 5-chloro-8-methoxyquinoline without excessive loss of chlorine. A preparative reduction at low temperature showed that it indeed was possible. Scheme 2 shows the carboxylation of 6-chloroquinoline.[3b]

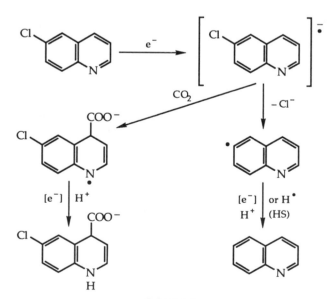

Scheme 2

Another way of determining the rate constant of a follow-up reaction is to measure the anodic peak current during the reverse sweep i^a_p and the cathodic peak i^c_p at different v. From the dependence of the ratio i^a_p/i^c_p on v the rate constant can be obtained by comparison with simulated values,[2a-c] when the order of the reaction is known.

In the reactions discussed so far the electrochemical step is a transfer of electrons between the electrode and the substrate. However, in some electrochemical reactions the heterogenous electron transfer occurs between the electrode and a species which exchanges electrons with the substrate in a homogeneous electron transfer reaction. In other cases the electrogenerated reagent reacts "chemically" with the substrate, e.g., in the electrocatalytic hydrogenation the electrogenerated hydrogen reacts with the substrate in much the same way as in a conventional hydrogenation.

Cyclic voltammetry may also be used to find the rate constant for the homogeneous electron exchange between the mediator and the substrate; the reactions are shown in eqs. (4 - 6)

$$A + e^- \rightleftarrows B \tag{4}$$

$$B + C \xrightarrow{k_5} A + D \tag{5}$$

$$D \rightarrow \text{products} \tag{6}$$

By measuring the peak hight of A without and with different concentrations of C at different sweep rates k_5 can be found from tables or curves of calculated values.[4a,b]

In some cases another technique, double step chronoampereometri, is preferred to investigate reaction mechanisms and obtain rate constants, but as the resulting curves do not give a qualitative picture of the events at first glance, it will not be discussed here.

When the relevant data, potentials, rate constants etc., have been obtained by electroanalytical experiments or taken from the literature, the data have to be translated into experimental conditions for the electrolysis. If the potentials have been obtained from the literature, there are two points one should notice.

1. Which sign convention is used?

2. Which reference electrode are the potentials referred to (and in which medium)?

In the older literature two sign conventions were used, the "American" one and the "European" one. The "American" sign convention was tied to the thermodynamic conventions and as potentials were given as oxidation potentials, they had the opposite sign of the "European" convention, which expressed the measurable potential difference between the standard hydrogen electrode and the system. According to the "Stockholm convention" all standard potentials should now be given as reduction potentials; the "American" and "European" conventions give the same sign for reduction potentials, so in modern literature (after 1960) there is, hopefully, no confusion about the sign of potentials.

Fig. 7 shows a schematic outline of an electrolytic cell with cathode, anode, and reference electrode; in Fig. 7b the potential distribution in the cell, in which a current i is running, is depicted. The ohmic resistance between cathode and anode is R.

At the cathode the potential difference ΔE_c indicates how much energy is involved in the transfer of electrons to the reduced species. This will differ according to the reaction in question. Between the cathode and the anode there is a potential drop, ΔE_R, due to the ohmic resistance of the solvent; this potential drop is proportional to the current and the

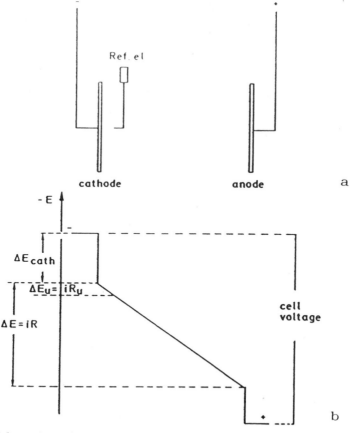

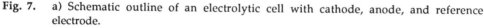

Fig. 7.   a) Schematic outline of an electrolytic cell with cathode, anode, and reference electrode.
b) Potential distribution in the cell in which a current i is running. The cell resistance is R. R_u is the uncompensated resistance.

ohmic resistance, and this energy is transformed into heat. From Fig. 7 it is also seen that the potential measured by the reference electrode is ΔE_c plus $\Delta E_{R_u} = iR_u$; some of this resistance can be eliminated by a feed back compensation, but not all of it. At the anode ΔE_A is a function of the anode reaction, and thus the cell voltage is $\Delta E_C + \Delta E_A + iR$. (The index u denotes "uncompensated").

In Fig. 8 a schematic circuit for controlled potential electrolysis is shown; besides a cathode, an anode, and a reference electrode it includes a voltmeter to measure the cathode potential, an ammeter for the current, and a coulometer for the amount of electricity. The potential at the cathode is adjusted by changing the cell voltage; it may be done manually, but generally a potentiostat is used.

A number of reference electrodes have been used, as a single one is not suitable under all conditions. A reference electrode is a half-cell that defines a potential to which all other measurements are referred. The primary standard is the standard hydrogen electrode (SHE), but as this electrode is inconvenient for practical work, other reference electrodes are used.

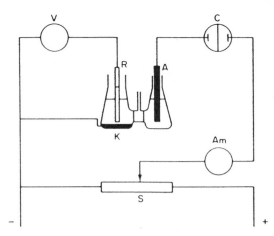

Fig. 8. Circuit for constant potential reduction: K, cathode; A, anode; R, reference electrode; V, potentiometer or pH meter; C, coulometer; Am, ammeter; S, voltage adjuster (from ref. 1a).

The ideal reference electrode is indefinitely stable, unpolarizable (i.e. the potential is independent of the current passing through the electrode), easy to handle, its potential is independent of temperature, it has a low resistance and does not cause contamination of the test solution.

In water the "saturated" calomel electrode (Hg_2Cl_2/Hg, SCE) or silver chloride/silver electrode are widely used; the aqueous SCE may also be used for measurements in other solvents, but it introduces an unknown liquid junction potential. When potentials in different solvents are compared, it has been recommended[5] to use the redox couples ferrocinium/ferrocene and/or bis(biphenyl)chromium(I)/bis(biphenyl)chromium(0); they behave reversibly in most solvents and their solvation energy is relatively low.

Another choice to make is the electrode material. For cathodes there are more possibilities than for anodes, which are more apt to corrode. In protic solvents (water, alcohols, acids) one may divide the materials according to their hydrogen overvoltage. Metals, such as Hg, Pb, and to a lesser degree Cd and Sn have a high hydrogen overvoltage which means that rather negative potentials may be employed before hydrogen evolution becomes a serious problem. For industrial use mercury is not tolerable and lead or lead alloys, possibly plated on a mechanically stronger support, may be used.

Sulfur-containing compounds may give problems if the substrate or its reduction products react with the electrode; this may be the case with mercury. Carbon electrodes may be the best choice for the reduction of most sulfur compounds.

A special way of introducing sulfur into a compound is to reduce the substrate at a cathode consisting of a mixture of graphite and sulfur.[6a,b] Thus reduction of the sulfur in the presence of a secondary amine yields a methyl dithiooxamide. The method may also be used to introduce selenium into organic compounds;[6c] tellurium is itself conducting enough and may be used without graphite.[6d]

Cathodes with low overvoltage (Pt, Pd, Ni) may be used for electrocatalytic hydrogenation; in aprotic medium hydrogen overvoltage is of minor importance.

For anodes, corrosion is an important consideration; nobel metals (Pt, Au) may be used in the laboratory, for industrial purposes rhodium plated on titanium (dimensionally stable anodes, DSA) is employed, e.g., for the production of chlorine. Graphite anodes are stable in many cases, and in basic, aqueous solution metals which form a protective oxide (Ni, Ti, Fe), are a possibility. Nickel, e.g., may be used for the oxidation of diacetone-L-sorbose (a step in the synthesis of ascorbic acid); the reaction is an indirect oxidation by the surface oxide $NiO(OH)$.

Sometimes corrosion of the anode is part of the desired reaction, e.g. in the electrochemical production of tetraalkyl lead. Magnesium or aluminum have also been employed as sacrificial anodes in reactions involving metal-organic compounds; an undivided cell may thus be used.[7a-c]

Recently chemically modified electrodes (CME) have been prepared with special catalytic properties;[8a] they have so far been useful mostly as indicator electrodes. Their stability is a problem for synthesis. Chiral induction with CME is an attractive possibility, but reproducibility and stability of the CME has been a problem.

The nature of the solvent is important for the course of electrolytic reactions; electrolysis can occur only in a medium which conducts the electric current. Factors, such as proton activity, usable potential range, dielectric constant, ability to dissolve electrolytes and substrates, ion-pair formation, accessible temperature range, vapor pressure, viscosity, toxicity and price must be taken into consideration when the choice of solvent is made.

Solvents may be divided in protic and aprotic solvents. If protons are desired or tolerable, water is the best solvent from an electrochemical point of view although the accessible potential range for anodic reactions is rather limited.

Several organic compounds are not water-soluble, and mixtures of water and organic solvents may be used. Another possibility is emulsions, sometimes in connection with phase-transfer reagents.

If protons are undesired, aprotic solvents are used, mostly dipolar, aprotic solvents, such as dimethylformamide (DMF), acetonitrile, and dimethylsulfoxide, but ethers may be employed when strong bases (e.g. metal-organic compounds) are involved; liquid ammonia is sometimes used for the production of solvated electrons.

In this connection it might be mentioned that electron transfer to a molecule makes it more basic/nucleophilic, and a loss of electrons makes it more acidic/electrophilic. As only the solvent/electrolyte system puts a limit on the strength of the base, which can be created electrochemically, reactions which require strong bases may be run by using electrogenerated bases (EGB).[8b]

The choice of electrolyte depends on properties such as its solubility, dissociation constant, mobility, discharge potential, and protic activity. The choice of anion is of most importance in anodic reactions. Unless the oxidized anion is desired in an indirect electrolytic oxidation, a difficultly oxidizable anion, such as perchlorate, tetrafluoroborate, hexafluorophosphate or nitrate, is chosen. In cathodic reactions the choice of cation has highest priority. Usually alkali metal ions or tetraalkylammonium ions are chosen, unless an acidic or buffered solution is desired.

A number of problems must be considered when designing a cell for electrolysis, such as potential distribution at the working electrode, position of the reference electrode (if used), ohmic resistance, mass and heat transfer, the need for a diaphragm for separating the anolyte from the catholyte, and the necessity of working in a closed system.

Table 1. Comparison of the Monsanto 1965 divided cell process with the recent Monsanto undivided cell process (from ref. 9)

	Divided cell	Undivided cell
Adiponitrile selectivity (%)	92	88
Inter-electrode gap (cm)	0.7	0.18
Electrolyte resistivity (Ω cm)	38*	12
Electrolyte flow velocity (m s^{-1})	2	1-1.5
Current density (A cm^{-2})	0.45	0.20
Voltage distribution (V)		
Estimated reversible cell voltage	-2.50	-2.50
Overpotentials	-1.22	-0.87
Electrolyte IR	-6.24	-0.47
Membrane IR	-1.69	—
Total	-11.65	-3.84
Energy consumption (kWh ton^{-1})	6700	2500

*Catholyte

The first choice is a divided/undivided cell. In an exploratory stage a separation of anolyte and catholyte is desirable. In aqueous media an ion-exchange membrane is usually chosen, in aprotic media some kind of porous material (e.g., fritted glass) is used. For industrial use undivided cells are preferred; data for the hydrodimerization of acrylonitrile to adiponitrile in a divided and undivided cell are compared in Table 1.[9] For relatively low-cost material, such as acrylonitrile, the decrease in power-consumption outweighs the small decrease in selectivity whereas this might not be the case for more expensive compounds.

A common laboratory cell is the so-called H-cell; one form of the type is shown in Fig. 9.[1a] It is a convenient all-round type, but the potential distribution at the working electrode is not particularly good. In Fig. 10 a "filterpress" (plate and frame) cell with diaphragm is depicted; the catholyte and anolyte are pumped through the cell as indicated. Cells of that type are commercially available both in laboratory size and for industrial use. In Fig. 11 an undivided plate and frame cell are shown. In this case one side of the electrode works as cathode, the other as anode.

Among the more specialized cells the "Swiss roll" cell (Fig. 12) has been used for the oxidation of diacetone-L-sorbose. Its special feature is the very large area of the electrode for a given volume; this is required as the reaction is a rather slow chemical oxidation of the substrate by the electrogenerated surface oxide [NiO(OH)].

For laboratory use, the cell design is usually not critical but for industrial cells most of the above mentioned factors must be taken into consideration. If one finds a reaction which might turn into a commercially useful process, it is advisable rather early in the investigation to employ a cell and other conditions, which mimic those of a large scale production, in order to be able to decide as early as possible whether the reaction has the qualities required for a commercial production.

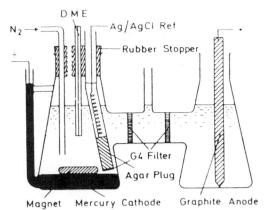

Fig. 9. Semimacroscale electrolytic cell (from ref. 1a).

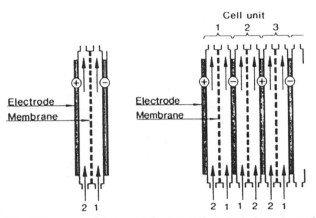

Fig. 10. Filter press (plate and frame) cell. 1. catholyte, 2. anolyte.

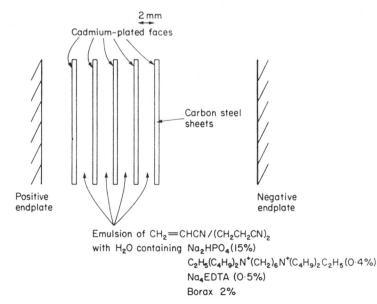

Fig. 11. The cell for the new Monsanto process for the hydrodimerization of acrylonitrile. (From ref. 9).

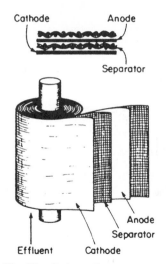

Fig. 12. The Swiss-roll cell (from ref. 9).

REFERENCES

1 a. H. Lund, *in*: "Organic Electrochemistry", 3. ed., H. Lund and M.M. Baizer, eds., Dekker, New York , Chapter 6, pp. 243 (1990).

   b. F. Goodridge and C.J.H. King, *in*: "Technique of Electroorganic Synthesis", Part I, N.L. Weinberg, ed., Wiley, New York, pp. 7 (1974).

2 a. A.J. Bard and L.R. Faulkner, "Electrochemical Methods", Wiley, New York (1980).

   b. O. Hammerich and V.D. Parker, *in*: "Organic Electrochemistry", 3. ed., H. Lund and M.M. Baizer, eds., Dekker, New York, Chapter 3, pp. 121 (1990).

   c. Southampton Electrochemistry Group, "Instrumental Methods in Electrochemistry", Wiley, New York (1985).

3 a. H. Lund, *Coll.Czech. Chem. Commun.* 30:4238 (1965).

   b. P. Fuchs, U. Hess, H.H. Holst, and H. Lund, *Acta Chem. Scand.* B35:185 (1981).

4 a. J.-M. Savéant and K.B. Su, *J. Electroanal. Chem.* 196:1 (1985).

   b. S.U. Pedersen, *Acta Chem. Scand.* A41:392 (1987) and references therein.

5. G. Gritzner and J. Kuta, *Pure Appl. Chem.* 56:461 (1984); *Electrochim. Acta* 29:869 (1984).

6 a. M.L. Contreras, S. Rivas, and R. Rozas, *J. Electroanal. Chem.* 177:299 (1984).

   b. G.Le Guillanton, Q.T. Do, and J. Simonet, *Tetrahedron Lett.* 27:2661 (1986).

   c. P. Jeroschewski, W. Ruth, B. Stübing, and H. Berge, *J. Prakt. Chem.* 324:787 (1982).

   d. G. Merkel, H. Berge, and P. Jeroschewski, *J. Pract. Chem.* 326:467 (1984).

7 a. O. Sock, M. Troupel, and J. Perichon, *Tetrahedron Lett.* 26:1509 (1985).

   b. J.C. Polest, J.Y. Nedelec, and J. Perichon, *Tetrahedron Lett.* 28:1885 (1987).

   c. S. Gambino, G. Filardo, and G. Silvestri, *J. Appl. Electrochem.* 12:549 (1982).

8 a. A. Diaz, *in*: "Organic Electrochemistry", 3rd ed., H. Lund and M.M. Baizer, eds., Dekker, New York, Chapter 33, pp. 1363 (1990).

   b. M.M. Baizer, *in*: "Organic Electrochemistry", 3rd. ed., H. Lund and M.M. Baizer, eds., Dekker, New York, Chapter 30, pp. 1265 (1990).

9. D. Pletcher, "Industrial Electrochemistry", Chapman and Hall, London (1982).

SINGLE ELECTRON TRANSFER IN ALIPHATIC NUCLEOPHILIC SUBSTITUTION

Henning Lund

Department of Organic Chemistry
University of Århus
8000 Århus C, Denmark

Since the classical studies by Ingold in the nineteen thirties the mechanism of the aliphatic nucleophilic substitution reaction has been described as a polar two-electron mechanism in which the nucleophile transfers two electrons to the new bond which is established to the electrophilic center. Depending on the reaction order the abbreviations S_N1 and S_N2 were used for first-order and second-order reactions, respectively.

The transition state (TS) for the S_N2–reaction was pictured as a five-coordinated central carbon with considerable bonding to both the attacking nucleophile and the leaving group, and with a change in spatial structure, in principle from an (R)-form to an (S)-form (or vice versa).

$$Nu^- \ + \ \overset{}{\underset{}{C}}-L \ \longrightarrow \ [Nu\cdots C\cdots L]^- \ \longrightarrow \ Nu-C \ + \ L^- \tag{1}$$

This reaction mechanism (together with S_N1 and hybrids between S_N1 and S_N2) was successful in explaining many experimental observations, and this success is part of the explanation why organic chemists for many years after tried to explain all reactions as polar events.

During the last 25 years, however, a number of nucleophilic reactions have been proposed to involve the transfer of a single electron rather than a pair of electrons. Some of the pioneering experiments were done by the groups of N. Kornblum[1a,b] and Glen A. Russell.[1c] It was observed that anions of aliphatic nitrocompounds were alkylated at oxygen by benzyl chloride whereas p-nitrobenzyl chloride alkylated the anion mainly at carbon. It was also found that the leaving group X in the alkylating agent $O_2NC_6H_4CH_2X$ influenced the C/O-alkylation ratio. Poor leaving groups favored C-alkylation whereas good leaving groups (e.g. I^-) favored O-alkylation.

The explanation for these and similar experiments was given independently by the N. Kornblum and G.A. Russell groups.[1] The reaction mechanism was later to be known as the $S_{RN}1$ reaction.

$$A^- + O_2NC_6H_4CH_2X \ \rightarrow \ A^\bullet + O_2NC_6H_4CH_2X]^{\bullet-} \tag{2}$$

Sulfur-Centered Reactive Intermediates in Chemistry and Biology, Edited by
C. Chatgilialoglu and K.-D. Asmus, Plenum Press, New York, 1990

$$O_2NC_6H_4CH_2X]^{\bullet-} \rightarrow O_2NC_6H_4CH_2{}^{\bullet} + X^- \tag{3}$$

$$O_2NC_6H_4CH_2{}^{\bullet} + A^- \rightarrow O_2NC_6H_4CH_2-A]^{\bullet-} \tag{4}$$

$$O_2NC_6H_4CH_2A^{\bullet-} + O_2NC_6H_4CH_2X \rightarrow O_2NC_6H_4CH_2A + O_2NC_6H_4CH_2X]^{\bullet-} \tag{5}$$

It is essential to have an electron-withdrawing substituent in the benzyl system, otherwise the rate of the electron transfer (ET) (eq. 2) is too slow to compete with the classical S_N2-reaction; even for the p-nitrobenzyl system the ET from the Li-salt of 2-nitropropane is relatively slow and cannot compete with the S_N2-reaction when X = I.

The $S_{RN}1$-reaction was different from the S_N2-reaction not only with regard to C/O-alkylation of nitronate ions, but also with respect to stereochemistry. The sterically hindered halide, p-nitrocumyl chloride, reacted readily with a number of anions; the Li-salt of 2-nitropropane was thus C-alkylated under conditions where an S_N1-reaction was unlikely. Furthermore, the reaction of chiral 2-nitro-2-(p-nitrophenyl)butane with sodium azide (6) yielded the racemic product.[1b]

$$\underset{O_2NC_6H_4}{\overset{C_2H_5}{\underset{}{\underset{CH_3}{\diagdown}}}}C-NO_2 + NaN_3 \xrightarrow{\text{HMPA}} NaNO_2 + \underset{O_2NC_6H_4}{\overset{C_2H_5}{\underset{CH_3}{\diagdown}}}C-N_3 + N_3-\underset{C_6H_4NO_2}{\overset{C_2H_5}{\underset{CH_3}{\diagup}}}C \tag{6}$$

Several systems react via a similar ET-chain reaction; thus 2-halo-2-nitropropanes, 2,2-dinitropropane or 2-arylsulfonyl-2-nitropropane undergo reactions with carbanions, such as anions from nitroalkanes, malonic esters, β-diketones and similar compounds. In some cases, however, the nitropropyl radical is reduced by the anion without a coupling to an anion radical.[2]

It is characteristic of all the alkylating species which react by an $S_{RN}1$-type reaction that they are substituted by one or two nitro groups or other electron-withdrawing groups. This is necessary in order to bring their reduction potential into a region sufficiently close to the oxidation potentials of the enolate (and other) anions; it also prolongs the life-time of the anion radicals involved.

Different evidence for the occurrence of radicals during the reaction between organo-metallic reagents and alkyl halides have been found. Thus t-butyl radicals have been detected by e.s.r. spectroscopy in the reaction between n-butyllithium and t-butyl iodide.[3a] Similarly, chemically induced dynamic nuclear polarization (CIDNP) has been employed to detect radicals in the reactions between ethyllithium and ethyl iodide in benzene[3b] and between butyllithium and iodobutane.[3c] However, the occurrence of radicals during a reaction does not necessarily mean that the main reaction proceeds through radical intermediates.

Evidence for species formed by ET was also found in the reaction between thiophenoxide and different butyl compounds.[4] Reaction of 2-butyl nosylate with lithium thiophenoxide in the presence of the spin trap phenyl-t-butylnitrone in THF gave the 2-butyl spin adduct nitroxide, and the phenylthiyl radical was detected by the styrene polymerization technique.

Before going further we should discuss electron transfer to alkyl halides. Aromatic anion radicals are one-electron donors; the reactions between aromatic anion radicals, prepared by the reaction of an alkali metal with the aromatic compund (A), and alkyl halides (BX) were investigated in the late sixties by several groups.[5a,b] It was concluded that the coupling products were formed after an inital electron transfer from the anion radical to the alkyl halide, e.g. (7)

$$2 \; \left[\text{(naphthalene)} \right]^{\bullet -} \; + \; RX \; + \; H^+ \; \longrightarrow \; \text{(R,H substituted)} \; + \; C_{10}H_8 \; + \; X^- \tag{7}$$

The reaction between electrochemically generated aromatic anion radicals and aromatic and aliphatic halides has been investigated since 1974.[6a-d] We found that the reaction with alkyl halides could be described by the following equations (8–11) followed by protonation of AB^- and B^-.

$$A \; + \; e^- \; \rightleftarrows \; A^{\bullet -} \tag{8}$$

$$A^{\bullet -} \; + \; BX \; \xrightarrow{k_9} \; A \; + \; B^{\bullet} \; + \; X^- \tag{9}$$

$$A^{\bullet -} \; + \; B^{\bullet} \; \xrightarrow{k_{10}} \; AB^- \tag{10}$$

$$A^{\bullet -} \; + \; B^{\bullet} \; \xrightarrow{k_{11}} \; A \; + \; B^- \tag{11}$$

The coupling reaction (eq. 10) and the reduction (eq. 11) are in competition which will be discussed later. When (10) predominates, the result is a substitution reaction; when (11) is the main reaction, a catalytic reduction of BX occurs.

Cyclic voltammetry (CV) is a valuable tool for the investigation of electron transfer reactions. Figure 1 shows CV-curves of anthracene in the presence of increasing concentrations of benzyl chloride (the latter is reduced at potentials more negative than that of anthracene). The current, which is proportional to the concentration of anthracene, increases on addition of benzyl chloride, which is due to the regeneration of anthracene in eqs. (9) and (11); the electrode thus "feels" a higher concentration of A than without BX. A functions only as an electron-carrier. In contrast to that the current becomes independent on added BX in the reaction between quinoline and butyl bromide (Fig. 2), because the catalyst, the anion, is eventually removed from the solution as the coupled product AB^-, eq. (10).

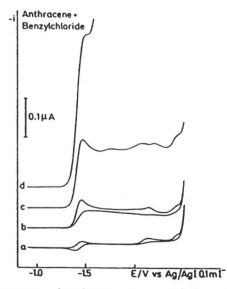

Fig. 1. Cyclic voltammograms of anthracene (3.4×10^{-4} M) in DMF/0.1 M TBAI in the presence of different concentrations of benzyl chloride. (a) 0 M; (b) 7.3×10^{-4} M; (c) 3.7×10^{-3} M; (d) 1.85×10^{-2} M benzyl chloride. Sweep rate 10 mV s^{-1} (from ref. 6c).

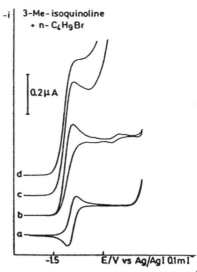

Fig. 2. Cyclic voltammograms of 3-methylisoquinoline (5.8×10^{-3} M) in DMF/0.1 M TBAI in the presence of different concentrations of butyl bromide. (a) 0 M; (b) 8×10^{-3} M; (c) 3.2×10^{-2} M; (d) 6.4×10^{-2} M butyl bromide. Sweep rate 10 mV s^{-1} (from ref. 6c.).

It was also shown[7] that enolate ions were good ET-reagents. When methyl 1-methyl-isonicotinoate (**1+**) was reduced in CV, two reversible waves were obtained (Fig. 3, trace b). The reduction is illustrated in eq. (12).

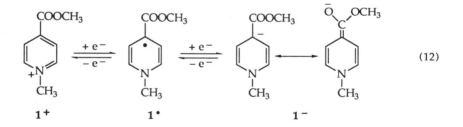

$$(12)$$

The CV of 1,2-dichloro-1,2-diphenylethane (Fig. 3, trace a) shows the 2-electron reduction to stilbene, and the reversible reduction of the latter.

Addition of 1,2-dichloro-1,2-diphenylethane to the solution gave an increase of the second wave (Fig. 3), indicating that a catalytic reduction of the dihalide took place and that the anion **1⁻** was a good reducing agent. A further advantage of **1⁻** is that the reversible oxidation potential of it is known from CV.

If **1+** is reduced to **1⁻** in the presence of an alkyl halide, e.g., *t*-butyl bromide, an alkylation takes place at C-4 in the pyridine ring (eq. 13). Furthermore, if **1⁻** reacts with *t*-butyldimethylsulfonium iodide, **1⁻** is *t*-butylated. Obviously this is not a classical S_N2-reaction; a methylation rather than a *t*-butylation would be expected. (The reaction is run under conditions where an S_N1-reaction is unlikely, *t*-BuBr does not react with iodide under these conditions.) The reaction was formulated as:

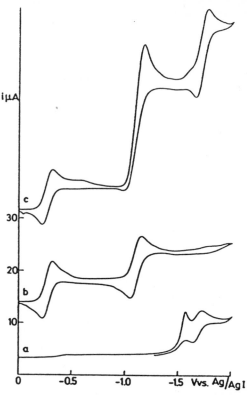

Fig. 3. Cyclic voltrammograms at a platinum electrode in DMF/TBAI; potentials *vs.* Ag/AgI, 0.1 M I⁻-reference electrode; v = 20 mV s⁻¹. a: 1.9 x 10⁻² M 1,2-dichloro-1,2-diphenylethane (**2**); b: 3.0 x 10⁻² M 1-ethyl-4-methoxycarbonyl-pyridinium iodide (**1⁺**); c: 5.6 x 10⁻² M **2** + 3.0 x 10⁻² **1⁺** (from ref. 7).

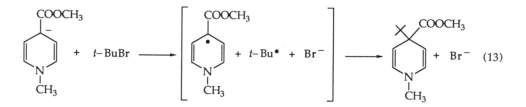

A similar mechanism has been proposed for the reaction between some lithium acetylides and 2-chloro-2-nitropropane or 2,2-dinitropropane.[2]

Several experimental facts thus indicate that ET plays a role in some aliphatic, nucleophilic substitutions. The question is then, is the rate of the ET-step high enough to account for the observed rate of the reaction? Two approaches have been employed to throw some light on this question, one based on thermodynamic data[8] and the Marcus theory for electron-transfer reactions, and a second experimental one[9] based on the assumption that aromatic anion radicals react with alkyl halides via ET as the rate-determining step; the data of the latter investigation are also treated assuming the Marcus theory being applicable to the electron transfer to alkyl halides, which probably is a dissociative ET.

73

The Marcus equation is concerned with the connection between the activation energy for the transfer of an electron and the free energy of the reaction. Whereas the Hammett equation is a linear free energy equation connecting the activation energy of a reaction with the change in free energy, the Marcus equation assumes a quadratic connection between $\Delta G^{\ddagger}$ and ΔG°. A simplified form of the Marcus equation, applicable for the transfer of an electron from an anion radical (or another anion) to an alkyl halide, is formulated in eq. (14)

$$\Delta G^{\ddagger} = \frac{\lambda}{4} (1 + \frac{\Delta G^{\circ}}{\lambda})^2 \tag{14}$$

In this equation λ is the total reorganization energy consisting of λ_i (the internal reorganization energy connected with changes in bond length and bond angles) and λ_s (the reorganization energy of the solvent connected with the electron transfer).

As a result of the thermodynamic approach Lennart Eberson[8] distinguished three groups within a number of reactions proposed to involve ET, one consisting of reactions in which ET might be possible, another for which ET seemed excluded by the thermodynamic data, and a third group consisting of boundary cases. Not all the relevant thermodynamic data were available and quite a number of them, including most of the λ-values, had to be estimated. ET was therefore excluded only for reactions for which the available data and estimated values indicated that the calculated rate constants were several orders of magnitude smaller than the experimentally determined rate constants.

The basis for the "experimental" approach was a measurement of the connection between the standard potential of an electron donor (stable anion radicals of aromatic and heteroaromatic compounds were chosen) and the rate of electron transfer to a given acceptor, an alkyl halide.[9a] When this connection was found, the rate of the substitution reaction between an anion and the same alkyl halide was measured; if the rate (k_{SET}) of ET from an anion radical with the same oxidation potential as the anion to the alkyl halide was the same as the rate (k_{SUB}) with which the anion reacted with the alkyl halide, then it was assumed that the rate-determining step in this aliphatic nucleophilic substitution was the transfer of an electron (Fig.4). In Table 1 some kinetic results are given, including k_{SUB}/k_{SET}; when discussing these data, it must be remembered that there are some assumptions involved. It is, e.g., assumed that the λ-value of the anion 1– is close to that of aromatic anion radicals ($\approx$ 10 kcal mol-1). Thus, due to the uncertainties of the method, when $0.2 < k_{SUB}/k_{SET} < 5$, the ratio is interpreted in terms of SET (single electron transfer) being the rate-determining step in the reaction.

The low values of k_{SUB}/k_{SET} are found for sterically hindered alkyl halides. The less hindered alkyl halides gave higher values, which has been interpreted as a stabilization of the transition state (TS) due to a bonding interaction in the TS. However, even in the reaction of 1– with ethyl bromide the stabilization of the TS corresponds to a bonding interaction of 4 - 5 kcal mol-1, much smaller than the bonding interaction in a "classical" S_N2-TS. The data were interpreted to suggest that the TS of the "classical" S_N2 and SET reactions were extremes and that different kinds of "hybrids" of these extremes existed, characterized by different degrees of bonding stabilization of the transition states ("SET with some S_N2-character, S_N2 with some SET-character"); this interpretation was preferred for an alternative one which explains the data as a competition between two different reaction paths, SET and S_N2.

Originally S_N2 meant all second order aliphatic nucleophilic substitutions, and the SET type would thus also be included. However, an essential characteristic of a "classical" S_N2-reaction is the inversion of the stereochemical arrangement at the central carbon atom; in a "pure" SET-reaction a razemization might be expected.

Table 1. Rate constants, k_{SUB}, for the reaction of different nucleophiles with some alkyl halides compared to k_{SET}. The nucleophiles were prepared by a two-electron reduction of the compounds shown below. Pe, Perylene, AQ, 9,10-anthraquinone.

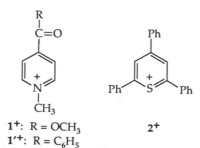 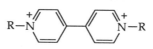

1⁺: R = OCH₃
1′⁺: R = C₆H₅
2⁺
BV²⁺: R = C₆H₅CH₂
MV²⁺: R = CH₃

Nucleophile	RX	E_{Nu}	$k_{SUB}/M^{-1}\,s^{-1}$	$k_{SET}/M^{-1}s^{-1}$	k_{SUB}/k_{SET}
1⁻	1-Adamantyl bromide	−1.13	1.5×10^{-2}	1.9×10^{-2}	0.8
—	Neopentyl bromide	—	2.9×10^{-2}	2.3×10^{-2}	1.3
—	t-BuBr	—	30	12	2.5
—	s-BuBr	—	480	2.8	170
—	n-BuBr	—	1420	3.5	400
—	CH₃CH₂Br	—	3052	1.2	2500
—	PhC(CH₃)(CH₂CH₃)Cl	—	218	171	1.3
—	PhCH(CH₃)Cl	—	2980	367	8.2
—	PhCH₂Cl	—	3.1×10^4	469	66
—	d,l-PhCHCl-CHClPh	—	376	693	0.55
—	meso-PhCHCl-CHClPh	—	948	1745	0.54
1′⁻	t-BuBr	−0.886	23	0.08	290
—	s-BuBr	—	84	0.02	4200
---	n-BuBr	—	141	0.07	2000
—	PhC(CH₃)(CH₂CH₃)Cl	—	21	3	7
—	PhCH(CH₃)Cl	—	107	3.4	33
—	PhCH₂Cl	—	540	6	90
—	PhCH(CH₃)Br	—	1.6×10^5	4.5×10^3	36
—	PhCH₂Br	—	1.2×10^6	4×10^3	300
2⁻	PhCH(CH₃)Br	−0.721	1.2×10^4	891	14
—	PhCH2Br	—	5.9×10^4	562	105
—	PhCH(CH₃)Cl	—	19.5	0.26	76
—	PhCH₂Cl	—	113	0.55	205
MV	PhCH(CH₃)Br	−0.350	54	4.5(18)	12(3)
—	PhCH₂Br	—	35	1.6(6.3)	22(6)
BV	PhCH(CH₃)Br	−0.287	137	1.6(6.3)	86(14)
—	PhCH₂Br	—	35	0.8(3.2)	44(10)
Pe²⁻	t-BuCl	−1.80	171	141	1.21
—	s-BuBr	−1.80	9.5×10^5	2.11×10^5	4.5
AQ²⁻	s-BuBr	−1.123	909	2.8	322

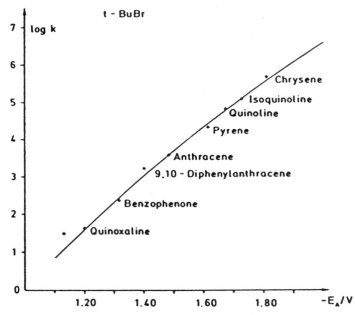

Fig. 4. Rate of electron transfer, k_{SET}, from some electrochemically generated anion radicals and 1^{-}(*) to 2-bromo-2-methylpropane (t-BuBr) in DMF/0.1 M TBABF$_4$ (from ref. 9a).

When bornyl bromide and isobornyl bromide was brought to reaction with the enolate anion 1^{-}, the same mixture of *exo*- and *endo*-substitution products were observed, eq. (15).[9b]

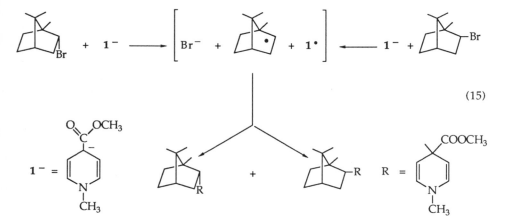

(15)

The interpretation suggested was that the radical obtained on ET from 1^{-} to the bornyl/isobornyl bromide was formed in a solvent cage and that it had time to isomerize before it coupled with the $1^{•}$-radical. In this reaction practically no bonding interaction existed in the TS between the radical from the donor ($1^{•}$) and the alkyl radical. The time between the creation of the two radicals and their coupling is, however, extremely short.

For the reaction between aromatic anion radicals and alkyl halides it was observed quite early[6] that a catalytic reduction of the alkyl halide [eq. (11)] or a coupling with the alkyl halide [eq. (10)] took place. By means of linear sweep voltammetry (LSV) and controlled potential electrolysis it was possible to determine whether reaction (10) or (11)

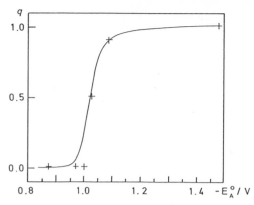

Fig. 5. q-Values of benzyl chloride vs. the redox potential $E°_A$ of some anion radicals. The data are tabulated in Table 2 (from ref. 9c).

was dominating or whether both were important.[9c] A dimensionless parameter q [= $k_{11}/(k_{10} + k_{11})$] indicated the competition, q = 0 for coupling and q = 1 for reduction. In Fig. 5, q for benzyl chloride is plotted as a function of $E^0(A/A^{•-})$, the oxidation potential of the anion radical, and a smooth S-shaped curve is obtained. The interpretation of the results is that when $E^0(A/A^{•-})$ is more negative than the potential $E^q_{1/2}$, where q = 1/2, then reduction takes place, but when $E^0(A/A^{•-})$ is less negative than $E^q_{1/2}$ then the anion radical is not able to reduce the radical and a coupling takes place. The knowledge of $E^q_{1/2}$ for a radical is directly useful for the planning of a synthesis; its relation to the thermodynamic interesting value E^0 depends a.o. on the value of $\lambda(R^{•}/R^-)$. The values of $E^q_{1/2}$ for some benzyl derivatives are given in Table 2 together with voltammetric half-wave potentials determined by phase-sensitive voltammetry of photochemically generated radicals.

Table 2. $E^q_{1/2}$ and $E_{1/2}$ values for some substituted benzyl radicals

BX	$-E^q_{1/2}/V$[a]	$-E_{1/2}/V$[b]
Benzyl chloride (1)	1.03	1.06[c]
1-Chloro-1-phenylethane (2)	1.20	1.21[c]
2-Chloro-2-phenylpropane (3)	1.20	1.34[c]
2-Chloro-phenylbutane (4)	1.20	—
Benzhydryl chloride (5)	0.70	0.75[c]
4-Methoxybenzyl choride (6)	1.10	1.36[c]
4-Chlorobenzyl chloride (7)	1.03	1.01[c]

[a] Versus Ag/AgI

[b] The $E_{1/2}$-values are corrected from vs. SCE to vs. Ag/AgI by adding +0.39 V.

[c] B. Sim, D. Griller, and D.D.M. Wayner, *J. Am. Chem. Soc.* 111:754 (1989).

A determination of such "practical redox potentials" as $E^q_{1/2}$ for radicals would help explaining many observations described in the literature and to bring the planning of new syntheses on a more rational basis. For instance, by having the knowledge of the oxidation potentials of the anions, the reduction potentials of the radicals, and the dimerization rates of the radicals involved it would be possible to predict in which cases the anion and the radical would couple to an anion radical and thus react by a chain reaction mechanism, or when the anion would be able to reduce the radical. An example relevant to this is the observation by Glen Russell[2] that when the 2-nitro-2-propyl radical reacts with certain anions, a coupling takes place, whereas others reduce the radical, eq. (16).

$$\overset{\bullet}{Me_2C}-NO_2 \;+\; A^- \quad \Bigg\langle \qquad \begin{array}{l} \longrightarrow \quad A-Me_2C-NO_2\Big]^{\bullet-} \\[1.5em] \longrightarrow \quad A^\bullet \;+\; Me_2C{=}NO_2^- \end{array} \tag{16}$$

The relationship between polar and electron–transfer pathways has been discussed by, a.o., S.S. Shaik and A. Pross.[10] They suggest that both polar and SET-reactions involve the "shift" of a single electron. They distinguish between "electron transfer" and "electron shift"; the former, in agreement with current usage, is the simple, elementary act of transferring an electron in a redox reaction, whereas "electron shift" describes a change in the position of an electron which is, in addition, coupled to bonding changes. In these terms the aliphatic nucleophilic substitution can be described by eq. (17).

$$Nu\,\updownarrow\!\uparrow \;+\; R\,\updownarrow\!\uparrow\, X \;\rightleftharpoons\; [\,Nu\,\updownarrow{\cdots}\uparrow R{\cdots}\updownarrow\!\uparrow\, X\,] \;\rightleftharpoons\; Nu\,\updownarrow\!\uparrow R \;+\; \updownarrow\!\uparrow X \tag{17}$$

It is attempted in eq. (17) to illustrate how a single electron is "shifted" from the nucleophile to the electrophile and the leaving group is "picking up" an electron. In Fig. 6 a reaction profile for an electron-transfer is depicted schematically according to the "configuration mixing" (CM) model which here is equivalent to the Marcus treatment. An electron-transfer requires that the donor-acceptor complex (DA) becomes isoenergetic with the resulting D^+A^- complex, and this involves changes in the solvation and geometry of the DA complex. According to the valence bond configuration mixing (VBCM) model the difference between a SET reaction and a classical S_N2-reaction is that the electron transfer in the latter is accompanied by a group transfer. In another language, mostly used in inorganic chemistry, the SET is an outer-sphere ET and S_N2 an inner-sphere ET.

The VBCM model "suggests that both SET and polar pathways (S_N2) involve a single electron shift in the transition state region; the factor that determines which particular pathway is followed in any given reaction is the feasibility of coupling of the two spin-paired electrons following the electron shift. Any factor (steric, electronic, or geometric) that operates so as to inhibit or hinder the coupling process will tend to favor a SET pathway over a polar one".[10a,b] As the TS of the polar reaction is stabilized by bond formation relative to the SET-TS the energy of the former TS should be lower than that of the latter, and the S_N2-reaction thus favored over SET, Fig. 7.

Inspection of Table 1 shows that alkyl halides with steric factors slowing down the classical S_N2 approach have a low k_{SUB}/k_{SET} ratio, as would also be expected intuitively.

Whereas the prediction by the VBCM model that the activation energy of the substitution will decrease if the electron-donating power of the nucleophile is increased ($\sim I_{Nu}$ decreased) or the electron-accepting power of the electrophile ($\sim A_{RS}$) is increased, is generally observed[9a] (also in agreement with what chemical intuition would expect), the suggestion that TS for a good electron donor will be located earlier on the reaction coordinate than a poor donor is not always met. This is illustrated in Fig. 8.

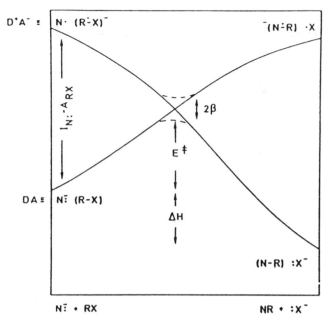

Fig. 6. State correlation diagram for an S_N2 reaction. The lower states are ground states of reactants and products, while the upper states are the corresponding charge transfer states. β is the degreee of avoided crossing. $\Delta E^{\neq}$ is the reaction barrier (from ref. 10a).

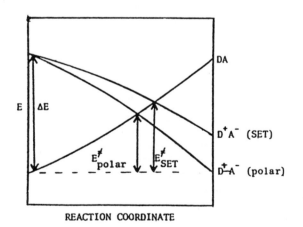

REACTION COORDINATE

Fig. 7. Schematic diagram illustrating the energy profiles for the SET and the polar pathways. $\Delta E^{\neq}$ for the polar process is lower than for the SET process (from ref. 10b).

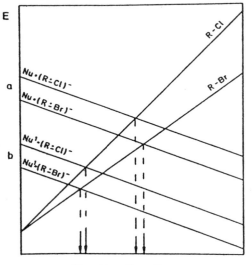

Reaction Coordinate

Fig. 8. Schematic energy diagram illustrating the effect of changing the acceptor from R–Cl to R–Br in two cases (a and b). The ionization potential of Nu, $I_{Nu\cdot}$ is higher than $I_{Nu}1$. The relative position of TS(RBr) and TS'(RCl) is changed from a to b (from ref. 9a).

Let us now summarize the working hypothesis of our group for the aliphatic, nucleophilic substitution: We consider three limiting cases corresponding to the classical S_N2, the classical S_N1, and the SET reaction.

a) When the steric hindrance is not prohibitive and the difference, ΔE, in oxidation potential of the nucleophile and reduction potential of the electrophile (RX) is not too great, the classical S_N2 (polar reaction) will take place.

b) If the steric conditions hinder the approach of the nucleophile to the central carbon atom and ΔE is not too large, then the SET reaction path is favored.

c) If steric factors inhibit the S_N2-reaction and ΔE is too large to allow SET to proceed with a reasonable rate, then the classical S_N1-reaction may come into play by creating a planar carbocation which also is a much better electron acceptor than RX; this might make ΔE sufficiently small for an electron shift to take place. Even then, if the E^0_{ox} of the nucleophile is too high (i.e. a very poor reducing agent with a low ability to donate an electron) the formation of a carbocation as electrophile might not be sufficient to set the energetics right for an electron shift. (In this connection it is noticeable that anions which are very difficult to oxidize such as perchlorate and tetrafluoborate ions are practically unreactive under S_N1-reaction conditions in aqueous solution).

The factors which control the reaction path of a nucleophilic substitution are not step functions, so a reaction does not have to follow one of the extreme cases but can be, e.g., "SET with some S_N2-character" or another hybrid.

Up till now we have discussed ET in the aliphatic, nucleophilic substitution, but what about ET in other organic reactions? The $S_{RN}1$-reaction has for several years been well established as a reaction path for the aromatic, nucleophilic substitution,[11a] and the influence of

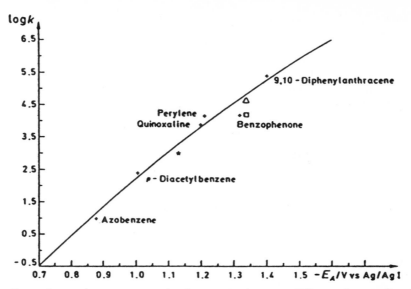

Fig. 9. Dependence of rate constants for the reaction between different electron donors and *meso*-1,2-dichloro-1,2-diphenylethane on the reversible redox potential of the donors; * 1^-; □ methyl isonicotinoate; Δ 10-nonyl-10-nonadecyl isonicotinoate (from ref. 9d).

the redox potentials of the participating species on the reaction has been thoroughly investigated by electrochemical methods.[11b]

The reductive elimination of vicinal halides is usually described as an atom-transfer reaction of S_N2-type, e.g., eq. (18).

$$I^- \quad I-C(RR_1)-C(R'R'')-I \longrightarrow I_2 + RR_1C=CR'R'' + I^- \tag{18}$$

It was found[9d] that the enolate ion 1^- reacted with 1,2-dichloro-1,2-diphenylethane (with formation of stilbene) with the same rate as anion radicals of aromatic compounds with the same redox potential (Fig. 9). This has been taken as evidence that the rate-determining step in the reductive elimination by 1^- is the transfer of an electron.

In the studies on the addition of Grignard reagents and lithium compounds to unsaturated compounds product analyses as well as different radical-detecting methods indicate the occurrence of radicals during some of these reactions. Similar results are found with other metalorganic reagents, and it has been suggested that ET is important in some electrophilic, aromatic substitutions, but a discussion of these reactions is outside the scope of this survey. It suffices to say that the problem ET/polar reaction seems to be of relevance to all organic reactions.[12]

ACKNOWLEDGEMENT

I would like to thank my collaborators, K. Daasbjerg, R. Fuhlendorff, T. Lund, D. Occhialini, and S.U. Pedersen for their dedicated work on SET reactions and related problems, and the Danish Science Research Council and Carlsberg Foundation for economic support.

REFERENCES

1.a. N. Kornblum, R.E. Michel, and R.C. Kerber, *J. Am. Chem. Soc.* 88:5660,5662 (1966).
  b. N. Kornblum, *Angew. Chem.* 87:797 (1975).
  c. G.A. Russell and W.C. Danen, *J. Am. Chem. Soc.* 88:5663 (1966).
2.   G.A. Russell, M. Jawdosiuk, and M. Makosza, *J. Am. Chem. Soc.* 101:2355 (1979).
3.a. G.A. Russell and G.W. Holland, *J. Am. Chem. Soc.* 91:3967 (1969).
  b. H.R. Ward, R.G. Lawler, and R.A. Cooper, *J. Am. Chem. Soc.* 91:746 (1969).
  c. A.R. Lepley and R.L. Landau, *J. Am. Chem. Soc.* 91:748 (1969).
4.   S. Bank and D.A. Noyd, *J. Am. Chem. Soc.* 95:8203 (1973).
5 a. J.F. Garst, *Acc. Chem. Res.* 4:400 (1971) and references therein.
  b. S. Bank and D.A. Juckett, *J. Am. Chem. Soc.* 89:7742 (1976).
6 a.. H. Lund, M.A. Michel, and J. Simonet, *Acta Chem. Scand.* B28:900 (1974).
  b. J. Simonet, M.A. Michel, and H. Lund, *Acta Chem. Scand.* B29:489 (1975).
  c. H. Lund and J. Simonet, *J. Electroanal. Chem.* 65:205 (1975).
  d. H. Lund, *J. Mol. Cat.* 38:203 (1986).
7.   H. Lund and L.H. Kristensen, *Acta Chem. Scand.* B33:495 (1979).
8.   L. Eberson, *Acta Chem. Scand.* B38:439 (1984).
9.a. T. Lund and H. Lund, *Acta Chem. Scand.* B40:470 (1986); B41:93 (1987); B42:269 (1988).
  b. K. Daasbjerg, T. Lund, and H. Lund, *Tetrahedron Lett.* 30:493 (1989).
  c. R. Fuhlendorff, D. Occhialini, S.U. Pedersen, and H. Lund, *Acta Chem. Scand.* B43 (1989), 803.
  d. T. Lund, S.U. Pedersen, H. Lund, K.M. Cheung, and J.H.P. Utley, *Acta Chem. Scand.* B41:285 (1987).
10a. A. Pross and S.S. Shaik, *Acc. Chem. Res.* 16:363 (1983).
  b. A. Pross, *Acc. Chem. Res.* 18:212 (1985).
11a. J.F. Bunnett, *Acc. Chem. Res.* 11:413 (1978).
  b. J.-M. Savéant, *Acc. Chem. Res.* 13:323 (1980).
12.   L. Eberson, Electron Transfer Reactions in Organic Chemistry, Springer Verlag, Berlin, Heidelberg (1987).

ELECTROCHEMISTRY AS A TOOL FOR THE STUDY OF REACTIVE INTERMEDIATES

George S. Wilson

Department of Chemistry
University of Kansas
Lawrence, KS 66045, USA

DETERMINATION OF KINETIC AND THERMODYNAMIC PARAMETERS

There are now available a wide range of electrochemical techniques which can provide fundamental information about both energetics and kinetics of thioether redox chemistry. Before discussing the various techniques, it is first appropriate to consider the kinds of reactions which thioethers are known to undergo and the information one desires about them. Equations 1 - 10 outline the type of reactions which can be studied by electrochemical methods. The discussion will be limited to oxidations and will not include multi-step reaction sequences leading to transitions from S-centered to C-centered radicals. These reactions have been extensively documented particularly in the pulse radiolysis literature.

a. One electron oxidation

$$R_2S \quad \overset{-e^-}{\rightleftarrows} \quad R_2S^{\bullet+} \qquad\qquad E_1^{0'} \qquad (1)$$

$$R_2S^{\bullet+} \quad \overset{-e^-}{\rightleftarrows} \quad R_2S^{2+} \qquad\qquad E_2^{0'} \qquad (2)$$

b. Dimer formation

$$2\,R_2S^{\bullet+} \quad \rightleftarrows \quad R_2S^+\!\cdot\!\cdot S^+R_2 \qquad\qquad \text{(diamagnetic)} \qquad (3)$$

$$R_2S^{\bullet+} + R_2S \quad \rightleftarrows \quad (R_2S\cdot^{\bullet}\!\cdot SR_2)^+ \qquad\qquad \text{(paramagnetic)} \qquad (4)$$

c. Adduct formation

$$R_2S^{\bullet+} + ROH \quad \rightleftarrows \quad R_2S\cdot^{\bullet}\!\cdot OR + H^+ \qquad (5)$$

$$R_2S^{\bullet+} + RNH_2 \quad \rightleftarrows \quad R_2S\cdot^{\bullet}\!\cdot NHR + H^+ \qquad (6)$$

d. Nucleophilic attack

$$R_2S^{2+} + H_2O \quad \rightarrow \quad R_2S=O + 2\,H^+ \qquad (7)$$

$$2\,R_2S^{\bullet+} + H_2O \quad \rightarrow \quad R_2S=O + R_2S + 2\,H^+ \qquad (7a)$$

Sulfur-Centered Reactive Intermediates in Chemistry and Biology, Edited by
C. Chatgilialoglu and K.-D. Asmus, Plenum Press, New York, 1990

e. Oxidation by atom transfer

$$R_2S + Br_2 \quad \rightarrow \quad R_2SBr^+ \; Br^- \quad \rightarrow \quad [R_2S^{2+} \; Br^-] \; Br^- \tag{8}$$

$$R_2S + OH^\bullet \quad \rightarrow \quad R_2S\bullet\cdot OH \quad \rightarrow \quad R_2S^{\bullet +} + OH^- \tag{9}$$

f. Further oxidation of sulfur

$$R_2S{=}O \; + \; H_2O \quad \xrightarrow{-2e^-} \quad R_2S\!\!\begin{smallmatrix}O\\\\O\end{smallmatrix} \; + \; 2\,H^+ \tag{10}$$

The simplest thioether redox reaction involves the loss of one-electron to form the cation radical (reaction 1) and this should serve as a measure of "ease of oxidation". This step could be followed by the loss of a second electron to form the dication. There are, to our knowledge, no examples of simple aliphatic thioether oxidations leading to a cation radical or dication which are sufficiently stable to exhibit reversible electron transfer. The most common technique now used to make such measurements is cyclic voltammetry. This technique is illustrated in Figure 1. A linearly-varying signal is applied to an electrode (typically platinum, gold, or glassy carbon) and the current is monitored as a function of the applied potential. The resulting current-voltage curve will reflect the electrochemistry of the substrate and of the products generated at the electrode surface. The example of Figure 1 is taken from the studies of Parker and co-workers[5] and involves the successive oxidation of thianthrene to the cation radical and dication. It will be noted that a peak corresponding to the reduction of the dication is observed. This means that on the time scale of the experiment the dication does not have time to react with water (reaction 7) a condition which is achieved by very careful drying of the solvent (acetonitrile). It is evident that by varying the rate of potential scan it is possible to change the time scale of the experiment over about five orders of magnitude.

By analyzing the current-voltage curves as a function of scan rate, it is possible to determine the rate constant for a chemical reaction following electron transfer (EC). Such a first

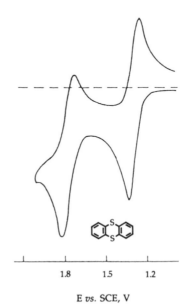

1.8   1.5   1.2

E vs. SCE, V

Fig. 1. Cyclic voltammogram for thianthrene in "dry" acetonitrile; negative currents indicate oxidation (from ref. 5)

order or pseudo-first order rate could be measured if the value is less than about 10^4 s^{-1} (half-life of about 60 µs). If the system approaches reversible behavior at high scan rates, then it may be possible to estimate the formal potential for the redox couple. More typically, however, oxidation is followed by a very rapid and irreversible chemical reaction leading to an oxidation peak with no corresponding reduction. The resulting products may have their own distinctive electrochemistry, for example, the oxidation (reaction 10) or reduction of the sulfoxide may be used to deduce the products formed.

The stability of aliphatic thioether cation radicals can be significantly enhanced by the formation of adducts (reactions 3 - 6). These serve to delocalize the positive charge on the sulfur. One of the best examples of this system is illustrated by the redox chemistry of 1,5-dithiacyclooctane (DTCO), first studied by Musker and co-workers[6] which yields both a stable cation radical and dication on oxidation with Cu(II) or NO$^+$. Figures 2 and 3 show the scan rate and concentration dependence of the cyclic voltammograms for DTCO oxidation carried out in our laboratories.[7] It is immediately evident that the system does not follow simple EC behavior and the fact that the current-voltage (CV) curves show strong concentration dependence is an immediate indication that higher order chemical reactions coupled to electron transfer are occurring. The oxidation may be formulated according to the following scheme:

$$DTCO \quad \rightleftarrows \quad DTCO^{\bullet+} + e^- \qquad E^{0'} = 0.335 \text{ V} \qquad (11)$$

$$DTCO^{\bullet+} \quad \rightleftarrows \quad DTCO^{2+} + e^- \qquad E^{0'} = 0.315 \text{ V} \qquad (12)$$

$$2\,DTCO^{\bullet+} \quad \overset{k_f}{\underset{k_b}{\rightleftarrows}} \quad (DTCO)_2^{2+} \qquad K \approx 5000 \qquad (13)$$

The dimer $(DTCO)_2^{2+}$ is not electroactive at the potential of the observed wave. It can be electrolytically consumed in three ways: 1. dissociation followed by oxidation to the dication, 2. dissociation followed by reduction to DTCO or 3. direct reduction to DTCO at −0.6 V. A peak is observed for this latter process which increases with increasing scan rate suggesting that process 3 occurs at the expense of process 2. It can also be demonstrated that the dication reacts slowly with water (k = 0.06 s^{-1}) to form the sulfoxide and this reaction becomes important at slow scan rates. The standard heterogeneous electron transfer rate constant can also be calculated and a value of 1.4×10^{-2} cm/s is obtained. This would correspond to a moderately fast reaction. The values of k_f and k_b are 8×10^4 M^{-1}s^{-1} and 15 s^{-1}, respectively. It is evident that extracting the kinetic and thermodynamic parameters from the experimental data involves assignment of 8 - 10 parameters. With such a large number of parameters it would seem that the data could be fitted to a wide variety of models. Fortunately in this case it is possible to fit the data under conditions where curve shape is sensitive to certain parameters but not others. One then goes to another set of conditions where other parameters play a dominant role. For example, values of $E^{0'}$ can be conveniently fixed at low scan rates where the heterogeneous rate constant does not have to be known exactly. Similarly, dimer formation is relatively unimportant at low concentration. Some parameters such as diffusion coefficients can be independently determined thus further reducing the dimensionality of the problem. A satisfactory set of parameters is obtained when a good fit occurs between theory and experiment over three orders of magnitude in scan rate and 1 - 2 orders of magnitude in concentration. It is clearly insufficient to rely on peak currents and peak potential shifts to establish the kinetic and thermodynamic properties of a system.

From the data mentioned above it is possible to conclude that the oxidation of DTCO occurs in two closely-spaced one-electron steps. In fact the oxidation from the cation radical to the dication occurs more easily than the first step, contrary to expectations. In the case of

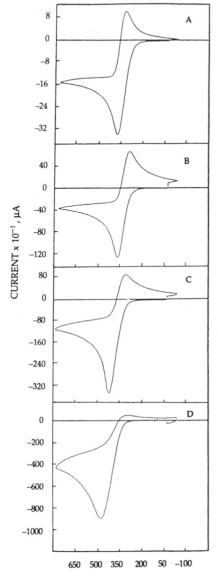

Fig. 2. Effect of scan rate on
voltammograms for 1 mM DTCO
in acetonitrile. (from ref. 7)
A– 0.01 v/s
B– 0.1 v/s
C– 1.0 v/s
D– 10 v/s

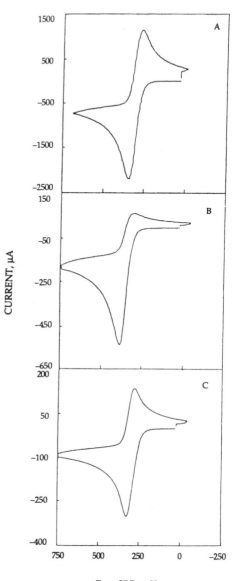

Fig. 3. Effect of concentration on
voltammograms for DTCO.
Scan rate: 0.1 v/s. (from ref. 7)
A– 2.0 mM DTCO
B– 0.5 mM DTCO
C– 0.1 mM DTCO

the cation radical, the species formed is in reality an adduct similar to that of reaction 4 resulting from a transannular interaction and the formation of a two center-three-electron bond. The electronic picture presented for this type of bond by Musker,[8] Asmus[9] and others suggests that the third electron should be in a sigma antibonding orbital (σ^*) which should have a destabilizing influence on bond strength. The removal of this electron when the dication is formed strengthens the S–S bond and this effect is energetically more important than the work required to remove an electron from a cationic species. The unexpected reversal of redox potentials is therefore appropriately explained. The three-electron bond energy is controlled by the geometry of the molecule which defines the non-bonding electron overlap. Thus the electrochemistry of related compounds such as the analogous 7-, 9- and 10-membered cyclic dithioethers does not exhibit this relatively straightforward behavior.

The behavior of DTCO is not typical of thioether oxidation. Most thioethers yield an initial single oxidation peak corresponding to an overall two-electron process. It is accordingly difficult to extract kinetic and thermodynamic information from such systems. The peak potential has no thermodynamic significance and the chemical reactions following electron transfer are too fast to be observed. Methods for extracting information from such systems will be discussed subsequently.

DETERMINATION OF "n" VALUES

In order to characterize the electron transfer reaction, it is necessary to determine the total number of electrons, n, per mole of substrate consumed. In general, there are two ways of accomplishing this: use of an electrochemical technique in which the observed current is controlled by diffusion of the substrate to the electrode surface (diffusion-controlled current) or exhaustive electrolysis of the substrate and measurement of the total charge required to accomplish the conversion. It should be emphasized that these two methods do not necessarily yield the same result. The time scales of the two experiments are quite different as are the concentrations of intermediates produced.

If methods based on the measurement of a diffusion-controlled current are used for the estimation of "n", then it is first necessary to demonstrate that the process in question is diffusion-controlled. In the case of cyclic voltammetry this consists in showing that the peak current is proportional to the square root of the scan rate.[1] Depending on the particular mechanistic scheme, appropriate theory is generally available to relate the current and the number of electrons.[1] A more reliable technique is to use a rotating disk electrode (RDE).[1] This technique has several advantages over cyclic voltammetry for this kind of measurement. Diffusion control is verified by changing the rotation speed of the electrode and measuring the steady-state current at a given applied potential. This is a much simpler measurement to make and the theory relating the current and n can often be made independent of the reversibility of the electron transfer process. Again using DTCO as an example, one can obtain an "n" value by applying a potential just beyond the oxidation peak. At low concentrations an apparent n=2 is obtained, whereas at high concentrations this value approaches n = 1. This is because the overall reactions in the two cases are:

$$DTCO \quad \rightleftarrows \quad DTCO^{2+} + 2e^- \qquad n = 2 \quad \text{(low concentration)} \qquad (14)$$

$$2\,DTCO \quad \rightleftarrows \quad (DTCO)_2^{2+} + 2e^- \qquad n = 1 \quad \text{(high concentration)} \qquad (15)$$

By contrast, at high concentration a controlled potential electrolysis yields an "n" value of 2 and the principal product is the sulfoxide resulting from the reaction of oxidized sulfur species with residual water in the non-aqueous solvent (acetonitrile). Detailed chemical analysis of the electrolysis products of thioether oxidation is very essential to understanding the redox chemistry. Two experimental difficulties peculiar to thioethers should be noted: 1. Oxidation often leads to electrode filming which blocks the surface and prevents further electron transfer from occurring. Using a large electrode, possibly with periodic cleaning, often alleviates this problem. 2. To analyze the products may require their separation from the supporting electrolyte. The product is often a sulfoxide which is quite polar and usually cannot be separated by a simple extraction. In the course of product analysis it is important to verify that all of the substrate has been consumed. It is often helpful to monitor the course of the electrolysis by running periodic cyclic voltammograms and by using spectral techniques such as uv-visible spectroscopy or EPR.

REDOX CATALYSIS

The direct oxidation of thioethers frequently occurs at potentials considerably higher than the hypothetical thermodynamic value. In the case of cyclic voltammetry, the observed peak potential, E_p, is related to $E^{0'}$ by the expression

$$E_p = E^{0'} - \frac{RT}{\alpha n_a F} [0.780 + \ln(\frac{D^{1/2}}{k^0}) + \ln(\frac{\alpha n_a Fv}{RT})^{1/2}] \tag{16}$$

where α is the electron transfer coefficient (usually 0.5), D, the diffusion coefficient, n_a, the number of electrons in the rate determining step, v, the scan rate, and k^0, the standard heterogeneous rate constant. It would appear from this expression that determination of k^0 and therefore $E^{0'}$ would be a simple matter, however the rate constant must be measured at $E^{0'}$ which is unknown. The "overpotential" which results from slow heterogeneous electron transfer causes a displacement of E_p to potentials which can be more than one volt positive of the reversible potential.

For both practical and fundamental reasons, redox catalysis can be useful in carrying out thioether oxidations. The catalyst in its reduced form, C_r, is a mediator of electron transfer which is converted into its oxidized form, C_o, at the electrode and diffuses out into solution where it reacts with the reduced form of the substrate, S_r (the thioether). The reaction sequence given below summarizes these reactions

$$C_r \rightarrow C_o + n e^- \tag{17}$$

$$C_o + S_r \rightarrow C_r + S_o \tag{18}$$

For reaction 18 to be energetically favorable, the $E^{0'}$ for the catalyst must be higher than that of the substrate. Since the substrate is oxidized with a high overpotential, it is often possible to observe the catalytic oxidation of the substrate at the potential corresponding to the catalyst which can be much lower than that of the direct oxidation. The above arguments assume that simple homogeneous electron transfer occurs in solution between the mediator and substrate (homomediation). It is also possible to form an intermediate adduct between the catalyst and substrate resulting in the oxidation of the substrate by atom transfer (heteromediation).[10] Oxidation of thioethers by this latter route is quite common and data analysis is complicated by the fact that the adduct formation must be taken into account.

Our interest in bromine catalyzed oxidation of thioethers was initiated by the coincidental presence of trace impurities of Br^- in the tetraalkylammonium tetrafluoroborate used as a supporting electrolyte for non-aqueous studies. The redox catalysis occurs at the potential corresponding to the oxidation of Br^- in acetonitrile (0.5 - 0.6 V vs. 0.1 M Ag/Ag^+ reference). The general reaction is

$$2\ Br^- \rightarrow Br_2 + 2e^- \tag{19}$$

$$R_2S + Br_2 \rightleftharpoons [R_2S^+ - Br]\ Br^- \tag{20}$$

88

Table 1. Peak Potentials for Thioether Oxidation[a]

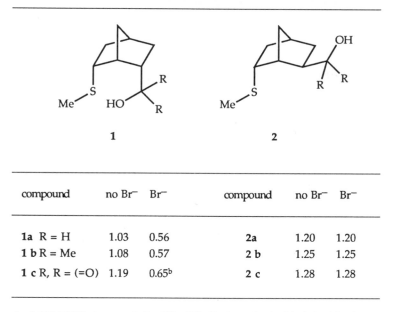

1   **2**

compound	no Br⁻	Br⁻		compound	no Br⁻	Br⁻
1a R = H	1.03	0.56		2a	1.20	1.20
1 b R = Me	1.08	0.57		2 b	1.25	1.25
1 c R, R = (=O)	1.19	0.65[b]		2 c	1.28	1.28

[a] 0.1 M LiClO$_4$ in acetonitrile, 100 mV/s, Pt electrode, Ag/Ag$^+$ 0.1 M reference.
[b] 2,6 di-(t-butyl)pyridine added as a base.

The bromosulfonium salt formed reacts with a nucleophile liberating a second mole of Br⁻ thus completing the catalytic cycle. Table 1 shows the peak potentials for a series of norbornane derivatives which are designed to demonstrate the influence of neighboring group participation. In the case of the endo-derivatives **1**, interaction between the S-methyl sulfur and the other endo nucleophilic substituent is possible whereas in the case of the corresponding exo-derivatives **2** such intramolecular interactions are not possible. The exo-derivatives show peak potentials in the 1.2 - 1.3 V range which is characteristic of the oxidation of "isolated" thioether sulfurs. The observed potentials are unaffected by the presence of Br⁻. By contrast the endo-derivatives show a shift to the 0.6 V region which is characteristic of bromide catalysis. In the case of **1c**, the endo-carboxylic acid, the shift occurs when base is added to form carboxylate.[11] Thus, although the potential to which the oxidation shifts is defined by the catalyst, the observed shift does serve as a test for neighboring group participation. Only the relatively small differences between the compound **1** (no Br⁻) and compound **2** (no Br⁻) values can be attributed to neighboring group participation (a subject dealt with in the articles by R.S. Glass in this book). Presumably efficient catalysis is dependent upon rapid decomposition of the bromosulfonium salt by a neighboring nucleophile.

Figure 4 shows the voltammogram for the Br⁻ catalyzed oxidation of **1b**. No discernible wave for Br⁻ oxidation alone can be observed at a concentration of 0.1 mM. On addition of the substrate, however, a diffusion-controlled wave which is proportional to substrate concentration is observed. It will be further noted that there is no wave at 1.08 V corresponding to direct oxidation of substrate and no wave corresponding to reduction of bromine suggesting that the catalytic reaction is rapid and complete. By measuring the catalytic enhancement as a function of the ratio of substrate to catalyst it is possible to show that significant amounts of catalysts are tied up as a catalyst substrate complex suggesting, as expected, that the process is not simple electron transfer. This complication has been treated theoretically by Nadjo and Su.[12] Even though the theoretical analysis of the system has proven difficult, redox catalysis is quite useful practically. Because the oxidation can be carried out at a lower potential and because substrate oxidation occurs in solution and not at

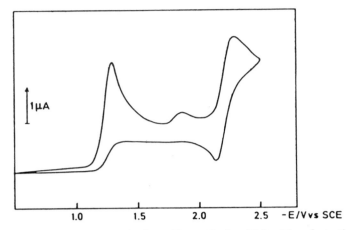

Fig. 4. Cyclic voltammogram for bromide catalysis of 1.0 mM endo-tertiary alcohol; 0.1 M LiClO$_4$ in MeCN; 0.1 mM LiBr. Scan rate : 100 mv/s

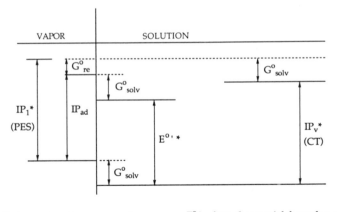

* Indicates parameter measured
IP$_v$ - vertical ionization potential
IP$_{ad}$ - adiabatic ionization potential

E$^{o\prime}$ - formal potential for redox process
G$^o_{re}$ - molecular reorientation energy
G$^o_{sol}$ - solvation energy

Fig. 5. Energetics of electron loss.

the electrode surface, electrode fouling is substantially reduced. This makes preparative scale electrolyses feasible provided that the catalyst can be continuously recycled.

Saveant and co-workers[13] have used redox catalysis as a means for estimating the $E^{0'}$ for the substrate. The basis for the method is the relationship (Marcus Theory) between the electron transfer cross reaction (reaction 18) and the difference in $E^{0'}$ between the substrate and a series of catalysts with known potentials. Because of the tendency for thioethers to form adducts, this approach has not so far proven useful for these systems.

CORRELATION OF PEAK POTENTIALS WITH SPECTROSCOPIC MEASURES OF ELECTRON LOSS

An estimate of the energetics of electron loss from a thioether often cannot be obtained directly from electrochemical data. This has inspired the correlation of electrochemical results with ionization potential estimates obtained from photoelectron spectroscopy (PES) and from the measurement of charge transfer complex spectral transitions. The energetic situation is shown in Figure 5. Several groups have reported good potentials determined by PES[14,15] even under conditions where the electron transfer is irreversible. Returning to eq. 16, one can see that if the peak potentials for all compounds are measured at the same scan rate and if the standard heterogeneous electron transfer rate is the same for all compounds, then a correlation between E_p and $E^{0'}$ might be expected. If the solvation energies are constant, then a correlation is possible. We have had limited success with this approach due partly to the fact that the above conditions cannot all be met and also the fact that the ionization potentials corresponding to electron loss of non-bonding electrons do not vary much with substituent.[16] This is, of course, a consequence of fact that the substituents only indirectly affect electron loss energy.

Despite the fact that thioether redox chemistry is very complicated and leads to irreversible oxidations often accompanied by electrode fouling, it is still possible to obtain useful information concerning the kinetics and thermodynamics of these reactions.

ACKNOWLEDGEMENT

The support of the National Institutes of Health (Grant No. HL 15104) is gratefully acknowledged.

REFERENCES

1. A.J. Bard and L.R. Faulkner, "Electrochemical Methods: Fundamentals and Applications", Wiley, New York (1980). Basic text covering the theory including especially the mathematics of diffusion.
2. A.J. Bard and H. Lund, eds., "Encyclopedia of the Electrochemistry of the Elements" (A Series), Marcel Dekker, New York (1978). Electrode potentials, rate constants, descriptive electrochemistry. Volume 12 deals specifically with organic compounds.
3. G. Dryhurst, "Electrochemistry of Biological Molecules", Academic Press, New York (1977). A good review which organizes the relevant biochemistry and electrochemistry.
4. D.T. Sawyer and J.L. Roberts, Jr., "Experimental Electrochemistry for Chemists", Wiley-Interscience, New York (1974). Gives extensive details on experimental techniques including cell design, solvent purification and reference electrodes.
5. O. Hammerich and V.D. Parker, Electrochim. Acta 18:537 (1973).
6. W.K. Musker, T.L. Wolford, and P.B. Roush, J. Am. Chem. Soc. 100:6416 (1978).

7. M.D. Ryan, D.D. Swanson, R.S. Glass, and G.S. Wilson, *J. Phys. Chem.* 85:1069 (1981).

8. T.G. Brown, A.S. Hirschon, and W.K. Musker, *J. Phys. Chem.* 85:3767 (1981).

9. K.-D. Asmus, *Acc. Chem. Res.* 12:436 (1979).

10. T. Shono, "Electroorganic Chemistry as a New Tool in Organic Synthesis", Springer-Verlag, Berlin, p. 114 (1984).

11. R.S. Glass, A. Petsom, M. Hojjatie, B.R. Coleman, J.R. Duchek, J. Klug, and G.S. Wilson, *J. Am. Chem. Soc.* 110:4772 (1988).

12. L. Nadjo, J.M. Saveant, and K.B. Su, *J. Electroanal. Chem.* 196:23 (1985).

13. C.P. Andrieux, J.M. Dumas-Bouchiat, and J.M. Saveant, *J. Electroanal. Chem.* 87:55 (1978).

14. L.L. Miller, G.D. Nordblom, and E.A. Mayeda, *J. Org. Chem.* 37:916 (1972).

15. P.G. Gassman and Y. Yamaguchi, *J. Am. Chem. Soc.* 101:1308 (1979).

16. B.R. Coleman, R.S. Glass, W.N. Setzer, U.D.G. Prabhu, and G.S. Wilson, *Adv. Chem. Ser.* 201:417 (1982).

ELECTROCHEMICAL REDUCTION OF SOME SULFUR COMPOUNDS

Henning Lund

Department of Organic Chemistry
University of Århus
8000 Århus C, Denmark

The electrochemistry of organic sulfur compounds with sulfur in the various oxidation states differs in many respects, and in this brief survey only a limited number of examples can be discussed. The reactions chosen are meant to illustrate the diversity of the reductions of sulfur compounds rather than to attempt a full coverage. A survey of the electrochemistry of organic sulfur compounds covering the literature up to 1975 is available;[1] in this paper we will mostly discuss reactions investigated at later times.

Elementary sulfur, dissolved in dimethylformamide (DMF), gives two quasi-reversible peaks in cyclic voltammetry (Fig. 1). The shape of the curves indicates that the heterogeneous rate constant for the electron transfer (ET) is higher for the second electron than for the first, and in this respect the reduction of S_8 is analogous to the reduction of cyclooctatetraene.

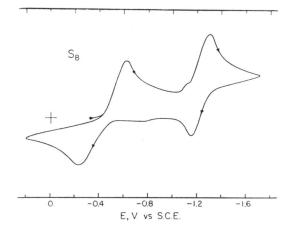

Fig. 1.   Cyclic voltammetry of S_8 in DMSO at an Au electrode, scan rate 0.1 Vs^{-1}. (M.V. Merritt and D.T. Sawyer, *Inorg. Chem.* 9:211 (1979).)

Sulfur-Centered Reactive Intermediates in Chemistry and Biology, Edited by
C. Chatgilialoglu and K.-D. Asmus, Plenum Press, New York, 1990

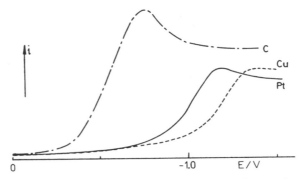

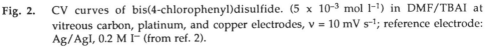

Fig. 2. CV curves of bis(4-chlorophenyl)disulfide. (5×10^{-3} mol l^{-1}) in DMF/TBAI at vitreous carbon, platinum, and copper electrodes, $v = 10$ mV s^{-1}; reference electrode: Ag/AgI, 0.2 M I^- (from ref. 2).

Reduction of sulfur in the presence of benzyl chloride gives mainly dibenzyl disulfide, regardless of whether the potential is adjusted to that of the first or second wave. Besides dibenzyl disulfide traces of dibenzyl polysulfides are isolated.

Another way of incorporating sulfur (or selenium) into organic compounds is to make an electrode of sulfur and graphite, as discussed in a previous lecture (Experimental Procedures in Electrolysis).

Electrochemical measurements of organic disulfides are in many cases complicated by reactions of the substrate or reduction products, $RS^{\bullet}$ or RS^-, with the electrode material. This is very pronounced at mercury electrodes, e.g., in classical polarography, but it is also evident at other metal electrodes (Fig. 2); usually glassy carbon electrodes give the least complications, but even here the rate constant for the heterogeneous ET is low, and the measured peak potential in CV is not the thermodynamic redox potential.[2]

The redox potential of disulfides may be estimated in different ways. It has been attempted[2] to use indirect electrolysis, using the reaction between electrogenerated anion radicals and aromatic disulfides. Measuring the rate of ET from stable anion radicals ($A^{\bullet -}$) of aromatic compounds to disulfides, establishes the connection between the rate of ET and the redox potential of the aromatic mediator, $E^0(A/A^{\bullet -})$. From this the E^0 of the disulfide could be obtained, if the disulfides electrochemically behave in a similar way as, for example, aryl halides. The reactions are (eqs. 1–5). (The disulfide anion radical is characterized by a three-electron bond for which a notation "$\bullet\bullet\bullet$" is also frequently used, see article by K.-D. Asmus in this book.)

$$A + e^- \rightleftarrows A^{\bullet -} \tag{1}$$

$$A^{\bullet -} + ArS\text{-}SAr \xrightarrow{k_2} A + [ArS\text{-}SAr]^{\bullet -} \tag{2}$$

$$[ArS\text{-}SAr]^{\bullet -} \xrightarrow{k_3} ArS^{\bullet} + ArS^- \tag{3}$$

$$A^{\bullet -} + ArS^{\bullet} \longrightarrow A + ArS^- \tag{4}$$

$$2\ ArS^{\bullet} \longrightarrow ArS\text{-}SAr \tag{5}$$

Formally the reactions (1)-(4) are analogous to those for the reduction of aryl halides,[3] but probably k_2 or k_3 are comparatively low. The estimated value for E^0 for diphenyl disulfide (≈ -0.9 V vs. SCE) is not much different from that estimated from the reaction between phenyl thiolate and some aromatic compounds, in which the occurrence of anion radicals, obtained by ET from the thiolate to the aromatic compound, was used for the estimation.

Aliphatic disulfides may also be reduced by anion radicals. If an anion radical of an activated olefin is used, thioalkylation is observed,[4a] probably as a result of a coupling between the anion radical and the alkylthio radical. The reaction between thiolate and activated olefin is slow compared to the electrochemical coupling.

ET from thiolate ions is also observed in the reduction of aliphatic disulfides in DMF in presence of dioxygen,[4b] in which sulfinates are formed according to reactions [eq.(6 – 9)].

$$RS-SR + 2 e^- \longrightarrow 2 RS^- \tag{6}$$

$$RS^- + O_2 \rightleftarrows RS^\bullet + O_2^{\bullet-} \tag{7}$$

$$RS^\bullet + O_2^{\bullet-} \longrightarrow RSO_2^- \tag{8}$$

$$RS^\bullet + RS^\bullet \longrightarrow RSSR \tag{9}$$

The ET takes probably place within a solvent cage, the alkylthio radical may either couple with the superoxide ion $O_2^{\bullet-}$ or diffuse from the solvent cage and form the disulfide. Preparatively, the only consequence of this is that the electron consumption is increased, but as all the disulfide eventually is reduced the yield of the sulfinate is high.

The reduction of carbon disulfide by sodium amalgam was originally believed[5a] to result in formation of tetrathiooxalate; later S. Wawzonek[5b] showed that the electrchemical reduction of CS_2 in DMF, after methylation of the product, yielded 4,5-bis(methylthio)-1,3-dithiole-2-thione (1), identical with the previously isolated product. It was suggested[5b] and later shown by HPLC[5c] that tetrathiooxalate was an intermediate which reacted further in fast reactions. It was thus realized,[6a,b] that only by somehow removing the tetrathiooxalate

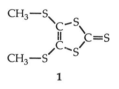

1

from the reaction mixture during the reduction it might be possible to isolate the elusive compound. By reduction of CS_2 in acetonitrile in the presence of tetraethylammonium[6a] or potassium[6b] ions a precipitate of the salt of tetrathiooxalate was obtained. The structure was ascertained[6b] by X-ray structure determination of the tetraphenylphosphcnium salt.

Wawzonek suggested originally[5b] that 1 was formed by reaction of tetrathiooxalate with CS_2 followed by elimination of S^{2-} [eq. (10)] and further reduction.

$$2 CS_2 \xrightarrow{2 e^-} \begin{matrix} CSS^- \\ | \\ CSS^- \end{matrix} \xrightarrow{CS_2} \begin{matrix} S{=}C{-}S \\ | \\ S{=}C{-}S \end{matrix} C \begin{matrix} S^- \\ S^- \end{matrix} \xrightarrow{-S^{2-}} \begin{matrix} S{=}C{-}S \\ | \\ S{=}C{-}S \end{matrix} C{=}S \xrightarrow[2) \text{ MeI}]{1) 2 e^-} 1 \tag{10}$$

Later it was suggested[5d] that carbon monosulfide, C=S, formed together with trithiocarbonate from tetrathiooxalate, was involved in the ring closure. It has, however, now been shown,[6c] that carbon monosulfide does not react with tetrathiooxalate under the reaction conditions, whereas a suspension of tetraethylammonium tetrathiooxalate reacted with CS_2 with, after methylation, formation of 1 and sulfur in high yield. Thus, the most probable reaction path for the formation of 1 is that originally suggested by Wawzonek, with the modification that the last step is a chemical reduction by S^{2-} or trithiocarbonate.[6c]

The electrochemical reduction of dialkyl trithiocarbonates,[7a] alkyl aryl trithio-carbonates,[7a] and diaryl trithiocarbonates,[7b,c] in DMF all yield tetrasubstituted tetrathio-ethylenes. Differences in the product distribution are mainly due to differences in the rate of the cleavage of the initially formed anion radical and to the ability of the substrates to alkylate or thioacylate intermediates.

The rate of the initial cleavage in the reduction [eq. (11)] was measured by cyclic voltammetry (CV).

$$RS-CS-SR + e^- \rightarrow [RS-CS-SR]^{\bullet-} \rightarrow RS-CS^{\bullet} + RS^- \qquad (11)$$

As discussed in an earlier lecture (Experimental Procedures in Electrolysis of Sulfur Compounds) the peak potential of a compound, the reduced species of which reacts fast in a first-order reaction, shifts to more negative values with increasing scan rates, $dE_p/d\log v \cong$ −30 mV, whereas a second order follow-up reaction shows $dE_p/d\log v = -19$ mV.

Fig. 3 shows the variation of the peak potential E_p for dimethyl trithiocarbonate as a function of the scan rate. At relatively low scan rate (until about 1 V s^{-1}) E_p becomes independent of the scan rate, and at the same time an anodic peak, due to the reoxidation of the anion radical on the reverse sweep, emerges. The sweep rate is then so fast that it "out-runs" the chemical follow-up reaction, the cleavage. Finally, at v higher than 100 V s^{-1} the sweep rate becomes fast compared to the heterogeneous ET and the reaction becomes quasi-reversible. From the sweep rate where the $dE_p/d\log v = -30$ mV line intersects with that of $dE_p/d\log v = 0$ the rate of the cleavage reaction can be calculated (for dimethyl trithiocarbonate $k \approx 30$ s^{-1}).

Reduction of dialkyl trithiocarbonates in DMF gives the products shown in eq. (12). The substrate, besides being reduced, thus reacts as an alkylating agent. By addition of RX to the catholyte the yield of tetraalkyl-thioethylene is increased.

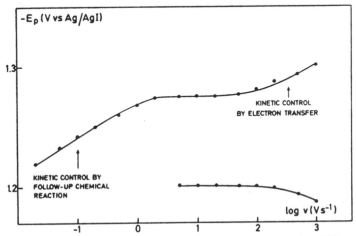

Fig. 3. Variation of the peak potential E_p with the scan rate for 2.59 mM dimethyl trithiocarbonate (3a) in DMF/0.1 M TBAI at an HMD-electrode. The upper curve depicts the reduction potentials; the lower one the reoxidation potentials (from ref. 7a).

$$(RS)_2CS \xrightarrow{\text{n e}^-} (RS)_2C=C(SR)_2 + HC(SR)_3 + RSCSS^- + RS^- \qquad (12)$$

The dimerization is probably not a reaction between two RSCS$^\bullet$ radicals, but rather a coupling between the radical and a radical anion, followed by elimination of RS$^-$ and further reduction. The formation of tri(alkylthio)methane may involve alkylation of the initial anion radical, followed by reduction and protonation.

Alkyl aryl trithiocarbonates behave analogously, but with formation of thioanisoles rather than tri(alkylthio)methane. The yields of tetrathioethylenes are in both cases only moderate, about 40%.

A certain interest exists in the synthesis of differently substituted tetrathioethylenes, especially with regard to the electrical properties of some of the derivatives. Carbon disulfide is an attractive starting material, but the yields from trithiocarbonates are moderate; 4,5-bis(alkylthio)-1,3-dithiole-2-thiones (**1**) might be employed, but again the reduction of **1** gives only about 40% tetrathioethylenes. A better way is to transform **1** into 4,5-bis(alkylthio)-1,3-dithiole-2-one (**2**). Electrochemical reduction of **2** yields tetrathioethylenes in almost quantitative yield, eq. (13).[7d]

$$\begin{matrix} RS-C-S \\ \quad\ \| \quad\ \ C{=}O \\ RS-C-S \end{matrix} \xrightarrow{2\,e^-} \begin{matrix} RS-C-S^- \\ \quad\ \| \\ RS-C-S^- \end{matrix} + CO \xrightarrow{2\,R'X^-} \begin{matrix} RS-C-SR' \\ \quad\ \| \\ RS-C-SR' \end{matrix} + 2\,X^- \qquad (13)$$

2

The reaction has the advantage that unsymmetrical tetrathioethylenes may be prepared.

Electrochemical reduction of benzenecarbodithioic esters (**3**) and benzenecarbothioic S-esters in DMF yields diphenylacetylene, eq. (14).[8a]

$$4\,C_6H_5CSSR + 6\,e^- \ \rightarrow\ C_6H_5C{\equiv}CC_6H_5 + 2\,C_6H_5CSS^- + 4\,RS^- \qquad (14)$$

3

A cyclic voltammetric investigation showed an irreversible peak followed by a smaller irreversible peak and a reversible redox reaction; the latter is due to diphenylacetylene, Fig. 4.

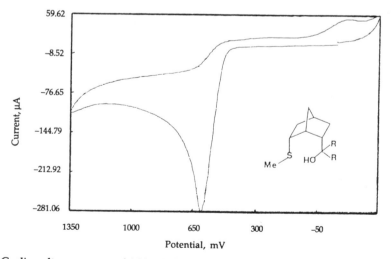

Fig. 4. Cyclic voltammogram of 1.83 mM benzyl benzenecarbodithioate in DMF/0.1 M TBAI at a Pt-electrode. Scan rate: 0.8 V s^{-1} (from ref. 8a).

The dependence of the peak potential E_p on the substrate concentration c was $dE/d\log c = 19$ mV which is consistent with the follow-up reaction being a second order reaction, most probably a dimerization of the primarily formed anion radical. A somewhat condensed reaction route is shown in eq. (15) - (19)

$$ArCSSR + e^- \rightarrow ArCSSR^{\bullet-} \tag{15}$$

$$\begin{array}{c} 2\ ArCSSR^{\bullet-} \rightarrow Ar-C(-S^-)-C(-S^-)-Ar \\ \ \ \underset{RS}{|}\qquad \underset{SR}{|} \end{array} \tag{16}$$

$$\begin{array}{c} Ar-\underset{RS}{\underset{|}{C}}(-S^-)-\underset{SR}{\underset{|}{C}}(-S^-)-Ar \rightarrow Ar-C(=S)-C(=S)-Ar\ +\ 2\ RS^- \end{array} \tag{17}$$

$$Ar-C(=S)-C(=S)-Ar \xrightarrow[2\ 3]{2\ e^-} \begin{array}{c} Ar-C=C-Ar \\ \underset{ArCSS}{|}\ \ \underset{SCSAr}{|} \end{array} +\ 2\ RS^- \tag{18}$$

$$\begin{array}{c} Ar-C=C-Ar \\ \underset{ArCSS}{|}\ \ \underset{SCSAr}{|} \end{array} +\ 2\ e^- \rightarrow ArC{\equiv}CAr\ +\ 2\ ArCSS^- \tag{19}$$

In DMF, in the presence of an alkylating agent, a derivative of a dithioketal of a ketone is formed on reduction of **3**. Benzenecarbodithioic methyl ester thus gives a dithioacetal of acetophenone.[8a] In protic medium a dithioacetal of benzaldehyd is formed.[8b] This is similar to the electrochemical reduction of thioamides.[8c]

Reduction of ethylene bis–dithiobenzoate in acetonitrile in the presence of methyl iodide yields a 1:1 mixture of meso- and d,l-2,3-bis(methylthio)-2,3-diphenyl-1,4-dithiane,[8b] eq. (20); the ring closure is analogous to the dimerization shown in eq. (16).

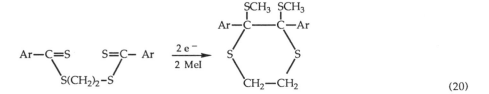

$$\tag{20}$$

S,S–Diaryl benzene-1,2-dicarbothioates (**4**) undergo an electrochemically induced rearrangement to 3,3-bis(arylthio)-phthalides (**5**).[8d] The primarily formed anion radical of **4** makes a ring closure and arylthiolate is expelled with the formation of the radical (**6**$^{\bullet}$). This reacts with the arylthiolate ion to the anion radial of **5**. The reduction potential of **5** is more negative than that of **4**, so the anion radical **5**$^{\bullet-}$ is able to reduce **4** to its anion radical **4**$^{\bullet-}$. A catalytic circle is thus established and the reaction may be regarded as an internal $S_{RN}1$ reaction, eq. (21).

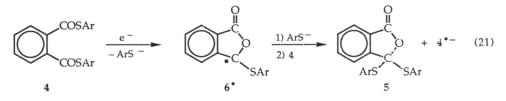

$$\tag{21}$$

The electrochemistry of sulfones has recently been reviewed.[9z] The most general reduction route of diaryl and alkylaryl sulfones in DMF is a cleavage of a carbon-sulfur bond, eqs. (22,23).

$$ArSO_2Ar + 2e^- + [H^+] \rightarrow ArH + ArSO_2^- \tag{22}$$

$$ArSO_2R + 2e^- + [H^+] \rightarrow RH + ArSO_2^- \tag{23}$$

The reduction potentials of the sulfones are rather negative, and highly unreducible supporting electrolytes must be employed. Unless activated, e.g. by double bonds, dialkyl sulfones are not reducible.

Alkylaryl sulfones are usually cleaved to arylsulfinic acids [eq. (23)]; this is in accordance with the results obtained by CV in DMF. The CVs of alkylaryl sulfones show generally, at moderately fast sweep rates, no oxidation peak on the reverse sweep indicating a fast cleavage of the anion radical, whereas the anion radical of diaryl sulfones cleave sufficiently slowly to give an anodic peak on the reverse sweep, even at relatively slow sweep rates (Fig. 5).

In unsymmetrical diaryl sulfones usually the better leaving group, that is the aryl group substituted with an electron-attracting group, will be cleaved as the hydrocarbon; similarly, a benzene ring substituted with an electron donating group will preferentially form the sulfinic acid. Thus reduction of (4-aminophenyl)phenyl sulfone in methanol yields 84% benzene and 9% aniline at a mercury cathode.[9b]

Substituents in the *ortho*-position change the situation, probably by steric inhibition of resonance. 2-(Aminophenyl)phenyl sulfone thus gives only 32% benzene but 68% aniline on reduction in methanol at a mercury cathode. Similarly, phenyl-(2,4,6-trimethylphenyl)-sulfone is reduced under similar conditions to 88% mesitylene.[9b]

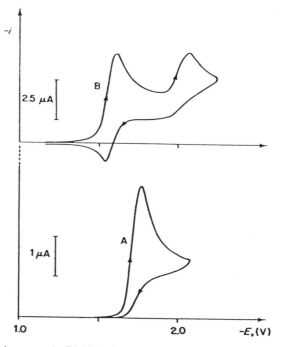

Fig. 5. Voltammetric curves in DMF in the presence of Bu_4NBF_4 (0.1 M), reference electrode $Ag/AgI/I^-$ (0.1 M), mercury stationary microelectrode: (A) $PhSO_2Me$ (10^{-3} M), sweep rate 500 mV s^{-1}. (B) $PhSO_2Ph$ (10^{-3} M), sweep rate 500 mV s^{-1} (from ref. 9a).

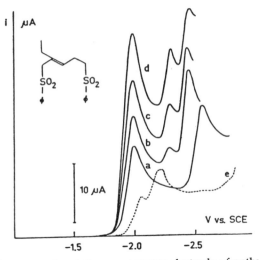

Fig. 6. Voltammetric curves at a stationary mercury electrode of anthracene in the presence of **7**. a: Anthracene (6.5×10^{-3} M); b, c, and d: a + **7** (2.3, 5.1, and 7.6×10^{-3} M, respectively). e: **7** without anthracene. Medium, DMF/1.4×10^{-1} M TBAP. Sweep rate 10 mV s^{-1} (from ref. 9c).

Sulfones may also be cleaved by indirect reduction using stable anion radicals A$^{\bullet-}$ as electron carriers. The rate of the electron transfer from an anion radical to a given sulfone depends on the redox potential of the electron carrier [$E^0(A/A^{\bullet-})$] (mediator); the more negative $E^0(A/A^{\bullet-})$ the faster is the ET. Similarly, a given mediator transfers electrons preferentially to the substrate with the least negative redox potential, i.e., the most activated, sulfone. This may be used preparatively.

2-Methyl-1,5-bis(benzenesulfonyl)pent-2-ene (**7**) has two benzenesulfonyl groups; one of them is separated from the double bond by one CH$_2$-group and the other by two. CV of **7** shows two poorly separated peaks (trace **e** in Fig. 6); trace **a** in Fig. 6 is the trace of the mediator anthracene. Addition of **7** to anthracene (traces **b, c,** and **d**) shows an increase in the peak height of anthracene and no peak corresponding to the first peak of **7**.

$$C_6H_5SO_2CH_2\underset{\underset{CH_3}{|}}{C}=CHCH_2CH_2SO_2C_6H_5$$

7

Reduction of anthracene in the presence of **7** gave a high yield of 2-methyl-5-benzene-sulfonyl-2-pentene together with minor amounts of the 1-isomer; in both cases the 1-benzene-sulfonyl group was cleaved. The anthracene anion radical is blue, but as long as some **7** was present the solution was colorless; however, after the consumption of 2 F mol^{-1} the blue color appeared and the reduction was stopped. The reactions are [eq. (24)][9c]

$$A + e^- \rightarrow A^{\bullet-}$$

$$2 A^{\bullet-} + C_6H_5SO_2CH_2C(CH_3)=CH(CH_2)_2SO_2C_6H_5 \xrightarrow{\ H^+\ }$$

$$\rightarrow C_6H_5SO_2^- + CH_3C(CH_3)=CH(CH_2)_2SO_2C_6H_5 \qquad (24)$$

The further reduction of the product by A$^{\bullet-}$ is too slow to interfere as long as **7** is present.

Electron transfer from sulfones may also take place; reduction of methyl styryl sulfone (8) in the presence of the more difficultly reducible *tert*-butyl bromide yields 3,3-dimethyl-2-phenyl methyl sulfone (9), 3,3-dimethyl-1-phenyl-1-butene (10), 4-*tert*-butylstyrene (11) and some 2-(4-*tert*-butyl)ethyl methyl sulfone (12).[9d]

$$C_6H_5CH(t\text{-}C_4H_9)CH_2SO_2CH_3 \qquad\qquad C_6H_5CH=CH-C(CH_3)_3$$
9   **10**

$$(CH_3)_3CC_6H_4CH=CH_2 \qquad\qquad (CH_3)_3CC_6H_4CH_2CH_4SO_2CH_3$$
11   **12**

The reactions may be formulated [eq. (25 – 27)]

$$8 + e^- \rightleftarrows 8^{\bullet -} \tag{25}$$

$$8^{\bullet -} + t\text{-}BuBr \rightarrow 8 + t\text{-}Bu^\bullet + Br^- \tag{26}$$

$$8^{\bullet -} + t\text{-}Bu^\bullet \rightarrow \text{products} \tag{27}$$

Coupling α to the benzene ring yields, after protonation, **9**, whereas coupling β to the ring yields an anion which eliminates methylsulfonate to form **10**.

Electrochemical reduction of β-ketosulfones (and β-ketosulfoxides) in DMF at a mercury electrode is a practical method of preparing methyl ketones [eq. (28)] from acid derivatives. Since the anodic reaction is the oxidation of the sulfinate to the sulfonate an undivided cell may be used.[9a]

$$ArCOCH_2SO_2CH_3 \xrightarrow[H^+]{2\,e^-} ArCOCH_3 + CH_3SO_2^- \tag{28}$$

In some cases a methyl(ene) group α to a sulfone is acidic enough to protonate products or intermediates, and in absence of added proton donors in aprotric media this might interfere with the reduction of the sulfone.

Esters of arylsulfonic acids are generally reducible, and this has been employed for protection of hydroxyl groups; the cleavage of tosylates[9b] and nosylates[10b] has especially been studied.

The arylsulfonate esters which carries no electron attracting groups in the aryl group are reducible at rather negative potentials; the primarily formed anion radical can, in principle, cleave between Ar and SO_2OR, between $ArSO_2$ and OR, and between $ArSO_2O$ and R. An O–R cleavage is unlikely in view of the observation that tosyl esters of chiral alcohols, where the tosyl ester is at the chiral centrer, are electrochemically cleaved with retention of the configuration.[9b] In general, the cleavage $ArSO_2$–OR is predominant [eq. (29)], but some ArH is observed in some cases.[10a]

$$Ar\text{-}SO_2\text{-}OR + 2\,e^- \rightarrow ArSO_2^- + RO^- \xrightarrow{H^+} ROH \tag{29}$$

It has been found that the cleavage of the primarily formed anion radical is the rate-determining step in an indirect reduction of tosylates in DMF when R is aliphatic, whereas with aromatic R, the cleavage is so fast that the homogeneous ET from the reduced mediator becomes rate-determining.

In the absence of protons the deprotection may be complicated by a nucleophilic attack of RO^- on the tosylate ester with formation of an ether, eq. (30).[10a]

$$ArSO_2\text{-}OR + RO^- \rightarrow ArSO_3^- + ROR \qquad (30)$$

Introduction of electron-attracting groups into the aryl group of the protecting sulfonate ester makes the reduction potential less negative, but also lowers the rate of the cleavage. Thus, esters of 4-nitrobenzenesulfonic acid (nosylates) are reduced at potentials about one volt less negative than tosylates, and a selective cleavage of nosylates in presence of tosylates can be achieved.[10b] Nosylates are reduced in two steps; in the second step a dianion is formed which cleaves rapidly.

In this brief survey several types of sulfur compounds have not been mentioned, e.g., sulfonium compounds, thioethers, sulfoxides, thiocyanates, and of the sulfur compounds included also only some aspects of the electrochemistry have been discussed, but hopefully the survey has given some impression of the diversity of the electrochemistry of organic sulfur compounds.

REFERENCES

1. J.Q. Chambers, in: "Encyclopedia of Electrochemistry of the Elements", A.J. Bard and H. Lund, eds., M. Dekker, New York, Vol. 13, p. 329 ff. (1978).
2. J. Simonet, M. Carriou, and H. Lund, Liebigs Ann. Chem. 1665 (1981).
3 a. C.P. Andrieux, C. Blocman, J.-M: Dumas-Bouchiat, and J.M. Savéant, J. Am. Chem. Soc. 101:3431 (1979).
   b. C.P. Andrieux, C. Blocman, J.-M. Dumas-Bouchiat, F.M´Halla, and J.-M. Savéant, J. Electroanal. Chem. 113:19 (1980).
4 a. C. Degrand and H. Lund, C.R. Acad. Sci. (Paris), Ser. C. 295 (1980).
   b. C. Degrand and H. Lund, Acta Chem. Scand. B33:512 (1979).
5 a. B. Fetkenheuer, H. Fetkenheuer, and H. Lecus, Ber. Dtsch. Chem. Ges. 60:2528 (1927).
   b. S. Wawzonek and S.M. Heilmann, J. Org. Chem. 39:511 (1974).
   c. P.R. Moses, R.M. Harnden, and J.Q. Chambers, J. Electroanal. Chem. 84:187 (1977).
   d. G. Bontempelli, F. Magno, G.A. Mazzocchin, and R. Seeber, J. Electroanal. Chem. 63:231 (1975).
6 a. P. Jeroschewski, Z. Chem. 21:412 (1981).
   b. H. Lund, E. Hoyer, and R.G. Hazell, Acta Chem. Scand. B36:207 (1982).
   c. E.K. Moltzen, A. Senning, R.G. Hazell, and H. Lund, Acta Chem. Scand. B40:593 (1986).
7 a. M. Falsig and H. Lund, Acta Chem. Scand. B34:545 (1980).
   b. M. Falsig, H. Lund, L. Nadjo, and J.-M. Savéant, Acta Chem. Scand. B34:685 (1980).
   c. M. Falsig, H. Lund, L. Nadjo, and J.-M. Savéant, Nouv. J. Chim. 445 (1980).
   d. M. Falsig and H. Lund, Acta Chem. Scand. B34:591 (1980).
8 a. M. Falsig and H. Lund, Acta Chem. Scand. B34:585 (1980).
   b. J. Voss, C. von Bülow, T. Drews, and P. Mischke, Acta Chem. Scand. B37:519 (1983).
   c. H. Lund, Coll. Czech. Chem. Commun. 25:3313 (1960).
   d. K. Praefcke, C. Weichsel, M. Falsig, and H. Lund, Acta Chem. Scand. B34:403 (1980).
9 a. J. Simonet, in: "The Chemistry of Sulphones and Sulphoxides", S. Patai, Z. Rappoport, and C.J.M. Stirling, eds., Wiley, New York, pp. 1001 (1988).
   b. H. Lund, in: "Organic Electrochemistry", 3. edition, H. Lund and M.M. Baizer, eds., Dekker, New York, pp. 1029 (1990).
   c. J. Simonet and H. Lund, Acta Chem. Scand. B31:909 (1977).
   d. C. Degrand and H. Lund, Nouv. J. Chim. 1:35 (1977).
10a. H.L.S. Maya, M.J. Medeiros, M.I. Montenegro, D. Court, and D. Pletcher, J. Electroanal. Chem. 164:347 (1984).
   b. D. Pletcher and N.R. Stradiotto, J. Electroanal. Chem. 186:211 (1985).

PHOTOCATALYTIC FORMATION OF SULFUR-CENTERED RADICALS BY

ONE-ELECTRON REDOX PROCESSES ON SEMICONDUCTOR SURFACES

Detlef Bahnemann

Institut für Solarenergieforschung GmbH
Sokelantstraße 5
3000 Hannover 1, F. R. Germany

INTRODUCTION

Free radical reactions cannot only be initiated and conveniently studied by radiation- or photo-chemical methods, but it is also possible to generate reactive free radical intermediates upon band-gap illumination of semiconducting materials. Depending on the individual material properties it is possible, for example, in an aqueous solvent to form primary radicals such as hydroxyl and sulfhydrol radicals or hydrated electrons. It is the microscopic structure of the semiconductor/electrolyte interface which determines the fate of these radical species: they may either diffuse into the bulk of the solution and initiate subsequent redox reactions like in a homogeneous system, or they are strongly adsorbed to the surface resulting in specific paths of the consecutive processes at the interface.

It will be the aim of this contribution to describe the pathways leading from the absorption of a photon in the semiconducting solid to the generation of primary and secondary radicals. Special emphasis will be laid on a detailed description of the interfacial region between the semiconductor and the aqueous electrolyte which is crucial for the reaction mechanisms observed. Following a brief description of the experimental methods employed to study mechanistic details as well as overall yields, examples will be given for free radical processes in heterogeneous systems involving sulfur-centered reactive intermediates. Finally perspectives for future research will be discussed in the light of the preceeding results and interpretations.

PHOTOPHYSICS AND PHOTOCHEMISTRY OF SEMICONDUCTOR PARTICLES

It is a characteristic property of semiconductors that they possess among others two distinct energy bands: a valence band with completely filled orbitals (resulting from the mixing of the Highest Occupied Molecular Orbitals "HOMO's" of the individual molecules forming the solid) and a conduction band which at 0 K does not contain any electrons in its orbitals (the latter can be regarded as the overlap of LUMO's, i.e. the Lowest Unoccupied Molecular Orbitals forming the solid).[1]

Figure 1 illustrates examples for the relative energetic positions of valence and conduction bands of several semiconductors on the usual electrochemical energy scale (i.e. vs. NHE).[2] It is ob-

Sulfur-Centered Reactive Intermediates in Chemistry and Biology, Edited by
C. Chatgilialoglu and K.-D. Asmus, Plenum Press, New York, 1990

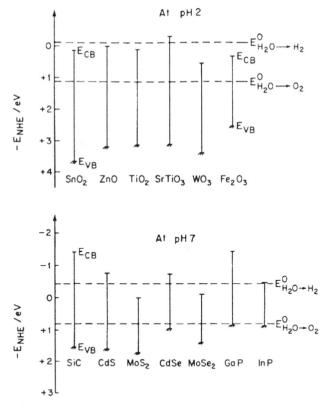

Figure 1. Band-edge positions of semiconductors in contact with aqueous electrolytes on the electrochemical scale of potentials relative to the normal hydrogen electrode (NHE) (taken from ref.[2]).

vious that band-gaps can vary between few tenths of an eV and several eV depending upon composition and structure of the solid. While the lower edge of the conduction band can be located at potentials as negative as -2.5 eV (vs. NHE), an energetic position exceeding $+3.0$ V (vs. NHE) for the upper end of the valence band is not unusual. The pH-dependence of the band positions of semiconductors immersed in aqueous solution will be discussed in a subsequent chapter [cf. eq. (10)]. Absorption of a photon with an energy $(E = h\nu = hc\lambda^{-1})$ exceeding the band-gap energy E_g of the semiconductor leads to the formation of an electron/ hole pair via

$$\text{semiconductor} \quad \xrightarrow{\quad h\nu \quad} \quad e^-_{CB} + h^+_{VB} \qquad (1)$$

where e^-_{CB} is an electron with an energy corresponding to the energy of the lower edge of the conduction band and h^+_{VB} is a hole with an energy level at the upper edge of the valence band. Any excess energy resulting from the photon absorption is transferred to heat by relaxation processes within the bands. The local electric field near the surface of a bulk semiconductor electrode normally results in the separation of the oppositely charged e^-_{CB} and h^+_{VB}. While one charge carrier is available for redox reactions at the semiconductor surface that with the opposite sign moves through the solid and is then reacting via the back contact e.g. to a metal electrode.[3]

The situation is different for a small semiconductor particle. Since the band bending and thus the electric field is negligable under these conditions, both charge carriers can easily reach the particle's surface.[34] There they are trapped in deeper electronic states. Such trap states are usually chemically identifyable. For example, a trapped electron in a metal oxide such as titanium dioxide, TiO_2, can be regarded as a Ti^{3+} ion (eq. 2),[5] whereas the corresponding hole is trapped as a surface bound hydroxyl radical, $^{\bullet}OH_S$ (eq. 3).[6]

$$e^-_{CB}\,(TiO_2) \quad \rightarrow \quad Ti^{3+}_S \tag{2}$$

$$h^+_{VB}\,(TiO_2) \quad \rightarrow \quad {}^{\bullet}OH_S \tag{3}$$

In analogy, an electron in a metal chalcogenide such as CdS is trapped as Cd^+ (eq. 4)[7] and a hole as a surface bound sulfhydrol radical, $^{\bullet}SH_S$ (eq. 5).[8]

$$e^-_{CB}\,(CdS) \quad \rightarrow \quad Cd^+_S \tag{4}$$

$$h^+_{VB}\,(CdS) \quad \rightarrow \quad {}^{\bullet}SH_S \tag{5}$$

While certain properties of these radical states on the surface of semiconductors such as redox potential and reactivity cannot automatically be taken from the corresponding intermediate known in homogeneous solution, similarities of their behavior are certainly to be expected. There is only a limited number of examples in the literature that surface trapped radicals can be directly detected: e.g., both, the Ti^{3+} state and the $^{\bullet}OH_S$ can be spectroscopically identified in colloidal suspensions of TiO_2.[9] While it is also possible to see a transient absorption change due to trapped holes in colloidal CdS particles (i.e. $^{\bullet}SH_S$),[10] the corresponding trapped e^-_{CB} (i.e. Cd^+_S) can only be measured indirectly, e.g., by their rapid reaction with oxidants such as methylviologen, MV^{2+} (1,1'-dimethyl-4,4'-bipyridylium chloride) (eq. 6).[7]

$$Cd^+_S \;+\; MV^{2+} \quad \rightarrow \quad Cd^{2+}_S \;+\; MV^{\bullet+} \tag{6}$$

It is interesting to note that in some cases also hydrated electrons, e^-_{aq}, are detected upon band-gap illumination of semiconductor particles, e.g., CdS, with a short laser pulse (eq.7).[10,11]

$$CdS \xrightarrow[\text{laser pulse}]{\quad h\nu > E_g \quad} e^-_{aq} \tag{7}$$

It has been shown, however, that this process requires the absorption of two photons for every ejected e^-_{aq} (i.e. a biphotonic process occurs).[10] The exciton (i.e. electron/hole pair) formed upon absorption of the first photon is further excited by an energy transfer from a second exciton thus generating an electron on a potential negative enough to facilitate its ejection into the aqueous phase. Since electron/hole pairs are trapped extremely fast, considerable yields of e^-_{aq} are only detected when the absorption of both photons occurs within a very short time. Indeed, quantum yields close to 50% have recently been observed for the e^-_{aq}-formation from colloidal CdS particles when fs-pulses were used.[12]

The above described ability of semiconductor particles to generate radical intermediates upon band-gap illumination is schematically summarized in Figure 2. Following their generation, trapped electrons and holes can then initiate subsequent electron transfer reactions. Selected one-electron redox potentials for a variety of compounds are given in Table 1.[13-23] Comparing these values with the band positions shown in Figure 1, it is evident that in principle almost every one-electron oxidation or reduction should be possible with the h^+_{VB} or e^-_{CB}, respectively. While the holes in semiconducting metal oxides are extremely good oxidants, electrons in metal chalcogenides are relatively strong reductants and may, in fact, efficiently reduce metal ions (or metals) adsorbed or incorporated at the semiconductor surface.

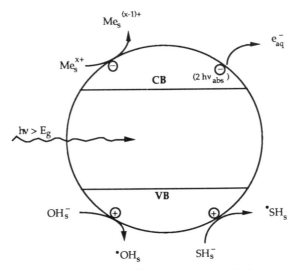

Figure 2. Schematic representation of possible primary free radical species generated on semi-conductor surfaces in contact with an aqueous electrolyte upon band-gap illumination.

Table 1. One-electron redox potentials E^1_7 for the reduction and the oxidation of various chemicals (reference NHE) (taken from refs.[13-23])

Reduction		Oxidation	
Compound	E^1_7 (vs. NHE),V	Compound	E^1_7 (vs. NHE),V
CH_3Cl	−2.000	C_6H_5OH	+0.900
CH_2Cl_2	−2.100	$p\text{-}CH_3OC_6H_4OH$	+0.600
$CHCl_3$	−1.440	$p\text{-}NH_2C_6H_4OH$	+0.410
CCl_4	−0.540	$p\text{-}NO_2C_6H_4OH$	+0.924
$C_6H_5CHCl_2$	−1.200	$p\text{-}ClC_6H_4OH$	+0.653
$C_6H_5CCl_3$	−0.300	$2,5\text{-}Cl_2C_6H_3OH$	+0.726
$(C_6H_5)_2CCl_2$	−0.470	$C_6H_5NH_2$	+0.625
$C_6H_4Cl_2$	−2.280	$p\text{-}NO_2C_6H_4NH_2$	+0.935
C_6Cl_6	−1.210	$p\text{-}ClC_6H_4NH_2$	+0.675
$C_6H_5NO_2$	−0.490	Cl^-	+2.590
Cu^+	+0.520	Br^-	+2.080
Fe^{3+}	+0.750	I^-	+1.310
Mn^{3+}	−0.240	SO_3^{2-}	+0.630
Na^+	−2.713	NO_2^-	+1.070
O_2	−0.330	H_2O	+2.820

Electron transfer reactions between charge carriers generated by light absorption within the semiconductor and solute molecules occur at the solid/liquid interface. It is, therefore, necessary to understand the surface chemistry of semiconductor particles in contact with the aqueous electrolyte. A schematic presentation of the cross section of the surface layer of a metal oxide is depicted in Figure 3.[24] In the presence of water, H_2O molecules will readily coordinate with free coordination sites of metal ions at the surface (Fig. 3b). Following an energetically favorable dissociative chemisorption of these water molecules, the surface of such a metal oxide is covered with metal hydroxide ($\equiv MeOH$) groups (Fig. 3c).

It is a well-known chemical principle that many metal hydroxides are amphoteric, i.e. they behave like a base in acidic solution and vice versa. Accordingly, the following protonation equilibria can be formulated:[24]

$$\equiv MeOH_2^+ \quad \rightleftarrows \quad \equiv MeOH \quad + \quad H^+ \tag{8}$$

$$\equiv MeOH \quad \rightleftarrows \quad \equiv MeO^- \quad + \quad H^+ \tag{9}$$

The corresponding equilibrium constants are given by

$$K^S_{a1} = \{\equiv MeOH\} [H^+] \{\equiv MeOH_2^+\}^{-1} \quad \text{and}$$
$$K^S_{a2} = \{\equiv MeO^-\} [H^+] \{\equiv MeOH\}^{-1},$$

where the notation { } refers to surface and [ ] to bulk concentrations. It is thus evident that depending on the pH of the solution, metal oxide particles can either be positively charged with ($\{\equiv MeOH_2^+\} > \{\equiv MeO^-\}$), neutral ($\{\equiv MeOH_2^+\} = \{\equiv MeO^-\}$) or negatively charged ($\{\equiv MeO^-\} > \{\equiv MeOH_2^+\}$). The characteristic pH of the Zero Point of Charge (i.e. where the particle's surface carries no net charge) pH_{zpc} is given by:[24]

$$pH_{zpc} = 1/2 \, (pK^S_{a1} + pK^S_{a2}) \tag{10}$$

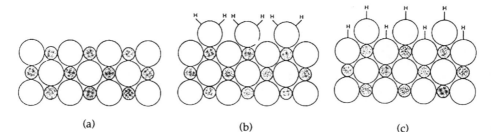

(a)   (b)   (c)

Figure 3.   Schematic representation of the cross section of the surface layer of a metal oxide. ●, metal ions; o, oxide ions. The metal ions in the surface layer (a) have a reduced coordination number. They thus behave as Lewis acids. In the presence of water the surface metal ions may first tend to coordinate water molecules (b). For most of the oxides dissociative chemisorption of water molecules (c) seems energetically favored. Oxide surfaces carry typically 4 - 10 hydroxyl groups per square nanometer. (taken from ref. [24]).

Table 2. pH-values at the zero point of charge for various metal oxide powders (taken from refs.[5, 24-26])

Material	pH_{zpc}
SiO_2	2.0
TiO_2	3.5 - 6.7
α-Fe_2O_3	6.5 - 8.5
ZnO	9.0 ± 0.3
α-Al_2O_3	9.1
CuO	9.5
MgO	12.5

Titration experiments are usually used to determine pH_{zpc} and to subsequently calculate pK^S_{a1} and pK^S_{a2}. Literature values for pH_{zpc} for some selected metal oxides are given in Table 2.[5, 24-26] It is evident that the zero point of charge can vary over a wide range. While some oxides such as SiO_2 or α-Al_2O_3 are negatively or positively charged, respectively, over most of the pH region, semiconducting oxides (e.g. TiO_2, α-Fe_2O_3 and ZnO) have a net surface charge that crucially depends on the pH of the surrounding solution.

It is certainly not surprising that the adsorption of solute molecules onto these oxide surfaces is strongly dependent on the latter's overall charge. While this conclusion seems to be trivial for ionic species, it normally requires rather extensive computer simulation routines to show that effects such as co-adsorption of "inert" ions also influence the adsorption of neutral molecules resulting in its significant pH-dependence.[27]

Since the reactivity of different molecules with e^-_{CB} or h^+_{VB} largely depends on their presence at the semiconductor's surface, i.e. their relative adsorption, it is crucial to understand the microstructure of this surface together with its acid/base properties to make any viable predictions.

It can readily be shown that the surface potential Ψ_o of a metal oxide particle in aqueous solution varies with pH via[24]

$$\Psi_o = 2.303 \, (RT/F) \, \{pH_{zpc} - pH\} \tag{11}$$

resulting in a "Nernstian" behavior of such particles. While Ψ_o does not contribute to the overall potential at pH = pH_{zpc}, the positions of conduction and valence band at the particle's surface are shifted by 59 mV (at 25°C) towards more negative values with every increasing pH unit. Thus, e^-_{CB} are better reductants at high pH whereas h^+_{VB} have their highest oxidation potential at low pH. Analogous considerations can be applied to other semiconductors in contact with an aqueous electrolyte. However, the situation is normally even more complicated since other equilibria have to be considered besides those given in equation (8) and (9). If like in the case of the chalcogenides the anion X (i.e. X = S, Se or Te) can also be protonated, the following equilibria cannot be neglected:

$$\equiv MeXH_2^+ \quad \rightleftarrows \quad \equiv MeXH \quad + \quad H^+ \tag{12}$$

$$\equiv MeXH \quad \rightleftarrows \quad \equiv MeX^- \quad + \quad H^+ \tag{13}$$

In these cases titration curves do not contain sufficient information to unravel all equilibrium constants or to even precisely yield pH_{zpc}. Furthermore, the pH dependence of the surface potential is normally not described by a simple relationship like that given in eq. (11). In most cases the experimental data is still too sparse to justify a detailed microscopic picture of the surface of non-oxide semiconductors in aqueous solution. Yet it should not be neglected that the surface composition of such materials also undergoes considerable changes upon varying the pH of the surrounding solution. Reaction rates and specificities of the electron transfer processes will therefore strongly depend upon pH, ionic strength and solute composition.

EXPERIMENTAL TECHNIQUES

Modern analytical techniques are generally used to analyze products of redox processes initiated by e^-_{CB} and h^+_{VB} which are generated upon band-gap illumination of aqueous suspensions of semiconductor particles. Ionic products can conveniently be detected by high pressure ion chromatography (HPIC), while capillary gas chromatography with FID (flame ionization detector) and ECD (electron capture detector) is used for the analysis of organic molecules - its combination with an ITD (ion trap detector) can be employed to identify unknown products by their mass spectra. HPLC and spectroscopic techniques are additional analytical means which can be useful for special applications. An elegant method has recently been introduced to study the formation of protons or hydroxyl ions during an illumination experiment in real time.[27] The "pH-stat" experiment employs an automatic titrator in combination with a sensitive pH-electrode and an auto-burette. The titrator is programmed to keep the pH constant and causes the burette to add base or acid to the illuminated solution as protons or hydroxyl ions, respectively, are formed. The volume of base added is a direct measure for the H^+_{aq} produced and vice versa and can be detected in real time.

Monochromatic illumination is a prerequisite when absolut quantum yields are required. Furthermore, it is extremely difficult to determine the amount of incident photons absorbed by a highly scattering suspension of particles. Therefore, it is most convenient to use suspensions of very small particles with an average diameter $d \leq \lambda_{ex}/20$ to avoid considerable scattering artefacts (λ_{ex} is the wavelength of excitation). Several methods have recently been developed to synthesize aqueous suspensions of such small colloidal particles.[28] Their stabilization is usually caused by a kinetic

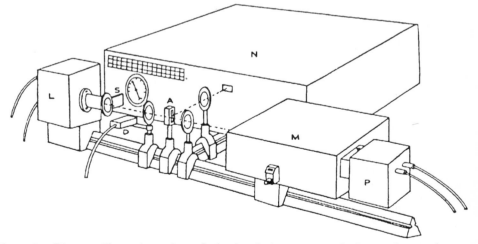

Figure 4. Diagram illustrating a laser flash photolysis set-up employing a nitrogen laser. L: xenon arc lamp housing with shutter (S), A: absorption cell holder, M: high radiance monochromator, D: photodiode, P: photomultiplier housing, N: 1MW nitrogen laser (Lambda Physik) (taken from ref.[32]).

hindrance of the coagulation due to surface charges high enough to ensure repulsion.[5,25] Reduction of the surface potential Ψ_o usually causes coagulation, thus the stability of colloidal suspensions is limited to a small pH regime. Stabilizing agents such as hexametaphosphate or polyvinylalcohol can be added yielding extremely stable colloids.[28] However, it should be pointed out that these surface active agents are strongly adsorbed and will, without doubt, change the surface chemistry of the particles to a considerable extend.

Optically transparent suspensions of small colloidal semiconductor particles are also ideal systems to be employed in time-resolved kinetic experiments.[9,10,29-31] Band-gap illumination is then performed with a short laser pulse (Figure 4);[32] artefacts due to laser light scattering do again pose no problem. Detection of intermediate free radical species is possible with time-resolved absorption spectroscopy perpendicular to the exciting beam. Often it will be convenient to concommittantly detect conductivity changes using AC- or DC-conductivity measurement techniques with µs or ns time resolution, respectively.[33,34] Transient optical changes in scattering samples containing bigger particles can be detected with a diffuse reflectance set-up that has recently been introduced.[35]

Typical results of these time-resolved laser studies are shown in Figures 5 and 6.[9] Figure 5 shows the transient absorption spectrum of surface trapped electrons (i.e. Ti_s^{3+}) in colloidal aqueous suspensions of TiO_2; the inset shows a typical absorption vs. time signal observed in the same solution. Under the given experimental conditions valence band holes are scavenged at the TiO_2 surface by polyvinylalcohol (PVA) molecules leaving the electrons behind (cf. eq. (2)).

$$TiO_2 \xrightarrow{\quad h\nu > E_g \quad} TiO_2(e^-_{CB}, h^+_{VB}) \tag{14}$$

$$PVA + h^+_{VB} \rightarrow PVA(-H)^\bullet + H^+_{aq} \tag{15}$$

When a good electron scavenger such as a platinum deposit is present on the TiO_2 particles instead of an alcohol molecule, the e^-/h^+ formation (eq. (14)) is followed by

$$(Pt)_x + e^-_{CB} \rightarrow (Pt)_x^- \tag{16}$$

leaving trapped holes behind (cf. eq. (3)). Their transient absorption spectrum is shown in Figure 6 together with an absorption/time signal as the inset. The pronounced difference of the absorption spectrum shown in Fig. 6 ($\lambda_{max} \approx 425$ nm) with that of the free hydroxyl radical in aqueous solution ($\lambda_{max} \approx 280$ nm) evinces that $^\bullet OH_S$ can certainly not be regarded as a free radical. However, its reactions can conveniently be studied from the kinetic characteristics of this absorption spectrum.

An example for the simultaneous detection of optical and conductivity changes is given in Figure 7.[9] The transient absorption spectrum of reduced methylviologen, $MV^{\bullet+}$, is detected immediately after the laser pulse admitted to an alkaline solution of colloidal TiO_2 particles (i.e. with MV^{2+} dications strongly adsorbed onto the negatively charged particle surface).

$$MV^{2+} + e^-_{CB} \rightarrow MV^{\bullet+} \tag{17}$$

Subsequently, a decrease in conductivity is observed (cf. Fig. 7, right) the kinetics of which obey a first order rate law with the rate constant increasing with OH^- concentration. These observations are interpreted by a reaction of the remaining h^+_{VB} with OH^-, i.e. the first step of water oxidation.

$$OH^- + h^+_{VB} \rightarrow {}^\bullet OH_S \tag{18}$$

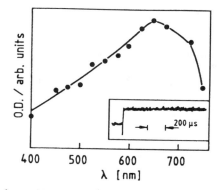

Figure 5. Transient absorption spectrum of a laser-flashed aqueous solution of 6.3×10^{-3} mol dm^{-3} TiO$_2$-colloid containing 5×10^{-3} mol dm^{-3} PVA (polyvinylalcohol) and time profile of the 650 nm-absorption signal (inset) at pH 10 (taken from ref.[9]).

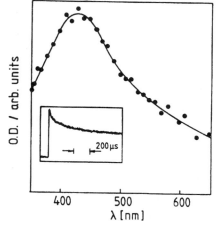

Figure 6. Transient absorption spectrum of a laser-flashed aqueous solution of TiO$_2$-Pt colloid (3.8×10^{-3} mol dm^{-3} TiO$_2$ and 1.6×10^{-5} mol dm^{-3} Pt) and decay of the 420 nm-absorption signal (inset) at pH 2.5 (taken from ref.[9]).

111

Ω^{-1}, $t_{1/2} = 8.5 \ \mu s$

600 nm

Figure 7. (left) Time profile of the 600 nm transient absorption and (right) the conductivity of a laser pulse illuminated 6.3×10^{-3} mol dm^{-3} TiO$_2$-colloid containing 5×10^{-5} mol dm^{-3} methylviologen at pH 10. (taken from ref.[9]).

The above examples illustrate the versatility of time-resolved measurements in colloidal suspensions of small semiconductor particles for a better understanding of charge transfer processes occurring at the interface of semiconductor electrodes and an aqueous phase.

SULFUR-CENTERED FREE RADICAL INTERMEDIATES

Oxidation of inorganic S (IV)

Sulfite molecules are readily oxidized when they are illuminated in aqueous suspension of semiconductor particles, e.g. Fe$_2$O$_3$ or TiO$_2$, with light energies exceeding the respective band-gap energy of the material.[26,36] Figure 8 shows the disappearance of SO$_3^{2-}$ (measured photometrically) as a function of illumination time ($\lambda_{ex} = 320$nm) in colloidal suspensions of hematite (α-Fe$_2$O$_3$).[26] While a linear (zero order) depletion of sulfite is observed in the presence of oxygen (lower curve in Fig. 8, air-saturated solution), the oxidation of SO$_3^{2-}$ starts with the same rate under anoxic conditions (upper curve in Fig. 8, N$_2$-sat.), but the reaction eventually slows down and ceases as 2×10^{-4}M SO$_3^{2-}$ are oxidized. At this point the catalyst, i.e. Fe$_2$O$_3$, is completely dissolved. Sulfate is the only oxidation product in the presence of air, but both SO$_4^{2-}$ and S$_2$O$_6^{2-}$ are found (by HPIC) in oxygen-free samples. The detection of dithionate evinces the intermediate formation of SO$_3^{\bullet-}$ radicals via

$$SO_3^{2-} + \ ^{\bullet}OH_S \ \rightarrow \ SO_3^{\bullet-} + \ OH^- \tag{19}$$

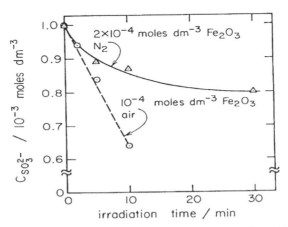

Figure 8. Photooxidation of S(IV) on colloidal α-Fe$_2$O$_3$ at pH 3.5 with $\lambda_{ex} = 320$ nm in the presence (----) and absence (——) of O$_2$. In the anoxic case, the reaction ceases upon stoichiometric reductive dissolution of the colloid to give Fe(II)$_{aq}$. (taken from ref.[26]).

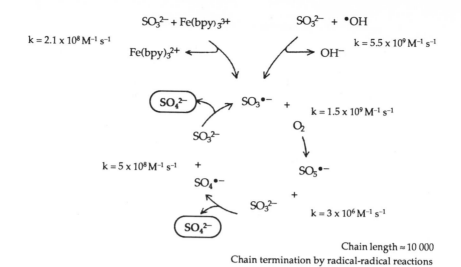

$$SO_3^{2-} + Fe(bpy)_3^{3+} \qquad\qquad SO_3^{2-} + {}^\bullet OH$$

$$k = 2.1 \times 10^8\,M^{-1}\,s^{-1}$$

$$Fe(bpy)_3^{2+} \qquad\qquad OH^- \qquad k = 5.5 \times 10^9\,M^{-1}\,s^{-1}$$

$$\boxed{SO_4^{2-}} \qquad SO_3^{\bullet-} \; + \qquad k = 1.5 \times 10^9\,M^{-1}\,s^{-1}$$

$$SO_3^{2-} \qquad\qquad O_2$$

$$k = 5 \times 10^8\,M^{-1}\,s^{-1} \quad + \qquad\qquad SO_5^{\bullet-}$$

$$SO_4^{\bullet-}$$

$$\qquad\qquad SO_3^{2-} \; +$$

$$\boxed{SO_4^{2-}} \qquad\qquad k = 3 \times 10^6\,M^{-1}\,s^{-1}$$

Chain length ≈ 10 000

Chain termination by radical-radical reactions

Scheme 1. Schematic representation of the free radical chain oxidation of S(IV) in homogeneous aqueous solution (Backström mechanism). Rate constants given are taken from the literature (refs.[38-40]).

It is well known that sulfite radicals react rapidly with molecular oxygen in homogeneous solution initiating a chain reaction called "Backström mechanism" (Scheme 1).[37-40] Chain lengths of more than 10,000 have been reported which should lead to enormously high "quantum yields" once the radicals are able to escape the particle's surface. However, the quantum yield (measured from the initial slope of the curves given in Fig. 8) does not change when N_2 is present instead of O_2. It is, therefore, concluded that the radicals remain on the semiconductor's surface where they undergo subsequent reactions before the products are released into the bulk. The following mechanism is suggested to explain the observations in anoxic solutions.

$$\alpha\text{-}Fe_2O_3 \quad \rightarrow \quad e^-_{CB} \quad + \quad h^+_{VB} \tag{20}$$

$$SO_3^{2-} \quad + \quad h^+_{VB} \quad \rightarrow \quad SO_3^{\bullet-}{}_{(S)} \tag{21}$$

$$SO_3^{\bullet-}{}_{(S)} \quad + \quad SO_3^{\bullet-}{}_{(S)} \quad \rightarrow \quad SO_3 \quad + \quad SO_3^{2-} \tag{22a}$$

$$SO_3^{\bullet-}{}_{(S)} \quad + \quad SO_3^{\bullet-}{}_{(S)} \quad \rightarrow \quad S_2O_6^{2-} \tag{22b}$$

$$SO_3 + H_2O \quad \rightarrow \quad 2H^+_{aq} \quad + \quad SO_4^{2-} \tag{23}$$

$$Fe^{3+}{}_{(S)} + e^-_{CB} \rightarrow \quad Fe^{2+}{}_{aq} \tag{24}$$

Thus, the reduction of surficial metal groups ($Fe^{3+}{}_{(S)}$) by conduction band electrons leads to the observed dissolution of the catalyst. Scheme 2 shows the mechanism proposed when O_2 is present in the system. $SO_3^{\bullet-}{}_{(S)}$ radicals formed in (21) are bound to the surface (most possibly to a positively charged ($\equiv FeOH_2^+$)-group) in close vicinity to a O_2 molecule. Thermodynamic calculations show that the one-electron redox potentials favor an electron transfer from $SO_3^{\bullet-}$ to O_2 (i.e. $E_7^1(O_2^{\bullet-}/O_2)$ = -0.33^{41} and $E_7^1(SO_3^{\bullet-}/SO_4^{2-}) = -0.62$ V[42] vs. NHE) as shown in Scheme 2. Adduct formation of $SO_3^{\bullet-}$ and O_2 like in the homogeneous case (cf. Scheme 1) is not possible on the particle's surface where both molecules are tightly bound. This explains the absence of free radical chain reactions in this case. This example most convincingly illustrates possible mechanistic differences between homogeneous and heterogeneous systems even when apparently the same intermediates are present.

113

Mechanistic Ideas

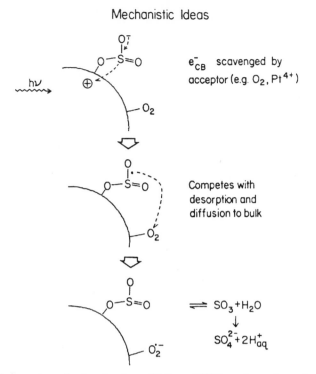

e^-_{CB} scavenged by acceptor (e.g. O_2, Pt^{4+})

Competes with desorption and diffusion to bulk

$\rightleftharpoons SO_3 + H_2O$

$SO_4^{2-} + 2H^+_{aq}$

Scheme 2. Proposed mechanism for the oxidation of S(IV) on the surface of colloidal metal oxide particles in the presence of molecular oxygen.

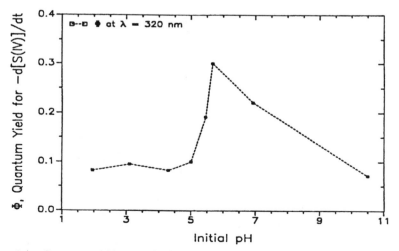

Figure 9. Quantum yield vs. pH for the photooxidation of 1.0×10^{-3} mol dm^{-3} S(IV) in a colloidal suspension of 0.1×10^{-3} mol dm^{-3} Fe$_2$O$_3$ colloid in the presence of 0.25×10^{-3} mol dm^{-3} O$_2$. (taken from ref.[26]).

Finally, it should be pointed out that O_2 also prevents dissolution of the catalyst since

$$O_2 \;+\; e^-_{CB} \;\rightarrow\; O_2^{\bullet -} \tag{25}$$

can compete successfully with (24). Therefore, a zero order depletion of sulfite (which is typical for catalytic processes) is observed (cf. Fig. 8) and the hematite particles remain unchanged. Figure 9 shows the observed pH dependence of the quantum yield for the photocatalytic sulfite oxidation when α-Fe_2O_3 colloids are the catalysts. A pronounced maximum is observed around pH 5.5. An extensive modelation of a hematite surface in the presence of SO_3^{2-} using a sophisticated computer code (SURFEQL) demonstrates that such a pH dependance can be readily explained.[26] Taking the sum of the concentrations of all iron/sulfite surface complexes plus the SO_3^{2-} in the electrical double layer at a distance of 0.3 nm from the particle surface, the curve $c = f(pH)$ exhibits almost identical shape and position as compared with the experimental results shown in Fig. 9. This can be taken as additional evidence for the importance of the surface chemistry for photocatalytic processes.

While similar observations are made when iron(III)oxide is replaced by colloidal titanium dioxide as the photoactive catalyst,[26] the situation changes drastically with a surface modified semiconductor. Recently, it has been possible to synthesize TiO_2 particles (P25, Degussa Co.) with cobalt tetrasulfophthalocyanine (CoTSP) groups chemically bound to their surfaces via a side chain of the dye (Scheme 3).[43] These modified particles are also extremely efficient for the oxidation of sulfite to sulfate upon band-gap illumination of the semiconductor[36]. As shown in Figure 10 as a function of SO_3^{2-} concentration and pH "quantum yields" well exceeding 100 are measured in these aqueous suspensions, evincing a considerable contribution of free radical chain reactions (cf. Scheme 1). Addition of ethanol, a typical radical scavenger, to these systems drastically reduces the observed yields of sulfite oxidation. Therefore it is concluded that also when TiO_2-CoTSP is the catalyst, $SO_3^{\bullet -}$ radicals formed in the initial oxidation step (cf. eq. (21)) do not find an O_2 molecule in a vicinity close enough to allow electron-transfer and will eventually leave the particle's surface to initiate a chain reaction in the homogeneous phase. The CoTSP-groups linked to the TiO_2 are good acceptors for e^-_{CB} and catalyze the formation of H_2O_2 from molecular oxygen thereby ensuring the stability of the catalyst.

Another example for electron transfer processes at semiconductor surfaces involving sulfur-centered free radical intermediates is shown in Figure 11[44]. Diethyldisulfide, EtSSEt, (qualitatively and quantitatively determined by capillary GC-FID) is readily formed when ethylmercaptane containing aqueous suspensions of cadmium sulfide (CdS) are illuminated with band-gap light. Similar observations are made when TiO_2 particles are used as photocatalysts[44].

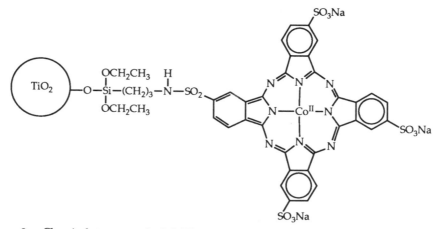

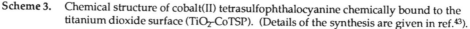

Scheme 3. Chemical structure of cobalt(II) tetrasulfophthalocyanine chemically bound to the titanium dioxide surface (TiO_2-CoTSP). (Details of the synthesis are given in ref.[43]).

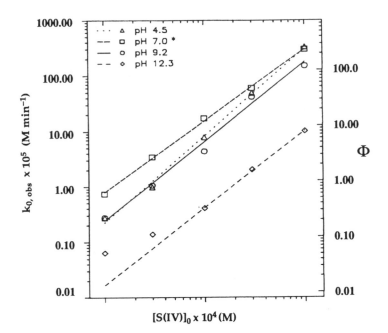

Figure 10. Dependence of the initial oxidation rate on the S(IV) concentration as a function of pH. Straight lines are least-squares fits with slopes indicating reaction orders in [S(IV)]. Concentrations are 50 mg/l for TiO_2-CoTSP and 1.2×10^{-3} mol dm^{-3} for O_2. Quantum yields are indicated. (*) Concentrations are 280 mg/l for TiO_2-CoTSP (taken from ref.[36]).

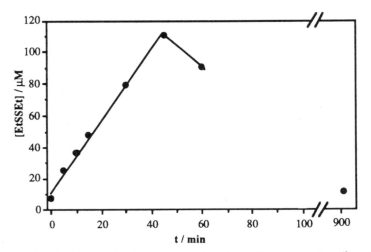

Figure 11. Yield of diethyldisulfide formation as a function of illumination time ($\lambda_{ex} = 480$ nm) of an air-saturated aqueous suspension of 1.0 g/l CdS powder and 1.1×10^{-3} mol dm^{-3} ethylmercaptane (taken from ref.[44]).

The proposed reaction mechanism is given in Scheme 4, which illustrates the formation of $CH_3CH_2S^\bullet$ radicals as first intermediates of the oxidation process. It is obvious from Fig. 11 that the photocatalytically formed product reaches a maximum concentration after approximately 90 min of illumination. The subsequent decrease of [EtSSEt] is explained by a competition with CH_3CH_2SH for the oxidizing equivalents:

$$CH_3CH_2SH + {}^\bullet OH_S / {}^\bullet SH_S \rightarrow CH_3CH_2S^\bullet + H_2O/H_2S \tag{26}$$

$$CH_3CH_2SSCH_2CH_3 + {}^\bullet OH_S / {}^\bullet SH_S \rightarrow products \tag{27}$$

The individual products of reaction (27) have not yet been identified.

Figure 12 shows the yield of disulfide formation in aqueous CdS suspensions containing ethylmercaptane as a function of the illumination wavelength[44]. It is evident that the photocatalytic activity ceases as the energy of the exciting light falls below the band-gap energy $E_g = 2.42$ eV of CdS.[44] It is interesting to note that CdS obviously possesses some catalytic dark activity for the EtSSEt-formation which is not observed for metal oxide semiconductors, e.g. TiO_2 or ZnO. A detailed study of the thermal oxidation of ethylmercaptane catalyzed by various low band-gap semiconductors which relates this catalytic activity to the different crystal structure of the materials has been published.[44]

initial process:

$$TiO_2 \xrightarrow{\ h\nu\ } TiO_2 (e^-, h^+)$$

$$TiO_2 (e^-, h^+) \longrightarrow TiO_2 (Ti^{III}, {}^\bullet OH_S)$$

$$CdS \xrightarrow{\ h\nu\ } CdS (e^-, h^+)$$

$$CdS (e^-, h^+) \longrightarrow CdS (Cd^I, {}^\bullet SH_S)$$

oxidative process:

$$CH_3CH_2SH + {}^\bullet OH_S / {}^\bullet SH_S \longrightarrow CH_3CH_2S^\bullet + H_2O/H_2S$$

$$2 CH_3CH_2S^\bullet \longrightarrow CH_3CH_2SSCH_2CH_3$$

reductive process:

$$O_2 + Ti^{III}/Cd^I \longrightarrow O_2^{\bullet -} + Ti^{IV}/Cd^{II}$$

$$O_2^{\bullet -} + 2 H^+_{aq} \longrightarrow H_2O_2 + O_2$$

overall:

$$2 CH_3CH_2SH \xrightarrow[TiO_2/CdS]{\ h\nu\ } CH_3CH_2SSCH_2CH_3$$

Scheme 4.   Proposed mechanism for the photocatalytic oxidation of ethylmercaptane in aerated aqueous suspensions with TiO_2- or CdS-particles as photocatalysts.

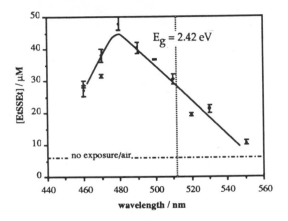

Figure 12.   Yield of diethyldisulfide formation as a function of illumination wavelength of an air-saturated aqueous suspension of 1.0 g/l CdS powder and 1.1 x 10⁻³ mol dm⁻³ ethyl-mercaptane (taken from ref.[44]).

Finally, it is interesting to note that the EtSSEt formation from ethylmercaptane in the photocatalytic system is supressed in the absence of molecular oxygen. The most likely explanation for this result is a reduction of the $CH_3CH_2S^\bullet$ radical formed in (26) by the conduction band electrons.

$$CH_3CH_2S^\bullet + e^-_{CB} \quad \rightarrow \quad CH_3CH_2S^- \tag{28}$$

The mercaptane then only promotes the recombination of e^-_{CB} and h^+_{VB} by the reaction sequence (26) and (28). Simular "short-circuiting" of semiconductor particles has before been reported for the inhibitory action of Fe^{3+} ions on the photocatalytic oxidation of methylviologen on TiO_2 colloids.   Molecular oxygen can, however, be replaced by other electron acceptors such as methylviologen without a significant reduction of the rate of mercaptane oxidation.[44]

SUMMARY AND OUTLOOK

It has been shown that oxidative and reductive free radical pathways can be initiated by one-electron transfer reactions on the surface of semiconductor particles following the band-gap illumination of the material. While the radical species formed can diffuse into the bulk solution to promote reaction pathways known from conventional, e.g., radiation chemical, free radical studies, many examples show that such radicals are stabilized on the surface giving them a residence time long enough to undergo further reactions before they are released. Thus, overall two-electron transfer processes are often encountered, chain reactions are supressed or the product distribution does not resemble that of the homogeneous case.

Inorganic ($SO_3^{\bullet-}$) and organic ($CH_3CH_2S^\bullet$) sulfur-centered radical intermediates are generated via oxidative attack when TiO_2, α-Fe_2O_3 or CdS particles are employed as photocatalysts. In most cases, the simultaneously formed conduction band electrons are scavenged by molecular oxygen or equivalent oxidants. In the absence of such oxidizing molecules reductive dissolution of the catalysts can pose a serious problem. Time resolved studies using a laser flash photolysis set-up have not yet been performed to study the mechanisms of these reactions. Optically transparent colloidal suspensions containing extremely small semiconductor particles are useful for this type of research since artefacts due to laser light scattering can be minimized.

Future research should furthermore concentrate on basic questions such as the redox potential of radicals at surfaces, e.g., $^\bullet OH_s$ or $^\bullet SH_s$, a better understanding of their chemical properties and the possibility to differentiate between reactions of h^+_{VB} and $^\bullet OH_s / ^\bullet SH_s$, respectively. Special emphasis should, however, also be directed towards more applied questions. Valence band holes or conduction band electrons can be extremely powerful oxidants or reductants, respectively, which could be utilized for the destruction of environmental pollutants. First attempts to employ illuminated aqueous suspensions of TiO_2 particles to the mineralization of halogenated hydrocarbons have been successful,[27,45-48] however, more systematic studies are required before technologically relevant solutions can be offered.

REFERENCES

1. P.W. Atkins, "Physical Chemistry", Oxford University Press, Oxford, U.K., p. 484 (1978).
2. H. Gerischer, *Topics in Appl. Physics* 31:115 (1979).
3. R. Memming, *Topics in Current Chemistry* 143:79 (1988).
4. G. Rothenburger, J. Moser, M. Grätzel, N. Serpone, and D.K. Sharma, *J. Am. Chem. Soc.* 107:8054 (1985).
5. C. Kormann, D.W. Bahnemann, and M. R. Hoffmann, *J. Phys. Chem.* 92:5196 (1988).
6. D.W. Bahnemann, Ch.-H. Fischer, E. Janata, and A. Henglein, *J. Chem. Soc., Faraday Trans. 1* 83:2559 (1987).
7. M. Gutiérrez and A. Henglein, *Ber. Bunsenges. Phys. Chem.* 87:474 (1983).
8. A. Henglein, *Ber. Bunsenges. Phys. Chem.* 86:301 (1982).
9. D. Bahnemann, A. Henglein, and L. Spanhel, *Faraday Discuss. Chem. Soc.* 78:151 (1984).
10. M. Haase, H. Weller, and A. Henglein, *J. Phys. Chem.* 92:4706 (1988).
11. Z. Alfassi, D. Bahnemann, and A. Henglein, *J. Phys. Chem.* 86:4656 (1982).
12. H. Weller, private communication (1988).
13. W.M. Latimer, "Oxidation Potentials", 2nd Edition, Prentice-Hall, New York, p. 38 (1952).
14. G. Charlot, A. Collumeau, and M. Marchon, "Selected Constants, Oxidation-Reduction Potentials of Inorganic Substances in Aqueous Solution", Butterworth Publ., London, (1971).
15. D. Meisel and P. Neta, *J. Am. Chem. Soc.* 97: 5198 (1975).
16. P. Wardman and E.D. Clarke, *J. Chem. Soc., Faraday Trans. 1* 72:1377 (1976).
17. M. v. Stackelberg and W. Stracke, *Z. f. Elektrochem.* 53:118 (1949).
18. S. Wawzonek and R.C. Duty, *J. Electrochem. Soc.* 188:1135 (1961).
19. J.C. Suatoni, R.E. Snyder, and R.O. Clark, *Anal. Chem.* 33:1894 (1961).
20. S. Steenken and P. Neta, *J. Phys. Chem.* 83:1134 (1979).
21. S. Steenken and P. Neta, *J. Phys. Chem.* 86:3661 (1982).
22. R.E. Huie and P. Neta, *J. Phys. Chem.* 89:3918 (1985).
23. A.J. Bard and H. Lund, "Encyclopedia of Electrochemistry of the Elements", Vol. XII, Marcel-Dekker, New York (1978).
24. W. Stumm and J.J. Morgan, "Aquatic Chemistry", J. Wiley & Sons, New York, pp. 599-640 (1981).
25. D.W. Bahnemann, C. Kormann, and M.R. Hoffmann *J. Phys. Chem.* 91:3789 (1987).
26. B.C. Faust, M.R. Hoffmann, and D.W. Bahnemann, *J. Phys. Chem.* 93: 6371 (1989).
27. C. Kormann, D.W. Bahnemann, and M.R. Hoffmann, *J. Phys. Chem.*, submitted.
28. A. Henglein, *Topics in Current Chemistry* 143:113 (1988).
29. D. Bahnemann, A. Henglein, J. Lilie, and L. Spanhel, *J. Phys. Chem.* 88:709 (1984).
30. D. Duonghong, J. Ramsden, and M. Grätzel, *J. Am. Chem. Soc.* 104:2977 (1982).
31. J. Moser and M. Grätzel, *J. Am. Chem Soc.* 105: 6547 (1983).
32. M. A. West, *Creat. Detect. Excited State* 4: 217 (1976).
33. G. Beck, *Int. J. Radiat. Phys. Chem.* 1:361 (1969).
34. E. Janata, *Radiat. Phys. Chem.* 16:37 (1980).

35. F. Wilkinson, C.J. Willsher, S. Uhl, W. Honnen, and D. Oelkrug, *J. Photochem.* 33:273 (1986).

36. A.P. Hong, D.W. Bahnemann, and M.R. Hoffmann, *J. Phys. Chem.* 91: 6245 (1987).

37. H.Z. Backström, *Phys. Chem.* 25B:122 (1934).

38. E. Hayon, E. Treinin, and J. Wilf, *J. Am. Chem. Soc.* 94:47 (1972).

39. R. Huie and P. Neta, *J. Phys. Chem.* 88:5665 (1984).

40. R. Huie and P. Neta, *EHP, Environ. Health Perspect.* 64:209 (1985).

41. B.H.J. Bielski, D.E. Cabelli, R.L. Arudi, and A.B. Ross, *J. Phys. Chem. Ref. Data* 14:1041 (1985).

42. D.W. Bahnemann and M.R. Hoffmann, *Proc. Electrochem. Soc.* 88-14:74 (1988).

43. A.P. Hong, D.W. Bahnemann, and M.R. Hoffmann *J. Phys. Chem.* 91:2109 (1987).

44. W. Hoyer, *Dissertation* TU Berlin, FRG D83, (1987).

45. D.F. Ollis, *Environ. Sci. Technol.* 19:480 (1985).

46. R.W. Matthews, *Wat. Res.* 2:569 (1986).

47. E. Pelizetti, M. Borgarello, C. Minero, E. Pramauro, E. Borgarello, and N. Serpone, *Chemosphere* 17:499 (1988).

48. C. Kormann, D.W. Bahnemann, and M.R. Hoffmann, *J. Photochem. Photobiol., A: Chemistry* 48:161 (1989).

APPLICATIONS OF PULSE RADIOLYSIS

FOR THE STUDY OF SHORT-LIVED SULPHUR SPECIES

David A. Armstrong

Department of Chemistry
University of Calgary
Calgary, Alberta, Canada, T2N 1N4

The first successful observations of free radical intermediates by spectroscopic techniques were made by Norrish and Porter using flash photolysis. Pulse radiolysis is the radiation-chemical analogue of flash photolysis. In it free radicals are created by the deposition of energy from a transient beam of high energy (0.6 - 10 MeV) electrons. These lose energy to electrons in the molecular orbitals of the target material due to repulsive coulombic interactions. The discrete energy losses result in excitations or ionizations of the molecules.[1] In water ionization dominates, with an average distance of about 100 nm between successive events. Secondary electrons produced in these ionizations have varying energies. They too cause ionization and excitation. Those of low energy (100 eV) deposit their energy within a relatively small volume, creating what is commonly called a spur with a number of ions, electrons, and free radicals in relatively close proximity. The distributions of radicals per spur (or of spur sizes) depends on the initial energy of the primary electrons and the medium irradiated. For 1 MeV electrons in water at 25 °C, about half the spurs contain only one ion pair, while for the others the number of reactive species varies up to about six.

Secondary electrons in water normally lose their energy in a time of about one psec, become localized, and then hydrated.[1] The hydrated electron, $e^-_{(aq)}$ has a characteristic spectrum which is developed at 10 ps, and the yield of hydrated electrons at this time is 0.5 μmoles J^{-1} of energy absorbed. On the ps time scale the positive ions undergo ion-molecule reactions and, like the electron, become solvated:

$$H_2O^+ + H_2O \rightarrow H_3O^+ + {}^\bullet OH$$

$$H_3O^+ + x\,H_2O \rightarrow H_3O^+_{(aq)}$$

and many of the excited molecules dissociate:

$$H_2O^* \rightarrow H^\bullet + {}^\bullet OH$$

From this point in time outward diffusion of the reactive species competes with geminate reactions within the spurs, and the spur structures are effectively dissipated at about 50 ns. The geminate reactions

$$e^-_{(aq)} + {}^\bullet OH \rightarrow OH^-$$

Sulfur-Centered Reactive Intermediates in Chemistry and Biology, Edited by
C. Chatgilialoglu and K.-D. Asmus, Plenum Press, New York, 1990

and

$$e^-_{(aq)} + H_3O^+_{(aq)} \rightarrow H^\bullet$$

cause the yield of electrons to fall from 0.50 to 0.28 µmoles J^{-1}. During the same time, the following reactions occur,

$$H^\bullet + H^\bullet \rightarrow H_2$$

$$^\bullet OH + {}^\bullet OH \rightarrow H_2O_2$$

giving rise to small yields of hydrogen molecules and hydrogen peroxide. The yields of these products and the reactive intermediates at 1 µs after passage of the primary electron are described by equation (1), where the coefficients are in units of µmoles J^{-1} of energy absorbed:

$$0.43\ H_2O \rightarrow 0.28\ e^-_{(aq)} + 0.05\ H^\bullet + 0.29\ {}^\bullet OH + 0.28\ H_3O^+_{(aq)} + 0.05\ H_2 + 0.07\ H_2O_2 \quad (1)$$

(For high energy x- or γ-ray irradiations electrons are set in motion in the medium by the photon. The energy deposition processes are then the same.[1]) As shown in equation (1), the major reactive species are the hydrated electron and the hydroxyl radical, with smaller numbers of H$^\bullet$ atoms.

PRODUCTION OF SECONDARY RADICALS

Because of the extremely fast time scale of the primary events in water, and the fact that the energy of the fast electrons is deposited in the molecules present in a solution on the basis of their relative concentrations, it is possible to add solute molecules at submolar concentrations without interfering with the production of primary water radicals in any significant way. However, when the solute concentrations are in excess of 1 mM, it may be necessary to make minor corrections to the yields in equation (1) to account for the fact that some radicals will react with the solute and not quite as many will undergo geminate reaction.[1]

A more interesting point is the fact that the hydroxyl radical is oxidizing, while the solvated electron and hydrogen atom are reducing in character. It is therefore often convenient to convert $^\bullet OH$ into the reducing species $^\bullet CO_2^-$ by the addition of sodium formate

$$^\bullet OH + HCO_2^- \rightarrow H_2 + {}^\bullet CO_2^- \quad (2)$$

and thus have only reducing species present, or to convert the hydrated electron into a hydroxyl radical via reaction (3):

$$e^-_{(aq)} + N_2O + H_2O \rightarrow {}^\bullet OH + OH^- + N_2 \quad (3)$$

and thus have mainly oxidizing species. By using suitable concentrations of HCO_2^- and N_2O, reactions (2) and (3) can be made to occur on a ns time scale. Other oxidizing solute radicals, such as $N_3^\bullet$, $Br_2^{\bullet-}$ and Tl^{2+}, can be created on a similar time scale by reacting $^\bullet OH$ with ca. 1 mM N_3^-, Br^- and Tl^+ (reference 1). These radicals (or $^\bullet CO_2^-$) may then react with a sulphur-containing molecule, added at a lower concentration (≤100 µM), on a 1 to 10 µs time scale.

At this point it is worth returning to a comparison between pulse radiolysis and the classical techniques of flash photolysis. Drawing on work in the published literature, it may be pointed out that the perthiyl radical can be produced by either technique, as illustrated by the following equations:

Pulse Radiolysis

$$^\bullet CO_2^- + RSSSR \xrightarrow{\text{ref 2a}} CO_2 + RSS^\bullet + RS^-$$

Flash Photolysis

$$RS_4R + h\nu \xrightarrow{\text{ref 2b}} 2RSS^\bullet$$

$$RSSCl + h\nu \xrightarrow{\text{ref 2b}} RSS^\bullet + Cl^\bullet$$

$$RSSR + h\nu \xrightarrow{\text{ref 2c}} RSS^\bullet + R^\bullet$$

Sometimes the flash photolysis technique provides an advantage, because by choice of wavelength a specific molecule can be excited to produce a desired radical species directly. On the other hand, there are occasions where the capability of creating either reducing or oxidizing species by pulse radiolysis can be an advantage. Thus, the two techniques have become complementary, and in a significant number of laboratories both are used with their relative advantages being exploited as the need arises. The major difference between the two techniques is that in radiolysis the energy is absorbed by the solvent (the major constituent present) and radicals formed from it then initiate reactions with the solute to create the species of interest. In flash photolysis this species is created by having the photon energy absorbed by a specific solute.

A TYPICAL APPARATUS

A schematic diagram of a typical pulse radiolysis apparatus is shown in Figure 1. The collimated beam of electrons is normally obtained from a linear accelerator or a Van de Graaff.[1] For reasonable penetration into water, the energy should be in the range of 2 to 10 MeV, although beams with energy as low as 1.5 MeV can be used. Beam intensities are usually designed to produce between 0.5 and 50 µM of radical intermediates per pulse. Most modern accelerators have pulse lengths which can be varied between 5 ns and 10 µs, but a number of specially designed accelerators are capable of producing pulses of a few ps duration. The solution to be irradiated is contained in a flushable optical cell situated in front of the electron beam. Ideally, the entire volume of the cell is uniformly irradiated. The cell may be fitted with electrodes, if changes in solution conductivity are to be measured, and various designs have been reported in the literature. Data from the light intensity monitor or conductivity bridge, acquired over the desired period of time, are normally processed on a micro-computer and stored in digital form for analysis. A triggering device

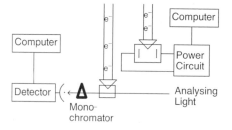

Fig. 1. Typical pulse radiolysis apparatus. For absorption studies the beam of electrons passes into the optical cell, which is traversed at right angles by the analysing light beam. The second cell (upper right) is for conductivity work.

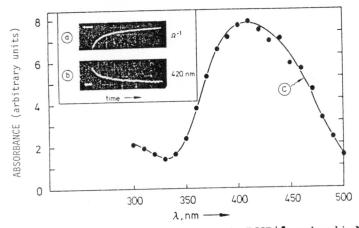

Fig. 2. Absorption spectrum and intensity time traces for RSSR$^{+\bullet}$ produced in N$_2$O-saturated 5 x 10^{-4} M (C$_2$H$_5$)$_2$S$_2$ at pH 4 by a 10 J kg^{-1} dose. (a) Conductivity change, 100 μs per division, (b) absorbance at 420 nm, 50 μs per division, (c) spectrum. Based on data from ref. 3b.

initiates data recording, and, after a suitable base line trace has been acquired, fires the electron beam pulse. A change in a signal is then observed and recorded. Traces of typical changes in absorption and conductivity are presented in Figure 2, inset. These traces can be analysed to give information on the kinetics of reactions of the reactive intermediates absorbing at the particular wavelength. Spectra can be made at different times by making observations at several wavelengths.

SPECTRUM AND REACTIONS OF PERTHYL RADICAL, RSS$^{\bullet}$

Chemical equations for the production of RSS$^{\bullet}$ were given above. The solid line in Figure 3 shows the spectrum of perthiyl radicals in water obtained by the pulse radiolysis method. The points are spectra obtained from two separate photochemical experiments. The

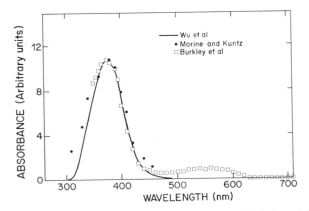

Fig. 3. Spectra of perthiyl radical from references 2a (———), 2b (□) and 2c (●). ε Values were reported to be 1630 ± 50^{2a} and ≈ 1700^{2c} in M^{-1} cm^{-1}.

124

R groups have different structure in each case, but the main absorption band is similar and peaks at 374 nm (ε_{max} = 1630 M^{-1} cm^{-1}). The weak band seen at 580 nm in cyclopentane (square symbols) was not reported in water.

Analysis of the kinetic traces and long lived absorbances in the pulse radiolysis[2a] showed that dimerization to tetrasulphide dominated for the decay of this species:

$$2 \text{ RSS}^\bullet \quad \rightarrow \quad \text{RSSSSR}$$

$$2k = (1.4 \pm 0.3) \times 10^9 \text{ M}^{-1}\text{s}^{-1}$$

Elimination of SS, if it occurs at all, takes place on a longer time scale. In this particular respect, perthiyl differs from its oxygen counterpart, the peroxyl radical, for which a tetroxide seems to be only an intermediate state, suffering immediate O_2 elimination. The difference has been attributed to the stronger bonding between the central S atoms in the tetrasulphide.[2a]

OXIDATIVE REACTIONS - RADICAL CATIONS OF DISULPHIDES AND THIOETHERS

Disulphides

Early studies of the reactions between $^\bullet$OH radicals and disulphides were interpreted in terms of the displacement reaction (4)

$$\text{RSSR} + {}^\bullet\text{OH} \quad \rightarrow \quad \text{RSOH} + \text{RS}^\bullet \tag{4}$$

However, work by Asmus and co-workers,[3] employing the conductivity technique, showed that ionic species were also produced, thus demonstrating that reaction (4) was an oversimplification of the mechanism.

The top oscilloscope trace, curve a, in Figure 2 shows the change in conductivity observed as a result of the reaction of $^\bullet$OH with $(C_2H_5)_2S_2$ at pH 4. The sharp drop in conductivity is due to the immediate removal of protons (reaction 6) by the OH$^-$ formed in reaction (5), and their replacement by the lower mobility species RSSR$^{\bullet+}$.

$$\text{RSSR} + {}^\bullet\text{OH} \quad \rightarrow \quad \text{RSSR}^{\bullet+} + \text{OH}^- \tag{5}$$

$$\text{OH}^- + \text{H}^+ \quad \rightleftarrows \quad \text{H}_2\text{O} \tag{6}$$

The simultaneous rapid growth in absorbance at 420 nm and its decay on the same time scale as the conductivity change, shown by trace b, indicates that the absorbing species is the radical cation. Its absorption peak is shown by curve c.

The yield of the cation calculated from the absolute magnitude of the conductivity change depended on the structure of the disulphide, lying in the region of 50% of $^\bullet$OH for reactions with low molecular weight alkyl disulphides. The remaining 50% of the reactions produce RSOH in reaction (4). Other oxidants, including Ag^{2+}, $Ag(OH)^+$, Tl^{2+}, $SO_4^{\bullet+}$, $Br_2^{\bullet-}$ and thioether cations such as $(CH_3)_2S^{\bullet+}$ have been shown to produce exclusively radical cations in their reaction with, e. g., $(CH_3S)_2$.

The decay reactions of RSSR$^{\bullet+}$ in acid or neutral systems are of second order, the two major processes suggested to be

$$\text{RSSR}^{\bullet+} + \text{RSOH} \quad \rightarrow \quad \text{RSSR} + \text{RSO}^\bullet + \text{H}^+ \tag{7}$$

$$2 \text{ RSSR}^{\bullet+} \quad \rightarrow \quad \text{RSSR} + \text{RSSR}^{2+} \tag{8}$$

The rate constant $2k_8$ is $(7.0 \pm 0.5) \times 10^8 \text{ M}^{-1}\text{s}^{-1}$, while k_7 is larger. In alkaline solution the decay become first order with rates proportional to the hydroxide ion concentration. This has been attributed to the formation of an OH adduct, which leads to

$$\text{OH}^- + \text{RSSR}^{\bullet+} \quad \rightarrow \quad [\text{RSSR(OH)}^{\bullet}] \quad \rightarrow \quad \text{products} \tag{9}$$

further products. Reactions (8) and (9) both exhibit the expected dependences of rate on ionic strength. Also, the rate constant of reaction (9) falls off with electron donating capacity of the R groups, showing that it is dependent on the charge density at the S centres (e.g., $k_9 = 1.8 \times 10^9 \text{ M}^{-1}\text{s}^{-1}$ for R = CH_3, and $3.9 \times 10^8 \text{ M}^{-1}\text{s}^{-1}$ for R = i-C_3H_7).

At very high pH a further rapid reaction leading to RS^- is seen. This has been attributed to:

$$[\text{RSSR(OH)}^{\bullet}] + \text{OH}^- \rightarrow \text{RS}^- + \text{RS(OH)}_2^{\bullet} \tag{10}$$

where $\text{RS(OH)}_2^{\bullet}$ is a hydrated form of the sulphinyl radical $\text{RSO}^{\bullet}$. The rate constant for this process is $(1.5 \pm 0.1) \times 10^8 \text{ M}^{-1}\text{s}^{-1}$. Finally, disulphide radical cations have been shown to act as oxidants, for example in reaction (11) ($k_{11} = (1.5 \pm 0.2) \times 10^{10} \text{ M}^{-1}\text{s}^{-1}$ for R = CH_3):

$$\text{RSSR}^{\bullet+} + \text{Fe(CN)}_6^{4-} \rightarrow \text{RSSR} + \text{Fe(CN)}_6^{3-} \tag{11}$$

As shown in Figure 2 (curve c) the absorptions of typical alkyl disulphide radical cations are broad, with λ_{max} near 400 nm and $\varepsilon = 2000 \text{ M}^{-1}\text{cm}^{-1}$. They are therefore relatively easy to observe. However, when attempts were made to produce them in the biologically relevant amino-containing disulphides of cysteine, penicillamine and other amino acid disulphides, they were completely unsuccessful.[4] Instead, weaker absorptions centered at 374 nm were observed. Also, unlike the $\text{RSSR}^{\bullet+}$ transient, these species exhibited no evidence of reaction with OH^- at high pH. The value of λ_{max} and the magnitude of the absorbances ($\varepsilon = 1600 \text{ M}^{-1}\text{cm}^{-1}$) corresponded to those of the perthiyl radical, and it would appear that the presence of the amino groups serves to catalyse the cleavages of C–S bonds.[4] Two possible mechanisms are given in Scheme 1. One of these is similar to that proposed for the decomposition of methionine radical cation by Asmus and co-workers (see below) in earlier work.

Thioether Systems

Very extensive studies of alkyl thioethers and cyclic dithia compounds have been made by Asmus and his co-workers.[5] Monosulphide radical cations $\text{R}_2\text{S}^{\bullet+}$ exhibit a λ_{max} around 300 nm. In aqueous solution they participate in the following equilibria:

$$\text{R}_2\text{S}^{\bullet+} \quad \rightleftarrows \quad \text{R–S–R(–H)}^{\bullet} + \text{H}^+$$

$$\text{R}_2\text{S}^{\bullet+} + \text{R}_2\text{S} \quad \rightleftarrows \quad \text{R}_2\text{S.}^{\bullet}\text{.SR}_2^+$$

The absorptions of the $\text{R}_2\text{S.}^{\bullet}\text{.SR}_2^+$ radical cation depend on the strength of the $\text{S.}^{\bullet}\text{.S}$ three-electron ($2 \sigma / 1 \sigma^*$) bond and are influenced by the structures of the R groups. In cyclic dithia compounds the degree of overlap of the sulphur orbitals has an important effect. The bands are broad and mainly in the visible (400 - 650 nm).

The monomolecular thioether radical cation $\text{R}_2\text{S}^{\bullet+}$ is a strong oxidant, while its dimer counter part $\text{R}_2\text{S.}^{\bullet}\text{.SR}_2^+$ is less strong. Examples of reactions are:

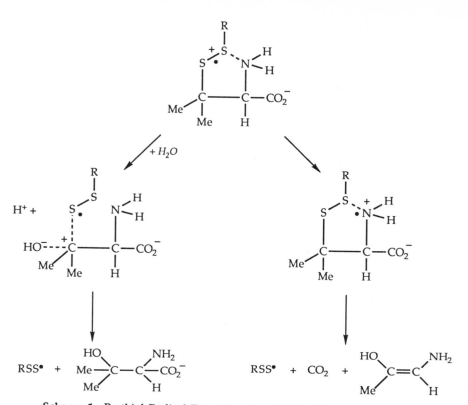

Scheme 1. Perthiyl Radical Formation from Amino-Disulphide Cations.

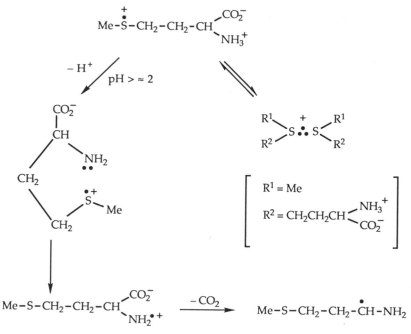

Scheme 2. Methionine Thioether Cation Equilibrium and Decomposition.

$$(CH_3)_2S^{\bullet+} + CysSH \rightarrow (CH_3)_2S + CysS^{\bullet} + H^+$$

$$k = 9.4 \times 10^9 \ M^{-1}s^{-1}$$

$$(CH_3)S.^{\bullet}.S(CH_3)^+ + CysS^- \rightarrow 2(CH_3)_2S + CysS^{\bullet}$$

$$k = 8.1 \times 10^9 \ M^{-1}s^{-1}$$

$$(CH_3)_2S.^{\bullet}.S(CH_3)_2^+ + CysSH \xrightarrow{\text{(slow)}} \text{products}$$

Very important to an understanding of the decomposition of certain biological systems are the intramolecular oxidations which can be initiated by $\gtrdot S^{\bullet+}$. The mechanism for de-carboxylation of methionine is a classic case. If the amino group is deprotonated and a rea-sonably unstrained ring can be formed to permit an $S\cdots N$ interaction, S to N charge transfer occurs. Carbon dioxide elimination then ensues and an α-amino radical is formed (Scheme 2).

SULPHYDRYL SYSTEMS AND DISULPHIDE REDUCTION

Many of the earliest studies of this kind were carried out by Henglein and his co-workers at the Hahn-Meitner-Institute in Berlin.[6] The pulse radiolysis of hydrogen-sulphide and simple mercaptans in nitrous oxide-saturated solutions at alkaline pH was shown to produce strong absorbances with maxima in the region of 400 to 450 nm. This was attributed to the occurrence of reaction (12):

$$^{\bullet}OH + RSH \rightarrow H_2O + RS^{\bullet} \tag{12}$$

However, the fact that the absorption depended on the concentration of thiolate ion (RS^- or HS^-) indicated that the absorbing species was not simply $RS^{\bullet}$. It was shown that the results could be explained quantitatively on the basis of the equilibrium (13):

$$RS^{\bullet} + RS^- \rightleftarrows RS.^{\bullet}.SR^- \tag{(13)/(-13)}$$

with the absorption being due almost entirely to the $RS.^{\bullet}.SR^-$ species. This assignment was subsequently confirmed by the fact that species with the same spectra could be created by the reaction of the hydrated electron with the corresponding disulphide.[7a]

$$RSSR + e^-_{(aq)} \rightarrow RS.^{\bullet}.SR^- \tag{14}$$

Quite extensive studies of the formation of these species from both mercaptans and disulphides were made during the 1970's at the Natick Laboratory and other institutions. A summary of some of the spectroscopic parameters and rate constants are given in Table 1. It has further been established that the sulphur-sulphur bond is $2\sigma/1\sigma^*$ in electronic nature as in the dimer thioether radical cations (see article by K.-D. Asmus in this book).

Table 1. Data for Disulphide Radical Anion $-RS^{\bullet}$ Equilibria (from refs. 7 and 8)

Thiol	λ(max)	ε(max)	$k_{13}/M^{-1}s^{-1}$	k_{-13}/s^{-1}
Cysteamine	410	8900	4.9×10^9	8×10^5
Cysteine	420	≥ 8800	3.1×10^9	3.2×10^5
Penicillamine	450	7400	2.8×10^9	1.5×10^6 (pH 7)
				6.5×10^6 (pH 10)

The absorbance of the thiyl (RS$^\bullet$) radical was shown to be much weaker than that of the disulphide anion radical and to have a maximum near to 340 nm.[7b,c,8] The decay of RS$^\bullet$ species followed second order kinetics with $2k_{15}$ being typically ca. 2×10^9 M^{-1}s^{-1} for uncharged species.[7b]

$$RS^\bullet + RS^\bullet \rightarrow RSSR \tag{15}$$

$$RS^\bullet + RS.^\bullet.SR^- \rightarrow RS^- + RSSR \tag{16}$$

When both RS$^\bullet$ and RS.$^\bullet$.SR$^-$ were present reaction (16) was also observed,[7c] and k_{16} was 6.4×10^9 M^{-1}s^{-1} for penicillamine at pH 9.

The work of Hayon and co-workers showed that protonation of RS.$^\bullet$.SR$^-$ led to a rapid loss of absorbance.[7a] They attributed this to the fact that the protonated form was unstable and dissociated rapidly, viz:

$$RS.^\bullet.SR^- + H^+ \rightarrow RSSR(H)^\bullet \rightarrow RSH + RS^\bullet \tag{17}$$

$$(k_{17} \text{ ca. } 5\times10^{10} \text{ M}^{-1}\text{s}^{-1})$$

However, it was found that the case of 1,3 and 1,4 disulphydryl-containing molecules was special. For lipoic acid and dithiothreitol, where the sulphydryls are separated by three and four carbon atoms respectively, the possibility for the existence of a five or six membered ring seems to stabilize the protonated disulphide radicals.[7,8,9] At this time it is not clear whether the resulting radical is of the sulphuranyl type or has a symmetrical structure, viz:

or

For practical purposes we shall write the structure as

The pk's of the dithiothreitol and lipoamide systems are quite similar:

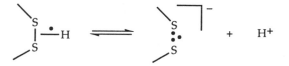

5.2 and 5.5, respectively.[8,9] The absorbance of the protonated lipoamide species with its five membered ring is notably stronger than that of the dithiothreitol. At present it is not clear whether this is due to a weaker absorbance coefficient of the ring form or to the fact that a

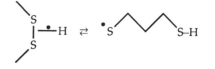

significant proportion of the radicals of the dithiothreitol system exist in the non-cyclic thiyl form[9].

The ability to form cyclic structures also leads to enhanced stability in disulphide anions. Thus the lipoate and dithiothreitol anion radicals at high pH appear to be stable to dissociation (i.e., ring opening), while the linear disulphide anions of the penicillamine system and other monothiols decomposed with typical rate constants of 10^6 s^{-1} (Table 1).

Typical spectra of $RS^\bullet$, $RS\bullet^\bullet SR^-$, and cyclic protonated forms of $RS\bullet^\bullet SR^-$ are shown in Figure 4. There is some evidence for small shifts in the wavelength of the absorption maxima and the absorbance coefficients with structure.

REDUCTION POTENTIALS

Reactions in which RSH molecules donate hydrogen atoms or thiolate (RS$^-$) ions donate electrons to free radicals constitute repair processes, which are believed to be involved in reducing the radiation sensitivity of living organisms. In some cases it has proven to be possible to set up an equilibrium between the sulphur donors and acceptor radicals, $A^\bullet$. Under those circumstances, if the reduction potential of the acceptor radical is known, then the reduction potential of the thiyl radical can be calculated from the standard free energy change of reaction, ΔG^0, calculated from the equilibrium constant. Thus for H atom transfer one has:

$$RS^\bullet + H^+ + e^- \rightleftarrows RSH \tag{18}$$

$$A^\bullet + H^+ + e^- \rightleftarrows AH \tag{19}$$

where equations (18) and (19) are the half cell reactions defining $E^0(RS^\bullet/RSH)$ and $E^0(A^\bullet/AH)$ respectively. The equilibrium (20)

$$RSH + A^\bullet \rightleftarrows RS^\bullet + AH \tag{(20)/(-20)}$$

must be reached sufficiently rapidly that other reactions of the radicals, including combinations and disproportionations, are negligible.

The course of the reaction is followed by measuring the change in absorbance due to one or more of the reacting species or products. The pseudo-first order rate constant for the approach to equilibrium with [A] small relative to [RSH] and [AH] is given by equation (21):

$$k_{obs} = k_{20}[RSH] + k_{-20}[AH] \tag{21}$$

The magnitude of k_{20} can be obtained from the slope of a plot of this first order rate constant vs. [RSH]. The magnitude of k_{-20} can be obtained from the intercept divided by the concentration of AH. The ratio of these two rate constants gives a value of the equilibrium constant. This constant can of course also be obtained from the equilibrium concentrations observed under different conditions.

The procedure described above is in fact a standard method for the determination of reduction potentials for transient radicals. One of the first applications to sulphur-centered species was made when Bonifacic and Asmus[10] determined $E^0(RSSR^{\bullet+}/RSSR)$ for several R species by equilibrations with $(SCN)_2^{\bullet-}$ and $I_2^{\bullet-}$. Their results are summarised in Table 2 along with data for $E^0(RS^\bullet, H^+/RSH)$, $E(RS^\bullet/RS^-)$ and the potentials of disulphide species. Values in parenthesis were calculated from gas phase bond dissociation energies and other data by a procedure to be explained in the second article of D. A. Armstrong in this book.

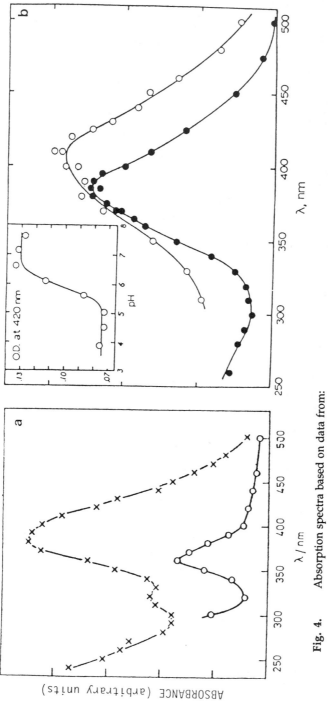

Fig. 4. Absorption spectra based on data from:

(a) ref. 9: RS$^{\bullet}$ from β-mercaptoethanol (o), and the protonated ring closed anion of dithiothreitol (x);

(b) ref. 7a: Deprotonated (o) and protonated (●) forms of the disulphide anion of lipoic acid. The ε_{max} values are, respectively, about 160, 430, 9000 and 6000 M^{-1} cm^{-1}.

Table 2. Summary of Reduction Potentials

Reaction	E^0/V
$CH_3SSCH_3^{\bullet+} + e^- \rightleftarrows CH_3SSCH_3$	1.39[a]
$HOOCLip(SS)^{\bullet+} + e^- \rightleftarrows HOOCLip(SS)$	1.13[a]
$^-O_2CLip(SS)^{\bullet+} + e^- \rightleftarrows {}^-O_2CLip(SS)$	1.10[a]
$RSSR + 2H^+ + 2e^- \rightleftarrows 2RSH$	0.08
$RSSR + e^- \rightleftarrows RS.^\bullet.SR^-$	−1.57 (−1.64)
$RS.^\bullet.SR^- + e^- \rightleftarrows 2RS^-$	0.57 (0.65)
$RS.^\bullet.SR^- + e^- + 2H^+ \rightleftarrows 2RSH$	1.72 (1.80)
$RS^\bullet + e^- \rightleftarrows RS^-$	0.77 (0.84)
$RS^\bullet + e^- + H^+ \rightleftarrows RSH$	1.34 (1.42)

[a] from ref. 10, all others from ref. 11a and 11b

( ) = calculated, ref. 11a. Standard states are 1 mole dm^{-3}.

Two standards were used with the RS$^\bullet$/RSH system: chloropromazine and phenol with E^0 values of 0.83 and 1.34 V, respectively.[11a] At pH $\geq$ 7 the RS$^\bullet$ radicals formed in reaction (22)

$$RSH + PhO^\bullet \rightleftarrows RS^\bullet + PhOH \tag{22}$$

also participated in the equilibrium (23)

$$RSH + RS^\bullet \rightleftarrows RS.^\bullet.SR^- + H^+ \tag{23}$$

Thus the observed overall equilibrium is:

$$2RSH + PhO^\bullet \rightleftarrows RS.^\bullet.SR^- + H^+ + PhOH \tag{24}$$

from which the E^0 found is for the half cell reaction:

$$RS.^\bullet.SR^- + e^- + 2H^+ \rightleftarrows 2RSH \tag{25}$$

The value of $E^0(RS^\bullet, H^+/2RSH)$ is found from this and the known value of ΔG^0 of (23).

The equilibria between the phenoxyl radical and both dithiothreitol, $D(SH)_2$, and lipoamide, $L(SH)_2$, were also studied, thus giving E^0 for the reactions:

$$LS_2^{\bullet-} + e^- + 2H^+ \rightleftarrows L(SH)_2$$

$$DS_2^{\bullet-} + e^- + 2H^+ \rightleftarrows D(SH)_2$$

These were similar to E^0 for reaction (25). The reduction potential of the disulphide, i.e., E^0 for:

$$RSSR + e^- \rightleftarrows RS.^\bullet.SR^-$$

in Table 2 was calculated from the difference between $E^0(RSSR, 2H^+/2RSH)$ and $E^0(RS\cdot\overset{\cdot}{\cdot}SR^-, 2H^+/2RSH)$.

Reaction (20) is the fundamental equation of free radical repair and it is important to examine the effect of R group structure on $E^0(RS^\bullet, H^+/RSH)$, since this potential determines the position of the equilibrium. It was shown that with aliphatic R groups, when electron-donating constituents such as CO_2^- or methyl groups were placed on carbon attached to the sulphydryl group, the reduction potential was decreased.[11b] Conversely, electron with-drawing groups tended to enhance it. However, in both cases the changes were small and within the overall estimated error for the potentials in Table 2, i.e., ±0.05 V.

The reliability of $E^0(RS^\bullet, H^+/RSH)$ can be assessed by making further equilibrium measurements and comparisons with other standards. A study of equilibrium (26)

$$RSH + {}^\bullet CO_2^- \rightleftarrows RS^\bullet + HCO_2^- \qquad (26)$$

has led to $E^0({}^\bullet CO_2^-, H^+/HCO_2^-) = 1.49$ V.[11c] Combination of this with the well established two electron reduction potential $E^0(CO_2^-, H^+/HCO_2^-)$ gives $E^0(CO_2/{}^\bullet CO_2^-) = 1.85$ V, in good agreement with Schwarz and Dodson's measurement of 1.90 V based on $E^0(Tl^+/Tl_{aq})$ as standard.[12] The utility of these potentials is discussed in the second paper of D. A. Armstrong in this book.

REFERENCES

1. I.G. Draganic and Z.D. Draganic, The Radiation Chemistry of Water, Academic Press, New York (1971).
2 a. Zhennan Wu, T.G. Back, R. Ahmad, R. Yamdagni, and D.A. Armstrong, *J. Phys. Chem.* 86:4417 (1982).
   b. T.J. Burkey, J.A. Hawari, F.P. Lossing, J. Lusztyk, R. Sutcliffe, and D. Griller, *J. Organic Chem.* 50:4966 (1985).
   c. G.H. Morine and R.R. Kuntz, *Photochem. and Photobiol.* 33:1 (1981).
3 a. H. Möckel, M. Bonifacic, and K.-D. Asmus, *J. Phys. Chem.* 78:282 (1974).
   b. M. Bonifacic, K. Schäfer, H. Möckel, and K.-D. Asmus, *J. Phys. Chem.* 79:1496 (1975).
   c. M. Bonifacic and K.-D. Asmus, *J. Phys. Chem.* 80:2426 (1976).
4. A.J. Elliot, R.J. McEachern, and D.A. Armstrong, *J. Phys. Chem.* 85:68 (1981).
5 a. M. Bonifacic, H. Möckel, D. Bahnemann, and K.-D. Asmus, *J. Chem. Soc. Perkin Trans.II* 675 (1975).
   b. K.-D. Asmus, *Acc. Chem. Research* 12:436 (1979).
   c. M. Bonifacic, J. Weiss, S.A. Chaudhri, and K.-D. Asmus, *J. Phys. Chem.* 89:3910 (1985).
6. W. Karmann, A. Ganzow, G. Meissner, and A. Henglein, *Int. J. Radiat. Phys. Chem.* 1:395 (1969).
7 a. M.Z. Hoffman and E. Hayon, *J. Am. Chem. Soc.* 94:7950 (1972).
   b. G.G. Jayson, D.A. Sterling, and A.J. Swallow, *Int. J. Radiat. Biol.* 19:143 (1971).
   c. J.W. Purdie, H.A. Gillis, and N.A. Klassen, *Can. J. Chem.* 51:3132 (1973).
8 a. J.E. Packer, *in:* "The Chemistry of the Thiol Group", Part 2, S. Patai, ed., Wiley, London, Chapter 11 (1974).
   b. C. von Sonntag and H.P. Schuchmann, *in:* "Chem. Ethers, Crown Ethers, Hydroxyl Groups and Their Sulphur Analogues", S. Patai, ed., Wiley, Chichester, England, Vol. 2 (1980).
9. M.S. Akhlaq, C. von Sonntag, *Z. Naturforsch. C.: Biosci.* 42:134 (1987).
10. M. Bonifacic and K.-D. Asmus, *J. Chem. Soc. Perkin Trans. II* 1805 (1986).

11a. P.S. Surdhar and D.A. Armstrong, *J. Phys. Chem.* 90:5915 (1986).

  b. P.S. Surdhar and D.A. Armstrong, *J. Phys. Chem.* 91:6532 (1986).

  c. P.S. Surdhar, S.P. Mezyk and D.A. Armstrong, *J. Phys. Chem.* 93:3360 (1989).

12. H.A. Schwarz and R.W. Dodson, *J. Phys. Chem.* 93:409 (1989).

STRUCTURE AND REACTION MECHANISMS IN SULPHUR-RADICAL

CHEMISTRY REVEALED BY E.S.R. SPECTROSCOPY

Bruce C. Gilbert

Department of Chemistry
University of York
Heslington, York YO1 5DD, U.K.

Free radicals derived from sulphur-containing compounds are believed to participate in a wide variety of reaction types, including those of synthetic or industrial importance, and they are increasingly implicated in a number of important biological processes. Although their mediation in chemical reactions has long been invoked, direct proof of their involvement - and their full characterization - has largely been the result of the relatively recent application of a number of spectroscopic techniques to the study of a variety of reaction systems.

This review will be particularly concerned with the use of e.s.r. spectroscopy to provide correct identification, as a basis for full structural study, of a range of sulphur-containing radical species, with special emphasis on sulphur-centred radicals [ranging from thiyl and related radicals (RS$^\bullet$) through oxygen-containing analogues RS(O)$_n{}^\bullet$ to sulphuranyl radicals (e.g. (RO)$_3$S$^\bullet$) and radical-ions]: it will be the intention to illustrate in each case how studies of g-values, hyperfine splittings, and line-widths provide crucial diagnostic information about spin density on sulphur and orbital occupancy (as indicated by g-values, for example) and conformational details and also how in some cases e.s.r. results complement information from other spectroscopic studies. Approaches to the generation and detection of many of the very short-lived species to be described have involved the use of fast-reaction techniques (e.g. radiolysis, photolysis, flow systems) which will also be discussed. Finally, examples will be discussed in which e.s.r. spectroscopy has been employed to reveal relevant details of reactions involving sulphur-conjugated radicals in which the steric and electronic character of the substituents [e.g. S(O)$_n$R, n = 0 - 2] plays an important role.

DETECTION AND CHARACTERIZATION OF SULPHUR-CENTERED RADICALS

Thiyl (RS$^\bullet$) and Related Species

Despite the continued and widespread interest in the use of e.s.r. spectroscopy for detection of sulphur-centred radicals, attention should be drawn to the particular problems which have been encountered in providing correct structural identification (and readily illustrated with reference to the problems concerning recognition of thiyl, RS$^\bullet$, species). Thus in sulphur-centered radicals, unlike nitrogen-centred species such as aminyls and nitroxides, splittings associated with the central atom (in principle observable for the isotope ^{33}S) are generally not available (or easily interpretable). Further, line-widths are often very large, with

Sulfur-Centered Reactive Intermediates in Chemistry and Biology, Edited by
C. Chatgilialoglu and K.-D. Asmus, Plenum Press, New York, 1990

considerable anisotropy (i.e. variation of position of absorption with orientation of the radical in the applied magnetic field) as is certainly the case for alkylthiyl radicals (which for this reason cannot be detected in fluid solution). In addition, proton splittings are often small or non-existent (in contrast with most carbon-centered species) and, until quite recently, the effect of changes in hybridization at sulphur on the g-value, splittings, and line-widths were not well understood.

Alkylthiyl radicals ($RS^{\bullet}$) appear to be characterized by a yellow colour, λ_{max} 400 nm, and - detectable by e.s.r. only in solid solution, following irradiation of photolysis of thiols or disulphides - an e.s.r. spectrum with, typically, an axial g-tensor ($g_{||} \approx 2.3$, $g_{\perp} \approx 2.0$) and splittings from hydrogen atoms on the carbon adjacent to sulphur [see e.g. 1].[1] The spectrum of rigid, randomly oriented $RS^{\bullet}$, as expected for a species in which the unpaired electron is localized on an atom with high spin-orbit coupling parameter (and which also has lone pairs of electrons), is found to be very broad, and at low concentrations would be expected to give only a weak asymmetric resonance (for the solid state) near $g = 2$:[2] however, narrowed parallel features (g ca. 2.16) have been detected in hydrogen-bonding solvents (RSH, ROH) and g_z values up to ca. 2.3 can be found in other matrices.

$$HO_2C-\overset{\overset{\displaystyle H}{|}}{\underset{\underset{\displaystyle ^+NH_3}{|}}{C}}-CH_2S^{\bullet}$$

a (β-H) 3.8, 1.4 mT

$g_{||}$ 2.29 $g_{\perp}$ 1.99

1

On the other hand, there is considerable controversy over the correct identification of other signals, detected in irradiated thiols[2] and disulphides,[1b,2,3] which possess an anisotropic g-tensor characterized by principal values near 2.00, 2.025 and 2.06 (with, in some cases, proton splittings). The spectra have been attributed both to adduct radicals $RS^{\bullet}SR_2$ (see e.g. ref. 1a, 4) and, in contrast,[1b,5] to perthiyl radicals $RS_2^{\bullet}$; mechanistic considerations appear to favour the former structure (though definite and unambiguous assignment does not appear to be possible at this stage). On the other hand, an e.s.r. spectrum attributed to $Bu^tSS^{\bullet}$ (obtained by photolysis of the thiosulphenyl chloride ^tBuSSCl) has been detected[6] in fluid solution (in methylbenzene at -86°C): this species has g_{iso} 2.025 (with line-width ΔH 0.4 mT) and decays via second-order kinetics with $2k_t = 2 \times 10^8$ dm^3 mol^{-1} s^{-1}. Photolysis of a sample of t-butylthiosuphenyl chloride in benzene at -176°C gives an anisotropic spectrum with g-values 2.059, 2.026, 2.001 (g_{av} 2.029), in reasonable agreement with those described above.

Indirect detection of alkylthiyl radicals may of course be accomplished via their addition to unsaturated compounds such as alkenes (e.g. maleic acid) and other spin-traps including, for example, the aci-anion of nitromethane[7] [a technique often used in flow-systems, photochemical reactions, or studies of biochemical systems; see reactions (1) and (2) and Figure 1]: (more examples will be discussed later).

$$HO^{\bullet} + HOCH_2CH_2SH \quad \rightarrow \quad H_2O + HOCH_2CH_2S^{\bullet} \qquad (1)$$

$$HOCH_2CH_2S^{\bullet} + {}^-CH_2NO_2 \quad \rightarrow \quad HOCH_2CH_2SCH_2NO_2^{\bullet-} \qquad (2)$$

On the other hand, some thiyl-substituted radicals in which the unpaired electron is delocalized away from the single sulphur generally have reduced g-values and isotropic spectra sharp enough to allow detection in fluid solution. These include, for example, $Et_2NS^{\bullet}$ **(2)**,[8] some dithiocarbamate radicals $R_2NC(S)S^{\bullet}$ (with g 2.015, line-width ca. 1 mT, and values of the enthalpy $\Delta H^{\ominus}$ and enthalpy of activation $\Delta H^{\ddagger}$ for formation from the corresponding thiuram disulphide of ca. 110 kJ mol^{-1}, with $2k_t$ 2×10^9 dm^3 mol^{-1} s^{-1})[9] and the analogous dialkoxy(thiophosphoryl)thio radicals $(RO)_2P(S)S^{\bullet}$ formed from photolysis of dimers or by oxidation of the corresponding thiol [e.g. **3**, see Figure 2][10]: for the latter, the

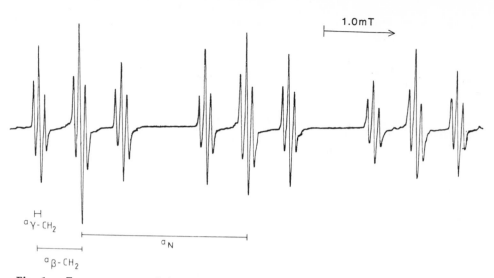

Fig. 1. E.s.r. spectrum of the radical anion $HOCH_2CH_2SCH_2NO_2^{\bullet-}$, obtained in the reaction of $^{\bullet}OH$ (from Ti^{III}/H_2O_2) and $HOCH_2CH_2SH$ in the presence of the aci-anion of nitromethane as a spin-trap (in a flow-system experiment).

unpaired electron is evidently shared between the two sulphur atoms in an in-plane orbital (and the increase in line-width with temperature reveals the occurrence of spin-rotation interaction).

$(CH_3CH_2)_2N{-}S^{\bullet}$

$a(N)$ 1.07, $a(4H)$ 0.61 mT
g 2.0156

2

$a(P)$ 2.460 mT
g_{iso} 2.0188 (2.0025, 2.0147, 2.037)

3

Sulphinyl Radicals (RSO•)

The difficulties involved in the correct identification of sulphur-centred free radicals, and hence the establishment of appropriate mechanistic conclusions, are illustrated by the reports of radicals with proton hyperfine splittings (up to *ca.* 1.0 mT) and g *ca.* 2.01 formed in the oxidation of thiols with Ce^{4+} in continuous-flow systems. Though these were original-ly assigned to the thiyl species themselves, it is now known, as a result of detailed structural and mechanistic studies, that these parameters characterize alkanesulphinyl radicals (RSO•, i.e. peroxy-radical analogues) whose properties have recently been reviewed.[11] Ta-ble 1 illustrates how their e.s.r. parameters compare with a range of other sulphur radicals.

The methanesulphinyl radical itself (MeSO•) has been unambiguously identified by e.s.r. spectroscopy following γ-irradiation of a single crystal of dimethyl sulphoxide;[12] from the hyperfine splittings [see 4] it can be concluded[11,12] that the unpaired electron lies mainly in a 3p-orbital on sulphur [72%, with small extent (0.65%) of 3s character]. Studies of the temperature dependence of the spectra suggest that there is a barrier of *ca.* 10 kJ mol⁻¹ for

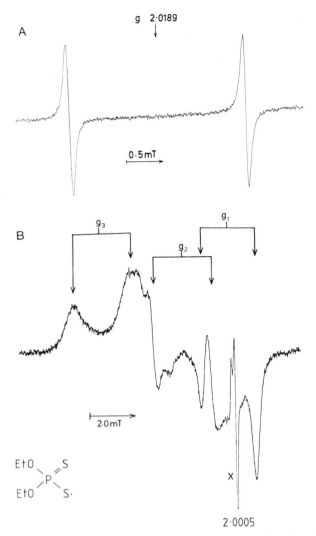

g 2·0189

A

0·5 mT

B

g_3

g_2

g_1

2·0 mT

$$\begin{array}{c} EtO \\ \diagdown \\ \diagup \\ EtO \end{array} P \begin{array}{c} \diagup S \\ \diagdown \\ S\cdot \end{array}$$

X

2·0005

Fig. 2. E.s.r. spectra of dialkoxy(thiophosphoryl)thio radicals showing hyperfine splitting by ^{34}P. (a) Isotropic spectrum of $(Pr^iO)_2P(S)S^{\bullet}$ in isopentane at 207 K (obtained in *in situ* photolysis of the corresponding disulphide). (b) Anisotropic spectrum of $(EtO)_2P(S)S^{\bullet}$ in a U.V.-irradiated polycrystalline sample of $(EtO)_2P(S)SH$ at 140 K. (X is a paramagnetic impurity in the e.s.r. cell).

rotation about C–S, with a preferred conformation [e.g. 5] in which one proton lies in the nodal plane of the p-orbital on sulphur. Analogous sulphinyl radicals have been generated in the liquid phase by the oxidation of thiols with $^{\bullet}$OH and with Ce^{4+},[7] by oxidation of Bu^tSOH with $^tBuO^{\bullet}$ at low temperatures[13] ($^tBuSO^{\bullet}$, shorter-lived than the isostructural $^tBuO_2^{\bullet}$ has g 2.0106, ΔH 0.19 mT), and by photolysis of sulphide chlorides.[14] These studies also establish that alkanesulphinyl radicals are characterized by a fairly large line-width (cf. $RO_2^{\bullet}$) and β-proton splitting [cf. 4]; the isotropic (solution) data for $MeSO^{\bullet}$ [g 2.010, a(3H) 1.15 mT at -113°C] are in excellent agreement with those for other radicals generated in rigid samples, but the radical shows marked spin-rotation broadening (cf. $RO_2^{\bullet}$) and is

Table 1. Isotropic E.s.r. Parameters for some Typical Aliphatic Sulphur-Centred Radicals

Type	Example	a(H) /mT		g
Sulphinyl	EtSO$^\bullet$	0.91	(2H)	2.0110
Alkoxysulphinyl	EtOSO$^\bullet$	0.040	(2H)	2.0049
Sulphonyl	MeSO$_2{}^\bullet$	0.095	(3H)	2.0049
'Dimer' radical-cation	Me$_2$S$\bullet^\bullet$·SMe$_2{}^+$	0.68	(12H)	2.0103
Disulphide radical-anion	$\begin{array}{c}\text{S}\,\overset{-}{\cdot\cdot}\,\text{S}\\/\qquad\backslash\\\text{CH}_2\qquad\text{CH}_2\\\qquad\text{CH}_2\end{array}$	0.68 0.10	(4H) (2H)	2.0132
Sulphuranyl	Me$\overset{\bullet}{\text{S}}(OEt)_2$	0.070 0.24	(3H) (4H)	2.0096
	$^\bullet$S(OMe)$_3$	0.17	(6H)	2.0068

undetectable in aqueous solution at room temperature. Some aliphatic sulphinyl radicals (e.g. PrSO$^\bullet$) show characteristic line-width alternation in the β-proton hyperfine splitting pattern which reveals[14] the occurrence of hindered rotation involving exchange (at an intermediate rate) between conformers 6 and 7 (confirmed by a recent solid-state study[15] in which analogous radicals were generated by irradiation of thiols in the presence of oxygen).

For aromatic analogues (see e.g. 8 and Figure 3), the hyperfine splittings from the ring protons confirm that these are delocalized p(π)-type species, with significant spin-density at the ortho- and para-positions of the ring.[14] The relative stability of sulphinyl species is

Me–$\overset{\bullet}{\text{S}}$=O

a (3H) 1.16 mT

A(^{33}S) −1.4, −2.1, +5.9

g 2.023, 2.011, 2.003

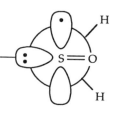

4

5

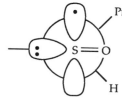

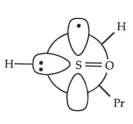

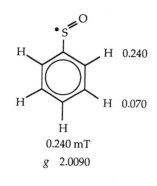

H 0.240

H 0.070

H
0.240 mT
g 2.0090

6

7

8

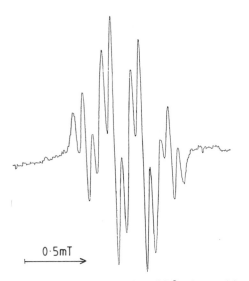

0·5mT

Fig. 3. E.s.r. spectrum of the sulphinyl radical $C_6H_5SO^\bullet$, obtained by *in situ* photolysis of benzenesulphinyl chloride at 200 K in diethyl ether.

apparently reflected in the ease of fragmentation, monitored by e.s.r., of radicals of the type $^\bullet CH_2CH_2S(O)R$ [cf. also $^\bullet CH_2CH_2S(O)_2R$], the lack of reactivity of $RSO^\bullet$ towards alkenes and spin-traps such as $CH_2=NO_2^-$ (contrast $RS^\bullet$, see above, and $RSO_2^\bullet$) and their ease of formation in systems in which sulphonyl radicals are generated (see later).[16] On the other hand, it should be noted that the rate-constants for self-reaction of sulphinyl radicals approach the diffusion-controlled limit; a value of $6 \times 10^7 \, dm^3 \, mol^{-1} \, s^{-1}$ at 173 K has been reported[13] for $^tBuSO^\bullet$ and a number of aromatic analogues have $2k_t$ *ca.* $5 \times 10^8 \, dm^3 \, mol^{-1} \, s^{-1}$ at 203 K.[17] The product from these reactions is generally believed[11] to be the corresponding thiolsulphinate (RSO_2SR).

Although space does not allow a detailed mechanistic treatment of all the reactions in which $RSO^\bullet$ and $ArSO^\bullet$ are formed, Schemes 1 and 2 outline how flow-system studies have revealed the ways in which attack of oxygen-centred radicals (e.g. $^\bullet OH$) at sulphur and further oxidation of thiyl radicals (as also with high-valent transition metals such as Ce^{4+}) give rise to sulphinyl (and, see later, sulphonyl) radicals. E.s.r. spectroscopy has also been employed[18] to characterize radical pathways in the thermal and photolytic decomposition of diaryl sulphoxides and S-aryl arenethiolsulphonates ($ArSO_2SAr$) (see Scheme 3). For the former, S–C cleavage leads directly to formation of $ArSO^\bullet$ (whose subsequent rapid reaction evidently involves, at least in part, disproportionation to $ArSO_2^\bullet$ and $ArS^\bullet$ via S–O coupling); for the latter substrate, this same pair of radicals, formed directly by homolysis, can evidently also revert to $ArSO^\bullet$ (again via S–O coupling). Direct and indirect (e.s.r.) evidence has been presented for the formation of α-disulphoxides $RS(O)S(RO)R'$ in the peroxidation of S-alkyl alkanethiolsulphinates [$RS(O)SR'$][15] (see also Scheme 3).

Spectroscopic studies (e.g. with microwave techniques) and calculations on the parent sulphinyl radical $HSO^\bullet$, summarized by Chatgilialoglu,[17] complement nicely the e.s.r. results described above. Other sulphinyl-type species detected by e.s.r. spectroscopy include a series of amino derivatives[20] [e.g. **9**, which has $2k_t$ $1.1 \times 10^9 \, dm^3 \, mol^{-1} \, s^{-1}$ at 163 K]; the analogous alkoxysulphinyl species [e.g. **10**] can be generated by the photolysis of the corresponding chlorosulphites $ROS(O)Cl$ and by the fragmentation of radicals $^\bullet CR^1R^2OS(O)OCHR^1R^2$ (derived by hydrogen-abstraction from the appropriate organic sulphites).[21] Photolysis of

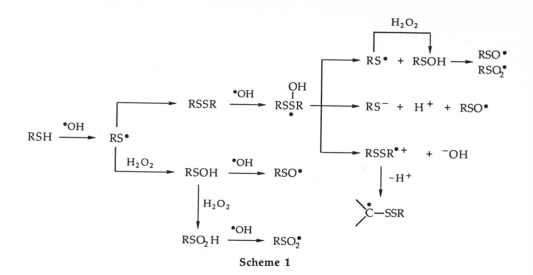

Scheme 1

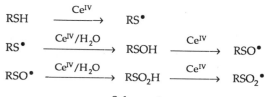

Scheme 2

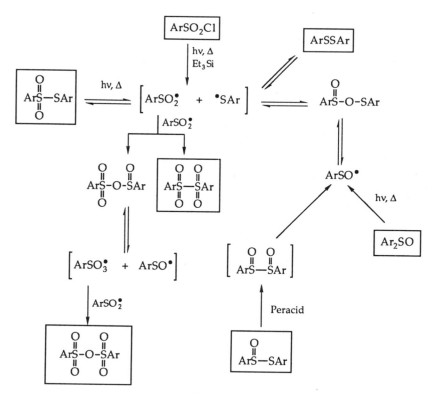

Scheme 3

dialkyl sulphites in the presence of di-t-butyl peroxide also gives rise to the detection of a signal (with g 2.0053) whose splitting varies with the nature of the solvent (and hence environment): evidence has been presented for the correct structural attribution to the hydroxysulphinyl radical **11**, formed from intermediate mono-alkyl sulphites $CHR^1R^2OS(O)OH$ formed *in situ*.

$$(CH_3CH_2)_2N\overset{\bullet}{S}O$$
$$a(N)\ 0.61,\ a(4H)\ 0.23\ mT$$
$$g\ 2.0060$$
9

$$CH_3CH_2O\overset{\bullet}{S}O$$
$$a(2H)\ 0.04\ mT$$
$$g\ 2.0049$$
10

$$HO\overset{\bullet}{S}O$$
$$a(H)\ 0.9\ mT$$
$$g\ 2.0053$$
11

Sevilla and his coworkers have recently reported[15c] the detection and characterization by e.s.r. of the cysteine thiyl *peroxyl* radical $(CysSOO^\bullet)$ formed by the reaction of cysteine thiyl radical $(CysS^\bullet)$, generated radiolytically, with dissolved molecular oxygen (and which is observed only when [RSH] is relatively low): $CysSOO^\bullet$ is violet (with λ_{max} 540 nm) and has an e.s.r. spectrum with (at 140 K) g_x 2.0027, g_y 2.0090 and g_z 2.035 (as well as ^{17}O splitting in a labelled sample, with A_x 7.8 mT and A'_x 6.2 mT at 160 K, the latter evidently being associated with the 'outer' oxygen). The thiylperoxyl radical $RSOO^\bullet$ is believed[15] to be the precursor of sulphinyl radicals generated in annealed irradiated solids from a variety of thiyl radicals in the presence of oxygen [via a mechanism which may involve reaction of R'SH with either $R'SO_2^\bullet$ or $RO_2^\bullet$, cf. reactions (3) and (4)]; it is also noteworthy that matrix photolysis (photobleaching) of $CysSOO^\bullet$ leads[15c] to the formation of the corresponding sulphonyl radical $CysSO_2^\bullet$ (with $g\ ca.\ 2.0053$).

$$R'S^\bullet + O_2 \quad \rightarrow \quad R'SO_2^\bullet \quad \overset{R'SH}{\rightarrow} \quad R'SO^\bullet + R'SOH \tag{3}$$

$$RO_2^\bullet + R'SH \quad \rightarrow \quad [ROOS(H)R']^\bullet \quad \rightarrow \quad R'SO^\bullet + ROH \tag{4}$$

Sulphonyl Radicals ($RSO_2^\bullet$)

During the oxidation of thiols and disulphides with $^\bullet OH$ or with high-valent transition-metal ions (e.g. Ce^{4+}) in continuous-flow e.s.r. studies, signals from $RSO^\bullet$ (g *ca.* 2.010) are often accompanied by signals from other sulphur-centred radicals with g *ca.* 2.005 and small hyperfine splittings (0.1 mT).[7] These are now understood to be the corresponding *sulphonyl* radicals, which may be generated in a variety of ways including direct photolysis of sulphonyl chlorides,[22] removal of the halogen from sulphonyl halides with, e.g., $Et_3Si^\bullet$, reaction of alkyl radicals with $SO_2^\bullet$, and hydrogen-atom abstraction from sulphenic acids (e.g. with $^tBuO^\bullet$, Ce^{4+},[23] or alkyl radicals[24]), these being formed *in situ*, for example, via addition of $^\bullet OH$ to sulphoxides[24] [see e.g. reactions (5) - (8)]. Mechanistic discussion will follow structural interpretations derived from e.s.r. spectra.

$$ArSO_2Cl \quad \rightarrow \quad ArSO_2^\bullet + Cl^\bullet \tag{5}$$

$$Et_3Si^\bullet + RSO_2Cl \quad \rightarrow \quad Et_3SiCl + RSO_2^\bullet \tag{6}$$

$$HO^\bullet + Me_2SO \quad \rightarrow \quad \overset{Me}{\underset{Me}{}}\overset{O}{S}\overset{}{OH} \quad \rightarrow \quad MeSO_2H + Me^\bullet \tag{7}$$

$$Me^\bullet + MeSO_2H \quad \rightarrow \quad CH_4 + MeSO_2^\bullet \tag{8}$$

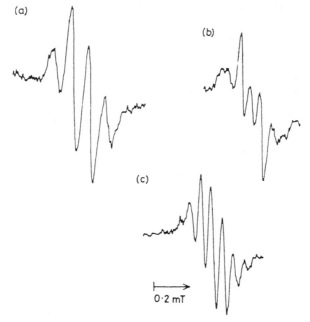

Fig. 4.   E.s.r. spectra from the pentamethylbenzenesulphonyl radical in methylbenzene,
showing evidence for restricted rotation about the C–S bond (see text) [(a) 193 K;
(b) 226 K; (c) 241 K].

The analysis of splittings and line-broadening (resulting from restricted rotation about
the C–S bond) have been used, in conjunction with INDO calculations, to establish that these
radicals are σ-species with the unpaired electron in an s-containing orbital located to
significant extent (ca. 50%) on the pyramidal sulphur atom[22a,d] (in agreement with solid
state data and with information on $HSO_2^\bullet$ and $FSO_2^\bullet$: see the review in ref. 25).   For
example, the largest splittings in $PhSO_2^\bullet$ [a(2H) 0.113 mT at ca. -80°] are derived from the
meta protons [with a(o-H) 0.033, a(p-H) 0.052 mT], which points to a structure (12) closely
resembling that of the (σ) benzoyl radical.[22a,d] Some mono-*ortho*-substituted radicals display
a marked preference for a conformation in which the oxygen atoms lie *anti* to the substituent
(and reveal that the *trans meta*-proton has the anticipated large splitting, see 13); for some
di-*ortho*-substituted radicals, barriers to rotation about C–S (14 ⇌ 15) have been determined
as ca. 22 kJ mol[-1].   Further evidence for such a structure derives from kinetic and u.v. spectroscopic studies of $ArSO_2^\bullet$, which indicate that there is little or no π-delocalization of the
unpaired electron into the ring (see ref. 22a, 25).

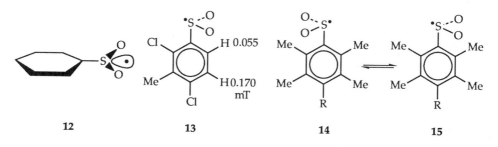

| 12 | 13 | 14 | 15 |

A conformational analysis has also been presented of a series of (pyramidal) aliphatic
sulphonyl radicals, for which it has been previously noted that a(γ-H) is typically greater

than $a(\beta\text{-H})$ [and $a(\delta\text{-H})$] (cf. $\text{PrSO}_2{}^{\bullet}$: a 0.070, 0.212, 0.070 mT, respectively for β, γ, δ protons at -40°C). This order, and the observed magnitudes, can be rationalized on the basis of INDO calculations and evidence from line-broadening resulting from restricted rotation about C_β–S and C_β–C_γ bonds (e.g. for $\text{CF}_3\text{CH}_2\text{SO}_2{}^{\bullet}$): the preferred conformation for $\text{RCH}_2\text{SO}_2{}^{\bullet}$ (16) and sample calculations indicating the angular dependence of $a(\beta\text{-H})$ (17) are shown. Spectra for some alkenesulphonyl radicals[22c] and some thiophen-2-sulphonyl radicals[25] may, in contrast to the aromatic analogues, indicate a preference for a π-type structure.

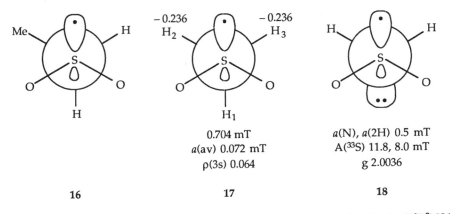

| | 16 | 17 | 18 |

Anisotropic and isotropic spectra of a series of aminosulphonyl radicals ($\text{R}^1\text{R}^2\text{NSO}_2{}^{\bullet}$) have been obtained[26] following irradiation and photolysis of some benzyl sulphenamides and by reaction of $\text{Et}_3\text{Si}^{\bullet}$ with aminosulphonyl chlorides $\text{R}^1\text{R}^2\text{NSO}_2\text{Cl}$. Analysis of the ^{33}S splittings indicates that these radicals are also essentially sulphur-centred, with over 50% of the unpaired electron on sulphur and pyramidal character at that atom (p/s ratio ca. 4.7: see e.g. 18), with a structure thus in-between $\text{RSO}_2{}^{\bullet}$ and $\text{SO}_3{}^{\bullet-}$; the g-values are considerably less than for the π-type species $\text{RSO}^{\bullet}$, $\text{R}_2\text{NS}^{\bullet}$, and $\text{R}_2\text{NSO}^{\bullet}$ (see above). Examples of the corresponding alkoxysulphonyl radicals $\text{ROSO}_2{}^{\bullet}$ (g 2.0032) have also been described.[25]

Kinetic-e.s.r. measurements have been employed[17] to establish that a variety of aromatic and aliphatic sulphonyl radicals undergo diffusion-controlled self-reaction in solution, with $2k_t$ typically ca. 10^9 dm^3 mol^{-1} s^{-1} at ca. 200 K (though the presence of two ortho substituents reduces this to ca. (2 - 4) x 10^8 dm^3 mol^{-1} s^{-1}). The observation of curved Arrhenius plots for some $\text{ArSO}_2{}^{\bullet}$, as well as the detection of sulphinyl radicals from sulphonyl precursors under steady-state conditions, is taken as evidence for the formation (and subsequent thermolysis) of an unstable intermediate sulphinyl sulphonate [$\text{RS(O)OS(O)}_2\text{R}$, ΔH ca. –20 kJ mol^{-1}] formed by S–O dimerisation (see Scheme 3: note that for $\text{ArSO}_2{}^{\bullet}$ it is estimated that ca. 50% of the unpaired electron density resides on sulphur, with each oxygen having ρ ca. 25%). The products reported from bimolecular reactions involving sulphonyl radicals (including disulphones, thiol-sulphonates and sulphonic acid anhydrides) have been interpreted[17] in terms of Scheme 3 and thermodynamic calculations based on the group additivity approach.

Attention is also drawn to mechanistic studies of the generation of $\text{RSO}_2{}^{\bullet}$ from $\text{RS}^{\bullet}$ (via $\text{RSO}^{\bullet}$) in reactions of thiols (and disulphides in some cases) with $^{\bullet}\text{OH}$ and Ce^{4+} (see Schemes 1 and 2), and to the findings that whilst the reaction of $\text{Me}^{\bullet}$ with SO_2 to give $\text{MeSO}_2{}^{\bullet}$ proceeds readily at room temperature,[24] stabilized radicals such as $^{\bullet}\text{CH}_2\text{CO}_2\text{H}$ (or benzyl[25]) do not undergo addition and that reaction between $^{\bullet}\text{CH}_2\text{OH}$ and SO_2 leads to electron-transfer (to give $^{+}\text{CH}_2\text{OH}$ and $\text{SO}_2{}^{\bullet-}$). The reader is referred to a recent review[25] on sulphonyl radicals for further discussion of C–S bond energies in $\text{RSO}_2{}^{\bullet}$, for kinetic details on the chlorine-atom abstraction reactions of alkyl and aryl radicals with RSO_2Cl, and for the participation of aromatic sulphonyl radicals in the copper-catalysed (CuCl) redox-transfer chain addition of $\text{ArSO}_2{}^{\bullet}$ to styrenes: other addition (and addition-elimination) reactions have been summarized.[25]

Sulphuranyl and Related Radicals

Radicals formed in the low-temperature photolysis of t-butyl methanesulphenate and of mixtures of, for example, dialkyl disulphides and di-t-butyl peroxide have g *ca.* 2.009, sharp lines, and smaller proton splittings than the corresponding RSO• radicals [e.g. *ca.* 0.6 mT from MeSOBu^t]. These are now believed to be correctly assigned to *sulphuranyl* radicals of the type **19** (nine-electron or "expanded octet" species), which are formed by addition of alkoxyl radicals (from the sulphenate or peroxides) to sulphenates (formed *in situ* from the reactions of thiols, disulphides or thioethers).[14] In an additional study[27a] detailed analysis of the e.s.r. spectra of a variety of examples reveal the occurrence of observable ring-proton splittings in C$_6$H$_5$S•(OEt)$_2$ (see **20**) and restricted rotation in CF$_3$S•(OBu^t)$_2$: these observations suggest that, at least for (**20**), the radical has a T-shaped structure with the unpaired electron in a sulphur orbital with considerable 3p-character (see also ref. 6b and 27b structure **21** for which ΔH values for S–S and S–O dimerization have been estimated[27b]). These assignments are based in part on the results of related earlier studies[28] involving the generation of trialkoxyl sulphuranyl radicals •S(OR)$_3$ by the photolysis of dialkyl sulph-oxylates[28]: for **22**, interaction with *two* groups of methyl-group protons is interpreted in terms of splittings from *apical* alkoxy groups in a quasi-trigonal bipyramidal radical. Labelling experiments establish that addition of •OMe to S(OMe)$_2$ is stereoselective, the incoming radical taking up an *apical* site.

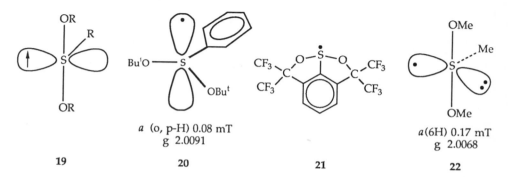

a (o, p-H) 0.08 mT
g 2.0091

a(6H) 0.17 mT
g 2.0068

19 **20** **21** **22**

Reactions of alkoxyl radicals (RO•) with R$_2$S probably involve the formation of a sulphuranyl radical ROS•R$_2$ and several analogues have been detected. For example, reaction of •OSiMe$_3$ with Me$_2$S gives[29] Me$_2$S•OSiMe$_3$, which has a(6H) 0.77 mT, g 2.0076, and CF$_3$S•SMe$_2$, from CF$_3$S• and Me$_2$S, has a(3F) 0.92, a(6H) 0.41 mT, g 2.0133 (related radicals include adducts from acyl radicals and thioethers). It is argued[30] that, in contrast to **20**, those radicals with *one* electronegative substituent possess a (σ^2 σ^{*1}) σ*-structure (see e.g. **23** and **24**) (akin to the 'three-electron bond' structures proposed for related species R$_2$S•∴•SR$_2^+$ and RS•∴•SR$^-$, and as described in the next section), rather than (π-type) sulphuranyl-structures.* The β-proton splittings from the S-alkyl groups probably arise mainly through a hyperconjugative mechanism and appear to depend on the spin density in thiol sulphur's 3p$_\alpha$-orbital (and the appropriate dihedral angle).

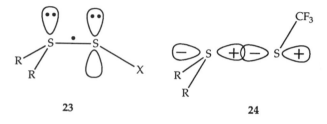

23 **24**

---

* The "•∴•" notation for 2σ/1σ* bonds has been introduced by K.-D. Asmus; see his article in this book.

The sulphuranyl radical **25** has been generated in low-temperature photolysis of *t*-butylperoxyl (2-methylsulphenyl)-benzoate in CH_2Cl_2 and may also be a σ*-species;[31] the sulphuranyl*oxyl* radical **26** has also been observed by e.s.r.[32]

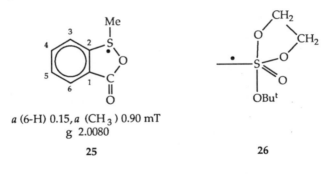

a (6-H) 0.15, *a* (CH$_3$) 0.90 mT
g 2.0080

25

26

Sulphur-Containing Radical Ions

Oxidation of some thioethers and β-hydroxysulphides with the hydroxyl radical provides evidence that these reactions proceed via electron abstraction from sulphur (probably via an •OH-adduct: see later) to give undetected (monomeric) radical-cations:[33] for example, oxidation of dimethyl sulphide at pH 1 gives an e.s.r. signal with splittings from twelve protons (0.68 mT), g 2.0103, assigned to **27**, formed from the intermediate radical-cation by reaction with another molecule of thioether [reaction (9)]; at higher pH •CH$_2$SMe is detected. Oxidation of, for example, 2-t-butylthioethanol gives, *inter alia*, signals from •CH$_2$SBu^t, •CH(CH$_2$OH)SBu^t, and in trapping experiments, •SBu^t and •SCH$_2$CH$_2$OH, all evidently derived from [Bu^{t}SCH$_2$CH$_2$OH]•$^+$.

$$Me_2S \xrightarrow[-e^-]{HO^\bullet} [Me_2S^{\bullet+}] \xrightarrow{Me_2S} (Me_2S^{\bullet}{\cdot}SMe_2)^+ \tag{9}$$

27

Radical **27** and its analogues, generated by a variety of le-oxidants, are generally described[34,35] as σ*-radicals (with the unpaired electron in a σ* orbital; *cf.* Cl$_2$•$^-$, RS•$^{\bullet}$•SR$^-$ and RS•$^{\bullet}$•SR$_2$ see earlier) as illustrated by structure **28** and the MO bonding scheme **29**, though other electronic structures have been suggested[36] (e.g. for some relatively long-lived intramolecular analogues formed from eight- or nine-membered cyclic dithioethers with e.g. NOPF$_6$ in MeCN, cf. **30**). Details of λ_{max} for σ → σ* absorptions and interpretation in terms of overlap between orbitals on the two sulphur atoms have been described.[34] Radicals detected in irradiated sulphides with, typically g 2.022, 2.011, 2.004 and with two 'sets' of β-proton splittings are also believed to be best assigned to the dimeric structure R$_2$S•$^{\bullet}$•SR$_2$+.[1a,2] Support for such an assignment comes from γ-irradiated N-acetyl-D,L-methionine [which has g 2.002, 2.013, 2.022 and A(^{33}S), from two *equivalent* sulphur atoms, 6.49, 1.84, 0.95 mT]: the orbital population on sulphur is calculated as 60% p, with a very small amount of s-character

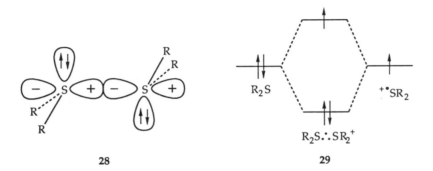

28

29

146

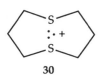

30

(suggesting a radical with *ca.* 50% of the unpaired electron in a p_z-type orbital on each sulphur). Further structural information is available from solution studies[36] which establish that the g-values are sensitive to R, the temperature, solvent and counter-ion, that $R_2S\cdot^\bullet\cdot SR_2^+$ (R=Pr^i) shows evidence for restricted rotation and that the ^{33}S splitting of *ca.* 3 mT is typical of structure **28**. Other *monomeric* radical-cations described include $But_2S^{\bullet+}$ [with g 2.0130, a(^{33}S) 3.25 mT at 200 K in CH_2Cl_2] and both $(Me_2N)_2S^{\bullet+}$ [g 2.0053, a(2N) 0.76, a(12H) 0.76 mT)] and $(Me_2N)_2SO^{\bullet+}$ [g 2.0037, a(2N) 1.30, a(12H) 1.24 mT].

Detailed pulse-radiolysis experiments have been described[38a] which lead to the distinction between hydroxyl adducts, radical-cations, and N–S and O–S bonded '3-electron' σ-radicals in the oxidation of methionine and some related molecules with $^\bullet OH$ and with metal ions. Features of particular interest include the ease of decarboxylation of methionine and its analogues (via prior oxidation at sulphur, see e.g. Scheme 4), the relative weakness of the S–O 3e-bond compared with the corresponding S–N (and S–S) 3e-bonds (as expected from the difference in electronegativities and in energy levels of the heteroatoms) and the relative ease of formation of such species if a five-membered ring can be achieved. In these terms some relevant e.s.r. observations may be rationalized.[38b]

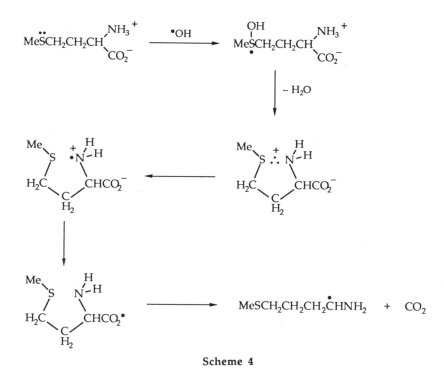

Scheme 4

Disulphide radical-anions are believed to possess the σ*-structure appropriate to the analogous 'dimer' radical-cations $R_2S\cdot^\bullet\cdot SR_2^+$ and 'adducts' $RS\cdot^\bullet\cdot SR_2$ [$MeS\cdot^\bullet\cdot SMe^-$ has $g_{||} \approx$ 2.020, $g_\perp \approx$ 2.0020, a(H) 0.5 mT][35]: these species may be formed by electron capture by disulphides in the solid or liquid phase (see e.g. **31** from lipoate ion) and by rapid cyclization[7] of a thiyl radical onto a thiolate. In contrast disulphide radical-cations, detected in the solid state, have, typically, g 2.032, 2.019, 2.004,[40] and examples derived from 1,2 dithiacycloalkanes (e.g. 32) have been detected in fluid solution.[41] They are, however, not σ* but π* radicals. The reader is referred elsewhere to the discussion of e.s.r. spectra[42] and assignment[34c,42] of other σ*-radical structures to electron-loss/halide adducts [e.g. $R_2S \cdot^\bullet \cdot Br$] and to evidence[42] from irradiated sulphoxides for the formation of monomer (33) and dimer (34) radical-cations.

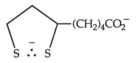

$a(\beta$-H) 0.78 (1), 0.435 (2) mT
$a(\gamma$-H) 0.145 (2) mT
g 2.0129

31

a(4H) 0.095 mT
a (^{33}S) 1.33 mT
g 2.0183

32

$Me_2SO^{\bullet+}$

a(6H) 1.2 mT
g 2.008, 2.008, 2.002

33

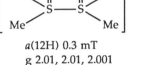

a(12H) 0.3 mT
g 2.01, 2.01, 2.001

34

MECHANISTIC AND STRUCTURAL INFORMATION ON SULPHUR-SUBSTITUTED RADICALS

In this final section, attention is focussed on the ways in which e.s.r. spectroscopy can provide insight and detailed information on the structure of sulphur-containing (rather than sulphur-centred) radicals, and how the steric, electronic, and polar factors which govern reactivity may be probed.

Structural Features in Sulphur-conjugated Radicals

The isotropic e.s.r. parameters given in Table 2 for some aliphatic sulphur-conjugated radicals $^\bullet C$–$S(O)_n$ establish[16] quite clearly that thiol-, sulphenyl-, and sulphonyl-substituted radicals are planar at the radical centre [with $a(\beta$-H) : $a(\alpha$-H) ca. 1.2 for radicals $^\bullet CHMeS(O)_n R$: contrast oxygen-substituted analogues, cf. $^\bullet CHMeOH$]. Use of the β-splitting in α-methyl-substituted radicals (in conjunction with Fischer's method, cf. ref. 16) shows, for example, that the ethylthio group in $^\bullet CHMeSEt$ removes ca. 22% of the spin from the adjacent tervalent carbon atom (cf. ca. 16% for a carbonyl group), which is of course reflected in the magnitude of the g-value and the substantial γ-proton splitting (reflecting the contribution to the resonance hybrid of the structure 36): it has also been noted that $a(\alpha$-H)

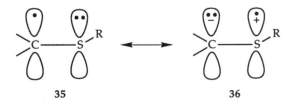

35   **36**

Table 2. E.s.r. Parameters for Some Sulphur-Conjugated Radicals

	$a(\alpha\text{-H})/mT$	$a(\beta\text{-H})$	$a(\gamma\text{-H})$	g
$^\bullet CH_3$	2.29			2.0025
$^\bullet CH_2Me$	2.22	2.71		2.0025
$^\bullet CH_2SMe$	1.65		0.36	2.0049
$^\bullet CHMeSEt$	1.70	2.10	0.15	2.0044
$^\bullet CH_2S(O)Me$	2.00			2.0025
$^\bullet CHMeS(O)Et$	2.02	2.53		2.0025
$^\bullet CH_2S(O)_2Me$	2.23		0.21	2.0025
$^\bullet CHMeS(O)_2Et$	2.16	2.73	0.21	2.0025
$^\bullet CHMeSO_3^-$	2.17	2.59		2.0025

is larger for radicals of the type $^\bullet CHMeSR$ than $^\bullet CH_2SR$, and it appears that the α-methyl group may serve to increase the compressional forces in the coplanar conformation of the $^\bullet C-S-R$ fragment, thereby reducing the effectiveness of the conjugation between the tervalent carbon and the sulphur (and also hence reducing the g-value). The delocalization represented below, also manifest via restricted rotation, clearly corresponds to a considerable radical stabilization (of ca. 25 kJ mol^{-1})[44] and weakening of the corresponding C–H bond, which is widely reflected in the reactivity shown in thio-ethers and other related compounds.[45]

In contrast, in the radical $^\bullet CHMeS(O)Et$ the sulphinyl substituent removes only ca. 6% of the unpaired electron from the tervalent carbon atom and both the sulphonyl substituent in $^\bullet CHMeS(O)_2Et$ and the sulphonate substituent in $^\bullet CHMeSO_3^-$ are totally ineffective at removing spin density (n.b. the γ-proton splitting observed for $^\bullet CHMeS(O)_2Et$ and some analogues is attributed to the operation of spin-polarization, rather than hyperconjugation from spin on sulphur). Corresponding conclusions have been reached for a series of *para*-substituted phenoxy radicals:[46] differences in *ring-proton* splittings are attributed, at least in part, to the consequences of the inductive effects of different substituents on electron distribution (e.g. between oxygen and the ring), and successfully modelled by HMO calculations in which h_{C-4} is varied [see the proton splittings in **37 - 41**].

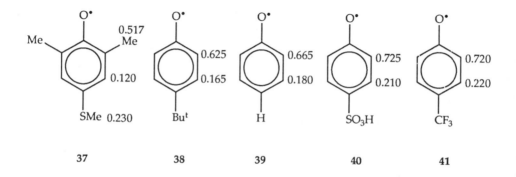

| 37 | 38 | 39 | 40 | 41 |

149

On the other hand, the observation of restricted rotation at low temperature [e.g. **42** and **43**] and a full analysis of $a(H)$ (and its temperature dependence) confirm the considerable extent of delocalization for thioether (but not sulphinyl or sulphonyl substituents):[46] for the interconversion of **42** and **43**, E_a is estimated as 37 kJ mol^{-1}.

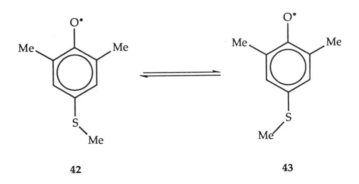

42   **43**

E.s.r. spectroscopy also provides further evidence for the rapid fragmentation of β-sulphur-substituted radicals, including the sulphinyl- and sulphonyl-substituted examples shown in reactions (10) and (11) (cf. evidence for the rapid, reversible addition of RS$^\bullet$ to alkenes[10]).

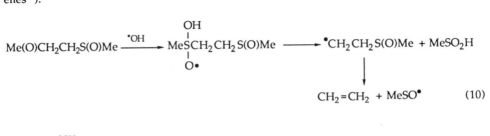

$$CH_2=CH_2 + MeSO^\bullet \qquad (10)$$

$$Pr_2SO_2 \xrightarrow{\ ^\bullet OH\ } Me\overset{\bullet}{C}HCH_2S(O_2)Pr \longrightarrow MeCH=CH_2 + PrSO_2^\bullet \qquad (11)$$

The failure of cyclic analogues of the latter (e.g. **44**) to undergo rapid fragmentation is perhaps best understandable in terms of their reluctance or inability to assume a conformation in which there is significant overlap between the half-filled orbital and the C–S bond to be broken.

44

Mechanistic Features

It is instructive briefly to compare the sites and rates of attack of oxygen and carbon-centred radicals at a variety of sulphur-containing species (see, e.g. ref. 16). For example, as we have seen above, the electrophilic oxygen-centred radicals $^\bullet OH$, $Bu^tO^\bullet$) usually attack at sulphur in thiols, disulphides, thioethers and sulphoxides, to give adducts (detectable in some cases) for which a variety of subsequent rapid reactions are possible, including fragmentation to give carbon - ($R^\bullet$, $^\bullet CH_2SR$) and sulphur-centred species ($RS^\bullet$, $RSO^\bullet$ etc.). In contrast, reaction of $^\bullet OH$ with sulphones proceeds via C–H abstraction at positions removed from the substituent (cf. RCO_2H); for example, reaction with Et_2SO_2 gives $^\bullet CH_2CH_2S(O)_2Et$. This illustrates the importance of polar effects in governing reaction direction.

150

On the other hand, though Me$^\bullet$ is relatively unreactive towards sulphoxides and sulphones, the phenyl radical reacts with diethyl sulphoxide and diethyl sulphone to give $^\bullet$CHMeS(O)Et and $^\bullet$CHMeS(O)$_2$Et, respectively: this apparently reflects the more nucleophilic character of the phenyl radical (rather than any contribution to radical stabilization in the resulting α-sulphur-substituted radical, for which we have argued[16] that there is little evidence for delocalization).

Radical stabilization presumably plays an important role in the conversion of, for example, $^\bullet$CH$_2$CH$_2$SMe into $^\bullet$CH$_2$SCH$_2$CH$_3$, which apparently proceeds via a novel 1,4-shift and which is revealed by e.s.r./photolysis studies.[47] A particular clear-cut example[48] is provided by the analogous rearrangement in which an intermediate vinyl radical (derived from addition of a thiyl radical to an alkyne) undergoes rapid transformation to the appropriate thio-substituted radical (see e.g. Figure 5 and Scheme 5). The reaction proceeds rapidly (10^5 s^{-1} at room temperature) when the resulting radical is stabilized by the presence of both sulphur and carboxyl substituents (otherwise a 1,5-shift is preferred). It has been argued that this is associated both with the exothermicity in the reaction (a C_{sp2}–H bond is formed at the expense of a C_{sp3}–H bond, and a stabilized radical is produced: each effect may contribute ca. 40 kJ mol^{-1} to $\Delta H^{\ominus}$) and geometric factors (long C–S bonds, of ca. 0.18 nm, together with small $\angle$CSC, ca. 100°) which allow a closest approach of the radical centre and δ-H of ca. 0.16 nm.

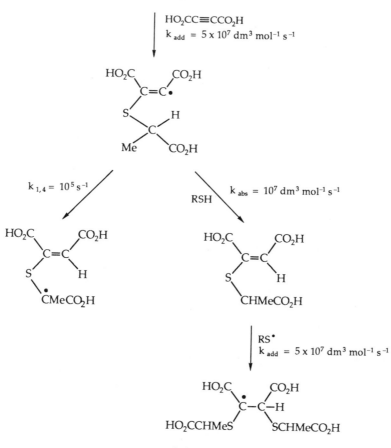

Scheme 5

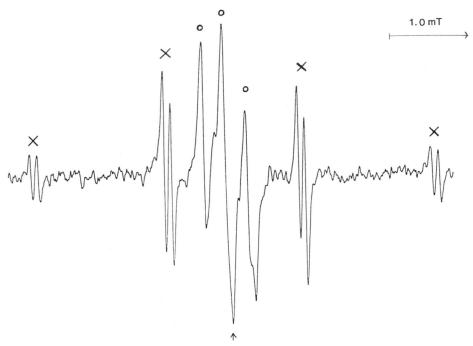

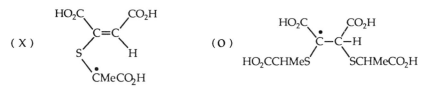

Fig. 5. E.s.r. spectra of sulphur-conjugated radicals obtained following reaction of •OH with MeCH(CO₂H)SH and HO₂CC≡CCO₂H (see Scheme 5).

ACKNOWLEDGEMENT

It is a great pleasure to thank many co-workers (to whom reference has been made in the text) for their collaboration via stimulating ideas, helpful suggestions and lively exchange of views. The financial support from the SERC is gratefully acknowledged.

REFERENCES

1.a. M.C.R. Symons, *J. Chem. Soc. Perkin Trans. 2* 1618 (1974).

  b. F.C. Adam and A.J. Elliot, *Canad. J. Chem.* 55:1546 (1977).

  c. H.C. Box and E.E. Budzinski, *J. Chem. Soc. Perkin Trans. 2* 553 (1976).

  d. K. Akasaka, *J. Chem. Phys.* 43:1182 (1965).

  e. W.W.H. Kou and H.C. Box, *J. Chem. Phys.* 64:3060 (1976).

2. D.J. Nelson, R.L. Petersen, and M.C.R. Symons, *J. Chem. Soc. Perkin Trans. 2* 2005 (1977).

3. H.R. Falle, Sir Frederick S. Dainton, and G.A. Salmon, *J. Chem. Soc. Faraday Trans 2* 72:2014 (1976).

4. R.L. Petersen, D.J. Nelson, and M.C.R. Symons, *J. Chem. Soc. Perkin Trans.* 2 255 (1978).
5. A.J. Elliot and F.C. Adam, *Canad. J. Chem.* 52:102 (1974).
6.a. J.E. Bennett and G. Brunton, *J. Chem. Soc. Chem. Commun.* 62 (1979).
  b. See also C. Chatgilialoglu, A.L. Castelhano, and D. Griller, *J. Org. Chem.* 50:2516 (1985).
7. B.C. Gilbert, H.A.H. Laue, R.O.C. Norman, and R.C. Sealy, *J. Chem. Soc. Perkin Trans.* 2 892 (1975): see also B.C. Gilbert and R.O.C. Norman, *Canad. J. Chem.* 60:12 (1982).
8. J.A. Baban and B.P. Roberts, *J. Chem. Soc. Perkin Trans.* 2 678 (1978).
9. P.J. Nichols and M.W. Grant., *Aust. J. Chem.* 36:1379 (1983).
10. B.C. Gilbert, P.A. Kelsall, M.D. Sexton, G.D.G. McConnachie, and M.C.R. Symons, *J. Chem. Soc. Perkin Trans.* 2 629 (1984).
11. C. Chatgilialoglu, *in:* "The Chemistry of Sulphones and Sulphoxides", S. Patai, Z. Rappoport, and C.J.M. Stirling, eds., John Wiley, Chichester, UK, p. 1081 (1988).
12. K. Nishikida and F. Williams, *J. Amer. Chem. Soc.* 96:4781 (1974).
13. J.A. Howard and E. Furimski, *Canad. J. Chem.* 52:555 (1974).
14. B.C. Gilbert, C.M. Kirk, R.O.C. Norman, and H.A.H. Laue, *J. Chem. Soc. Perkin Trans.* 2 497 (1977).
15a. S.G. Swarts, D. Becker, S. DeBolt, and M.D. Sevilla, *J. Phys. Chem.* 93:155 (1989).
  b. D. Becker, S.G. Swarts, M. Champagne, and M.D. Sevilla, *Int. J. Radiat. Biol.* 53:767 (1988).
  c. M.D. Sevilla, M. Yan, and D. Becker, *Biochem. Biophys. Res. Commun.* 155:405 (1988).
16. P.M. Carton, B.C. Gilbert, H.A.H. Laue, R.O.C. Norman, and R.C. Sealy, *J. Chem. Soc. Perkin Trans.* 2 1245 (1975).
17. J.E. Bennett, G. Brunton, B.C. Gilbert, and P.E. Whittall, *J. Chem. Soc. Perkin Trans.* 2 1359 (1988).
18. B.C. Gilbert, B. Gill, and M.D. Sexton, *J. Chem. Soc. Chem. Commun.* 78 (1978); C. Chatgilialoglu, B.C. Gilbert, B. Gill, and M.D. Sexton, *J. Chem. Soc. Perkin Trans.* 2 1141 (1980).
19. F. Freeman and C.N. Angeltakis, *J. Amer. Chem. Soc.* 105:4039 (1983), 104:5766 (1982); B.C. Gilbert, B.Gill, and M.J. Ramsden, *Chem. and Ind. (Lon)* 283 (1979); see F. Freeman, *Chem. Rev.* 84:117 (1984).
20. J.A. Baban and B.P. Roberts, *J. Chem. Soc. Perkin Trans.* 2 678 (1978).
21. B.C. Gilbert, C.M. Kirk, and R.O.C. Norman, *J. Chem. Research* (S) 173, (M) 1974 (1977).
22a. C. Chatgilialoglu, B.C. Gilbert and R.O.C. Norman, *J. Chem. Soc. Perkin Trans.* 2 770 (1979).
  b. A.G. Davies, B.P. Roberts, and B.R. Sanderson, *J. Chem. Soc. Perkin Trans.* 2 62 (1973).
  c. C. Chatgilialoglu, B.C. Gilbert, and R.O.C. Norman, *J. Chem. Soc. Perkin Trans.* 2 1429 (1980).
  d. M. Geoffroy and E.A.C. Lucken, *J. Chem. Phys.* 55:2719 (1971).
23. M. McMillan and W.A. Waters, *J. Chem. Soc. (B)* 422 (1966).
24a. B.C. Gilbert, R.O.C. Norman, and R.C. Sealy, *J. Chem. Soc. Perkin Trans.* 2 308 (1975).
  b. B.C. Gilbert, R.O.C. Norman, and R.C. Sealy, *J. Chem. Soc. Perkin Trans.* 2 303 (1975).
25. C. Chatgilialoglu, *in:* "The Chemistry of Sulphones and Sulphoxides", S. Patai, Z. Rappoport, and C.J.M. Stirling, eds., John Wiley, Chichester, UK, p. 1089 (1988).
26. C. Chatgilialoglu, B.C. Gilbert, R.O.C. Norman, and M.C.R. Symons, *J. Chem. Research* (S) 185, (M) 2610 (1980).
27a. W.B. Gara, B.P. Roberts, B.C. Gilbert, C.M. Kirk, and R.O.C. Norman, *J. Chem. Research* (S) 152, (M) 1748 (1977).
  b. C.W. Perkins, R.B. Clarkson, and J.C. Martin, *J. Amer. Chem. Soc.* 108:3206 (1986).
28. J.S. Chapman, J.W. Cooper, and B.P. Roberts, *J. Chem. Soc. Chem. Commun.* 835 (1976); 228 (1977).
29. W.B. Gara and B.P. Roberts, *J. Organometallic Chem.* 135:C20 (1977).
30. J.R.M. Giles and B.P. Roberts, *J. Chem. Soc. Chem. Commun.* 623 (1978); *J. Chem. Soc. Perkin Trans.* 2 1497 (1980); see also E. Anklam and P. Margaretha, *Res. Chem. Int.* 11:127 (1989).

31. C.W. Perkins, J.C. Martin, A.J. Arduengo, W.W. Lau, A. Alegria, and J.K. Kochi, *J. Am. Chem. Soc.* 102:7753 (1980).

32. W.B. Gara, B.P. Roberts, C.M. Kirk, B.C. Gilbert, and R.O.C. Norman, *J. Magn. Res.* 27:509 (1977).

33a. B.C. Gilbert, D.K.C. Hodgeman, and R.O.C. Norman, *J. Chem. Soc. Perkin Trans. 2* 1748 (1973).

 b. B.C. Gilbert, J.P. Larkin, and R.O.C. Norman, *J. Chem. Soc. Perkin Trans. 2* 272 (1973).

 c. See also B.C. Gilbert and P.R. Marriott, *J. Chem. Soc. Perkin Trans. 2* 191 (1980).

34a. K.-D. Asmus, D. Bahnemann, Ch.-H. Fischer, and D. Veltwisch, *J. Amer. Chem. Soc.* 101:5322 (1979).

 b. See also K.-D. Asmus, H.A. Gillis, and G.E. Teather, *J. Phys. Chem.* 82:2677 (1978).

 c. M. Göbl, M. Bonifacic, and K.-D. Asmus, *J. Amer. Chem. Soc.* 106:5984 (1984).

 d. K.-D. Asmus, *Acc. Chem. Res.* 12:436 (1979).

35. W.B. Gara, J.R.M. Giles, and B.P. Roberts, *J. Chem. Soc. Perkin Trans. 2* 1444 (1979).

36. W.K. Musker and T.L. Wolford, *J. Amer. Chem. Soc.* 98:3055 (1976); W.K. Musker, T.L. Wolford, and P.B. Roush, *J. Amer. Chem. Soc.* 100:6416 (1978); see also G.S. Wilson, D.D. Swanson, J.T. Klug, R.S. Glass, M.D. Ryan, and W.K. Musker, *J. Amer. Chem. Soc.* 101:1040 (1979).

37. A. Naito, S. Kominami, K. Asaka, and H. Hatano, *Chem. Phys. Lett.* 47:171 (1977).

38a. K.-D. Asmus, M. Göbl, K.-O. Hiller, S. Mahling, and J. Mönig, *J. Chem. Soc. Perkin Trans. 2* 641 (1985).

 b. J. Mönig, M. Göbl, and K.-D. Asmus, *J. Chem. Soc. Perkin Trans. 2* 647 (1985).

 c. M.J. Davies, B.C. Gilbert, and R.O.C. Norman, *J. Chem. Soc. Perkin Trans. 2* 731 (1983).

39. T. Gillbro, *Chem. Phys.* 4:476 (1974); J. H. Hadley and W. Gordy, *Proc. Nat. Acad. Sci., U.S.A.* 71:4409 (1974).

40. H.C. Box, H.E. Freund, K.T. Lilga, and E.E. Budzinski, *J. Phys. Chem.* 74:40 (1970).

41. H. Bock and U. Stein, *Angew. Chem. Int. Ed.* 19:834 (1980); H. Bock, U. Stein, and A. Semkow, *Chem. Ber.* 113:3208 (1980).

42. M.C.R. Symons and R.L. Petersen, *J. Chem. Soc. Faraday Trans. II* 210 (1979).

43. M.C.R. Symons, *J. Chem. Soc. Perkin Trans. 2* 908 (1976).

44. See D. Griller, D. Nonhebel, and J.C. Walton, *J. Chem. Soc. Perkin Trans. 2* 1817 (1984).

45. O. Ito and M. Matsuda, Chemical Kinetics of Sulphur-containing Radicals, *in*: "Chemical Kinetics of Small Organic Radicals", Z.B. Alfassi, ed., CRC Press, Boca Raton, Florida, Vol. III (1988).

46. B.C. Gilbert, P. Hanson, W.J. Isham, and A.C. Whitwood, *J. Chem. Soc. Perkin Trans. 2* 2077 (1988).

47. L. Lunazzi, G. Placucci, and L. Grossi, *J. Chem. Soc. Perkin Trans. 2* 703 (1981); *Tetrahedron* 39:159 (1983).

48. B.C. Gilbert, D.J. Parry, and L. Grossi, *J. Chem. Soc. Faraday Trans. I* 83:77 (1987).

SULFUR-CENTERED THREE-ELECTRON BONDED RADICAL SPECIES

Klaus-Dieter Asmus

Hahn-Meitner-Institut Berlin
Bereich S, Abteilung Strahlenchemie
Postfach 39 01 28,
1000 Berlin 39, F. R. Germany

We all know that it is not possible to stabilize a He_2 molecule. Any interaction of the filled He 1s orbitals would lead to the establishment of bonding σ and antibonding σ^* energy levels, both doubly occupied. As σ^*, for quantum mechanical reasons, is raised a little bit more than σ is lowered relative to the original atomic energy levels the net result of this electronic σ/σ^* configuration would be a slight repulsion of the two atoms. The situation becomes significantly different though if one of the antibonding σ^* electrons is removed.

$[\,He_2\,]$ He_2^+ $(He\cdot\!\cdot\!\cdot He)^+$

The combined effect of two bonding and one antibonding electrons results in net bonding with an energy of approximately half of a normal two-electron σ bond.

Experimentally the He_2^+ ion is a well characterized species and can be generated via ionization of a helium atom followed by association with an neutral atom, i.e.

$$He^+ \quad + \quad He \quad \rightarrow \quad He_2^+ \tag{1}$$

These gas phase, most often mass spectrometry studies have been corroborated by many theoretical calculations.

Although some notations have been suggested and used for such *two-center-three-electron* $(2\sigma/1\sigma^*)$ *bonds* none of them has commonly been accepted, presumably because they did not express convincingly enough the particular electronic situation in this kind of bond. With the discovery of numerous examples of such species in the condensed phase, involving in particular group V-VII elements, during the last 10-15 years it seemed appropriate to identify this type of bonding situation by a more specific symbol. We have introduced the "$\cdot\!\cdot\!\cdot$" notation as shown in the $(He\cdot\!\cdot\!\cdot He)^+$ example. In our mind it shows the different character of the two bonding $vs.$ the one antibonding electrons and the relative weakness of the bond

Sulfur-Centered Reactive Intermediates in Chemistry and Biology, Edited by
C. Chatgilialoglu and K.-D. Asmus, Plenum Press, New York, 1990

by inserting a somewhat wider space between the two interacting atoms. It seems to be widely accepted by now and therefore will be used throughout this article.

The identification and physico-chemical characterization of $2\sigma/1\sigma^*$ three-electron bonded species has particularly benefitted from the availability of sensitive, time-resolved techniques[1] such as the radiation chemical method of pulse radiolysis or the photochemical laser flash equipments. Valuable information has also been gathered from ESR measurements under steady-state flow conditions[2] as well as in low temperature solid matrices.[3-5] In addition, excellent high level calculations have provided a very good theoretical description of these species.[6-9] (See also articles by T. Clark in this book).

As of today, cationic, neutral, and anionic radical species of the general structures are known with X and Y representing heteroatoms of group V-VII (e.g. N, P, O, S, Se, Cl, Br, I).

$$X \mathbin{.\!.\!.} X \rceil^{+,\,0,\,-} \qquad\qquad\qquad X \mathbin{.\!.\!.} Y \rceil^{+,\,0,\,-}$$

A prominent element among those which are particularly susceptible for the establishment of $2\sigma/1\sigma^*$ bonds and well suited for experimental investigations is sulfur. This article will focus mainly on $(>\!S\!\mathbin{.\!.\!.}S\!<)^+$ type radical cations, but a number of $S\mathbin{.\!.\!.}X$ bonded neutral and cationic species with X = Cl, Br, I, and O will also be discussed.

At this point it may be appropriate to briefly mention two other types of three-electron bonds the $2\sigma/1\sigma^*$ should not be mistaken with. Oxidation of a π-bond, for example, results in a radical cation with a $2\sigma/1\pi$ three-electron bond. This bond contains no antibonding electrons, and the total bond strength exceeds that of a 2σ bond by the energy of half a π-bond. Oxidation of an organic disulfide , RSSR, to its radical cation $(RSSR)^{\bullet+}$ yields a species in which the unpaired electron from the oxidized sulfur interacts with the unbound p-electron pair of the second sulfur. This establishes a $2\pi/1\pi^*$ bond on top of the already existing σ bond. The overall bond strength of this five-electron $(2\sigma/2\pi/1\pi^*)$ bond also exceeds that of the normal 2s bond by *ca* half a π bond, i.e. the $(RSSR)^{\bullet+}$ assumes a partial π-character. While these two examples describe a one-and-a half bond situation with respect to bond strength there is, however, also another possibility for an overall half-bond, namely, the one-electron (1s) bond which exists, for example, in the H_2^+ molecular cation. All the various odd-electron bonds are depicted in the following structures for comparison.

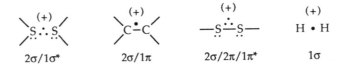

DIMER SULFIDE RADICAL CATIONS AND RELATED SPECIES

One-electron oxidation of an organic sulfide, R_2S, yields a radical cation, $R_2S^{\bullet+}$, with the unpaired electron being located in the sulfur p-orbital. This species is usually highly reactive and generally has a strong tendency to stabilize by association with a free p-electron pair of an unoxidized sulfur from a second sulfide molecule. The p-orbital interaction results in the generation of σ and σ^* energy levels and the establishment of the $2\sigma/1\sigma^*$ three-electron bond, similar to that shown above for the He_2^+ ion. The radical cation thus formed from sulfides is characterized by a sulfur-sulfur bond in which the antibonding electron exerts enough bond weakening to facilitate easy redissociation.

$$
\begin{array}{c}
(+) \\
R\!-\!\overset{\displaystyle .}{\underset{\displaystyle ..}{S}}\!-\!R
\end{array}
\quad + \quad
R\!-\!\overset{\displaystyle ..}{\underset{\displaystyle ..}{S}}\!-\!R
\quad \rightleftharpoons \quad
\overset{\displaystyle R}{\underset{\displaystyle R}{\diagdown}}\!\!S \mathbin{.\!.\!.} S\!\!\overset{\displaystyle R}{\underset{\displaystyle R}{\diagup}}
\qquad (2)
$$

As a result $R_2S^{\bullet+}$ and $(R_2S\therefore SR_2)^+$ establish an equilibrium. It may be noted that the total number of electrons located around each of the sulfur atoms amounts to nine, i.e. one above the magic figure of eight.

Most of the studies on these sulfide-derived three-electron bonded species have been carried out in aqueous solutions and using hydroxyl radicals, generated by radiolysis or Fenton-type chemistry, as oxidants. (The $^\bullet$OH radical induced process, incidentally, proceeds via an adduct as primary intermediate, which subsequently eliminates OH^- to yield the radical cation; this aspect shall, however, not be discussed in further detail in this article). In principle, any one-electron oxidant with high enough redox potential (> 1.5 V), as well as electro- or photochemical methods may be used to generate such sulfur-centered radical cations. Also, their formation is not limited to aqueous solutions and seems possible in any environment, including hydrocarbon solutions.

One of the classical methods to study radicals and particularly their structure is electron spin resonance (ESR)[2-5] (see also article by B. C. Gilbert in this book). However, this technique has some obvious limitations with respect to sulfur-centered radicals owing to the lack of nuclear spin on the main sulfur isotope ^{32}S. Nevertheless, ESR has provided some essential information on several $(R_2S\therefore SR_2)^+$ type radical cations. For example, in case of the completely methylated species, $(Me_2S\therefore SMe_2)^+$, a spectrum could be resolved which showed the presence of twelve equivalent protons demonstrating the high degree of symmetry in this species. ESR also revealed the antibonding character of the unpaired electron.

However, it has been another physical property, namely, optical absorption which has provided a still deeper insight into the nature of such three-electron bonded systems. Generally, they exhibit broad and strong absorption bands with maxima in the visible, easily detectable by, for example, the radiation chemical technique of pulse radiolysis.[10-12] (For a detailed description of this time-resolved detection method see article by D. A. Armstrong in this book).

Optical absorption spectra of $(R_2S\therefore SR_2)^+$ radical cations

Oxidation of dimethylsulfide, Me_2S, (e.g. by $^\bullet$OH radicals in pulse irradiated aqueous solutions) yields a strong transient absorption which is unambiguously attributable to the $(Me_2S\therefore SMe_2)^+$ radical cation. (Identification is based not only on the optical characteristics, as discussed below, but also on other parameters including time-resolved conductivity measurements).[10] The complete visible spectrum of this three-electron bonded species is shown in Fig. 1. It is characterized by the lack of structure and its broadness (halfwidth ca 1 eV). The extinction coefficient of $\approx 6 \times 10^3$ M^{-1} cm^{-1} at the 465 nm maximum indicates that the transition is an allowed one.

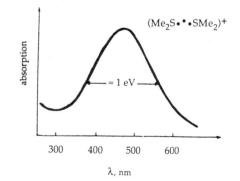

Fig. 1   Optical absorption spectrum of $(Me_2S\therefore SMe_2)^+$

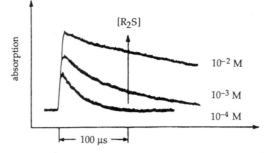

Fig. 2. Absorption *vs.* time traces recorded at 465 nm upon pulse radiolysis of N_2O saturated, pH ≈ 4, aqueous solutions of 10^{-2}, 10^{-3}, and 10^{-4} M Me_2S, respectively.

Three individual absorption *vs.* time traces recorded at 465 nm upon pulse radiolysis of N_2O saturated, pH ≈ 4 solutions of three different Me_2S concentrations, respectively, are shown in Figure 2. They reveal two significant informations. First, the signal height after the (*ca* 2 μs) pulse increases with sulfide concentration, and this reflects the influence of the latter on the equilibrium

$$(Me_2S\text{·}^{\cdot}\text{·}SMe_2)^+ \quad \rightleftarrows \quad Me_2S^{\bullet +} \quad + \quad Me_2S \tag{3}$$

Second, the lifetime of the three-electron bonded species increases with increasing sulfide concentration, i.e. from ≈ 30 μs at 10^{-4} M to almost 1 ms at 10^{-2} M Me_2S. This finding shows that $(Me_2S\text{·}^{\cdot}\text{·}SMe_2)^+$ is kinetically more stable than $Me_2S^{\bullet +}$ and, in addition, it demonstrates the influence of equilibrium (3). Qualitatively, the same holds for practically all other $(R_2S\text{·}^{\cdot}\text{·}SR_2)^+$ as well.

The optical absorption of $(R_2S\text{·}^{\cdot}\text{·}SR_2)^+$ is due to a transition between the uppermost doubly occupied orbital which is essentially the σ energy level disturbed by the non-bonding sulfur electrons, and the singly occupied σ* energy level which is practically not influenced by other orbitals.

This picture, simplified in Figure 3 from a theoretical study by Clark,[6a] has been evaluated for the simplest of these dimer radical cations, namely, $(H_2S\text{·}^{\cdot}\text{·}SH_2)^+$. The energy difference $a_g(\sigma - n^-) / b_u(\sigma^*)$ has been calculated to *ca* 3.25 eV, corresponding to an optical transition band with maximum at 380 nm. Experimentally, this species could be generated via pulse irradiation of very acid (pH < 2) solutions of high H_2S concentrations[13] via the overall reaction sequence:

$$H_2S \quad + \quad {}^{\bullet}OH / H^{\bullet} \quad \rightarrow \quad HS^{\bullet} \quad + \quad H_2O / H_2 \tag{4}$$

$$HS^{\bullet} \quad + \quad H_2S \quad + \quad H^+ \quad \rightarrow \quad (H_2S\text{·}^{\cdot}\text{·}SH_2)^+ \tag{5}$$

The measured absorption spectrum peaks at 370 nm in excellent agreement with Clark's calculations.

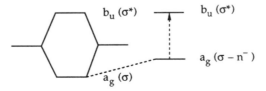

Fig. 3. Simplified MO energy level scheme for the optical transition in $2\sigma/1\sigma^*$ bonded species (see ref. 6a).

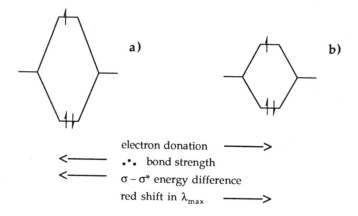

electron donation ———————>

<——————— $\cdot\,\dot{}\,\cdot$ bond strength

<——————— $\sigma - \sigma^*$ energy difference

red shift in λ_{max} ———————>

Fig. 4. Simplified MO energy level scheme for a stronger (**a**) and a weaker (**b**) $2\sigma/1\sigma^*$ bonded species such as $(H_2S\cdot\,\dot{}\,\cdot SH_2)^+$ and $(Me_2S\cdot\,\dot{}\,\cdot SMe_2)^+$, for example.

The underlying $2\sigma/1\sigma^*$ concept, and the consequences based on it have been substantiated by a variety of experiments, as will be presented and discussed in the following sections.

Effect of the substituents "R" on λ_{max}

Considering the two experimental results mentioned so far, i.e. λ_{max} of 465 and 370 nm for $(Me_2S\cdot\,\dot{}\,\cdot SMe_2)^+$ and $(H_2S\cdot\,\dot{}\,\cdot SH_2)^+$, respectively, it is noted that they differ significantly from each other. Somehow this must be connected with methyl being substituted by hydrogen at the sulfur atoms. This can be rationalized in terms of the comparatively higher release of electron density from the methyl groups into the three-electron bond, and there most likely into the antibonding orbital. This, in turn, results in bond weakening and a corresponding smaller difference between the σ and σ^* energy levels. Electron donation by substituents thus causes S$\cdot\,\dot{}\,\cdot$S bond weakening and is associated with a red-shift in optical absorption as schematically depicted in Figure 4 for a stronger (**a**) and a weaker (**b**) three-electron bond.[11,12,14]

This theoretical expectation is fully corroborated by the experimental results. The λ_{max} listed in Table 1 reveal, in fact, quite a dramatic effect of the substituents on the optical transition energies. The combined effect of four *tert*-butyl groups does not even allow stabilization of the sulfur-sulfur three-electron bond anymore. In this case (i.e. upon oxidation of *t*-Bu_2S) only the monomer *t*-$Bu_2S^{\cdot+}$ radical cation ($\lambda_{max} \approx 310$ nm) can be detected.

Table 1. Optical Absorption Maxima for $(R_2S\cdot\,\dot{}\,\cdot SR_2)^+$ Radical Cations[12]

R	λ_{max}, nm
H	370
Me	465
Et	485
i-Pr	555
t-Bu	not stable

A quantitative picture on the effect of electron donating by the substituents on the optical properties of $(R_2S\textbf{.}\text{•}\textbf{.}SR_2)^+$ radical cations is given by the linear free energy correlation shown in curve (a) of Figure 5.[12] An excellent straight line is obtained between the energy of the optical transition (which seems proportional to the S.•.S bond strength) and the electron induction (given in terms of Tafts σ* parameter) for all radical cations with *unbranched* substituents R. For *branched* R the respective points fall below this straight line indicating relatively too much red-shift in λ_{max}, i.e. too low S.•.S bond strength. This deviation from linearity is attributed to steric hindrance which is assumed to exert an additional S.•.S bond weakening effect.

No such steric effect is apparent in another type of 2σ/1σ* three-electron bonded species which has only one substituent bound to each sulfur, namely, the disulfide radical anion, $(RS\textbf{.}\text{•}\textbf{.}SR)^-$. This species exists in equilibrium with the thiyl radical, $RS^\bullet$,

$$(RS\textbf{.}\text{•}\textbf{.}SR)^- \quad \rightleftarrows \quad RS^\bullet \quad + \quad RS^- \tag{6}$$

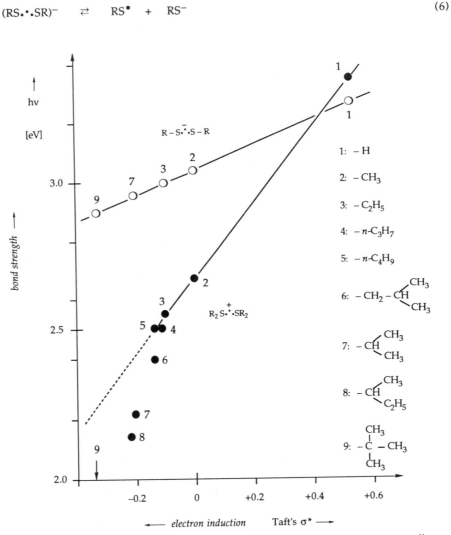

Fig. 5. Linear free energy correlation between the optical transition energy (in terms of hv) (proportional to S.•.S bond strength) and Tafts inductive σ* parameter (electron induction by R) for $(R_2S\textbf{.}\text{•}\textbf{.}SR_2)^+$ radical cations and $(RS\textbf{.}\text{•}\textbf{.}SR)^-$ radical anions.[12]

Table 2. Optical Absorption Maxima, λ_{max}, for Radical Cations with Mixed Substitution

radical cation	λ_{max}, nm
$(Me_2S\bullet\!\therefore\!\bullet SH_2)^+$	420
$(Me_2S\bullet\!\therefore\!\bullet SMe_2)^+$	465
$(Me_2S\bullet\!\therefore\!\bullet SEt_2)^+$	480
$(Me_2S\bullet\!\therefore\!\bullet Si\text{-}Pr_2)^+$	530
$(Me_2S\bullet\!\therefore\!\bullet St\text{-}Bu_2)^+$	545
$(Me_2S\bullet\!\therefore\!\bullet SMe,t\text{-}Bu)^+$	495
$(Me,t\text{-}BuS\bullet\!\therefore\!\bullet SMe,t\text{-}Bu)^+$	510
$(Me,t\text{-}BuS\bullet\!\therefore\!\bullet St\text{-}Bu_2)^+$	> 550

and can be generated either by one-electron reduction of the correponding disulfide or by association of a thiyl radical with a thiolate anion. As can be seen from Fig. 5b an excellent straight line is obtained in the linear free energy correlationship for the entire series of disulfide radical anions, i.e. for R varying from H to *tert*-butyl, indicating that no steric effect seems to be in operation for these species.

Radical cations with mixed substitution

Formation of symmetrically mixed sulfide radical cations, $(R'R''S\bullet\!\therefore\!\bullet SR'R'')^+$, is achieved by one-electron oxidation of the corresponding mixed thioethers, R'R''S.[15] Asymmetric $(R'_2S\bullet\!\therefore\!\bullet SR''_2)^+$ radical cations can be generated not only statistically in mixtures of the two corresponding sulfides but also exclusively via H-atom induced reduction of a sulfoxide R'_2SO in the presence of small amounts of the conjugate R''_2S sulfide and at very acid conditions (pH < 2).[16] The overall reaction

$$R'_2SO \;+\; H^\bullet \;+\; H^+ \;\rightarrow\; R'_2S^{\bullet+} \;+\; H_2O \tag{7}$$

provides the uncomplexed radical cation $R'_2S^{\bullet+}$ in the absence of the corresponding sulfide. This may then associate with any sulfide added to the solution, as formulated, for example, for $Me_2S^{\bullet+}$ generated from dimethylsulfoxide in its reaction with diethylsulfide.

$$Me_2S^{\bullet+} \;+\; Et_2S \;\rightleftarrows\; (Me_2S\bullet\!\therefore\!\bullet SEt_2)^+ \tag{8}$$

Table 2 summarizes some instructive data on the absorption maxima of such mixed $2\sigma/1\sigma^*$ radical cations.[15,16]

The optical results reflect again the influence of the substituents. Generally, the mixed radical cations exhibit absorption maxima lying between those for the corresponding systems with only one kind of substituents (see Table 1 and Figure 5 for comparison). The various mixed methyl/*tert*-butyl species, for example, show an increasing red-shift with increasing degree of *tert*-butyl substitution. Noteworthy is perhaps the marked difference in λ_{max} between $(t\text{-}Bu_2S\bullet\!\therefore\!\bullet SMe_2)^+$ and $(Me,t\text{-}BuS\bullet\!\therefore\!\bullet SMe,t\text{-}Bu)^+$. The former absorbs at significantly longer wavelengths indicating a comparatively lower strength of the three-electron bond. The reason for this is considered to be steric hindrance. It appears that the presence of two bulky *tert*-butyl substituents on one side of the S$\therefore$S bridge distorts the orbital symmetry to become less favourable for p-orbital overlap.

Cyclic systems and influence of structure

Sulfur-sulfur interaction may be achieved not only *inter*molecularly as in $(R_2S. \cdot .SR_2)^+$ but also *intra*molecularly if both sulfur atoms are located within the same molecule.[9,11,17-21] In this case structural parameters play, in fact, the decisive role for the stability and optical properties of the three-electron bonded species. The two extreme situations in this respect are depicted in Figure 6.

A molecular geometry which allows optimal p-orbital interaction to yield a $2\sigma/1\sigma^*$ bond between the two sulfur atoms is shown in Figure 6a. In reality such a configuration seems closely approached in the radical cation obtained upon one-electron oxidation of 1,5-dithiacyclooctane, **1**, and was first described in ESR work by W. K. Musker.[21] It exhibits an optical absorption with λ_{max} at 400 nm and costitutes the most blue-shifted example except for the all-hydrogen substituted $(H_2S. \cdot .SH_2)^+$. Stabilization is facilitated by the establishment of two five-membered rings on either side of the $S. \cdot .S$ bridge.

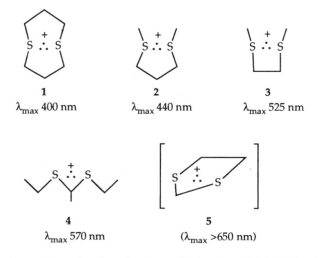

1	2	3
λ_{max} 400 nm	λ_{max} 440 nm	λ_{max} 525 nm

4
λ_{max} 570 nm

5
$(\lambda_{max} > 650$ nm$)$

The other extreme is given for the radical cation of 1,3-dithiacylopentane (**5**). In this rigid molecule the sulfur p-orbitals are aligned more or less perpendicular to the ring plane (only very slight envelope structure), and this prevents any appreciable overlap for σ-bond formation. Experimentally only a very weak and red-shifted absorption (>650 nm) is indicated upon oxidation of this compound, and assignment to a three-electron bonded radical cation remains, in fact, ambiguous.

The difference between **1** and **5** becomes also apparent in the kinetic (and most likely also thermodynamic) stabilities. While the radical cation **5** has a life-time, if any, of less than 0.2 μs species **1** lives much longer and under pulse radiolysis conditions (*ca* 10^{-6} M radical concentration) decays predominantly by second order kinetics on the ms time scale.

a

b

Fig. 6. Most (a) and least (b) favorable alignment of sulfur p-orbitals for establishment of a $2\sigma/1\sigma^*$ three-electron bond.

Table 3. λ_{max} for some selected five- and six-membered, *intra*molecular S·⁺·S bonded radical cations, and $(R_2S\cdot^+\cdot SR_2)^+$ type *inter*molecular radical cations [12,22]

R	$\begin{matrix} R \quad + \quad R \\ S \cdot\cdot S \\ \text{(5-ring)} \end{matrix}$	$\begin{matrix} R \quad + \quad R \\ S \cdot\cdot S \\ \text{(6-ring)} \end{matrix}$	$(R_2S \therefore SR_2)^+$
Me	440 nm	475 nm	465 nm
i-Pr	455	510	555
t-Bu	460	545	not stable

Radical cations **2-4** are generated from open chain dithia compounds and represent a series with decreasing extent of possible sulfur-sulfur p-orbital interaction. (The sulfur p-orbitals are practically perpendicular to the C–S–C plane). Accordingly the alignment of the p-orbitals relative to each other gradually changes from the picture given in Figure 6a to that in Figure 6b. As expected, the most blue-shifted optical absorption (440 nm) is observed for the radical cation **2** from 2,6-dithiaheptane which can assume a five-membered ring configuration. Steric strain in the 4-membered (**3**) and three-membered (**4**) radical cations obtainable from 2,5-dithiahexane and 3,5-dithia-4-methylheptane, respectively, results in considerably weaker S·⁺·S bonds and considerably red-shifted absorptions (525 nm and 570 nm, respectively). These considerations hold for all *intra*molecular radical cations generated from dithia compounds R–S–(CH$_2$)$_n$–S–R or *transannular* sulfur-sulfur interaction upon oxidation of cyclic dithia compounds (as in **1**) with the most favourable situation generally given for n = 3.[11,17,19,22]

Sensitivity of optical data

The high sensitivity with which structural parameters and substituents are reflected in the optical absorptions are illustrated in the following two example series.[22] Table 3 compares the λ_{max} for some *intra*molecular S·⁺·S bonded, 5- and 6-ring radical cations with Me, *i*-Pr, and *t*-Bu substituents with the optical data on the corresponding *inter*molecular dimeric radical cations, $(R_2S\cdot^+\cdot SR_2)^+$. Very little effect of the substituents R are found for the five-membered ring structures with $\Delta\lambda_{max}$ 20 nm between R = Me and *t*-Bu, respectively. For the six-membered ring species this difference amounts already to 70 nm which, however, is still much smaller than for the dimer species obtained from the monosulfides.

This conclusion is further substantiated by the findings on 2-substituted-1,3-dithia-cyclopentanes.[23] While the radical cation derived from 1,3-dithiacyclopentane (**6**) which contains two hydrogen atoms at the 2-position is very unstable if formed at all (λ_{max} > 650 nm), the analogue radical cation generated upon oxidation of 1,3-dithia-2,2-dimethylcyclopentane (**7**) exhibits a pronounced absorption at 610 nm and a considerable kinetic and thermodynamic stability (substitution at the 4-position does not exert any such effect).

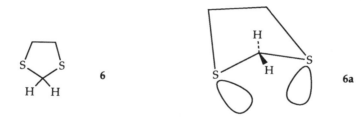

As electron induction by the methyl groups would rather result in abond weakening, and substitution in the 4-position does not lead to any such stabilization the observed effect in the 2-substituted species is considered to have steric reasons. It has been proposed that the shift in λ_{max} reflects the *gem-dialkyl* effect causing a structural distortion which favours p-orbital interactions with respect to $2\sigma/1\sigma^*$ bond formation.[23] This is depicted in pictures 6a and 7a.

PHOTOCHEMICAL GENERATION OF S∴S BONDED RADICALS

The formation of $2\sigma/1\sigma^*$ three-electron bonded radical cations is not limited to radiation chemical methods but may be initiated by any other suitable chemical or electrochemical means as well. Time-resolved photochemical methods, for example, have proven to be very useful in this sense, and two interesting mechanisms shall briefly be outlined.

Photolysis of sulfonium salt peresters subjected to laser flashes (248 or 254 nm) yield *intra*molecular radical cations via the reaction sequence[24]

$$R - S - \cdots - S^+ - CH_2 - CH_2 - CO - OO - t\text{-}Bu$$
$$\mid$$
$$R$$

$$\downarrow \quad h\nu, -t\text{-}BuO^{\bullet}$$

$$R - S - \cdots - S^+ - CH_2 - CH_2 - CO - O^{\bullet}$$
$$\mid$$
$$R$$

$$\downarrow \quad -CO_2$$

$$R - S - \cdots - S^+ - CH_2 - CH_2^{\bullet}$$
$$\mid$$
$$R$$

$$\downarrow \quad -C_2H_4$$

$$R - S - \cdots - S^{\bullet+} - R \quad \rightarrow \quad \left[\begin{array}{c} R \quad\quad R \\ \diagdown S \therefore S \diagup \end{array} \right]^+ \quad (9)$$

Photolysis of aliphatic dithia compounds (248 nm KrF laser) leads to alkylthio substituted thiyl radicals (among other radical species) which can cyclized to generate neutral $2\sigma/1\sigma^*$ three-electron bonded radicals.[25]

The most stable radical in this series is again that with n = 3, i.e. the one which can assume a five-membered ring structure. The substituents R exhibit hardly any influence for this case ($\lambda_{max} = 395$ nm for R = Me, 400 nm for R = Et and Pr).

$$R - S - (CH_2)_n - S - R \quad \xrightarrow{h\nu} \quad R^\bullet \quad + \quad {}^\bullet S - (CH_2)_n - S - R$$

$$\downarrow$$

$$(10)$$

It may be appropriate to point out that the λ_{max} of such $(R_2S.^\bullet.SR)$ type radicals cannot be compared with the optical data of the $(R_2S.^\bullet.SR_2)^+$ radical cations or $(RS.^\bullet.SR)^-$ radical anions on an *absolute* basis. Owing to the number and quality of unbound electron pairs at sulfur the overall electronic structure is different in any of these series. Within each group, however, all the parameters affecting the *trends* in λ_{max} and stability apply as discussed in detail in connection with the S.$^\bullet$.S bonded radical cations.

THERMODYNAMIC AND KINETIC STABILITY OF S.$^\bullet$.S BONDS

An exact evaluation of thermodynamic and kinetic parameters for S.$^\bullet$.S bonds is a relatively complex task and shall not be presented here in any detail. Most of the experimentally determined data have been revealed from kinetic analysis of the $(R_2S.^\bullet.SR_2)^+$ radical cation decay as a function of sulfide concentration (see traces in Figure 2), and deliberate acceleration of the $R_2S^{\bullet+}$ deprotonation by addition of a proton acceptor such as, for example, $(HPO_4{}^{2-})$.[14] The reactions considered are:

$$R_2S^{\bullet+} \quad + \quad R_2S \quad \underset{\overset{\longleftarrow}{k}}{\overset{\overset{\longrightarrow}{k}}{\rightleftarrows}} \quad (R_2S.^\bullet.SR_2)^+ \tag{2a}$$

$$R_2S^{\bullet+} \quad + \quad HPO_4{}^{2-} \quad \rightarrow \quad >C^\bullet - S - R \quad + \quad H_2PO_4{}^- \tag{11}$$

$$R_2S^{\bullet+} \quad \rightarrow \quad >C^\bullet - S - R \quad + \quad H^+ \tag{12}$$

Absolute rate constants have been determined for all individual reactions and also for the equilibrium constant for two sulfides, Me_2S and $i\text{-}Pr_2S$. Table 4 summarizes the most interesting data, i.e. the rate constant for the dissociation of the three-electron bond, $\overset{\longleftarrow}{k}$, and the stability (equilibrium) constant $K = (\overset{\longrightarrow}{k}/\overset{\longleftarrow}{k})$. It also contains the reaction enthalpies for the transition state of the S.$^\bullet$.S dissociation, as evaluated from the temperature dependence of the decay, and the S.$^\bullet$.S bond energies estimated therefrom.[14]

The results substantiate the electronic influence of the substituents with the stability constant of the *iso*-propyl substituted species being lower than that of the methylated one by almost three orders of magnitude. The stability constant, furthermore, is controlled practically only by the rate constant for the S.$^\bullet$.S dissociation.

Based on the transition state enthalpies and some further assumptions (practically no activation energy for the forward reaction in equilibrium (2a) and negligible entropic contributions) the S.$^\bullet$.S bond energies could be estimated (although only within relatively large error limits).[14] The bond energy of $(Me_2S.^\bullet.SMe_2)^+$ thus derived amounts to 100 - 120 kJ mol^{-1} agrees very well with a theoretically calculated value[8] of *ca* 100 kJ mol^{-1} and is only slightly higher than the *ca* 80 kJ mol^{-1} which could be estimated from time-resolved Raman resonance experiments.[26] In any case, the bond strength of the three-electron bond seems to be somewhat lower than that of a normal 2σ sulfur-sulfur bond in an organic disulfide. The strongest three-electron bond within this series is expected for the $(H_2S.^\bullet.SH_2)^+$ as discussed already in connection with the optical data. Theoretical calculations gave a value of *ca* 120 - 130 kJ mol^{-1} for the heat of formation of this species,[6] which very nicely fits into the general trend.

Table 4 Thermodynamic and kinetic data for stability of S∴S bond in $(R_2S∴SR_2)^+$ radical cations. (* see comment on re-evaluation in text)

R	$k^{\leftarrow}$, [s^{-1}]	$K = k^{\leftarrow} \times k^{\rightarrow}$ [M^{-1}]	$\Delta H^{\neq}$ [kJ mol^{-1}]	D (S∴S) [kJ mol^{-1}]	ref.
i-Pr	5.6×10^6	5.4×10^2	17	80 - 90	14,*
Me	1.5×10^4	2.0×10^5	57	100 - 120	14,*
				80	26
				100	8
H				120 - 130	6

It should be mentioned at this point that our published values on the bond energies of the $(i$-Pr$_2$S∴S-Pr$_2)^+$ and $(Me_2S∴SMe_2)^+$ radical cations were lower by *ca* 50 kJ mol^{-1}. They were based on an early estimate of *ca* 20 kJ mol^{-1} for the bond energy of $(R_2S∴OH_2)^{+}$.[14] This latter value has now been corrected to *ca* 70 kJ mol^{-1} (see article on "The Electronic Properties of Sulfur Containing Substituents ...." by T. Clark in this book), and the difference must be added to the old values.

CHEMICAL CONSEQUENCES

Stabilization of an oxidized sulfur center in an S∴S bond generally results in a lower reactivity as compared to the uncomplexed >S$^{•+}$ radical cation. Reactions usually proceed via redissociation of the three-elctron bond. This is illustrated by an example on the oxidation of *tert*-butyl mercaptane (*t*-BuSH) by the radical cations from dimethylsulfide. The formation of the *t*-BuS$^{•}$ radical via

$$Me_2S^{•+}$$

$$+ \ Me_2S \ \updownarrow \qquad + \ t\text{-BuSH} \quad \rightarrow \quad t\text{-BuS}^{•} \ + \ H^+ \ + \ (2) \ Me_2S \qquad (13)$$

$$(Me_2S∴SMe_2)^+$$

occures with apparent rate constants which depend on the Me$_2$S concentration, ranging from 2.6×10^6 M^{-1} s^{-1} at 10^{-2} M Me$_2$S to 2.4×10^8 M^{-1} s^{-1} in the absense of Me$_2$S.[27] (In the latter case the Me$_2$S$^{•+}$ is generated by reduction of DMSO, as described in an earlier section of this article).

Generally, all $2\sigma/1\sigma^*$ three-electron bonded species exhibit lower redox potentials than their single-electron conjugate radicals. This is not only true for the oxidizing power of $(R_2S∴SR_2)^+$ relative to $R_2S^{•+}$ as shown in the above example. It become even more apparent in the RS$^{•}$/(RS∴SR)$^-$ equilibrium (6) where RS$^{•}$ is a mild oxidant (E° ≈ 0.75 V) while (RS∴SR)$^-$ is quite a strong reductant (E° ≈ −1.6 V ; see articles by D.A. Armstrong in this book).

In the absence of suitable redox partners the predominant fate of the radical cations appears to be deprotonation to yield the α-thioalkyl type radicals, as shown in reaction (12), or radical-radical disproportionation. The latter occurs particularly in those cases where deprotonation cannot take place or is too slow to compete with this second order process. The rate constant for deprotonation is strongly dependent on the molecular structure of the radical

(a) $(R_2S \cdot \cdot SR_2)^+$ **(b)** $R_2S \cdot \cdot X$

Fig. 7. Schematic picture of MO levels in an energetically symmetric (**a**), and an asymmetric (**b**) system.

cation or, to be more specific, on the relative alignment of the singly occupied sulfur p-orbital with the C–H σ-bond to be cleaved. A very fast deprotonation is observed, for example, for the i-Pr$_2$S$^{\bullet +}$ radical cation ($k = 2.7 \times 10^5$ s^{-1}).[14] As depicted in structure **8** this is presumably facilitated by a parallel alignment of the sulfur-p- and C–H-σ-orbitals. In the radical cation from 9-thia-bicyclo-[3.3.1]-nonane (**9**) the neighbouring C–H bonds are perpendicular to the sulfur p-orbital. This practically prevents deprotonation, and **9** decays by a purely second order process under pulse radiolysis conditions ($\approx 10^{-5}$ M radical concentration).[28]

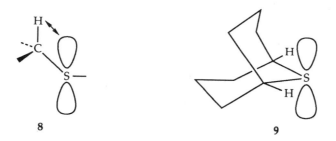

8   9

FORMATION OF $R_2S \cdot \cdot X$ RADICALS (X = HALOGEN)

Stabilization of an oxidized sulfur atom can, in principle, be achieved by other hetero atoms as well, provided it offers, like sulfur, a free (preferably p-) electron pair. Numerous examples have been identified and described by now with X being an atom from group V-VII of the periodic table. The thermodynamic stability of an S$\cdot \cdot$X bond is generally not as high as that of a symmetric S$\cdot \cdot$S system since the energy levels of S and X are usually different and consequently the gain in bond energy upon interaction becomes smaller. This can be rationalized from the corresponding MO diagrams in Figure 7. The gain in bond energy is clearly smaller in the asymmetric case.

One of the well investigated combinations is sulfur-halogen interaction, in particular in neutral $R_2S \cdot \cdot X$ radicals.[19,29] For X = iodine and bromine such species can be generated via oxidation of the sulfide and subsequent association of $R_2S^{\bullet +}$ with the halide anion, or ligand displacement as shown in reaction scheme (14). Alternatively, the halide is oxidized first to the $X_2^{\bullet -}$ radical anion which then enters a ligand exchange equilibrium (15).

$$R_2S^{\bullet +} \quad + \quad X^- \quad \rightleftarrows \quad R_2S \cdot \cdot X$$

$$\uparrow\downarrow \; + Me_2S \qquad\qquad\qquad\qquad\qquad\qquad \text{for } X = I \text{ and } Br \qquad (14)$$

$$(R_2S \cdot \cdot SR_2)^+ \quad + \quad X^- \quad \rightleftarrows \quad R_2S \cdot \cdot X \quad + \quad R_2S$$

$$X_2^{\bullet -} \quad + \quad R_2S \quad \rightleftarrows \quad R_2S \cdot \cdot X \quad + \quad X^- \qquad\qquad \text{for } X = Br \text{ and } Cl \qquad (15)$$

Table 5. Optical absorption maxima, λ_{max}, and stability constants for $Me_2S.\cdot.X$ with respect to ionic, K_{16}, and atomic dissociation, K_{17}.[14,29]

$Me_2S.\cdot.X$	λ_{max}, nm	K_{16}	K_{17}
$Me_2S.\cdot.I$	410	$\ll 10^{-6}$	4.4×10^{-5}
$Me_2S.\cdot.Br$	400	8.2×10^{-6}	4.2×10^{-10}
$Me_2S.\cdot.Cl$	390	1.2×10^{-2}	$\ll 10^{-10}$
- - - - - - - -	- - - - -	- - - - - -	- - - - - -
$Me_2S.\cdot.F$		$\gg 1$	$\lll$

This mechanism applies for bromine and chlorine systems. Table 5 summarizes some interesting data concerning $Me_2S.\cdot.I$, $Me_2S.\cdot.Br$, and $Me_2S.\cdot.Cl$, and includes a couple of reasonable extrapolations for a potential fluorinated species, $Me_2S.\cdot.F$. The stability constants K_{16} and K_{17} refer, respectively, to ionic dissociation

$$Me_2S.\cdot.X \quad \rightleftarrows \quad Me_2S^{\bullet+} \quad + \quad X^- \qquad\qquad (16)$$

and atomic dissociation

$$Me_2S.\cdot.X \quad \rightleftarrows \quad Me_2S \quad + \quad X^\bullet \qquad\qquad (17)$$

The λ_{max} are relatively close together and lie in between those for $(Me_2S.\cdot.SMe_2)^+$ and $X_2^{\bullet-}$ (the latter is also a $2\sigma/1\sigma^*$ three-electron bonded species). The equilibrium constants for the ionic and atomic dissociation have been recalculated from our earliest results but using more recent and probably most accurate values on the $(Me_2S.\cdot.SMe_2)^+ \rightleftarrows Me_2S^{\bullet+} + Me_2S$ equilibrium. The data are most reasonable not only with respect to trends but also to the absolute values. Thus ionic dissociation increases with increasing difference in electronegativity between S and X. Accordingly, $Me_2S.\cdot.Cl$ is still stabilized over the respective ions (and thus observable as such). The fluorinated species, by extrapolation, should, however, not be stable anymore which is in agreement with the complete failure to verify a $Me_2S.\cdot.F$ experimentally.

Particularly interesting and fully supportive of the considerations presented so far are the calculated equilibrium constants for the atomic dissociation K_{17}. This process leading to halogen atoms appears to be practically unimportant (relative to the ionic dissociation) for all halogens except iodine. The three-electron bond in $Me_2S.\cdot.I$ seems to be preferably split by leaving two of the three electrons at sulfur. This becomes understandable in terms of iodine being the only halogen atom whose electronegativity is close and, in fact, even slightly higher than that of sulfur.

STABILIZATION OF $R_2S(O).\cdot.X$ RADICALS

Recently it has been possible to characterize a particularly stable sulfur-chlorine three-electron bonded radical, $R_2S(O).\cdot.X$.[30] Its formation can be achieved via one-electron oxidation of a sulfoxide to the sulfoxide radical cation, $R_2S(O)^{\bullet+}$, in the presence of chloride ions (in acid solutions of pH < 4-5):

$$R_2S(O)^{\bullet+} \quad + \quad Cl^- \quad \rightleftarrows \quad R_2S(O).\cdot.Cl \qquad\qquad (18)$$

It is not possible to stabilize the corresponding $R_2S(O) \cdot \cdot \cdot Br$ or $R_2S(O) \cdot \cdot \cdot I$ species. This can be rationalized in terms of the MO pictures given in Figure 7. The energy levels of the sulfoxide function is expected to lie at a considerably lower value than that of the sulfide function, and it would seem that $>S(O) \cdot \cdot \cdot Cl$ interaction resembles the "symmetric" picture in Figure 7 much more than the "asymmetric" situation prevailing for $>S \cdot \cdot \cdot Cl$. In case of the corresponding iodine species the situation appears to be reverse. In this case, the more "symmetric" situation is given for $R_2S \cdot \cdot \cdot I$, and $R_2S(O) \cdot \cdot \cdot I$ is the species with the least favourable electronic conditions for stabilization of the three-electron bond.

S.•.X BONDED RADICAL IONS

Sulfur-halogen $2\sigma/1\sigma^*$ three-electron bonds are not limited to neutral species but have also been found to be stabilized as radical ions. The number of anionic species is still very low and only some $(RS \cdot \cdot \cdot Br)^-$ have been observed.[31] Radical cations have, however, been detected much more frequently. They include *intra*molecular bromine (10) and iodine (11) as well as *inter*molecular iodine (12) species.[32-34]

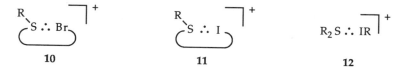

$$\begin{array}{ccc} \textbf{10} & \textbf{11} & \textbf{12} \end{array}$$

With regard to optical absorptions, influence of substituents and molecular structure all observed trends follow the same rules as outlined for the $(>S \cdot \cdot \cdot S<)^+$ radical cations. The most stable $(>S \cdot \cdot \cdot X-)^+$ radical cations are *intra*molecular species (10 and 11) which are able to attain five-membered ring structures (see also structures 1 and 2). Particularly interesting are the $S \cdot \cdot \cdot Br$ bonded systems since they represent rare examples of bromine based cationic species in organic chemistry.

Pulse radiolysis could also answer the question where in a sulfur-halogen organic compound, e.g. in $CH_3S(CH_2)_3I$, the primary oxidation takes place. By comparison with the physico-chemical, particularly pH characteristics of the all-sulfur- and all-iodine-centered radical cations, $(>S \cdot \cdot \cdot S<)^+$ and $(-I \cdot \cdot \cdot I-)^+$, it seems to be the sulfur atom which is primarily attacked.[33]

SULFUR-OXYGEN INTERACTION

Many examples have been found and described for the interaction between sulfur and other hetero atoms, particularly nitrogen. Corroborating results exist also for $X \cdot \cdot \cdot X$ and $X \cdot \cdot \cdot Y$ bonded systems not involving sulfur at all. In principle, the same considerations apply for their characterization as for the sulfur-centered species. Therefore, they shall not be presented and discussed in this article, except for some final remarks on sulfur-oxygen interactions. Establishment of a three-electron bond between these two elements of very different electronegativity is expected to lead to a rather asymmetric MO energy diagramme, exemplified in Figure 7 (b), and the $S \cdot \cdot \cdot O$ bond strength should, accordingly, be relatively low.

Nevertheless, the discussion of such $S \cdot \cdot \cdot O$ bonds is most relevant since, as mentioned before, the major part of knowledge on $2\sigma/1\sigma^*$ three-electron systems in the condensed phase stems from investigations in aqueous solutions. A most pertinent question in this context is concerned with the possibility of an interaction between an oxidized sulfur and the oxygen of a water molecule and the stability of the radical cation formed in reaction (19).

$$R_2S^{\bullet+} \quad + \quad H_2O \quad \rightleftarrows \quad (R_2S \cdot \cdot \cdot OH_2)^+ \tag{19}$$

Clark has estimated the three-electron bond strength to *ca* 70 kJ mol^{-1} (see article on "The Electronic Properties of Sulfur Containing Substituents ...." by T. Clark in this book) This would, most reasonably, identify the "molecular" radical cation also as a defined species composed by association of two hetero atoms. The magnitude of the bond strength would also nicely explain why in case of the all-*tert*-butyl substituted sulfide no $(t\text{-}Bu_2S\cdot^\bullet\cdot St\text{-}Bu_2)^+$ dimer complex could be detected. Its S$\cdot^\bullet\cdot$S bond strength as extrapolated from Figure 5a and Table 2 would be smaller than that for the $(t\text{-}Bu_2S\cdot^\bullet\cdot OH_2)^+$ radical cation and, indeed, only the latter can be observed ($\lambda_{max} \approx 310$ nm). Generally, any $R_2S^{\bullet+}$ in aqueous (and probably all hetero atom containing) solutions should be viewed as the monosolvated three-electron bonded species.

Much more stable sulfur-oxygen interaction can be achieved via *intra*molecular association in sterically rigid and structurally defined molecules as in the following two examples 13 and 14.[35,36]

13   14

They are obtained upon oxidation (which takes place at the sulfur atom) of *endo*-2-(2-hydroxy-2-methyl-ethyl)-*endo*-6-(methylthio)-bicyclo[2.2.1]heptane and *endo*-2-(carboxyl)-*endo*-6-(methylthio)-bicyclo[2.2.1]heptane, respectively. Optical absorptions peak near 400 nm. The sulfur-oxygen stabilization energies in 13 and 14 rely on the knowledge of the $(R_2S\cdot^\bullet\cdot OH_2)^+$ bond energy and, with Clark's new value, can be estimated to ca 90-100 kJ mol^{-1}.

CONCLUSIONS

2σ/1σ* Three-electron bonded radical species from sulfur organic compounds constitute significant and interesting intermediates. They exhibit unique physical and chemical properties, and are particularly accessible for investigation by using time-resolved optical detection techniques such as pulse radiolysis or laser-flash photolysis. Their stability is mainly controlled by the substituents at sulfur and structural parameters. Bond energies seem to range from about half to one third of that for a normal 2σ bond.

For sulfur interaction with other hetero atoms the thermodynamic stability generally decreases with increasing difference between the electronegativity of the two associating centers. As a consequence the distribution of the three electrons involved becomes less and less symmetric, and the spin is accordingly expected to be increasingly localized at the more electropositive atom. This raises, of course, the question as to what extent we are still allowed to view this eventually only very weak interaction in terms of the 2σ/1σ* concept. Clearly, the sulfur-oxygen examples mentioned in this article display all characteristic features of an S$\cdot^\bullet\cdot$O bond. However, a sulfur-centered σ-radical ($>S^\bullet-$ O$-$)[37], a coulombic interaction (e.g. $>S^{\bullet+} \leftrightarrow {}^-OOC-$ as in 14),[36] or π-delocalization (in case of aromatic substituents)[38] may equally be forwarded for the most appropriate description of one or the other property. In that sense sulfur-oxygen (and in general X-Y) interactions lead to the limits of the 2σ/1σ* concept. This seems best verified in systems of high symmetry and strong hetero atom interaction. There is no question that such species play a most important and interesting role in free radical and particularly redox chemistry of organic sulfides and related compounds.

ACKNOWLEDGEMENT

Most of the experimental work presented and discussed in this article has been performed in the Radiation Chemistry laboratories of the Hahn-Meitner Institute Berlin, often in close collaboration and contact with renowned colleagues and research groups all over the world. I would like to express my sincere thanks to all of those who experimentally and/or intellectually have contributed to the picture of sulfur-centered three-electron bonds as it exists now. Although their names are documented in the numerous publications it seems more than appropriate to name at least those who were directly involved in the work performed at our institute and which is reported in this article (in alphabetical order) : Elke Anklam, Detlef Bahnemann, Marija Bonifacic, Shamim A. Chaudhri, Christian H. Fischer, Manfred Göbl, Roland Goslich, Kamal Kishore, Sabine Mahling, Hermann J. Möckel, Jörg Mönig, Hari Mohan, and Dieter Veltwisch. I also very much appreciated the most fruitful collaboration with Richard S. Glass (University of Arizona at Tucson) and George S. Wilson (University of Kansas at Lawrence) and their students on the structurally defined norbornane derivatives. Special thanks are also due to Tim Clark (University of Erlangen-Nürnberg) and Steven Nelsen (University of Wisconsin at Madison) who, through their open and lively discussion with us, contributed significantly to our understanding of the three-electron systems and thus have been a great stimulus for our investigations.

Finally, it is my great pleasure to acknowledge the financial support provided by the "Deutsche Forschungsgemeinschaft", the "Internationales Büro der KfA Jülich" (German-Yougoslav science exchange programme), and (in connection with the collaboration with R. S. Glass and G. S. Wilson) by the National Institute of Health and NATO-Science.

REFERENCES

1. K.-D. Asmus, in "Methods in Enzymology", L. Packer (ed.), Academic Press, New York, Vol. 105, p. 167 (1984); and Vol. 186, p. 168 (1990).
2. B.C. Gilbert, D.K.C. Hodgeman, and R.O.C. Norman, *J. Chem. Soc. Perkin Trans. 2* 1748 (1973).
3. W.B. Gara, J.R.M. Giles, and B.P. Roberts, *J. Chem. Soc. Perkin Trans. 2* 1444 (1980).
4. R.L. Petersen, D.J. Nelson, and M.C.R. Symons, *J. Chem. Soc. Perkin Trans. 2* 225 (1978).
5. L. Bonazzola, J.P. Michaut, and J. Roncin, *Can. J. Chem.* 66:3050 (1988).
6 a. T. Clark, *J. Comput. Chem.* 2:261 (1981).
   b. T. Clark, *J. Am. Chem. Soc.* 110:1672 (1988).
7. P.M.W. Gill and L. Radom, *J. Am. Chem. Soc.* 110:4931 (1988).
8. A.J. Illies, P. Livant, and M.L. McKee, *J. Am. Chem. Soc.* 110:7980 (1988).
9. T. Momose, T. Susuki, and T. Shida, *Chem. Phys. Letters* 107:568 (1984).
10. M. Bonifacic, H. Möckel, D. Bahnemann, and K.-D. Asmus, *J. Chem. Soc. Perkin Trans. 2* 675 (1975).
11. K.-D. Asmus, *Acc. Chem. Res.* 12:436 (1979).
12. M. Göbl, M. Bonifacic, and K.-D. Asmus, *J. Am. Chem. Soc.* 106:5984 (1984).
13. S.A. Chaudhri and K.-D. Asmus, *Angew. Chem.* 93:690 (1981), *Angew. Chem. Int. Ed. (Engl.)* 20:672 (1981).
14. J. Mönig, R. Goslich, and K.-D. Asmus, *Ber. Bunsenges. Phys. Chem.* 90:115 (1986).
15. S.A. Chaudhri and K.-D. Asmus, unpublished results.
16. S.A. Chaudhri, M. Göbl, T. Freyholdt, and K.-D. Asmus, *J. Am. Chem. Soc.* 106:5988 (1984).
17. K.-D. Asmus, D. Bahnemann, Ch.-H. Fischer, and D. Veltwisch, *J. Am. Chem. Soc.* 101:5322 (1979).
18. D. Bahnemann and K.-D. Asmus, *J. Chem. Soc. Chem. Commun.* 238 (1975).
19. K.-D. Asmus, D. Bahnemann, M. Bonifacic, and H. A. Gillis, *Faraday Discussions* 63:213 (1977).

20. K.-D. Asmus, H.A. Gillis, and G.G. Teather, *J. Phys. Chem.* 82:2677 (1978).

21a. W.K. Musker and T.L. Wolford, *J. Am. Chem. Soc.* 98:3055 (1976).

  b. W.K. Musker, T.L. Wolford, and B.P. Roush, *J. Am. Chem. Soc.* 100:6416 (1978).

  c. W.K. Musker, *Acc. Chem. Res.* 13:200 (1980).

22. E. Anklam, K.-D. Asmus, and H. Mohan, *J. Phys. Org. Chem.* 3:17 (1990).

23. M. Bonifacic and K.-D. Asmus, *J. Org. Chem.* 51:1216 (1986).

24. E. Anklam, R.S. Glass, K.-D. Asmus, unpublished results.

25. E. Anklam and S. Steenken, *J. Photochem. Photobiol.* A43:233 (1988).

26. R. Wilbrandt, N.H. Jensen, P. Pagsberg, A.H. Sillesen, K.B. Hansen, and R.E. Hester, *J. Raman Spectrosc.* 11:24 (1981).

27. M. Bonifacic, J. Weiß, S.A. Chaudhri, and K.-D. Asmus, *J. Phys. Chem.* 89:3910 (1985).

28. S.F. Nelsen and K.-D. Asmus, unpublished results.

29. M. Bonifacic and K.-D. Asmus, *J. Chem. Soc. Perkin Trans.* 2 758 (1980).

30. K. Kishore and K.-D. Asmus, *J. Chem. Soc. Perkin Trans.* 2 2079 (1989).

31. J.E. Packer, *J. Chem. Soc. Perkin Trans.* 2 1015 (1984).

32. E. Anklam, H. Mohan, and K.-D. Asmus, *J. Chem. Soc. Chem. Commun.* 629 (1987).

33. E. Anklam, H. Mohan, and K.-D. Asmus, *J. Chem. Soc. Perkin Trans.* 2 1297 (1988).

34. E. Anklam, H. Mohan, and K.-D. Asmus, *Helv. Chim. Acta* 70:2110 (1987).

35 R.S. Glass, M. Hojjatie, G.S. Wilson, S. Mahling, M. Göbl, and K.-D. Asmus, *J. Am. Chem. Soc.* 106:5382 (1984).

36. S. Mahling, K.-D. Asmus, M. Hojjatie, R.S. Glass, and G.S. Wilson, *J. Org. Chem.* 52:3717 (1987).

37. C. Chatgilialoglu, A.L. Castelhano, and D. Griller, *J. Org. Chem.* 50:2516 (1985).

38. C.W. Perkins, J.C. Martin, A.J. Arduengo, W. Lau, A. Alegria, and J.K. Kochi, *J. Am. Chem. Soc.* 102:7753 (1980).

RADICAL CATIONS OF CONJUGATED CYCLIC DISULFIDES

Glen A. Russell and W.C. Law

Department of Chemistry
Iowa State University
Ames, Iowa 50011, USA

Persistent radical cations of heterocycles containing two conjugated sulfur atoms are readily detected by ESR spectroscopy. These systems are often more persistent than their acyclic analogues because scission of a sulfur-sulfur bond does not lead to irreversible decomposition. The cation radicals formed from **1-10** will be discussed in terms of their formation, conformations and ESR spectroscopy. The α-substituted derivatives of **1-10** will be labeled as **a**, monomethyl; **b**, dimethyl; **c**, butane-1,4-diyl; **d**, monophenyl; **e**, pentane-1,3-diyl; **f**, ethane-1,2-diyl.

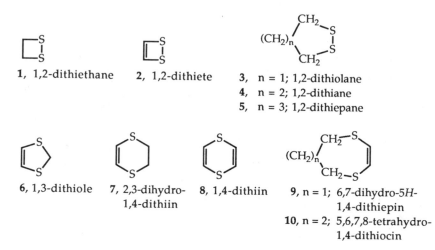

1, 1,2-dithiethane **2, 1,2-dithiete**

3, n = 1; 1,2-dithiolane
4, n = 2; 1,2-dithiane
5, n = 3; 1,2-dithiepane

6, 1,3-dithiole **7, 2,3-dihydro-1,4-dithiin** **8, 1,4-dithiin**

9, n = 1; 6,7-dihydro-5H-1,4-dithiepin
10, n = 2; 5,6,7,8-tetrahydro-1,4-dithiocin

Radical cations of **3-10** can be prepared by oxidation of the preformed heterocycles using reagents as simple as H_2SO_4 or Al_2Cl_6/CH_2Cl_2. The preparation of radical cations in the series **1** and **3** is best performed by oxidations of the 1,2- or 1,3-dithiols. Radical cations in series **1** and **2** can also be prepared by the reaction of S_2Cl_2 or S_8 in the presence of Al_2Cl_6 or $SbCl_5$, usually in CH_2Cl_2 solution.[2]

Thiete radical cations slowly decompose to the 1,4-dithiin radical cations at room temperature although in the presence of an excess of the sulfurizing agent they can often be observed free of the dithiin radical cation. In the case of benzodithiete[•+] decomposition leads to thianthrene[•+], a species also formed by the reaction of diphenyl disulfide with Al_2Cl_6.[2]

Sulfur-Centered Reactive Intermediates in Chemistry and Biology, Edited by
C. Chatgilialoglu and K.-D. Asmus, Plenum Press, New York, 1990

Cyclic conjugated disulfide radical cations are often the final and/or most persistent paramagnetic species formed upon oxidation in acidic solutions of non-conjugated compounds. Scheme 1 gives some typical examples.

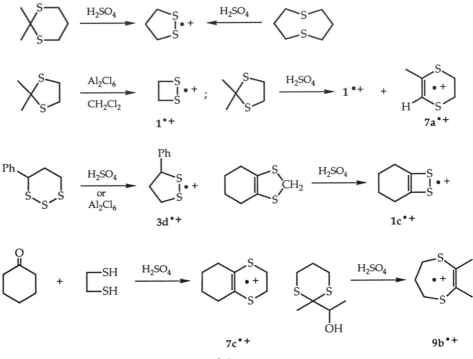

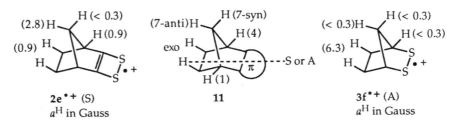

Scheme 1

For π spin systems such as **(2,6-10)**ˑ⁺, the symmetry of the SOMO (or at least the relative phases of the coefficient in the SOMO at C(2) and C(3) that the spin label shares with the bicyclo[2.2.1]hept-2-ene system) can be experimentally determined from the ESR spectra of the bicyclic radical cation **11**.[3] For symmetric SOMO's (defined by the plane of the dotted

line in **11** and passing through C(7)), it is observed that a^H (7-anti) > a^H (7-syn) or a^H (5,6-exo) and that $a^H_{1,4} \approx a^H_{5,6\text{-exo}} >> a^H_{5,6\text{-endo}}$. For antisymmetric SOMO's, a^H (7-anti) ≈ a^H (7-syn) > $a^H_{5,6\text{-exo}}$ while $a^H_{1,4}$ is variable but often ≈ 0).

SATURATED CYCLIC 1,2-DISULFIDE RADICAL CATIONS

Treatment of ethane-1,2-dithiol or 2,3-dihydro-1,4-dithiin with H_2SO_4 forms 1,2-thietane ˑ⁺ (**1**ˑ⁺) as a transient intermediate which is slowly replaced by the very stable Δ²,²'-bi-1,3-dithiolanylidene radical cations (reaction 1).[4] With mono- or 1,2-disubstituted

$$\boxed{\begin{array}{c}\text{—SH}\\[-2pt]\text{—SH}\end{array}} \quad \text{or 7} \quad \xrightarrow{\text{H}_2\text{SO}_4} \quad \mathbf{1}^{\bullet+} \quad \longrightarrow \quad (1)$$

ethane-1,2-dithiols the initially formed 1,2-dithietane$^{\bullet+}$ in H_2SO_4 is further oxidized to the dithiete$^{\bullet+}$ which slowly decomposes to the 1,4-dithiin$^{\bullet+}$ (reaction 2).

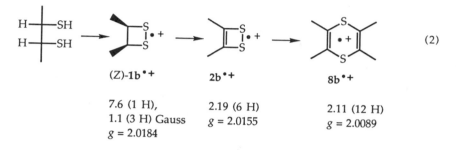

(2)

(Z)-1b$^{\bullet+}$	2b$^{\bullet+}$	8b$^{\bullet+}$
7.6 (1 H),	2.19 (6 H)	2.11 (12 H)
1.1 (3 H) Gauss	$g = 2.0155$	$g = 2.0089$
$g = 2.0184$		

Treatment of ethanedithiols with Al_2Cl_6/CH_2Cl_2 at 25 °C for 10-30 min leads to persistent $\mathbf{1}^{\bullet+}$ free of further oxidation products. The ESR spectra from –90 to 25 °C indicate that $\mathbf{1a}^{\bullet+}$, (Z)-$\mathbf{1b}^{\bullet+}$, (E)-$\mathbf{1b}^{\bullet+}$, (Z)-$\mathbf{1c}^{\bullet+}$ or (E)-$\mathbf{1c}^{\bullet+}$ are locked in a single non-planar conformation where quasi-axial hydrogen atoms (position 1 in $\mathbf{12}^{\bullet+}$) have a large hyperfine splitting constant (hfsc) of 6-8 Gauss and quasi-equatorial hydrogen atoms (position 2 in $\mathbf{12}^{\bullet+}$) have values of $a^H < 0.5$ Gauss.

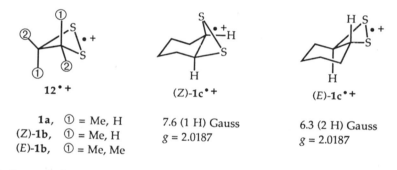

$\mathbf{12}^{\bullet+}$	(Z)-$\mathbf{1c}^{\bullet+}$	(E)-$\mathbf{1c}^{\bullet+}$
	7.6 (1 H) Gauss	6.3 (2 H) Gauss
	$g = 2.0187$	$g = 2.0187$

$\mathbf{1a}$, ① = Me, H
(Z)-$\mathbf{1b}$, ① = Me, H
(E)-$\mathbf{1b}$, ① = Me, Me

For $\mathbf{1a}^{\bullet+}$ the methyl substituent prefers the quasi-axial position to give a single hfsc of 8.5 Gauss. For (Z)-$\mathbf{1b}^{\bullet+}$ there is a single α-hydrogen with a large hfsc but for (E)-$\mathbf{1b}^{\bullet+}$ both methyls occupy the axial positions and no hfs was resolved ($\Delta H_{1/2} = 2.5$ Gauss), $g = 2.0185$. The Z and E cyclohexane derivatives $\mathbf{1c}^{\bullet+}$ are consistent with this assignment. The lack of hfs for the quasi-equatorial hydrogens in $\mathbf{12}^{\bullet+}$ is probably a result of the antisymmetric SOMO and the cancellation of 1,2 and 1,3 hyperconjugation and homohyperconjugation interactions.[4] Radical cation (Z)-$\mathbf{1c}^{\bullet+}$ is also conveniently prepared from cyclohexene by reaction with S_2Cl_2 or S_8 in the presence of Al_2Cl_6/CH_2Cl_2 at 25 °C. However, (E)-$\mathbf{1c}^{\bullet+}$ must be prepared by oxidation of the trans dithiol.[1]

Cooling a solution of the $\mathbf{1}^{\bullet+}$ in CH_2Cl_2 (4 equivalent hydrogens at 25 °C) gives line broadening and coalescence at $\approx$ –35 °C to yield, at –50 °C, a sharp triplet, $a^H = 6.8$ (2 H), < 0.6 (2 H) Gauss. From line broadening in the fast exchange mode above –35 °C, a value of $\Delta H^{\ddagger} = 3.3$ kcal, $\Delta S^{\ddagger} = -12$ cal/° is found for the ring flip of $\mathbf{12}^{\bullet+}$.

1,2-Ditholane radical cation ($\mathbf{3}^{\bullet+}$) also shows a strong conformational preference with an appreciable barrier to conformational interconversion.[5] The monophenyl derivative, $\mathbf{3d}^{\bullet+}$ (see Scheme 1), shows a single populated conformation up to 60 °C in CH_2Cl_2 but now the

3d•+

14.4 (1 H), 11.9 (1 H), 4.7 (1 H), 1.2 (1 H) Gauss

$g = 2.0172$

substituent prefers the quasi-equatorial position. For unsubstituted **3**, readily prepared from propane-1,3-dithiol or by the reactions of Scheme I, 4 equivalent α-hydrogen atoms are observed at 25 °C with $a^H = 10.0$ Gauss, $g = 2.0182$. Cooling in CH_2Cl_2 results in line broadening and coalescence with $\Delta H^{\ddagger} = 4.4$ kcal, $\Delta S^{\ddagger} = -4.5$ cal/° (previously reported $\Delta H^{\ddagger} = 8$ kJ/mol)[5] to give at –90 °C a triplet of triplets, $a^H = 16.25$ (axial), 3.9 (equatorial, $a(^{33}S) = 13.3$ Gauss.[2] The value of a^S indicates that the unpaired electron is in a p-orbital on sulfur since in a s-orbital a value of 670 Gauss would be observed for a full electron spin. Usually the value of a^S is given by 33 ρ_s (in Gauss) where ρ_s is the sulfur (p) spin density (c_s^2). For **3** a spin density on each sulfur of ≈ 0.4 is indicated. For **1•+** the value of a^S is similar (14 Gauss) indicating that in conformation **12** that electrons are still in p-orbitals of sulfur.

Radical cations are easily prepared from **4** to **5** by treatment with H_2SO_4 or Al_2Cl_6/CH_2Cl_2. For the parent system the α-hydrogen atoms are time-averaged by ring flip at 25 °C ($a^H ≈ 9.5$(4 H) Gauss, $g = 2.0182$-5) but for 4,4,6,6-tetramethyl-1,2-dithiepane, the radical cation is in a locked conformation ($\Delta H^{\ddagger} > 7$ kcal) at 25 °C with hfsc from only the two quasi-axial hydrogen atoms.[5] The 3-methyl-1,2-dithiane radical cation (**4a•+**) is time-averaged at 25 °C, $a^H = 8.0$ (1 H), 9.9 (1 H), 12.6 (1 H) Gauss but at –70 °C the half-chair conformation is frozen with the quasi-axial hydrogen having $a^H = 12.0$ (1 H), 12.5 (1 H) Gauss and the single quasi-equatorial hydrogen having $a^H = 4.0$ Gauss. A locked conformation is also observed for the radical cation of *trans*-2,3-dithiadecalin (**13•+**) and 1,4-dithiatetralin (**14•+**). Radical cation **14•+** is amazingly stable conformationally since only a sharp triplet is observed at 80 °C. The α,α-dideuterio derivative of **14•+** gives $a^H = 15.8$, $a^D = 2.42$ Gauss and thus confirms the structure.

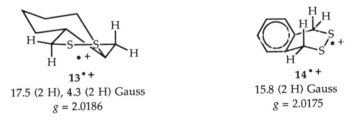

13•+

17.5 (2 H), 4.3 (2 H) Gauss

$g = 2.0186$

14•+

15.8 (2 H) Gauss

$g = 2.0175$

Bicyclic derivatives of **4•+** such as **3f•+** (see introductory section) or **4f•+** have boat conformations for the 6-membered rings. Again the α-equatorial hydrogen atoms are not seen but long range splittings can be seen for β-hydrogen atoms not in the nodal plane of the π-system.

4f•+

3.8 (4 H) Gauss

$g = 2.0186$

d₂-4f•+

d_2-4f•+

3.8 (2 H) Gauss

$g = 2.0186$

Radical cations **3f•+** and **4f•+** can be prepared from the *cis*-dithiols in H_2SO_4 although the parent disulfides cannot be isolated.

THE DITHIETE AND 1,4 DITHIIN RADICAL CATIONS

The 1,2-dithiete radical cation is a rather ubiquitous species since it is formed in many oxidizing systems.[6,7] 1,2-Disubstituted thietes are stable with the substituents CF_3, t-Bu or $-CMe_2(CH_2)_3CMe_2-$ and these compounds in H_2SO_4 or Al_2Cl_6/CH_2Cl_2 form the radical cations readily (for 3,4-di-*tert*-butyl-2 the first vertical ionization potential is ≈ 7.95 eV),[7] usually accompanied by the 1,4-dithiin radical cation. Some of the other routes to dithiete radical cations are shown in Scheme 2. In addition 1,3-dithioles without substituents at C(2) are degraded to the dithiete by concentrated H_2SO_4 (Scheme 1).

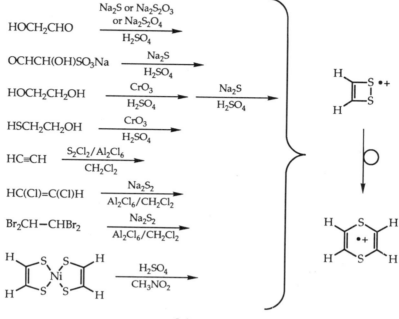

Scheme 2

The unsubstituted dithiete radical cation has $a^H = 2.75$ (2 H), $a(^{33}S) = 8.5$ Gauss with $g = 2.0153$ while the 1,2-dimethyl derivative has $a^H = 2.15$ (6 H), $a^C(\text{vinyl}) = 1.2$, $a^S = 7.6$ Gauss, $g = 2.0148$. A wide variety of substituted 1,2-dithiete radical cations and the corresponding *cis*-1,2-semidiones, *e.g.* **15a,b** and **16a,b** have a constant ratio of a^H for the vinyl hydrogen atoms, or for the α-hydrogen atoms of the substituent, of 0.30-0.40.[6] Even the long range splittings of **2e**$^{\bullet+}$ and **17**$^{\bullet-}$ are in a similar ratio. Since the *cis*-1,2-semidione has a carbonyl-carbon spin density of 0.25, the carbon spin density in the dithiete radical cation must be no greater than 0.1 and hence, $\rho_s \approx 0.4$. With CF_3 substituents the values of a^F are much greater for the semidione (10.9 Gauss) than for the dithiete (1.35 Gauss) reflecting the greater hyperconjugation for an electronegative substituent when attached to a negatively charged spin center.[6] However, HMO calculations with $\beta(SS) = 0.4 \beta(CC)$ or RHF/STO-3G (with UHF 3-21G optimized geometry) for **2**$^{\bullet+}$ indicate that the SOMO coefficients should be nearly 0.5 at each atom,[8] i.e., $\rho_c = \rho_s = 0.25$ and this is in agreement with the low value of a^S which is comparable to other highly delocalized cyclic disulfide cations, e.g., **18**$^{\bullet+}$ and **19**$^{\bullet+}$ (compare with a^S for **1**$^{\bullet+}$ and **3**$^{\bullet+}$ of 14.0 and 13.3 Gauss).[7]

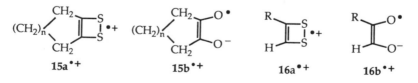

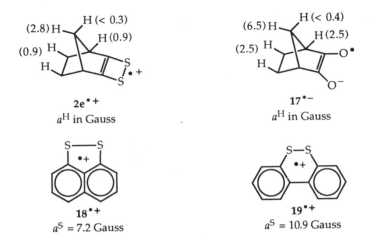

2e•+

a^H in Gauss

17•−

a^H in Gauss

18•+

$a^S = 7.2$ Gauss

19•+

$a^S = 10.9$ Gauss

The MO calculations predict that the SOMO should be S as defined in structure **2e•+** (see Introduction). One might expect that with $\beta(CC) > \beta(CS)$ that the aS orbital would be lower in energy than the sA orbital. The main factor leading to the higher energy of the aS

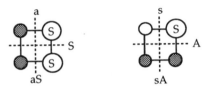

orbital is that $\beta(SS) << \beta(CC)$, i.e., the magnitude of the interaction between the two sulfur atoms is much weaker than between the two carbon atoms. This leads to a smaller energy gap between the symmetric and antisymmetric local SS π-orbitals than between the corresponding local CC orbitals. When the local orbitals of the same symmetry interact with each

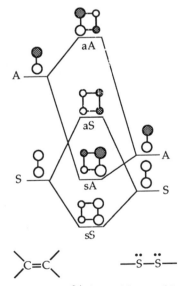

Fig. 1. Molecular orbitals for **2•+** formed by combination of local –SS– are >C=C< π-orbitals.

other, the resulting aS orbital becomes higher in energy than the sA molecular orbital (Figure 1).[9] The dithiete radical cation appears to be a perfectly behaved π-radical when compared to the 1,4-dithiin$^{\bullet+}$ ($8^{\bullet+}$) where HMO calculations predict the spin densities quite well ($a^H = -23\ \rho_c$, $a^H_{CH_3} = 28\ \rho_c$), Table 1. It appears that both systems have similar π-electron spin densities at carbon of about 0.1.

It is not obvious why $2^{\bullet+}$ appears to be a simple π-radical but with unexpectedly low carbon spin density. It is true that HMO calculations using $\beta(CC) = \beta(SS) > \beta(CS)$ will give a carbon spin density of ≈ 0.1 but this value of $\beta(SS)$ is unrealistic. Possibly the dithiete radical cation is delocalized but non-planar like $1^{\bullet+}$, but the value of a^C and a^S indicates little s-character in the atomic orbitals involved. Another possibility is that the structure is misassigned. An equilibrium such as reaction 3 could perhaps explain the low carbon spin density but the missing sulfur spin density and g-value is hard to understand. (However, Me$_2$SSMe$_2^{\bullet+}$ has a fairly low g-value of 2.0103). The ^{33}S labeling experiments of Bock et al.[2]

$$\underset{20a^{\bullet+}}{\left[\!\!\!\begin{array}{c}\end{array}\!\!\!S^1\overset{\bullet+}{\rule{0.8cm}{0.4pt}}S^2\right]} \rightleftarrows 2^{\bullet+} \rightleftarrows \underset{20b^{\bullet+}}{\left[\!\!\!\begin{array}{c}\end{array}\!\!\!S^2\overset{\bullet+}{\rule{0.8cm}{0.4pt}}S^1\right]} \tag{3}$$

definitely require two equivalent sulfur atoms and so the thiirene radical cation can be excluded (of course, the value of a^S also excludes this species). The postulation of $20^{\bullet+}$ does provide a reasonable explanation of the conversion of the dithiete$^{\bullet+}$ to the 1,4-dithiin$^{\bullet+}$, e.g., reaction (4).

$$2\ \overset{+}{\left[\!\!\!\begin{array}{c}\end{array}\!\!\!S\!-\!\overset{\bullet}{S}\right]} \xrightarrow{-S_2} 2\ \left[\!\!\!\begin{array}{c}\end{array}\!\!\!S^{\bullet+}\right] \longrightarrow \left[\begin{array}{c}S\\ \\S\end{array}\right]^{2+} \longrightarrow 8^{\bullet+} \tag{4}$$

Table 1. Hydrogen Hyperfine Splitting Constants of 1,2-Dithiete and 1,4-Dithiin Radical Cations (in Gauss)

R¹, R²	R¹⎥⎥S / R²⎥⎥S $^{\bullet+}$	R¹,S,R¹ / R²,S,R² $^{\bullet+}$
H, H	2.75 (2 H)[a]	2.84 (4 H)[b]
Me, Me	2.17 (6 H)[c]	2.11 (12 H)[d,e]
Et, Et	2.00 (4 H)	1.46 (8 H)
-(CH$_2$)$_3$-	5.37 (4 H)	4.75 (8 H)
-(CH$_2$)$_4$-	3.04 (4 H)	2.92 (8 H)
H, CH$_3$	3.2 (1 H), 1.9 (3 H)	2.85 (2 H), 2.45 (6 H)[f]

[a] $a^S = 8.5$ [b] $a^S = 9.8$ [c] $a^S = 7.6$, a^C (vinyl) = 1.2.
[d] For the dioxin analogue, $a^H = 4.06$ (12 H), a^C (vinyl = Me) = 2.3.
[e] a^C (vinyl = Me) = 1.4.
[f] For the isomer with methyls in the 2,5-positions, $a^H = 3.50$ (2 H), 1.75 (6 H).

1,3-Dithioles

1,3-Dithiole radical cations are relatively unstable and readily decompose to the 1,2-dithiete radical cation, particularly in H_2SO_4 and when C(2) is unsubstituted. However, decomposition is also observed in Al_2Cl_6/CH_2Cl_2 and when C(2) is disubstituted. In the benzo-1,3-dithiole series this decomposition is observed with the 2-methyl and 2,2-dimethyl derivatives, reaction 5, although $21^{\bullet+}$ can be observed free of $22^{\bullet+}$. In the presence of

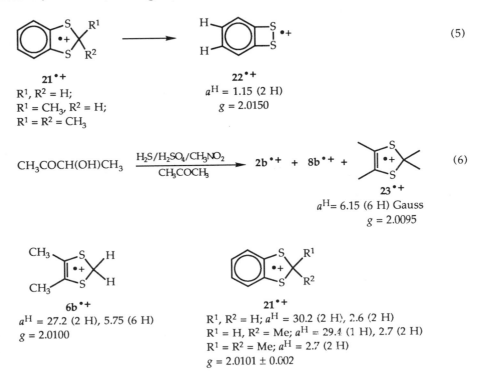

$$\text{(5)}$$

21$^{\bullet+}$
$R^1, R^2 = H;$
$R^1 = CH_3, R^2 = H;$
$R^1 = R^2 = CH_3$

22$^{\bullet+}$
$a^H = 1.15$ (2 H)
$g = 2.0150$

$$CH_3COCH(OH)CH_3 \quad \xrightarrow[\quad CH_3COCH_3 \quad]{H_2S/H_2SO_4/CH_3NO_2} \quad 2b^{\bullet+} + 8b^{\bullet+} + \quad \text{(6)}$$

23$^{\bullet+}$
$a^H = 6.15$ (6 H) Gauss
$g = 2.0095$

6b$^{\bullet+}$
$a^H = 27.2$ (2 H), 5.75 (6 H)
$g = 2.0100$

21$^{\bullet+}$
$R^1, R^2 = H; a^H = 30.2$ (2 H), 2.6 (2 H)
$R^1 = H, R^2 = Me; a^H = 29.4$ (1 H), 2.7 (2 H)
$R^1 = R^2 = Me; a^H = 2.7$ (2 H)
$g = 2.0101 \pm 0.002$

acetone, reaction of $CH_3COCH(OH)CH_3$ with Na_2S in H_2SO_4/CH_3NO_2 forms a mixture of $2b^{\bullet+}$, $8b^{\bullet+}$ and 2,2,4,5-tetramethyl-1,3-dithiole radical cation (**23**), reaction 6.[9] 4,5-Dimethyl-1,3-dithiole (**6b**) with Al_2Cl_6/CH_2Cl_2 at $-70\,°C$ forms $6b^{\bullet+}$ where the large hyperfine splitting for the methylene hydrogen atoms at C(2) is a result of the symmetrical MO and the hyperconjugative interaction, $a^H(C-2) = Q^H \cdot {}_{SCH}(c_1 + c_3)^2\cos^2\theta$.[4,10] A similar effect is observed for **21** at $-60\,°C$. No effect of temperature is observed for $6b^{\bullet+}$ or $21^{\bullet+}$. The 1,3-dithiole radical cation has a rather high carbon spin density of ≈ 0.20 (sulfur spin density ≈ 0.30) which leads to a value of $Q^H \cdot {}_{SCH} \approx 30$ Gauss (observed for $CH_3S^{\bullet}O$, 22.6 Gauss),[11] rather similar to $Q^H \cdot {}_{CCH} \approx 56$ Gauss for neutral radicals and ≈ 80 Gauss for radical cations. The sulfur atom is nearly as good as a carbon atom for $p\sigma$ hyperconjugative interactions.[4]

The cyclohexene derivative $6c^{\bullet+}$ shows similar hfs at C(2) but now at $-90\,°C$ the half-chain conformation of the cyclohexene ring is frozen (coalescence temperature $\approx -80\,°C$). Decomposition of $6c^{\bullet+}$ yields $2c^{\bullet+}$ which also possesses a conformationally mobile cyclohexene ring (Scheme 3). At low temperatures both systems demonstrate that for the α-hydrogen atoms in the cyclohexene ring a^H (axial) $\approx 2\,a^H$ (equatorial) as found previously for the semidione radical ion.[12] From selective line broadening above the coalescence temperature values of $\Delta H^{\ddagger} = 6.2$ and 6.0 kcal/mol and $\Delta S^{\ddagger} = 5.6$ and 3.8 cal/° are found for $6c^{\bullet+}$ and $2c^{\bullet+}$, respectively (cyclohexene-1,2-semidione has $\Delta H^{\ddagger} = 4.0$ kcal/mol, $\Delta S^{\ddagger} = 1.0$ cal/°).[12]

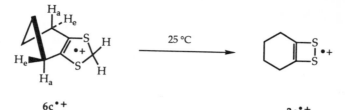

25 °C

6c•+

a^H (–30 °C, CH₂Cl₂) = 26.6 (2 H), 8.2 (4 H)
  g = 2.0101
a^H (–90 °C, CH₂Cl₂) = 26.2 (2 H), 11.0 (2 H),
  5.5 (2 H) Gauss

2c•+

a^H (25 °C, CH₂Cl₂) = 3.04 (4 H) Gauss
  g = 2.0155
a^H (–90 °C, CH₂Cl₂) = 4.02 (2 H),
  2.01 (2 H) Gauss

Scheme 3

2,3-Dihydro-1,4-dithiins

Radical cation 7•+ is unstable in H₂SO₄ and decomposes to 2•+ but 7•+ can be prepared by Al₂Cl₆/CH₂Cl₂ at –60 °C. Radical cations 7•+ or 7b•+ are conformationally mobile. However, for 7b•+ or 7e•+ in H₂SO₄, coalescence of the quasi-axial and equatorial hydrogen atoms was not observed until ≈ 70 °C ($\Delta G^{\ddagger}$ ≈ 9 kcal/mol). For 7b•+ values of $\Delta H^{\ddagger}$ = 2.3 kcal/mol and $\Delta S^{\ddagger}$ = –20 cal/° have been measured.[3]

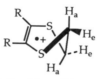

7•+, R = H; 7b•+, R = Me; 7e•+, R = cyclopentane-1,3-diyl

7•+, a^H (–80°C, CH₂Cl₂) = 8.15 (2 H), 3.4 (2 H), 2.7 (2 H) Gauss, g = 2.0092
7b•+, a^H (–80°C, CH₂Cl₂) = 5.7 (6 H), 6.9 (2 H), 2.1 (2 H) Gauss, g = 2.0080
7b•+, a^H (–10°C, H₂SO₄) = 5.6 (6 H), 7.0 (2 H), 2.1 (2 H) Gauss, g = 2.0080
7e•+, a^H (70°C, H₂SO₄) = 5.0 (1 H), 3.6 (4 H), 1.6 (4 H) Gauss, g = 2.0075

The cyclohexene rings in (24-26)•+ are in rapid conformational equilibrium even at –95 °C in CH₂Cl₂ and only four equivalent α-hydrogen atoms are seen. The 2,3-dihydro-1,4-dithiin ring is firmly locked in a half-chair structure by the cyclohexane chair conformation in 24•+ and 25•+ or in a boat conformation by the norbornane ring in 26•+.

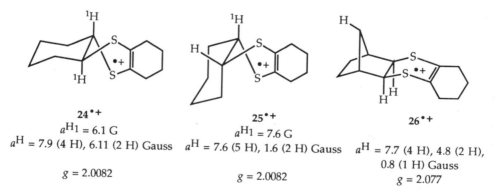

24•+
$a^H{}_1$ = 6.1 G
a^H = 7.9 (4 H), 6.11 (2 H) Gauss

g = 2.0082

25•+
$a^H{}_1$ = 7.6 G
a^H = 7.6 (5 H), 1.6 (2 H) Gauss

g = 2.0082

26•+
a^H = 7.7 (4 H), 4.8 (2 H),
  0.8 (1 H) Gauss
g = 2.077

In 27•+ presumably both rings are in half-chain conformations. One ring is conformationally mobile at low temperatures but the other ring does not begin to undergo inversion until

≈ -10 °C. By analogy with $7b^{\bullet+}$, it appears that the dihydrodithiin ring is conformationally more stable and this is consistent with $\Delta G^{\ddagger} \approx 9$ kcal for the ring flip of $7b^{\bullet+}$ at the coalescence temperature whereas $\Delta H^{\ddagger}$ for the cyclohexene ring flip is 5-6 kcal/mol.[13]

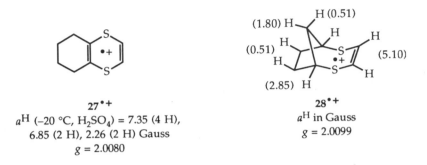

27$^{\bullet+}$
a^H (−20 °C, H_2SO_4) = 7.35 (4 H),
6.85 (2 H), 2.26 (2 H) Gauss
$g = 2.0080$

28$^{\bullet+}$
a^H in Gauss
$g = 2.0099$

6,7-Dihydro-5H-1,4-dithiepin and 5,6,7,8-tetrahydro-1,4-dithiocin derivatives

2,3-Dimethyl-6,7-dihydro-5H-1,4-dithiepin (**9b**) is converted to the radical cation with either H_2SO_4 or Al_2Cl_6/CH_2Cl_2. Below −60 °C in CH_2Cl_2 the α-hydrogen atoms have hfsc of $a^H = 5.3$ (2 H) and 1.85 (2 H) Gauss. Coalescence occurs at ≈ -60 °C with $\Delta H^{\ddagger} = 5.8$ kcal/mol, $\Delta S^{\ddagger} = -9.5$ cal/°.

Radical cation **28**$^{\bullet+}$ is both a dithiepien at a dithiocin derivative. It is formed from the parent bicyclic compound by Al_2Cl_6/CH_2Cl_2.

The dihedral angle dependence of the hfsc for saturated hydrogen atoms α to the sulfur atoms in **9b**$^{\bullet+}$ and **28**$^{\bullet+}$ supports the assumption of a strong pσ hyperconjugative interaction in these systems.

1,4-Dithiins

Both the mono- and disubstituted dithiin with the cyclopentane-1,4-diyl substituent have been prepared (Scheme 4). The hfsc demonstrate that the SOMO of **8**$^{\bullet+}$ (as well as **7**$^{\bullet+}$) has coefficients of the same sign at C(2) and C(3).

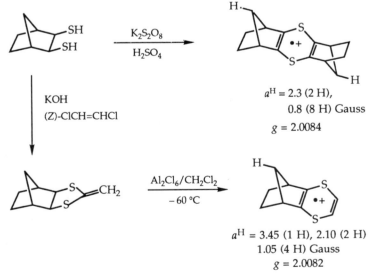

$a^H = 2.3$ (2 H),
0.8 (8 H) Gauss
$g = 2.0084$

$a^H = 3.45$ (1 H), 2.10 (2 H)
1.05 (4 H) Gauss
$g = 2.0082$

Scheme 4

ACKNOWLEDGMENT

This work was supported by grants from the National Science Foundation. Collaboration with Prof. Chen Zhixing, Zhongshan University, on calculations related to the structure of $2^{•+}$ is gratefully acknowledged.

REFERENCES

1. G.A. Russell and W.C. Law, *Heterocycles* 24:321 (1986).
2. H. Bock, P. Rittmeyer, and U. Stein, *Chem. Ber.* 119:3766 (1986).
3. G.A. Russell, W.C. Law, and M. Zaleta, *J. Am. Chem. Soc.* 107:4175 (1985).
4. G.A. Russell and M. Zaleta, *J. Am. Chem. Soc.* 104:2318 (1982).
5. H. Bock and U. Stein, *Angew. Chem. Inter. Ed. Engl.* 19:834 (1980).
6. G.A. Russell, R. Tanikago, and E.R. Talaty, *J. Am. Chem. Soc.* 94:6125 (1972).
7. H. Bock, P. Rittmeyer, A. Krebs, K. Schütz, J. Voss, and B. Köpke, *Phosphorus and Sulfur* 19:131 (1984).
8. Z. Chen, H. Ge, G.A. Russell, and W.C. Law, *Phosphorous and Sulfur* 35:121 (1988).
9. G.N. Schrauzer and H.N. Rakinowitz, *J. Am. Chem. Soc.* 92:5769 (1970).
10. D.H. Whiffen, *Mol. Phys.* 6:223 (1963).
11. K. Nishikida and F. Williams, *J. Am. Chem. Soc.* 96:4781 (1974).
12. G.A. Russell, G.R. Underwood, and D.C. Lini, *J. Am. Chem. Soc.* 89:6636 (1967).
13. F.A.L. Anet, and M.Z. Haq, *J. Am. Chem. Soc.* 87:3147 (1965).

BEAM EXPERIMENTS AS A MEANS FOR THE GENERATION OF

ELUSIVE SULFUR-CONTAINING MOLECULES

Detlev Sülzle, Thomas Drewello, and Helmut Schwarz

Institut für Organische Chemie
Technische Universität Berlin
1000 Berlin 12, F.R. Germany

INTRODUCTION

Quite a few molecular species can only be generated in the gas phase because in solution or even in the solid state *intermolecular* interactions will make them unstable and lead to isomerization, polymerization or dissociation.[1-3] The diluted gas phase which exists in a mass spectrometer is ideally suited for the study of isolated molecules. First of all, as the experiments are conducted in the absence of any counterions and solvents one does not have to account for any possible intermolecular effects. Secondly, the past ten years have witnessed an impressive development in mass spectrometric methodologies[4-8] for the tailor-made generation and characterization of solitary ions with unusual structures and/or intriguing properties. Among the species studied are not only singly-charged ions, but also di- and polycations, which are produced from a high kinetic energy beam of mono-cations via charge-stripping mass spectrometry (CSMS),[9-12] as well as unique neutrals. The latter can easily be generated and characterized by the promising technique of neutralization/reionization mass spectrometry (NRMS).[13-18] Without going into technical details, the two experiments consist of the following steps:

CSMS: Monocations, m_1^+, formed in the ion source of a multi-sector mass spectrometer, are accelerated to several keV kinetic energy and mass-selected by means of a magnetic field B before colliding with a collision gas, preferably oxygen, in a collision cell located in the field-free region between B and the electric sector E. Charge-stripping peaks due to reaction (1) are obtained by scanning the electric sector voltage (E) around E/2, where E represents the voltage required to transmit stable m_1^+ ions. The displacement of the peak from 0.5 E reflects the translational energy loss (Q_{min}) in process (1) and corresponds to the ionization energy (IE) of m_1^+.[10-12] The ionization energies thus obtained can be compared with theoretically evaluated energies. Further characterization can be obtained by performing high level *ab initio* MO calculations.[12,18]

$$m_1^+ + target \longrightarrow m_1^{2+} + target + e^- \qquad (1)$$

$$m_1^+ \xrightarrow[\text{neutralization}]{G_1} m_1 \xrightarrow[\text{reionization}]{G_2} m_1^+ \longrightarrow F_i^+ \qquad (2)$$

NRMS: The starting point is, again, a mono-cation, which after acceleration to typically 8 keV translational energy is mass selected and subjected to a sequence of collision

Sulfur-Centered Reactive Intermediates in Chemistry and Biology, Edited by
C. Chatgilialoglu and K.-D. Asmus, Plenum Press, New York, 1990

experiments. The ion beam is first reduced by electron transfer from a suitable gas G_1 (Xe or metal atoms [eq. 2]). All ions which have survived the neutralization step are deflected away from the molecular beam by a charged deflector electrode which is located between the two collision cells. The only species which enter the second collision cell are therefore the neutral molecule m_1 (having the momentum of m_1^+) and their uncharged dissociation products m_i (having the momentum m_i/m_1). The fast moving neutrals are reionized by collision with a target gas G_2 in the second chamber. This oxidation step should be performed in such a way that not only reionized molecular species m_1^+ but also charged, structure-indicative fragments are formed. The mass spectrum of the latter may serve to characterize m_1^+. Upon its generation in cell 1 it takes ca. 10^{-6} s for m_1 to reach cell 2. The presence of a recovery signal m_1^+ in the NR mass spectrum therefore signifies that the neutral m_1 has a lifetime $>10^{-6}$ s. For further details, in particular for the suitability of various collision gases or the problem of generating ground *vs.* excited states, see refs. 13 - 17.

In the present contribution a few examples from our laboratory are discussed to indicate the scope and limitations of both CSMS and NRMS as a means to generate elusive sulfur-containing species which, due to *inter*molecular reactions, are difficult if not impossible to make in the condensed phase.

DISCUSSION OF SELECTED EXAMPLES

Sulfurous Acid (H_2SO_3)

Sulfurous acid, H_2SO_3, has in common with carbonic acid, $(HO)_2CO$, and carbamic acid, H_2NCO_2H, that general textbook knowledge takes it to be non-existent in the free state. Recently we were able to demonstrate[19-21] that this instability does not reflect an intrinsic property of these molecules but is rather determined by the "environment" (acid/base or solvent catalysis provoke a rapid decomposition). In the gas phase all three neutral molecules as well as their cation radicals are perfectly stable species, separated by significant barriers from both unimolecular dissociations and interconversion to isomers. Indeed, carbonic acid **1** and carbamic acid **2**, were generated by thermolysis of NH_4OCOX (X = OH, NH_2) according to eq. 3; both species could be oxidized by electron impact ionization to the likewise stable cation radicals $1^{\bullet+}$ and $2^{\bullet+}$, respectively, whose neutralization was accomplished using NRMS.

$$\text{NH}_4\text{OCOX} \xrightarrow[-\text{NH}_3]{\Delta} \underset{X}{\overset{\text{HO}}{>}}\text{C=O} \underset{\text{reduction}}{\overset{\text{oxidation (70 eV)}}{\rightleftharpoons}} \underset{X}{\overset{\text{HO}}{>}}\text{C=O}^{\bullet+} \qquad (3)$$

1: X = OH $1^{\bullet+}$: X = OH
2: X = NH_2 $2^{\bullet+}$: X = NH_2

3   **5**$^{\bullet+}$ **5**   (4)

4   **6**$^{\bullet+}$

186

For the generation of **5** and **5$^{\bullet +}$** a different approach had to be used. The obvious route to **5**, namely via thermolysis of ammonium hydrogen sulfite, is thwarted by the fact that anhydrous NH_4HSO_3 is unknown. On the other hand, dissociative ionization of **3** and **4** (eq. 4) was demonstrated[21] (using labeling studies in conjunction with collisional activation (CA) experiments[11]) to result in the exclusive formation of **5$^{\bullet +}$** (but *not* **6$^{\bullet +}$**); neutralization of **5$^{\bullet +}$** furnished the long-sought after neutral sulfurous acid **5**. According to HF 6-31G*//6-31G* + ZPVE calculations **5** is 17 kcal/mol more stable than the as yet unknown **6**; the barrier for the (hypothetical) reaction **5 $\rightarrow$ 6** was calculated to 88 kcal/mol. Similarly, at this level of theory **5$^{\bullet +}$** is predicted to be 37 kcal/mol more stable than **6$^{\bullet +}$** with a barrier of 82 kcal/mol for the 1.2-H migration of reaction **5$^{\bullet +}$ $\rightarrow$ 6$^{\bullet +}$**.

The HSO$_3$$^{\bullet}$ Radical

The key intermediate in the production of "acid rain" by atmospheric oxidation of sulfur dioxide, SO_2, which accounts for essentially all antropogenic sulfur emitted, is $HSO_3{}^{\bullet}$. While *ab initio* MO calculations leave no doubt that the "isolated" $HSO_3{}^{\bullet}$ radical should be a stable molecule, until recently the free molecule remained undetected in the gas phase. However, $HSO_3{}^{\bullet}$ is easily accessible by neutralization of its ionic counterpart **8$^+$**; the latter is formed by electron impact ionization of methane sulfonic acid **7** (eq. 5).

In addition, the NRMS experiment[22] provide evidence for the MO based prediction[23] that the two major unimolecular decomposition modes of **8** correspond to the extrusion of HO$^{\bullet}$ and O, respectively.

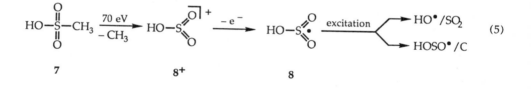

$$\text{(5)}$$

       7   **8$^+$**   **8**

Ethylenedithione (SCCS), its Mono-Oxygen Analogue OCCS, and Butatrienedithione S=(C$_4$)=S

The considerable recent interest in linear (or quasi-linear) molecules of the general structure X(C_n)Y (X,Y = lone electron pair, H_2, O, S), containing poly-cumulated double bonds (n = 2, 3, 4, etc.), is partly due to the particularly interesting spectroscopic and chemical properties of these species; in addition, some of these cumulenes are also believed to play a crucial role in the genesis of interstellar organic molecules. Unfortunately, quite many of these molecules cannot even be produced as a transient due to their high reactivity towards *inter*molecular reactions. This holds true in particular for cumulenes having an *even* carbon number. In contrast to their well-known *odd*-numbered analogues, they are generally believed to be significantly less stable.

In spite of this pessimistic prognosis, we were recently able[24] to generate and characterize in the gas phase the parent molecule, ethylenedithione S=C=C=S (**11**), by using NRMS (eq. 6) thus confirming a theoretical prediction of Schaefer et al.[25] Both **11** (as a triplet) and **11$^{\bullet +}$** correspond to the global minima of the respective C_2S_2 and $C_2S_2{}^{\bullet +}$ potential energy surfaces. As indicated in eq. 6, **11** and **11$^{\bullet +}$** possess quite close geometries; this could explain the unusually high fraction of C_2S_2 "survivor" ions observed in the NR spectrum of $C_2S_2{}^{\bullet +}$.

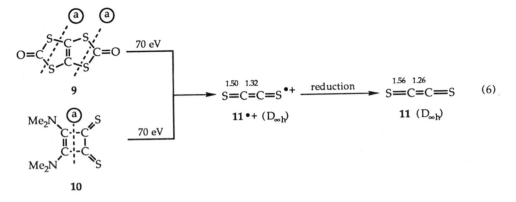

$$S=C=C=S^{\bullet+} \xrightarrow{\text{reduction}} S=C=C=S \quad (6)$$
$$\text{11}^{\bullet+} (D_{\infty h}) \qquad \text{11} (D_{\infty h})$$

For thioxaethyleneone, O=C=C=S, the "best" precursor turned out to be **12**; electron impact ionization of **12** (eq. 7) afforded *inter alia* a $C_2OS^{\bullet+}$ species whose collisional activation (CA) mass spectrum (with ionic fragments of $C_2S^{\bullet+}$, $C_2O^{\bullet+}$, $CS^{\bullet+}$, $S^{\bullet+}$, and $CO^{\bullet+}$) is only compatible with the connectivity $OCCS^{\bullet+}$.[26] Neutralization/reionization of the latter gives rise to stable **13**, and the experimental data are in accordance with *ab initio* MO calculations.[26] The latter indicate that **13** is formed in its triplet state. While spin-allowed dissociation of $^3(OCCS)$ is hampered by a significant endothermicity of 44 kcal/mol (6-31G*//6-31G* + ZPVE), the singlet electromer would spontaneously fall apart to CO and CS.

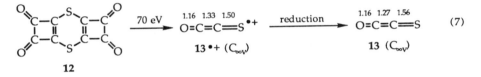

$$O=C=C=S^{\bullet+} \xrightarrow{\text{reduction}} O=C=C=S \quad (7)$$
$$\text{13}^{\bullet+} (C_{\infty v}) \qquad \text{13} (C_{\infty v})$$

The next higher, *even*-numbered homologue of **11**, i.e. the butatrienedithione **16**, is accessible from several precursors.[27] Dissociative ionization (70 eV) of **12, 14** and **15** (eq. 8) generates *inter alia* an intense signal at m/z 112 ($C_4S_2^{\bullet+}$). Again, the CA spectrum of mass-selected $C_4S_2^{\bullet+}$ (Figure 1) clearly points to an unbranched connectivity $S=(C_4)=S^{\bullet+}$ (**16**$^{\bullet+}$), and the NRMS study (Figure 2) lends support for the existence of a stable neutral counterpart **16**. The experimental findings are, again, supported by *ab initio* MO calculations.

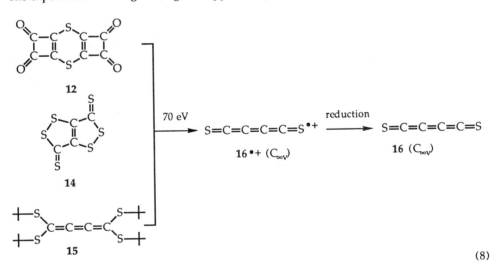

$$S=C=C=C=C=S^{\bullet+} \xrightarrow{\text{reduction}} S=C=C=C=C=S$$
$$\text{16}^{\bullet+} (C_{\infty v}) \qquad \text{16} (C_{\infty v})$$

$$(8)$$

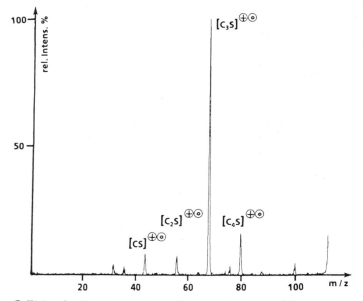

Fig. 1. Collisional activation spectrum of mass-selected $C_4S_2^{+\bullet}$ generated from **14**.

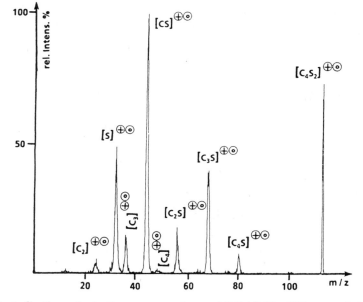

Fig. 2. Neutralization-reionization mass spectrum of $C_4S_2^{+\bullet}$ (Xe, 90% transmission (T)//O_2, 90% T).

In an extensive combined experimental/theoretical study,[28] evidence was presented for the existence of several C_xS_y species including also the corresponding cation ($\bullet +$) and anion radicals ($^{-\bullet}$). For these species the experimental data and MO results suggest connectivities as shown in eq. 9; for the neutral molecules, 6-31G* calculations indicate the triplet states to be as if not more stable than the singlet electromers.

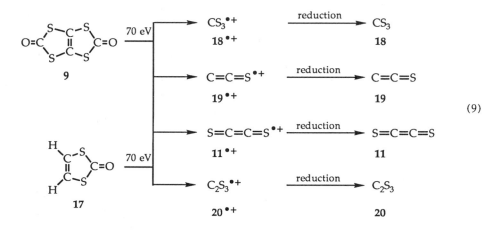

(9)

Doubly-Charged Sulfur-Containing Cations

The method of choice to generate from a beam of *solitary* monocations the corresponding isolated dications is charge stripping mass spectrometry (CSMS).[9-12] In fact, literally speaking hundreds of quite exotic dications were studied over the last decade,[12,29-35] and the species were rightly described as an "emerging class of remarkable molecules".[30] Salient features of these species, including stable molecules as small as He_2^{2+},[36] are the following ones: (1) Although most dications are thermochemically extremely unstable towards charge separation reactions (coulomb explosion), these reactions are very often prevented by significant barriers thus kinetically stabilizing the dications. (2) However, the prospects of generating in solution any of the small organic dications are often extremely remote. This is due to the fact that proton transfer from the dication to the solvent shell or addition of negatively charged species, like electrons, to the dication will occur avidly. (3) The relative stabilities of dicationic isomers are frequently reversed when compared with their neutral or mono-charged counter-parts. (4) Significant structural changes are observed which often favor anti-van't Hoff geometries.

In this contribution only a few sulfur-containing examples will be discussed; they may further illustrate the extraordinary potential of CSMS to generate and characterize dications in the gas phase.

It was recently demonstrated[37,38] in line with theoretical finding,[39-41] that systems with three-electron/two-center (3e/2c) sulfur-sulfur bonds (21) are indeed viable species even under isolated conditions. (For an early review and leading references on the generation of these species in solution see, for example, ref. 42; see also article by K.-D. Asmus in this book). For $[(CH_3)_2\bullet^\bullet S(CH_3)_2]^+$ the S–S bond order was determined to ca. 25 kcal/mol[38,43] in excellent agreement with *ab initio* predictions.[40,41] The 3e/2c bond strength is *inter alia* dependent upon the overlap between the two centers and is also subject to substituent effect (with electron releasing groups lowering the bond strength).[42] The fact that one electron of the 3e/2c bond of 21 occupies a σ^* orbital indicates several features:

(i) Oxidation of 21 to generate a dication 22 (eq. 10) should be a relatively facile process being accompanied with a shortening of the respective bond. This is, indeed, found experimentally.

(ii) As the orbital energy of the σ* orbital of a 3e/2c system is quantitatively correlated with the σ/σ* separation[39-42] which itself is indicated by the λ_{max} data for the electronic transition σ → σ*, one should expect a qualitative correlation between the λ_{max} data and the orbital energies of σ*. The latter are, of course, related to the ionization energy of the reaction 21 → 22, which itself can be obtained from a charge stripping experiment.

$$(10)$$

21   22

Indeed, this expectation is born out by recent gas-phase experiments.[44] For the monocations 23 - 28 it is observed that with decreasing λ_{max} indicating a larger σ/σ* separation, the ionization energy for the process 21 → 22 decreases (Chart 1). The detailed analysis[44] further suggests a reexamination of the λ_{max} data for the dithia-cyclopentane systems 29 - 32.

Chart 1

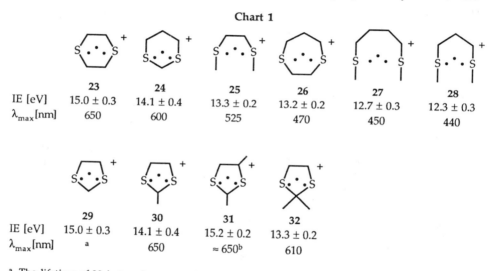

	23	24	25	26	27	28
IE [eV]	15.0 ± 0.3	14.1 ± 0.4	13.3 ± 0.2	13.2 ± 0.2	12.7 ± 0.3	12.3 ± 0.3
λ_{max}[nm]	650	600	525	470	450	440

	29	30	31	32
IE [eV]	15.0 ± 0.3	14.1 ± 0.4	15.2 ± 0.2	13.3 ± 0.2
λ_{max}[nm]	a	650	≈ 650[b]	610

a The lifetime of 29 is too short to permit an accurate determination of λ_{max} (λ_{max} is presumably >700 nm[44]).

b This value is associated with a large error due to low concentration of 31 in solution.[44]

ACKNOWLEDGEMENTS

We are indebted to the "Deutsche Forschungsgemeinschaft", the "Fonds der Chemischen Industrie", and the "Gesellschaft von Freunden der Technischen Universität" for the generous support of our work. The long-time, fruitful collaboration with Dr. J.K. Terlouw, Utrecht, is particularly appreciated.

REFERENCES

1. C. Wentrup, "Reactive Molecules: The Neutral Reactive Intermediate in Organic Chemistry", Wiley, New York (1984).
2. W.W. Duley and D.A. Williams, "Interstellar Chemistry", Academic Press, London (1984).

3. G. Winnewisser and E. Herbst, *Top. Curr. Chem.* 139:119 (1987).

4. M.T. Bowers (ed.), "Gas Phase Ion Chemistry", Vol. 1 - 3, Academic Press, Orlando (1984).

5. F.W. McLafferty (ed.), "Tandem Mass Spectrometry", Wiley, New York (1983).

6. J.H. Futrell (ed.), "Gaseous Ion Chemistry and Mass Spectrometry", Wiley, New York (1986).

7. P. Ausloos and S.G. Lias (eds.), "Structure/Reactivity and Thermochemistry of Ions", Reidel, Dordrecht (1987).

8. J.L. Holmes, *Org. Mass Spectrom.* 20:169 (1985).

9. R.G. Cooks, T. Ast, and J.H. Beynon, *Int. J. Mass Spectrom. Ion Phys.* 11:490 (1973).

10. T. Ast, *Adv. Mass Spectrom.* 8A:555 (1980).

11. K. Levsen and H. Schwarz, *Mass Spectrom. Rev.* 2:77 (1983).

12. W. Koch, F. Maquin, D. Stahl, and H. Schwarz, *Chimia* 39:376 (1985).

13. C. Wesdemiotis and F.W. McLafferty, *Chem. Rev.* 87:485 (1987).

14. J.K. Terlouw and H. Schwarz, *Angew. Chem. Int. Ed. Engl.* 26:805 (1987).

15. J.L. Holmes, *Mass Spectrom. Rev.* 8:513 (1989).

16. J.K. Terlouw, *Adv. Mass Spectrom.* 11:984 (1989).

17. J.L. Holmes, *Adv. Mass Spectrom.* 11:53 (1989).

18. H. Schwarz, *Pure Appl. Chem.* 61:685 (1989).

19. J.K. Terlouw, C.B. Lebrilla, and H. Schwarz, *Angew. Chem. Int. Ed. Engl.* 26:354 (1987).

20. K.J. van den Berg, C.B. Lebrilla, J.K. Terlouw, and H. Schwarz, *Chimia* 41:122 (1987).

21. D. Sülzle, M. Verhoeven, J.K. Terlouw, and H. Schwarz, *Angew. Chem. Int. Ed. Engl.* 27:1533 (1988).

22. H. Egsgaard, L. Carlsen, H. Florencio, T. Drewello, and H. Schwarz, *Chem. Phys. Lett.* 148:537 (1988).

23. S. Nagase, S. Hashimoto, and H. Akimoto, *J. Phys. Chem.* 92:641 (1988).

24. D. Sülzle and H. Schwarz, *Angew. Chem. Int. Ed. Engl.* 27:1337 (1988).

25. G.P. Raine, H.F. Schaefer III, and R.C. Haddon, *J. Am. Chem. Soc.* 105:194 (1983).

26. D. Sülzle, J.K. Terlouw, and H. Schwarz, *J. Am. Chem. Soc.* in press.

27. D. Sülzle and H. Schwarz, *Chem. Ber.* 122:1803 (1989).

28. D. Sülzle and H. Schwarz, in preparation.

29. W. Koch and H. Schwarz in ref. 7, p. 413.

30. P.v.R. Schleyer, *Am. Chem. Soc. Div. Pet. Chem. Prep.* 28:413 (1983).

31. P.v.R. Schleyer, *Adv. Mass Spectrom.* 10:287 (1985).

32. T. Ast, *Adv. Mass Spectrom.* 10:471 (1985).

33. L. Radom, P.M.W. Gill, M.W. Wong, and R.H. Nobes, *Pure Appl. Chem.* 60:183 (1988).

34. K. Lammertsma, *Rev. Chem. Interm.* 9:141 (1988).

35. K. Lammertsma, P.v.R. Schleyer, and H. Schwarz, *Angew. Chem. Int. Ed. Engl.* 28:1321 (1989).

36. M. Guilhaus, A.G. Brenton, J.H. Beynon, M. Rabrenovich, and P.v.R. Schleyer, *J. Chem. Soc. Chem. Commun.* 210 (1985).

37. T. Drewello, C.B. Lebrilla, H. Schwarz, L.J. de Koning, R.M. Fokkens, N.M.M. Nibbering, E. Anklam, and K.-D. Asmus, *J. Chem. Soc. Chem. Comm.* 1381 (1987).

38. A.J. Illies, P. Livant, and M.L. McKee, *J. Am. Chem. Soc.* 110:7980 (1988).

39. N.C. Baird, *J. Chem. Ed.* 54:291 (1977).

40. T. Clark, *J. Am. Chem. Soc.* 110:1672 (1988).

41. P.M.W. Gill and L. Radom, *J. Am. Chem. Soc.* 110:4931 (1988).

42. K.-D. Asmus, *Acc. Chem. Res.* 12:436 (1979).

43. J. Mönig, R. Goslich, and K.-D. Asmus, *Ber. Bunsenges. Phys. Chem.* 90:115 (1986)

44. T. Drewello, C.B. Lebrilla, K.-D. Asmus, and H. Schwarz, *Angew. Chem. Int. Ed. Engl.* 28:1275 (1989).

DICATIONS FROM CYCLIC POLYTHIOETHERS

Hisashi Fujihara and Naomichi Furukawa

Department of Chemistry
University of Tsukuba
Tsukuba, Ibaraki 305, Japan

Recently, several procedures for formation of the cation radical of 1,5-dithiacyclooctane and related compounds, and isolation of their dications have been presented.[1-4] These new sulfur species are of particular interest and become attractive if one could prepare the derivatives bearing multithia centers (tri-, tetra, etc.) since their cation radicals or dications once formed could play an important role in the development of a new field in organosulfur chemistry. Clear cut examples of transannular interactions between more than three sulfur atoms in multithia compounds have been hitherto unknown, except for our recent results.[6,7] This paper presents the following topics. 1. The first characterization of the crystal structure and reactivity of dithia dication salt.[5] 2. The first observation of transannular sulfur-sulfur bond formation of a new cyclic tris-sulfide and the isolation of its dicationic salt.[6,7] 3. A novel $p\pi$-(S^+-S^+) interaction in a new double bond bridged dibenzodithiocin.[8]

CRYSTAL STRUCTURE AND REACTIVITY OF DITHIA DICATION

The reaction of 1,5-dithiacyclooctane 1-oxide (1) with $(CF_3SO_2)_2O$ afforded the corresponding dithia dication salt, 1,5-dithioniabicyclo-[3.3.0]octane-bis(trifluoromethanesulfonate) (2) which was characterized by ^1H and ^{13}C NMR, and X-ray crystallographic analysis (Scheme 1 and Figure 1). The X-ray structure of 2 indicates that the S(1)–S(2) length is 2.121 Å which is only slightly longer than the normal S–S single bond (2.08 Å) in disulfides. It is interesting to note that very strong interaction was observed between the S^+ of the dication and the oxygen atoms of the triflate counter anions. The $S^+\cdots O$ distance is 2.682 Å which is remarkably shorter than the van der Waals' contact of 3.35 Å. The angle of S–S$\cdots$O is 176.3°.

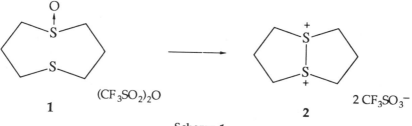

Scheme 1

Sulfur-Centered Reactive Intermediates in Chemistry and Biology, Edited by
C. Chatgilialoglu and K.-D. Asmus, Plenum Press, New York, 1990

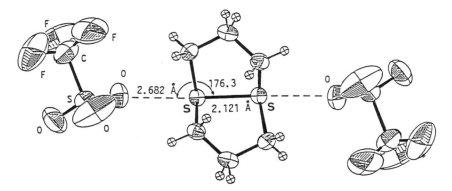

Fig. 1. X-Ray Structure of 1,5-Dithioniabicyclo-[3.3.0]octane-bis(trifluoromethane-sulfonate) (**2**).

Dithia dication **2** acts both as an oxidizing agent and electrophile. Surprisingly, when the dithia dication (**2**) was treated with carbanions (R⁻Li⁺) or alkoxides (RO⁻), only redox reactions took place. No hydrogen atom abstraction from **2** by strong bases was observed, although normal sulfonium salts bearing α-protons easily undergo deprotonation to give the sulfur ylides.

DITHIA DICATION IN CYCLIC TRIS-SULFIDE

The transannular bond formation between the three sulfur atoms, *i.e.*, dication 6 of 1,11-(methanothiomethano)-5H,7H-dibenzo[b,g][1,5]dithiocin (**3**) was observed in the reaction of sulfide **3** and its sulfoxides (**4** and **5**) with concentrated H_2SO_4 by ¹H and ¹³C NMR spectroscopy (Scheme 2).

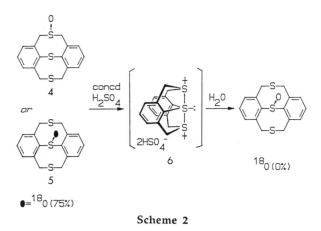

Scheme 2

A new type of oxygen-transfer reaction via the dication **6** was found by the hydrolysis of the H_2SO_4 solutions of **3-5**. Meanwhile, the reaction of the sulfoxide **5** with $(CF_3SO_2)_2O$ gave the dicationic salt **6a** (X = $CF_3SO_3^-$) [mp 134 - 135°C (decomp); ¹H NMR (CD$_3$CN) δ 4.65, 5.20 (ABq, *J* = 17 Hz, 8H), 7.39 - 7.98 (m, 6H); ¹³C NMR (CD$_3$CN) δ 42.5, 131.4, 137.2, 139.8; field-desorption (FD) mass spectrum, *m/z* 601 (MH⁺), 451 (M–OTf⁺), 302 (M–2OTf⁺), 151 (doubly charged cation)]. This is the first example for the dication with *hypervalent bond* of the central sulfur atom.

The electrochemical oxidation of a new aliphatic cyclic tris-sulfide, 3,7,9-trithia-bicyclo[3.3.1]nonane (7) shows the existence of sulfur-sulfur interaction. Namely, the peak potentials (Ep) for 7 and for the cyclic bis-sulfide 1,4-dithiacyclohexane (10) showed the following values: 7, 0.53 V and 10, 1.32 V vs. Ag/0.01 M AgNO$_3$. Simple sulfides normally exhibit a peak potential of ca. 1.2 to 1.5 V (Ag/Ag$^+$). Interestingly, comparison of the tris-sulfide 7 with the bis-sulfide 10 shows a peak potential 800 mV more cathodic for the former, so that 7 should be oxidized more readily. Thus, the trithia compound 7 exhibits a large negative potential shift which is undoubtedly related to transannular interaction. The reaction of the tris-sulfide 7 with concentrated H$_2$SO$_4$ shows the formation of the intrabridged dithia dication 8 which was characterized by 500 MHz ^{1}H NMR spectroscopy (Scheme 3). In contrast bis-sulfide 10 did not react with concentrated H$_2$SO$_4$. Thus, the tris-sulfide 7 is easily oxidized to the dication 8 which is remarkably stabilized by transannular S–S interaction.

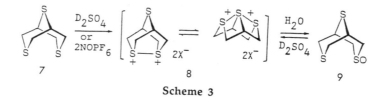

Scheme 3

TRANSANNULAR pπ(>S$^+$–$^+$S<) INTERACTION

A new double bond-bridged dibenzodithiocin, 1,11-(etheno)-5H,7H-dibenzo[b,g][1,5]di-thiocin (11) has been prepared to examine the intramolecular interaction between the dithia dication and a double bond. Sulfide 11 was treated with CF$_3$CO$_2$D, however, the recovered 11 did not contain the deuterium. Interestingly, on dissolution in concentrated D$_2$SO$_4$ (98%) at room temperature, sulfide 11 resulted in a deep-red colored solution. The 500 MHz ^{1}H NMR spectrum of the D$_2$SO$_4$ solution of 11 shows the formation of dithia dication 12 (Scheme 4).

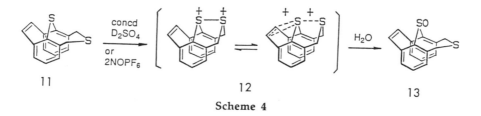

Scheme 4

Furthermore, the transannular interaction between the double bond and dithia dication in structure 12 is supported by the ^{13}C NMR spectroscopic data indicating that the olefinic carbon signal is shifted towards downfield. The D$_2$SO$_4$ solution of 11 was very stable as evidenced by NMR spectroscopy. Hydrolysis of the D$_2$SO$_4$ solution of 11 led to the sulfoxide 13 in a good yield and none of sulfide 11 was obtained. No addition products to the double bond or incorporation of deuterium were observed at all. The reaction of 11 with NO$^+$PF$_6^-$ (2 equiv) as one electron-oxidizing agent gave the dicationic salt 12 which was reacted with H$_2$O to afford the sulfoxide 13. The reaction of sulfide 11 in concentrated D$_2$SO$_4$ may proceed through the initial formation of the monocationic species such as the cation radical by an electron-transfer or the sulfidonium cation of 11, which is subsequently converted into the dithia dication 12. This finding is quite different from that of the bis-sulfide 5H,7H-di-benzo[b,g][1,5]dithiocin (14) which is unstable in concentrated H$_2$SO$_4$. The ^{1}H and ^{13}C NMR spectra of the D$_2$SO$_4$ solution of 14 showed complex signals. These results demonstrate that the dication 12 is remarkably stablized by transannular participation of the double bond to the (>S$^+$–$^+$S<) dicationic center, namely pπ-(>S$^+$–$^+$S<) interaction.

REFERENCES

1. K.-D. Asmus, *Acc. Chem. Res.*12:436 (1979).
2. W.K. Musker, *Acc. Chem. Res.* 13:200 (1980).
3. R.S. Glass, M. Hojjatie, A. Petsom, G.S. Wilson, M. Göbl, S. Mahling, and K.-D. Asmus, *Phosphorus and Sulfur* 23:143 (1985).
4. H. Fujihara and N. Furukawa, *J. Mol. Struct. (Theochem)* 186:261 (1989).
5 a. H. Fujihara, R. Akaishi, and N. Furukawa, *J. Chem. Soc. Chem. Commun.* 930 (1987).
   b. F. Iwasaki, N. Toyoda, R. Akaishi, H. Fujihara, and N. Furukawa, *Bull. Chem. Soc. Jpn.* 61:2563 (1988).
   c. H. Fujihara, R. Akaishi, and N. Furukawa, *Chem. Lett.* 709 (1988).
   d. H. Fujihara, R. Akaishi, and N. Furukawa, *Bull. Chem. Soc. Jpn.* 62:616 (1989).
   e. H. Fujihara, R. Akaishi, and N. Furukawa, *J. Chem. Soc. Chem. Commun.* 147 (1989).
6. H. Fujihara, J.-J. Chiu, and N. Furukawa, *J. Am. Chem. Soc.* 110:1280 (1988).
7. H. Fujihara, R. Akaishi, and N. Furukawa, *Tetrahedron Lett.* 30:4399 (1989).
8. H. Fujihara, J.-J. Chiu, and N. Furukawa, *Tetrahedron Lett.* 30:2805 (1989).

SULFENYLIUM CATION CARRIERS IN CHEMISTRY AND BIOCHEMISTRY

Giorgio Modena and Lucia Pasquato

Centro Meccanismi di Reazioni Organiche, C.N.R.
Dipartimento di Chimica Organica
Via Marzolo 1, 35131 Padova, Italy

Sulfenylium cations **1** have often been invoked as intermediates in reactions of sulfenyl derivatives.[1] Formally they are related to sulfenic acids **2**, as acylium ions or carbonium ions are related to carboxylic acids or alcohols, respectively.

$$R\text{–}S^+ + H_2O \ \rightleftarrows \ R\text{–}S\text{–}OH_2{}^+ \ \rightleftarrows \ R\text{–}S\text{–}OH + H^+ \tag{1}$$
$$\quad \mathbf{1} \qquad\qquad\qquad\qquad\qquad\qquad \mathbf{2}$$

Although serious doubts have been advanced[2,3] on the presence of these intermediates in solution the possibility for some of them to exist under very specific conditions cannot entirely be ruled out.[4] In the gas phase the situation is somewhat different as ions of mass/charge ratios corresponding to RS^+ are detected, e.g. in mass spectroscopy,[5] by ionization of appropriate precursors. However, for aliphatic derivatives and in particular for the ion with a mass equal to H_3CS^+ it has been shown that the structure is that of protonated thioformaldehyde, $(CH_2{=}SH^+)$ (**3**), which probably derives from the molecular ion by a concerted mechanism . At least, it has been proven that this is the pathway of lowest energy.[6] The instability of the sulfenylium cations, which in the case of aromatic derivatives may perhaps be partially compensated by resonance, should be related not only to the ease of rearrangement to protonated or alkylated thiocarbonyl compounds but also to more general factors. When a sulfenylium cation is generated from a closed shell precursor, it should be in the singlet state configuration. This is expected to be higher in energy than the corresponding triplet. *Ab intio* calculations have suggested that there is a significant spacing of the energy levels as indicated in Figure 1.[7]

Some years ago we studied this problem,[8,9] as far as condensed phase is concerned, and the results may be summarized as follows. ^{1}H–NMR spectra of either aromatic or aliphatic sulfenyl chlorides in liquid sulfur dioxide show, at low temperature, two groups of signals of equal intensity on addition of excess boron trifluoride (Figure 2) which gradually coalesce at higher temperature. The coalescence temperature increases with increasing concentration of boron trifluoride. The same trend is caused by decreasing the concentration of the Lewis acid at a fixed temperature. This conclusion was further substantiated by the observation that one of the ethyl groups of the ethylthioethylchlorosulfonium cation has a diasterotopic methylene and hence is linked to a tricoordinate sulfur atom.[8]

The rationale that we proposed, and which was later substantiated by further evidence, is condensed in the dimerization-ionization equilibrium reported in eq. 2. The Lewis acid has the function of binding the halide ion thereby shifting the equilibrium to the right. The use of a stronger Lewis acid such as antimony pentachloride makes the "dimerization

Sulfur-Centered Reactive Intermediates in Chemistry and Biology, Edited by
C. Chatgilialoglu and K.-D. Asmus, Plenum Press, New York, 1990

197

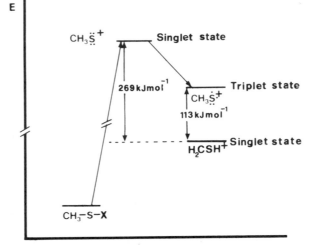

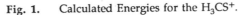

Fig. 1. Calculated Energies for the H_3CS^+.

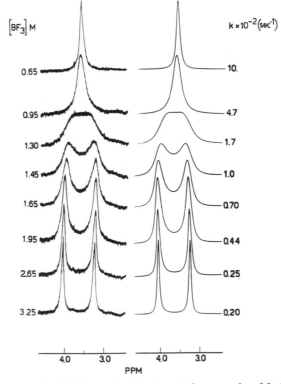

Fig. 2. Experimental (90 MHz) and Calculated Spectra for Methane Sulphenyl Chloride (0.6 M) in SO_2 at –12°C in the Presence of BF_3. (Reprinted with permission from G. Capozzi, V. Lucchini, G. Modena, and F. Rivetti, *J. Chem. Soc. Perkin Trans.* 2 361 (1975). Copyright by the Royal Society of Chemistry).

198

reaction" irreversible. However, no evidence of further ionization, i.e. formation of sulfenylium cations, was observed. The same conclusions were reached by reacting methanesulfenyl chloride with silver tetrafluoroborate. It was observed that only one half of the chlorine may be released as chloride ion. We are confident that much of the chemistry of sulfenyl halide, and sulfenyl derivatives in general, is based on equilibria like that given in eq. 2.

$$2\ RSCl\ +\ BF_3\ \rightleftarrows\ R-\overset{+}{\underset{Cl}{S}}-S-R\ +\ ClBF_3^- \tag{2}$$

It follows that it is not proper to talk about the chemistry of sulfenylium cations but rather of sulfenylium cation carriers since the cation is always associated with a base (or a nucleophile) and hence every reaction may be dealt with as a displacement of the bonded base by a stronger one (eq. 3). The dimerization equilibrium (2) discussed above also represents a sulfenylium cation transfer from one base to another (eq. 4).

$$RSX\ +\ Nu\ \rightleftarrows\ RS-Nu\ +\ X \tag{3}$$

$$2\ RSCl\ \rightleftharpoons\ \left[R-\overset{}{\underset{Cl}{S}}:\curvearrowright\overset{R}{\overset{|}{S}}-Cl \right]\ \rightleftharpoons\ R-\overset{+}{\underset{Cl}{S}}-S-R\ +\ Cl^- \tag{4}$$

Such a displacement, as proven for some specific systems, can be considered to occur with the same stereochemistry as the S_N2 aliphatic substitution, i.e. the three atoms which undergoe the coordination changes have to be collinear.[10] The nucleophile (or base) may either be negatively charged or neutral, giving rise to reagents (sulfenylium cation carriers) either neutral or positively charged. A set of the most common and utilized reagents are reported below:

Neutral carriers: $RS-Hal, RS-OSO_2R^1, RS-SOR, RS-SO_2R, RS-SR^1, RS-OR^1$

Charged carriers: $R\overset{+}{S}(Hal)SR, R\overset{+}{S}(SR)_2, R_2\overset{+}{S}SR, H_2\overset{+}{N}(R)SR$

At least one negatively charged compound is also known, i.e. $RS-SO_3^-$.

The reactivity or, in more general terms, the behavior of the sulfenyl derivatives depends also on the nature of the R group. A rather homogeneous set of reagents are those where R is equal to an aliphatic, like methyl and ethyl, or simple aromatic residues as phenyl, tolyl, chlorophenyl, etc. However, decreased reactivity is observed when the sulfur

Table 1. Relative Rates of Addition of Alkane Sulfenyl Chlorides to 1-Hexyne at 25 °C in Ethyl Acetate and in Chloroform

RSCl R =	ethyl acetate $k \times 10^2$ (M^{-1} s^{-1})	chloroform $k \times 10^2$ (M^{-1} s^{-1})
CH_3	16.2	153
C_2H_5	8.66	47.1
$n\text{-}C_3H_7$	6.25	35.8
$i\text{-}C_3H_7$	0.493	2.12
$i\text{-}C_4H_9$	3.11	18.9
$neo\text{-}C_5H_{11}$	0.153	0.96

is bonded to a secondary and particularly to a tertiary carbon atom because of steric effects in the substitution reactions (Table 1).[11] On the other hand, 2-nitro- and 2,4-dinitrobenzene-sulfenyl chloride are about 10^4 times less reactive than 4-nitrobenzene derivatives towards many substrates, owing presumably to some kind of specific interaction.[12] Of course, greater structural modifications affect the reactivity more stronlgy, and this is particularly true when the sulfur is linked to other heteroatoms. We shall discuss the chemistry of these compounds only briefly though.

CHEMICAL PROPERTIES

As already said, every reaction of the sulfenylium cation carrier amounts to a nucleophilic displacement. In most of the cases the reaction mechanism may be assimilated to that of the classical aliphatic S_N2 reactions.[10-12] With most of the sulfenylium cation carriers listed above the reactions with either n- or π-nucleophiles are fast and lead to a variety of compounds of major or minor interest. Reversibility or irreversibility of the reaction as well as the position of the equilibria depend on the relative basicity or thiophilicity of the entering and leaving groups in the specific reaction medium. The most common reagents used, whenever consecutive or side reactions do not require the use of *ad hoc* compounds, are the sulfenyl chlorides. Sulfenyl bromides, to a minor extent, and iodides, only in special cases, are also employed.

Generally, sulfenyl halides are prepared by reaction of the halogen with either the corresponding thiols or disulfides. As equation 7 is very fast the two preparative methods are undistinguishable in most of the practical cases. However equation 6, which is irreversible when X is equal to chloride ion, is largely or totally shifted to the left in the case of iodination. Indeed, sulfenyl chlorides may be "titrated" with iodide ion on the basis of equation 6 (from right to left). It follows that sulfenyl iodides may be obtained only from thiols (eq. 5) and under the condition that eq. 7 is slow (high dilution, secondary or better tertiary alkyl-thiols, physically separated thiol groups in matrices, etc.). The behavior of bromine is, of course, intermediate between that of chlorine and iodine.

$$RSH \ + \ X_2 \ \rightarrow \ R - \overset{+}{\underset{X}{S}} - H \ + \ X^- \ \rightarrow \ RSX \ + \ HX \tag{5}$$

$$R-S-S-R \ + \ X_2 \ \rightleftarrows \ R - \overset{+}{\underset{X}{S}} - S - R \ + \ X^- \ \rightleftarrows \ 2 \ RSX \tag{6}$$

$$RS-X \ + \ RSH \ \rightarrow \ R-S-S-R \ + \ HX \tag{7}$$

With most of the n-nucleophiles the reactions do not exhibit particular features and generally go to completion; with neutral reagents carrying mobile hydrogens, buffers are needed to drive the reaction to completion and/or to avoid side reactions. Few typical examples are reported below (eq. 8 - 15).

$$RS-X \ + \ CN^- \ \rightarrow \ RSCN \ + \ X^- \tag{8}$$

$$RS-X \ + \ HNR^1R^2 \ \overset{B}{\rightarrow} \ RS-NR^1R^2 \ + \ BH^+ \ + \ X^- \tag{9}$$

$$RS-X \ + \ HSR^1 \ \overset{B}{\rightarrow} \ RS-SR^1 \ + \ BH^+ \ + \ X^- \tag{10}$$

$$\text{RS–X} \quad + \quad \underset{R^2}{\overset{S}{\triangle}} \quad \rightarrow \quad \text{RS–SCH}_2\text{–CH(X)–R}^2 \tag{11}$$

$$\text{RS–X} \quad + \quad \overset{O}{\triangle} \quad \rightarrow \quad \text{RS–O–CH}_2\text{CH}_2\text{X} \tag{12}$$

$$\text{RS–X} \quad + \quad \text{R}^1\text{–S–R}^2 \quad \rightarrow \quad (\text{R}^1)_2\overset{+}{\text{S}}\text{–S–R} \quad + \quad \text{X}^- \rightarrow \quad \text{R}^1\text{S–SR} \quad + \quad \text{R}^1\text{X} \tag{13}$$

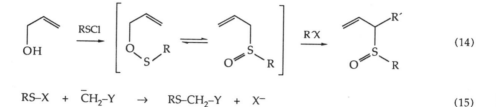

$$\text{RS–X} \quad + \quad \overline{\text{C}}\text{H}_2\text{–Y} \quad \rightarrow \quad \text{RS–CH}_2\text{–Y} \quad + \quad \text{X}^- \tag{15}$$

The reaction of sulfenyl chlorides with allylic alcohols[13](eq. 14) is particularly interesting since the sulfenate ester sulfoxide equilibrium allows selective functionalization which found several applications in organic synthesis. Also, the reaction with stabilized carbanions[14] (eq. 15) or enolates represents an entrance to more complex compounds because of the broad variety of transformation of the sulfur functionality.

The reaction with π-nucleophiles, in particular with alkenes, have a more complex and rich chemistry. The general scheme of the reaction is reported below:

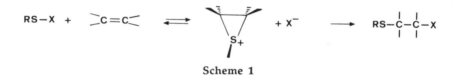

Scheme 1

Under normal conditions the reaction goes to completion and it is, in practice, irreversible. The intermediate bridged ion (thiiranium ion) never reaches concentrations high enough to be detected and it may be considered as an unstable intermediate at low stationary concentration. There is, however, strong evidence that the bridged ion is always formed along the reaction coordinate, as the addition is stereospecific under almost every condition. Furthermore, studies[15,16] in systems where thiiranium ions are long-living species have shown a rigorous preservation of the olefinic structure and no evidence of isomerization is detected, at least not for acyclic compounds at low temperature. Indeed, freezing of the reaction at the bridged cation stage is relatively easy. It is sufficient to use a non-nucleophilic solvent (usually chlorinated hydrocarbons or, better, liquid sulfur dioxide) and to trap the nucleophile present in the sulfenylium cation carrier used for the reaction by appropriate reagents. In most cases the thiiranium ions are stable enough to be detected by physical methods, in other cases products from rearrangements and decompositions are observed. Several thiiranium ions have been isolated as salt of non-nucleophilic anions. Two rather general procedures have been developed to isolate thiiranium ions. One method is the reaction between an appropriate sulfenylium cation carrier and the suitable alkene at low temperature. Several sulfenyl derivatives have been used which do not release a nucleophile strong enough to react with the thiiranium ions, trinitrobenzenesulfonate of sulfenyl[17] is the prototype of these reagents. Other weakly basic sulfonates as well as association of sulfenyl chlorides and silver salts of non-nucleophilic anions, such as silver tetrafluoroborate, have also been employed.[18] Perhaps a better method is based on the use of cationic derivatives such as the dimethylthiomethyl sulfonium salt of not nucleophilic anions ($SbCl_6^-$, BF_4^-, etc.).[16,19] The other method is based on the reaction of the β-thioalkyl chloride with strong lewis acids like SbF_5, $SbCl_5$, $AgBF_4$ in non-nucleophilic solvents.[20,21] Of course, the stability of the thiiranium ions (*versus* nucleophilic attack as well as *versus* rearrangements and decomposition) also plays a very significant role. It is enhanced by alkyl substitution at ring carbons.

In the thiiranium ion, the substitution pattern at carbons and sulfur affects both the balance between attack at sulfur relative to carbon, and between the two ring carbons (Scheme 2). It also has an influence on the ease of skeleton rearrangements, as will be discussed below.

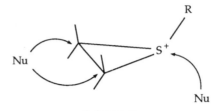

Scheme 2

Whenever the nucleophile is a base weak enough to yield a reactive sulfenylium cation carrier, like the chloride ion, attack at sulfur becomes a *virtual* reaction, evidence for which may be provided by carrying out the reaction in the presence of another olefin. This lead to the transfer of the sulfenylium derivative from one double bond to another one (Scheme 3).[22]

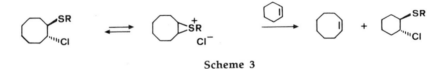

Scheme 3

The balance between Markovnikov (M) and *anti*-Markovnikov (AM) adducts of sulfenyl chlorides is rather important from several points of view including synthetic applications. Under conventional conditions, and with aliphatic alkenes, the AM products dominate and may even be the only product formed. However, the use of solvents which interact more strongly with nucleophiles, or the presence of carbon substitutents which stabilize adjacent positive charges, and electronegative substitutents at sulfur which indirectly destabilize the intermediate, may gradually change the M/AM ratio. They may even cause a complete inversion so that only products with M structure are formed (Table 2). A reasonable rationale for this trend would be the multivariational character of the system.

Under a formal point of view, thiiranium ions are the valence isomers of the corresponding open carbonium ions (Scheme 4). Even though it is a *virtual* equilibrium in most of

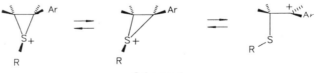

Scheme 4

Table 2. Product Distribution for the Addition of Sulfenyl Chlorides to Some Alkenes

System			CH$_2$Cl$_2$	
			M	AM
CH$_3$SCl	+	CH$_3$CH=CH$_2$	18	82
CH$_3$SCl	+	(CH$_3$)$_2$CHCH=CH$_2$	6	94
CH$_3$SCl	+	(CH$_3$)$_2$C=CH$_2$	20	80
CH$_3$SCl	+	C$_6$H$_5$CH=CH$_2$	98	2
CH$_3$SCl	+	(C$_6$H$_5$)$_2$C=CH$_2$	90	10
CH$_3$SCl	+	ClCH=CH$_2$	23	77
C$_6$H$_5$SCl	+	CH$_3$CH=CH$_2$	32	68
C$_6$H$_5$SCl	+	(CH$_3$)$_2$C=CH$_2$	13	87
CH$_3$C(O)SCl	+	(CH$_3$)$_2$C=CH$_2$	68	32

the cases, a real equilibrium may be established when the open ion is very efficiently stabilized by appropriate substituents.

When the substituent at sulfur destabilizes the positive charge on the adjacent atom, more positive charge will be transferred to the carbons. If they cary different substituents also the charge density at the carbons will be more differentiated, the thiiranium ion may not be any longer symmetric as far as geometry and charge distribution are concerned and the open carbonium ion structure may be reached as a limiting case.

Finally, solvation of the counteranion decreases its nucleophilicity and hence transition states with more advanced bond breaking will be favoured, and consequently more M products will be formed via transition states with less advanced bond formation. On the contrary, stronger nucleophiles will favor processes with more advanced bond formation which are more sensitive to steric repulsion and by consequence AM products will be formed preferentially.

The complexity and the interconnections among the factors affecting the reactivity and, in particular the regiochemistry, are such to make a detailed and possibly quantitative analysis rather difficult, however the general trends observed seem to be in line with this hypothesis. The problem of the balance between rearranged and unrearranged products, particularly when the double bond is part of a system favouring skeleton rearrangements is even more complex. That is the case for sulfenyl chloride addition to norbornene which in tetra-chloromethane and with benzene or tolylsulfenyl chloride yields the unrearranged *anti*-adduct, whereas with 2,4-dinitrobenzenesulfenyl chloride, particularly in polar solvents, products are formed with rearrangement of the skeleton .[23] (Scheme 5).

It is suggested that the same factors affecting the M/AM balance play a major role also in the ratio between rearranged and unrearranged products. An important role is played by the solvent which modifies the activity of the nucleophile and hence the rate of quenching of the intermediate. The longer the life-time is of the bridged ion, the higher are the chances for rearrangements to occur. This is also suggested by our recent study[24] on the reaction of *cis*- and *trans*-1,2-di-*t*-butylethylene. The *cis*-isomer irreversibly adds 4-methyl-benzenesulfenyl chloride to give the expected adduct.[25] The *cis*-thiiranium ion, obtained in

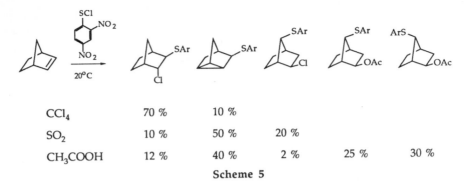

CCl₄	70 %	10 %		
SO₂	10 %	50 %	20 %	
CH₃COOH	12 %	40 %	2 %	25 %	30 %

Scheme 5

the usual way with the methylthio(bismethylthio)sulfonium salt, is stable almost indefinitely. The *trans*-isomer in methylene chloride gives an adduct which is, however, unstable.[24,25] By addition of liquid sulfur dioxide it reverts to the *trans*-thiiranium ion which slowly rearranges to the thietanium ion (Scheme 6).[24]

Under these conditions no *cis-trans* isomerization was observed and we have been able to show, by isotope labeling of one of the two *t*-butyl groups and preparing the two isomeric *trans*-2-adamantyl-3-*t*-butyl-1-methyl-thiiranium ions, that the rearrangement involves the group *cis* to the S-methyl group, and that it occurs in the thiiranium ion but not in the open carbonium ion.[24] Besides the absence of any cross-over between the *cis*- and *trans*-isomers, the stereochemistry of the thietanium ion formed (*t*-butyl and 3-methyl groups in *trans* relationship) rules out the possibility that the 1,2-methyl shift occurs in the open ion unless methyl migration is faster than rotation around the single bond (Scheme 6).[24]

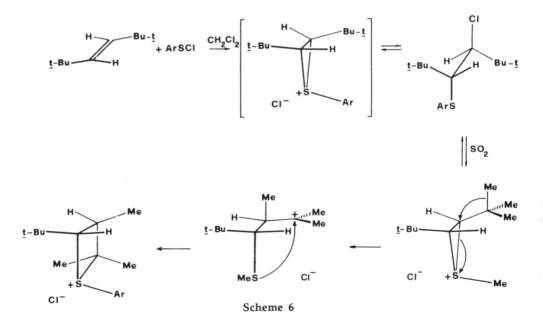

Scheme 6

The methodology developed to obtain stable or at least long-living thiiranium ions, *i.e.* to produce them from sulfenylium cation carriers which transfer the sulfenylium cation without releasing an active nucleophile, allowed the synthesis of novel compounds and better synthetic methods for several compounds of interest. In the presence of excess lithium perchlorate[26] the highly nucleophilic chloride ion is very likely substituted by the much less reactive perchlorate anion, slowing down the rate of adduct formation, and thus favouring rearrangements. Eventually this leads also to the formation of the highly reactive alkyl perchlorates. Another application of reactions of $R_3S_2^+$ and $R_3S_3^+$ with alkenes in conjunction with a weak nucleophile allows the synthesis of fluoro derivatives[27], and if the nucleophile is part of the substrate the synthesis of heterocyclic compounds.[28]

An original approach to thiirane (episulfuranes) synthesis has been developed by Capozzi[29] taking advantage of the high reactivity of silicon derivatives with nucleophiles. Silylsulfenyl chlorides are rather unstable species and present the reactivity features of the sulfenyl chloride function as well as of the lability of the silicon-sulfur bond *versus* nucleophiles. As shown in Scheme 7, trimethylsilylsulfenyl chloride reacts with alkenes to give the episulfide. Very likely the reaction proceeds in the usual way, but the S-silyl thiiranium is easily desililated by bromide ion. This kind of attack at the excocyclic atom bonded to sulfur is very rare and seems specific for silicon.

$$Me_3Si-S-SiMe_3 + Br_2 \xrightarrow[-78^{\circ}C]{-Me_3SiBr} Me_3Si-S-Br$$

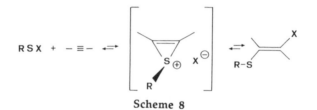

$+ Me_3SiBr$

Scheme 7

The addition of sulfenyl chlorides to the carbon-carbon triple bond has much the same features as the addition to double bonds. The reaction mechanism is also very much alike (Scheme 8) with the same *anti*-stereochemistry and the prevalence of AM over M regiochemistry.

Scheme 8

Even the relative reactivity of olefins and alkynes (k_o and k_a) toward electrophiles are fairly similar, quite contrary to what is observed for halogen addition where reactivity ratios k_o/k_a as large as 10^8 have been measured.[30,31] Some data are reported in Table 3.

Table 3. Relative Reactivity of Alkenes and Alkynes toward Different Electrophiles

	k_o/k_a		
	H^+	RSX	Br_2
n-BuCH=CH$_2$/n-BuC≡CH	$3.5^a /5.5^b$	8.4×10^1 c	5×10^7 d
$trans$-EtCH=CHEt/EtC≡CEt	$16^a /2.6^b$	1.5^c	
cis-EtCH=CHEt/EtC≡CEt	13.9^a	1.4×10^1 c	3.7×10^5 e
PhCH=CH$_2$/PhC≡CH	$0.65^a /3.8^f$	1.86×10^2	2.6×10^3 e

a 48% H$_2$SO$_4$, 25 °C.	b CF$_3$COOH, 60 °C.	c C$_2$H$_2$Cl$_4$, 25 °C.
d CF$_3$COOH, 25 °C.	e CH$_3$COOH	f HCl–CH$_3$COOH, 25 °C.

Also in these reactions the intermediate bridged cations, i.e. the thiirenium ions may be observed by physical methods and, under favourable conditions, isolated as salts of non-nucleophilic anions. The procedures to obtain thiirenium ions either in solution or as isolated salts are the same as previously reported for the thiiranium ions although the method based on the action of Lewis acids on the adducts has a rather limited applicability because of the interference of the double bond.[11] The stability of thiirenium ions, too, is strongly dependent on the substitution pattern, with carbon alkylation by bulky groups being very effective. The 2,3-di-*t*-butyl-1-methyl-thiirenium tetrafluoroborate or hexachloroantimonate are, in fact, so stable that a complete structure has been obtained.[32]

Comparison of the relative rates of addition to double and triple carbon-carbon bond in sulfenyl chlorides with those of two typical electrophilic addition reactions, hydration and bromination, suggests that thiirenium ions are relatively more stabilized by bridging than bromirenium ions. It was indeed suggested that the rather small reactivity ratios for protonation of double and triple carbon-carbon bonds in solution reflects the almost equal basicity of the two π-systems, whereas the large reactivity ratios observed for bromination are related to a "faster than expected" bromination of alkenes and are due to a lowering of the energy of the intermediate, and also of the transition state, because of bridging. This stabilization is fully effective in the case of addition to alkenes but almost totally ineffective in the case of addition to alkynes. In fact, bromirenium ions are destabilized by electron-electron repulsion between the π-electrons of the carbon-carbon double bond and the unshared electron pairs on the heteroatom.

The rather high reactivity of sulfenyl chloride with acetylenes suggests that a significant fraction of the energy lowering by bridging is preserved, and consequently the anti-aromatic character of thiirenium ions is partially removed. The structures of the thiirenium ion 4, as well as theoretical studies[33,34] offer a key to the explanation of this behavior.

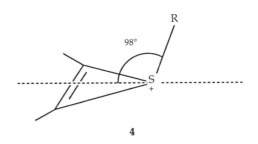

4

The small angle between the molecular plane and the *exo*-cyclic substituent at sulfur suggests that the fourth, doubly occupied, sulfur orbitals be almost in the molecular plane and hence almost orthogonal with the π-electrons of the double bond, making the destabilizing interaction rather weak.

In thiirenium ions there are three sites for nucleophlic attack: the sulfur and the two ring carbons. The observed general trends indicate that the same factors affecting the nucleophilic reactions on thiiranium ions are operative also in the case of thiirenium.

Even though systematic studies have not been carried out thiirenium ions seem to be less stable than thiiranium ions, and the nucleophilic attack at sulfur to be relatively more important in the former than in the latter.

As far as the product formation process is concerned there is one point to be emphasized. The nucleophilᵉ attacks the "ethylenic" carbon from the side opposite to the carbon-sulfur bond which constitutes a process of substitution with complete inversion of configuration at the ethylenic carbon. (Scheme 9).

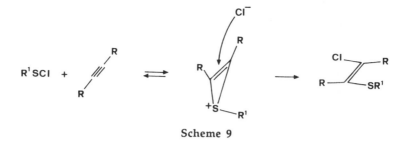

Scheme 9

It can be appreciated that this is not the normal stereochemistry of substitution at such a center. In fact, only in a few cases it has been observed a "partial inversion" of configuration in nucleophilic substitution at a vinylic center.[11,35]

Until now bicyclic thiirenium ions, for which rearrangements should be important, have not been studied. However, occurrence of this mode of reaction has been observed as a rather slow process for the 2,3-di-*t*-butyl-1-chlorobenzene thiirenium ion which undergoes 1,2-methyl shift[36] in much the same way as observed for 2,3-*trans*-di-*t*-butylthiiranium, as discussed above.

The chemistry developed for obtaining thiirenium ions may be applied for several kinds of cyclofunctionalization whenever the acetylenic function is associated with a nucleophilic function in a molecule.

The double and differentiated functionalization with defined stereochemistry obtained by addition of sulfenyl derivatives to acetylenic compounds have been exploited in several ways. An example is reported below.[37]

$$\text{PhS} \diagdown \diagup_{\text{Br}} \xrightarrow[\text{NiCl}_2(\text{dppe})]{n\text{-}C_{13}H_{27}\text{MgBr}} \text{PhS} \diagdown \diagup_{C_{13}H_{27}} \xrightarrow[\text{NiCl}_2(\text{dppe})]{n\text{-}C_8H_{17}\text{MgBr}} C_8H_{17} \diagdown \diagup_{C_{13}H_{27}}$$

(dppe = $Ph_2PCH_2CH_2PPh_2$)

(16)

Finally, one further point, common to addition reactions to alkenes and to alkynes, has to be discussed briefly. Several authors have speculated on the possibility of intervention of sulfurane, like intermediates 5, along the reaction path and on the possibility that the addition products may be formed directly from this intermediate (Scheme 10, analogous scheme may be written for addition to acetylenes).[16] However, there is no experimental evidence at all on the existence of such an intermediate, nor is there a need of it to explain either the kinetic or the nature of products. It may rather validate the hypothesis that the thiiranium ion formation is preceeded by a π-complex, as it has been suggested for bromination (and halogenation), whose dissociation leads to the ion pair (intimate, solvent separated, etc.) which may either revert to reagents, or collapse to products, or have a finite life-time so that other transformations may occur (Scheme 10).

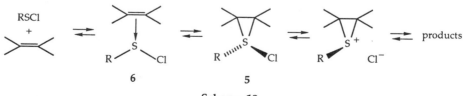

Scheme 10

Unless the sulfurane **5** (and the π-complex **6**) is directly detected or its existence is proven to be a necessary requirement to explain the product structures the problem remains an open one, more semantic than real in character.[15]

SOME APPLICATIONS IN BIOCHEMISTRY

In the area of bioorganic chemistry the sulfenylium ion chemistry appears limited at first glance, however, the ubiquitous presence of sulfur compounds suggests that much more has still to be learned in this area.

Furthermore, disulfides, which are very mild reagents but very sensitive to combined electrophlic-nucleophilic catalysis,[38] may play a role in the synthesis of complex molecules, *viz.*

$$\text{RS–SR} + \text{E} + \text{Nu} \rightarrow \text{R–}\overset{\overset{\text{E}}{|}}{\underset{\underset{\text{Nu}}{|}}{\text{S}}}\text{–S–R} \rightarrow \text{RSE} + \text{RSNu} \tag{17}$$

There are several fields in which sulfenylium derivatives play already an established role as, for example, in peptide synthesis and in the protection-deprotection sequence of amino and thiol functionalities.[39]

As far as the use of sulfenyl derivatives as reagents in peptide synthesis is concerned, the role of o-nitrobenzenesulfenyl chloride (**7**) as protecting group of the amino functionality may be mentioned.[39] It joins a number of features which makes this reagent very attractive. Its low reactivity, as compared with other arene or alkane sulfenyl chlorides, allows the use of it in protic solvents, with which it reacts only very slowly, while it is reactive enough to give complete and rapid transformation of the amino function in sulfenamide. The sulfenamides, in turn, are easily deblocked by the action of hydrochloric acid, better, in the presences of thiols (thioglycolic acid, thiophenol, etc.). In the latter case the less reactive disulfide is formed, thus avoiding some side reactions of **7** with sensitive amino acid residues (*e.g.* tryptophane).

An interesting modification of reagent **7** has been introduced recently by substituting the chloride ion as sulfenylium ion carrier by the saccharin anion, which is easily obtained from the same **7** and saccharin.[40] The new reagent is sufficiently active with amino-acids but its reactivity is low enough to be stored for longer time without appreciable decomposition.

The known properties of sulfonium and thiosulfonium ions as alkylating reagents and as precursor of ylides found application also in protein chemistry as for example in the deblocking of the *t*-butylcysteine[41] (see Scheme 11) or in the cleavage of proteins at the level of methionine.[42]

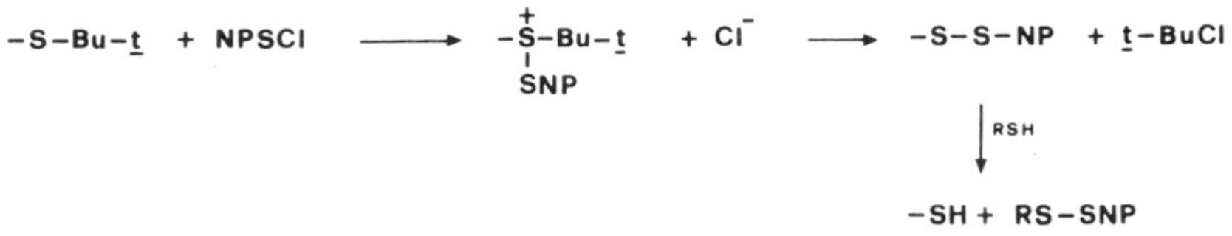

$$\text{–S–Bu–}\underline{t} + \text{NPSCl} \longrightarrow \text{–}\overset{+}{\underset{\underset{\text{SNP}}{|}}{\text{S}}}\text{–Bu–}\underline{t} + \text{Cl}^- \longrightarrow \text{–S–S–NP} + \underline{t}\text{–BuCl}$$

$$\downarrow \text{RSH}$$

$$\text{–SH} + \text{RS–SNP}$$

Scheme 11

The suggestion that sulfenylium cation carriers may play a role in the oxidative phosphorilation was advanced.[43] Even though such hypothesis has been abandoned in more recent times in favor of "membrane potential pumps" theories,[44] the possibility that sulfenylium cation carriers still play a role in the enzymatic processes might, nonetheless, be valid.

In the early seventies several authors suggested that one route of oxidative phosphorilation of ADP to ATP might involve a sulfenylium cation transfer to ADP or to inorganic phosphate to form a mixed anhydride which would then couple with inorganic phosphate or ADP, respectively. Indeed, by oxidizing thiols or disulfides in dry solvents (pyridine) with bromine or iodine sizable quantities of ATP were formed (up to 15% in the iodine-disulfide reaction).[43] The reaction was formulated in several ways including the generation of sulfenyl bromide and, respectively, iodide or the thio-halosulfonium salt.

Even though other sulfenylium ion carriers more compatible with the biological system might be suggested nowadays (eqs. 18 and 19), these hypotheses have the drawback that either thiols or disulfides are oxidized to the sulfenic acid level. It follows that the oxidative phosphorilation would be a pulsed or cyclic reaction in which an oxidative step has to be followed by a reductive one to restore the initial situation.

$$RS-SR \ + \ RSH \quad \underset{-H^+}{\overset{-2e^-}{\rightarrow}} \quad (RS)_2^+ -SR \tag{18}$$

$$RS-SR \quad \overset{-e^-}{\rightarrow} \quad (RS-SR)^{\bullet +} \tag{19}$$

It is now generally accepted that the oxidative phosphorilation reaction, *i.e.* the condensation of ADP with inorganic phosphate, is driven by some form of acid catalysis through pH gradients caused by a parallel oxidative process.[44] From a chemical point of view, this hypothesis is certainly reasonable if the local activity of H^+ in the hydrophobic cavity of the enzyme rises enough to form the conjugate acid of ADP, or of the inorganic phosphate, in kinetically signifcant amounts (Scheme 12, eq. a). However, it may be argued whether the acid catalysis might be associated with a disulfide catalysis as in Scheme 12, eqs. b-e.

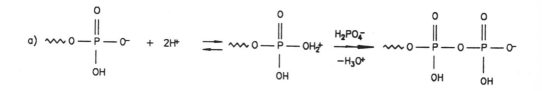

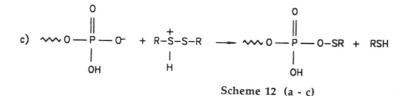

Scheme 12 (a - c)

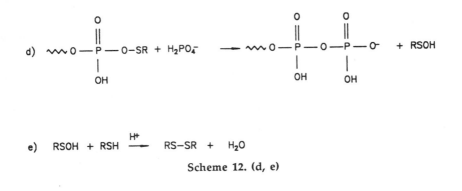

d) $\sim\sim$O—$\overset{\displaystyle O}{\underset{\displaystyle OH}{\overset{\displaystyle \|}{P}}}$—O–SR + $H_2PO_4^-$ $\longrightarrow$ $\sim\sim$O—$\overset{\displaystyle O}{\underset{\displaystyle OH}{\overset{\displaystyle \|}{P}}}$—O—$\overset{\displaystyle O}{\underset{\displaystyle OH}{\overset{\displaystyle \|}{P}}}$—O⁻ + RSOH

e) RSOH + RSH $\xrightarrow{\text{H}^+}$ RS–SR + H_2O

Scheme 12. (d, e)

The hypothesis concerning this aspect is based on the well-known acid-base catalysis in the reactions of disulfides[38] and on the possibility that the acid catalysis manifests itself in causing the internal redox process of disulfide where one sulfur is reduced to the thiol and the other one raised to the sulfenylium cation level. This scheme does not alter the overall state of oxidation of the disulfide which is regenerated via the last equation of Scheme 12. It is still an acid catalyzed reaction, but mediated by the intervention of the disulfide.

Of course, the hypothesis reported above is highly speculative but it might overcome the difficulty to accept the intervention of the conjugate acid of ADP or of the phosphoric acid both of which are weak bases.

Experimental evidence is still totally lacking. However, with the methodologies of study now available verification seems to be within reach.

REFERENCES

1. N. Kharasch, *J. Chem. Educ.* 33:585 (1956).
2. E.A. Robinson and S.A.A. Zaidi, *Can. J. Chem.* 46:3927 (1968).
3. G.K. Helmkamp, D.C. Owsley, W.M. Barnes, and H.N. Cassey, *J. Am. Chem. Soc.* 90:1635 (1968).
4. K.C. Malhotra and J.K. Puri, *Indian J. Chem.* 9:1409 (1971).
5. J.J. Butler, T. Baer, and S.A. Evans, *J. Am. Chem. Soc.* 105:3461 (1983), and references therein.
6. J.D. Dill and F.W. McLafferty, *J. Am. Chem. Soc.* 101:6526 (1979).
7. M. Roy and T.B. McMahon, *Org. Mass Spectrom.* 17:392 (1982).
   R.H. Nobes, W.J. Bouma, and L. Radom, *J. Am. Chem. Soc.* 106:2774 (1984).
8. G. Capozzi, V. Lucchini, G. Modena, and F. Rivetti, *J. Chem. Soc., Perkin Trans.* 2 361, 900 (1975).
9. G. Capozzi, O. DeLucchi, V. Lucchin, and G. Modena, *Synthesis* 677 (1976).
10. E. Ciuffain and A. Fava in, *Progrm. Phys. Org. Chem.* 6:81 (1968).
11. G. Capozzi, V. Lucchini and G. Modena, *Rev. Chem. Intermediates* 2:347 (1979).
12. G.H. Schmid and D.G. Garratt, *in:* "The Chemistry of Functional Groups: The Chemistry of Double-Bonded Functional Groups", Supplement A, Part 2, S. Patai, ed., Wiley, London, pp. 828 (1977).
13. S. Braverman, *in:* "The Chemistry of Sulphones and Sulphoxides", Chapt. 14, S. Patai, Z. Rappoport, and C.J. Stirling, eds., Wiley, London, p. 717 (1988).
14. R.M. Williams and W.H. Rastetter, *J. Org. Chem.* 45:2625 (1980).
15. G. Capozzi, G. Modena, and L. Pasquato, *in:* "The Chemistry of Sulfenic Acids, Esters and their Derivatives", S. Patai, ed., in press.

16. G. Capozzi and G. Modena, in: "Studies in Organic Chemistry. 19. Organic Sulfur Chemistry. Theoretical and Experimental Advances", F. Bernardi, I.G. Csizmadia, and A. Mangini, eds., Elsevier, New York, Chap. 5, p. 246 (1985).

17. D.J. Pettitt and G.K. Helmkamp, J. Org. Chem. 35:2006 (1970).

18. A.S. Gybin, W.A. Smit, V.S. Bogdanov, M.Z. Krimer, and J.B. Kalyan, Tetrahedron Lett. 21:383 (1980).

19. G. Capozzi, O. De Lucchi, V. Lucchini, and G. Modena, J. Chem. Soc. Chem. Commun. 2603 (1975).

20. V. Lucchini, G. Modena, T. Zaupa, and G. Capozzi, J. Org. Chem. 47:590 (1982).

21. A.S. Gybin, W.A. Smit, M.Z. Krimer, N.S. Zefirov, L.A. Novgordtseva, and N.K. Sadovaja, Tetrahedron 36:1361 (1980).

22. D.C. Owsley, G.K. Helmkamp, and S.N. Spurlock, J. Am. Chem. Soc. 91:3606 (1969).

23. W.A. Smit, N.S. Zefirov, and I.V. Bodrikov, in: "Organic Sulfur Chemistry", R.Kh. Friedlina et al., eds., Pergamon Press, New York, p. 159 (1981).

24. V. Lucchini, G. Modena, and L. Pasquato, J. Am. Chem. Soc. 110:6900 (1988).

25. C.L. Dean, D.G. Garratt, T.T. Tidwell, and G. Schmid, J. Am. Chem. Soc. 96:4958 (1974).

26. N.S. Zefirov, A.S. Koz'min, and V.V. Zhdankin, Tetrahedron 38:291 (1982) and references therein.

27. G. Haufe, G. Alvernhe, D. Anker, A. Laurent, and C. Saluzzo, Tetrahedron Lett. 29:2311 (1988).

28. F. Capozzi, G. Capozzi, and S. Menichetti, in: "Reviews on Heteroatom Chemistry", Vol. 1, S. Oae, ed., A. Ohno, N. Furukawa,eds., Myu, Tokio, p. 178 (1988).

29. F. Capozzi, G. Capozzi, and S. Menichetti, Tetrahedron Lett. 29:4177 (1988).

30. G.H. Schmid, in: "The Chemistry of the Carbon-Carbon Triple Bond. Part 1: Electrophilic Additions to Carbon-Carbon Triple Bonds" S. Patai, ed., Wiley, London, p. 306 (1978).

31. G. Melloni, G. Modena, and U. Tonellato, Acc. Chem. Res. 14:227 (1981).

32. R. Destro, T. Pilati, and M. Simonetta, Nouv. J. Chim. 3:533 (1979).

33. Y. Kikuzono, T. Yamabe, S. Nagata, H. Kato, and K. Fukui, Tetrahedron 30:2197 (1974).

34. J.W. Gordon, G.H. Schmid, and I.G. Csizmadia, J. Chem. Soc., Perkin Trans. 2 1722 (1975).

35. Z. Rappoport, Acc. Chem. Res. 14:7 (1981).

36. V. Lucchini, G. Modena, and L. Pasquato, unpublished results.

37. F. Naso, Pure Appl. Chem. 60:79 (1988).

38. J.L. Kice, in: "Sulfur in Organic and Inorganic Chemistry", Chapter 6, A. Senning, ed., M. Dekker, New York, p. 153 (1971).

39. T.W. Green, in: "Protective Groups in Organic Synthesis", J. Wiley, New York, p. 214, 283 (1981).

40. S. Romani, G. Bovermann, L. Moroder, and E. Wuensch, Synthesis 512 (1985).

41. J.J. Pastuszak and A. Chimiak, J. Org. Chem. 46:1868 (1981).

42. I.J. Galpin and D.A. Hoyland, Tetrahedron 41:895, 901 (1985).

43. E. Baeuerlein, in: "Glutathione", L. Flohe, H.Ch. Beuhoer, H. Sies, H.D. Waller, and A. Wandel, eds., G. Thieme, Stuttgart, p. 44 (1974).

44. E.L. Smith, R.L. Hill, I.R. Lehmon, R.J. Lefkowitz, P. Handler, and A. White, in: "Principles of Biochemistry, General Aspects", 7th Ed., McGraw-Hill, p. 348 (1983).

NEIGHBORING GROUP PARTICIPATION: GENERAL PRINCIPLES AND APPLICATION TO SULFUR-CENTERED REACTIVE SPECIES

Richard S. Glass

Department of Chemistry
The University of Arizona
Tucson, AZ 85721, USA

GENERAL PRINCIPLES

There are many ways in which a substituent can affect the reactivity of a functional group, such as electronic effects (resonance, hyperconjugation, inductive, and field effects), steric effects, conformational effects, and neighboring group participation. This review focuses on one of these ways: neighboring group participation.[1]

In neighboring group participation, the substituent forms a bond with the reactive center in a transition state or reactive intermediate. The nature of this interaction is as varied as that between functional groups on separate molecules. It includes nucleophilic participation, electrophilic participation, acid catalysis, base catalysis, and free radical participation. Although neighboring group participation is obviously an effect of general applicability in chemical reactions, its tenets were most extensively developed in studies on nucleophilic substitution reactions of alkyl halides and sulfonates and such reactions of carboxylic acids and their derivatives. These tenets will be outlined and then applied to the redox chemistry of organothioethers.

To discern the magnitude of the effect of neighboring group participation in the transition state for the rate determining step, two quantitative treatments are commonly used.[2] In both of these methods it is critical that the rate determining step for the reactions being compared is the same otherwise the comparison provides only a limit at best. If the reaction mechanisms are different for the two reactions, then there is no point in comparing them.

In the first treatment, the rate of reaction with the compound with a neighboring group is compared with that of a model compound devoid of the neighboring group, but otherwise the same. As an example, the rate constant for acetolysis of **1** is 10^{11} times greater than that for **2**

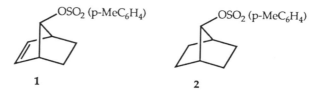

$$\text{1} \qquad\qquad \text{2}$$

Sulfur-Centered Reactive Intermediates in Chemistry and Biology, Edited by
C. Chatgilialoglu and K.-D. Asmus, Plenum Press, New York, 1990

under the same conditions. The difference between these two compounds is that there is a neighboring double bond in **1** but not in **2**. The π-electrons of the double bond in **1** are well oriented for overlap with the developing empty p-orbital on heterolysis of the C(7)–O bond, i.e. cleavage of the C–O bond is facilitated by backside displacement by the π-electrons. The limitations in such comparisons is that a neighboring group has multiple effects which may not all be compensated for in the model compound.

In the second method for quantitatively assessing neighboring group participation, the rate or equilibrium constant for reaction of the molecule containing the neighboring group is compared with that for the analogous intermolecular reaction. This ratio has the units of concentration and is known as the "effective" concentration or molarity. It is in essence the concentration at which the intermolecular reaction would have to be run in order to have the same rate as the intramolecular one, whether physically possible or not. For example, the "effective" concentration obtained by comparing the acid catalyzed rate of lactonization of **3**

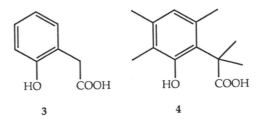

3   4

with the intermolecular reaction of phenol with acetic acid is 6×10^4 M. Such comparison between the rate of acid catalyzed lactonization of **4** with the intermolecular esterification of phenol with acetic acid provides an "effective" concentration of 3×10^{15} M. This enormous effect is clearly not due solely to neighboring group participation, but to conformational and steric effects as well.

Consideration of the energetics of these reactions provides insight into the origin of such large effects.[2] In the ensuing brief discussion differences in free energies are discussed. It is to be understood that for rates of reactions, the free energy differences are between the transition state and reactants, and for equilibria the differences are between products and starting materials. Let us compare the energetics in the gas phase between (1) two monofunctional molecules in equilibrium with their adduct containing a bond between their functional groups and (2) one bifunctional molecule in equilibrium with itself with a bond formed intramolecularly between its functional groups comparable to that formed in the intermolecular case. In case (1), three degrees of translational freedom and three degrees of rotational freedom are lost and six degrees of vibrational freedom are gained. In case (2), no net degrees of translational, rotational, or vibrational freedom are lost or gained. For typical values for translation, rotation, and new vibrational modes of high frequency, the two-molecule-association has a substantially higher free energy than one molecule interacting with itself. This entropic factor could contribute a factor of up to 10^7 - 10^9 favoring the intramolecular case (2) over the intermolecular case (1). This factor decreases as the importance of low energy vibrations increases, i.e. as the "looseness" of the transition state increases. This large entropic advantage of case (2) over case (1) in the gas phase also appears to apply to reactions in solution.

In addition to the discussed entropic effects, resonance, inductive, field, electrostatic, and solvation effects contribute to the differences between cases (1) and (2) as do bond stretching, bond angle bending, torsion, zero point energies, and van der Waals interactions which can be conveniently treated using force-field or molecular mechanics calculations. These factors may result in case (2) being less favorable than case (1) even though case (2) is favored by the entropic factors already discussed.

An alternative view of the origin of the advantage of *intra*molecular over *inter*molecular reactions has been cogently presented by Menger.[3] The attractive idea which was proposed is that intramolecular reactions are favored because the reactive groups are held within a "critical distance" longer than the corresponding intermolecular ones. Although this concept is intuitively appealing and certainly plays a role in neighboring group participation, it does not appear to provide the basis for a complete quantitative understanding of this effect.[4]

Comparison of an intramolecular case (2) reaction with others of this case can conveniently be done in the following terms. In such reactions, a ring is formed. The loss of entropy on cyclization follows the order of ring size: six-membered ring > five-membered ring > four-membered ring > three-membered ring. Ring strain follows the opposite order. The interplay of these effects is illustrated in the solvolysis of $X(CH_2)_nY$, where X = MeO, OH, O^-, NH_2, PhNH, or Cl and Y = halide or sulfonate. The rates of these reactions indicate that participation leading to ring formation follows the order five-membered ring > six-membered ring > three-membered ring. There is also an electronic effect disfavoring cyclization to the three-membered ring as well. The electron-withdrawing neighboring group disfavors heterolysis of the C–Y bond. The order of ring formation is a balance of factors and the result is not always the same with different neighboring groups. For example, solvolysis of $X(CH_2)_nY$, where X = PhS and Y = Cl, the order is three-membered ring > five-membered ring > six-membered ring > four-membered ring.

There are two other important effects, well-recognized in solvolysis reactions, that mediate the importance of neighboring group participation. (1) There may be stereoelectronic effects in the reaction that preclude or, at least disfavor, intramolecular reactions because the required geometry cannot be achieved.[5] (2) It has been shown that the extent of neighboring group participation depends on the need for it. That is, in solvolysis reactions, the less stable a cationic center the greater the extent of participation by a neighboring group.

Experimental evidence for neighboring group participation involves any or all of the following three general categories. (1) The rates or equilibria for the reaction are enhanced with the neighboring group relative to the comparison reaction as already discussed. (2) An intermediate with a bond between the neighboring group and reaction center can be isolated or trapped. (3) The products of the reaction in the presence of the neighboring group are different than those in its absence and the formation of these products can be reasonably rationalized by bond formation between the substituent and reaction center. The stereochemistry of the reaction products, the presence of rearrangement products, and isotopic labeling studies are all commonly used to detect neighboring group participation.

Discussion of examples of neighboring group participation can be cogently organized on the basis of the identity of the center undergoing reaction, the nature of the reaction, and the substituent participating in this reaction. The most detailed studies, as already pointed out, have been made on nucleophilic substitution reactions at a carbon center. A wide range of neighboring groups, which are often organized in terms of the class of electrons used in the participation: nonbonding (n) or bonding π-, or σ-electrons, have been so studied. In the ensuing discussion neighboring group participation by thioether sulfur in solvolysis and electrophilic addition reactions with carbon as the reactive center will be presented first. Then neighboring group participation by thioether sulfur in homolytic perester decomposition with oxygen as the reaction center will be discussed. Finally consideration of participation by a variety of functional groups in the oxidation of thioethers, in which sulfur is the center of reaction, will be considered.

Since examples illustrating neighboring group participation by thioether sulfur in solvolysis reactions have been exhaustively reviewed,[1,6] only a few examples have been selected here to illustrate the application of the general categories of experimental evidence for neighboring group participation. Studies on the hydrolysis of mustard gas, 1,5-dichloro-3-thiapentane (2,2'-dichloroethylsulfide), $(CH_2CH_2Cl)_2$, have been seminal in the development of neighboring group participation. The key moiety here is $RSCH_2CH_2Cl$ which forms a thiiranium salt (episulfonium salt) as a result of neighboring group participation in the rate determining step. The kinetic evidence for this is exemplified by a hundred-fold increase in the rate constant for hydrolysis of $PhS(CH_2)_2Cl$ over that for 1-chlorohexane.

In addition to kinetic evidence product studies also provide convincing evidence for neighboring group participation by thioether sulfur. Acetolysis of chloromethylthiirane **5** results in rearrangement to give 3-acetoxythietane **7**. This rearrangement provides evidence for neighboring thioether participation to afford bicyclic sulfonium salt **6** which then is

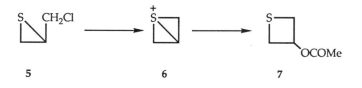

| 5 | 6 | 7 |

attacked by acetate to yield the observed product. Acetolysis of 2-endo-chloro-7-thiabicyclo[2.2.1]heptane **8** provides the corresponding acetate **10** with *retention* of configuration. This

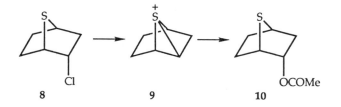

| 8 | 9 | 10 |

stereochemistry is unusual because nucleophilic attack on carbocation **11** should occur predominantly from the sterically less hindered exo rather than endo direction. However,

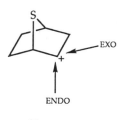

11

neighboring group participation by the thioether sulfur to give tricyclic sulfonium salt **9** would account for preferential attack from the endo direction. The nonbonding electrons on the sulfur atom participate in the heterolysis of the carbon-chlorine bond in **8** by backside (exo) attack on the carbon atom. Acetate attack on tricyclic sulfonium salt **9** at C(2) must now occur from the side opposite to that of the sulfur bonding electrons i.e. endo.

Neighboring thioether participation occurs on electrophilic addition to alkenes. Unactivated alkenes do not ordinarily react with iodine, but 6-thiahept-1-ene reacts readily with iodine to form cyclized sulfonium salt **12** thereby providing compelling evidence for

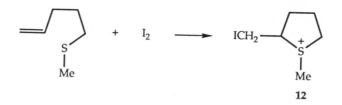

12

neighboring group participation.[7,8] Similar participation transpires on reaction of 5-methyl-ene thiocane with iodine and the structure of the resulting sulfonium salt **13** was determined

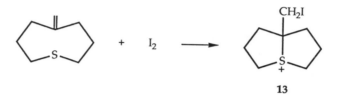

13

by x-ray crystallographic methods.[8] In both of these cases, a five-membered ring is formed in preference to a six-membered ring and iodocyclization happens rather than thioether oxidation.

Peresters **14**, R = Ph or Me, undergo decomposition by a predominantly free radical pathway 2.8×10^4 and 1.8×10^4 times faster, respectively, than t-butylperbenzoate.[9] Since

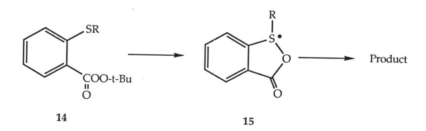

14   **15**

t-butyl O-t-butylperbenzoate does not undergo accelerated decomposition, neighboring group participation by the thioether moiety leading to **15**, instead of a steric acceleration, was inferred. Radical **15** could be generated by photolysis of **14**, R = Me at low temperature.[10] Interpretation of the well-resolved ESR spectrum of this species unequivocally established its overall structure and even provided considerable insight into its bonding. This example provides excellent evidence for neighboring group participation by thioether sulfur in homolytic perester decomposition. An important feature of thioether sulfur is that it, in principal, can interact with more than one reaction center simultaneously. To provide experimental evidence on this point, the decomposition of bis-perester **16** was studied. It decomposed almost 80

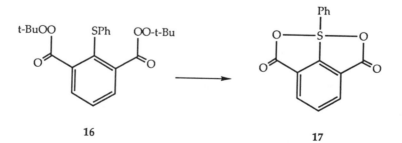

16   **17**

times faster than perester **14**, R = Ph (40 times faster after statistical correction because **16** has two perester moieties and **14**, R = Ph only one). The enthalpy of activation is lower in **16** than **14**, R = Ph, but the activation entropy is also lowered by 8.2 eu as expected for the more highly ordered transition state in **16** with simultaneous interaction of all three groups.

NEIGHBORING GROUP PARTICIPATION ON OXIDATION AT SULFUR

This next section focuses on neighboring group participation by a variety of functional groups in the oxidation of thioethers. Not only are the concepts of neighboring group participation extended to a new type of reaction, but a second-row atom, i.e. sulfur, rather than a first-row atom, i.e. carbon, is the reaction center. With these substantial changes, the only point which can unambiguously be extrapolated from the voluminous prior studies, is that there should be an entropic advantage to intramolecular rather than intermolecular interaction. Of particular interest is that there are bonding possibilities for sulfur not available to carbon. Such bonding possibilities may have important consequences in neighboring group participation at sulfur as a reaction center which engender results unprecedented in reactions at a carbon center. Among these may be unusual stereoelectronic effects and multicenter interactions. This last consequence has already been encountered in the unique three-group interaction in the homolytic decomposition of bis-perester **16**. An understanding of these effects may render intelligible the basis for long-range electron-transfer in biological redox proteins and provide the framework for rational design of new materials with novel electron-transfer properties, i.e. semiconductors, organic metals, superconductors, and photoconductors.

There are many ways to oxidize thioethers. For the purposes of this review such reactions will be organized as follows: (1) Thioethers may be oxidized to the corresponding sulfoxides by formally a two-electron oxidation coupled with an atom transfer. A simple example of this reaction is the oxidation of thioethers with peracids. The mechanism for this reaction involves only the transition state depicted below in which an oxygen atom is transferred from the peracid to the sulfur atom of the thioether.

$$R_2S + R'CO_3H \longrightarrow R_2S\text{---}O \cdots \overset{H\text{-----}O}{\underset{O}{\cdots}} \overset{\parallel}{C}\text{---}R' \longrightarrow R_2SO + R'COOH \tag{1}$$

(2) Thioethers may be oxidized by electron transfers to electron-acceptors without concomitant atom transfer. Such literal (outer-sphere) electron-transfers occur stepwise as depicted below.

$$R_2S \overset{-e^-}{\rightarrow} R_2S^{\bullet+} \overset{-e^-}{\rightarrow} R_2S^{2+} \tag{2}$$

(3) Thioethers may be oxidized by formally one-electron oxidation coupled to atom transfer. Such reactions in general form are presented in eq. (3). The previously discussed decomposition of perester **14** to give radical **15** is related to this reaction in that it involves formally

$$R_2S + X^{\bullet} \rightarrow R_2S^{\bullet}{\cdot}X \tag{3}$$

a one-electron oxidation of sulfur coupled with oxygen transfer. Heterolysis of the S–X bond in $R_2S^{\bullet}{\cdot}X$ affords $R_2S^{\bullet+}$ and X^-. This two-step process generates the same products as reaction (2) in which $X^{\bullet}$ is the one-electron acceptor. The difference between these two processes is that R_2SX is an intermediate in reaction (3) followed by heterolysis, but a transition state in reaction (2) with $X^{\bullet}$ acting as the electron acceptor. In reactions (2) and (3) the penultimate stable oxidation products may be sulfoxides or other products.

Oxidation of thioethers with halogens in the presence of water provides the corresponding sulfoxides. This reaction has been very extensively studied. In particular,

218

neighboring group participation has been demonstrated by rate enhancements, isolation of intermediates, and product studies. Oxidation of thioethers by aqueous halogen may be complicated because there may be buffer-dependent and buffer-independent pathways as well as acid-base equilibria and equilibria involving halide ions preceding the rate-determining step, and neighboring group participation with or without sulfurane formation. The mechanistic steps of relevance to this discussion are outlined below.

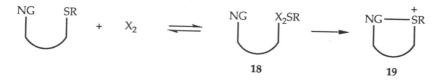

18   19

The halogen reacts with the thioether moiety to generate a complex **18** which may be a molecular complex, $R_2S \cdots X-X$, a halosulfonium halide, $R_2SX^+ X^-$, or trigonal bipyramidal dihalosulfurane. The neighboring group then forms a bond to the sulfur atom of this complex to produce **19** which in turn rapidly reacts with water to ultimately yield the corresponding sulfoxide. Neighboring group participation in heterolysis of the S–X bond in a halosulfonium halide resembles nucleophilic displacement of halide in alkyl halides with neighboring groups. However, owing to the possibility for sulfur to form a sulfurane such hypervalent species may be intermediates in the conversion of **18** into **19**.

Participation by Nitrogen

An important example of neighboring group participation in the oxidation of a thioether by halogen is the oxidation of methionine, $MeSCH_2CH_2CH(NH_3^+)CO_2^-$, by aqueous iodine.[11] The pH-rate profile for this oxidation demonstrates that the unprotonated amino group is involved in this reaction as required for neighboring group participation. Furthermore, the corresponding sulfoxide is formed in this reaction, but at high pH an intermediate designated "dehydromethionine" can be isolated. This intermediate has structure **20** in

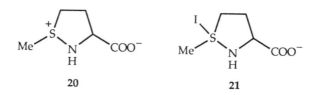

20   21

which there is an S–N bond as required for participation by the amino group in oxidation at sulfur.[12] Evidence has also been presented[13] that the formation of this intermediate is preceded by iodosulfurane **21**.

Comparable results to those obtained with methionine were found for primary amine **22**, R = R' = H, and secondary amine **22**, R = Me, R' = H.[14b,c] That is, all three compounds are

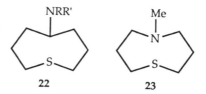

22   23

oxidized with comparable rates and faster than simple thioethers. In addition, the corresponding azasulfonium salts with S–N bonds have been isolated and their structures established by x-ray techniques. Tertiary amines **22**, R = R' = Me[14c] and $Me_2N(CH_2)_3SMe$[14a] do not show accelerated rates of oxidation by aqueous iodine, but both exhibit pH rate profiles suggesting neighboring group participation. However, tertiary amine, 5-methyl-5,1-azathio-cane, **23**, shows a spectacular rate enhancement for aqueous iodine oxidation of 10^5 relative to that of thiocane.[15]

Participation by neighboring nitrogen incorporated into heterocyclic rings has been reported. Although 2-[(2-methylthio)ethyl]pyridine does not show an accelerated rate of oxidation by aqueous iodine, the pH-rate profile for this reaction suggests neighboring group participation by the pyridine nitrogen forming a five-membered ring with an N–S bond.[14a] Participation by the neighboring nitrogen in imidazole and benzimidazole thioethers resulting in five-or six-membered rings with N–S bonds results in rate accelerations of 10^2 - 10^6 in aqueous iodine oxidations.[16] In addition, the pH-rate profiles for these oxidations provide additional evidence for such participation.

Aqueous iodine oxidation of thioether-amide **24** also occurs with neighboring group

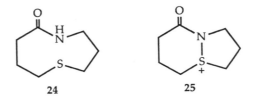

24   25

participation by the nitrogen atom.[17] This oxidation proceeds about 10^2 times faster than oxidation of a dialkyl thioether. Iodine oxidation of **24** under anhydrous conditions provided amidosulfonium salt **25** whose structure was proven by a single crystal x-ray structure analysis.

Participation by Oxygen

Neighboring alcohol participation in thioether oxidation of **26**, R = Ph or H by t-butyl hyprochlorite was inferred from the stereochemistry of the products.[18] Ordinarily,

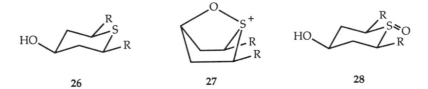

26   27   28

t-butyl hyprochlorite oxidation of thianes affords predominantly the sulfoxide stereoisomer which results from introduction of the oxygen from the axial direction. The predominant product formed by t-butyl hyprochlorite oxidation of **26** was **28** in which the oxygen is introduced from the equatorial direction. This result is accounted for by formation of the corresponding chlorosulfonium salt from **26**, in which the chlorine is introduced from the equatorial direction, and then neighboring hydroxyl group participation with inversion of configuration at sulfur to give **27**. Attack by t-butoxide at sulfur with inversion of configuration and tBu–O cleavage results in the formation of **28**.

Aqueous iodine oxidation of hydroxy-thioether **29** is faster than that of thiocane under comparable conditions. In addition, alkoxysulfonium salt **30** with an S–O bond could be isolated and its crystal and molecular structure determined by x-ray methods. These data demonstrate neighboring alcohol participation in this oxidation.[14c,19]

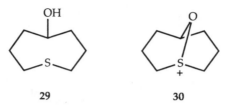

29   30

Neighboring alcohol participation occurred in the bromide-catalyzed electrochemical oxidation of thioethers **31**, R = H and **31**, R = Me.[20] In this electrochemical system, the

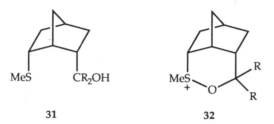

31   32

bromide ion concentration is 3×10^{-5} M. The anodic current associated with oxidation of such a small amount of bromide is miniscule compared with that for oxidation of thioethers at a concentration of 10^{-3} M. If the thioether is devoid of a suitable neighboring group or the substituent is geometrically precluded from participation, i.e. 2-exo-substituent, then the oxidation of the thioether is observed at a peak potential in the range for ordinary thioethers (1.2 - 1.7 V). However, for thioethers **31**, R = H and **31**, R = Me, which have neighboring hydroxyl groups, the oxidation occurs with a peak potential of approximately 0.6 V, which is the peak potential for oxidation of bromide to bromine. In this oxidation once the bromine is formed it oxidizes the thioether so rapidly with respect to the electrochemical time scale that the reformed bromide can be reoxidized and oxidize more thioether. This catalyzed oxidation occurs sufficiently rapidly that the currents measured for this oxidation are those expected for the direct oxidation of the thioether. Further evidence that this remarkable acceleration of bromine oxidation of the thioether is due to neighboring group participation is that alkoxysulfonium salts **32**, R = H, Me are isolated after controlled potential electrolysis of the corresponding alcohols at 0.60 - 0.65 V.

Neighboring ether participation in the oxidation of thioethers by halogens has not been reported. Indeed, aqueous iodine oxidation of 5-oxa-1-thiocane **33** is 40 times *slower*

$$OCOC_6H_33,5\text{-}(NO_2)_2$$

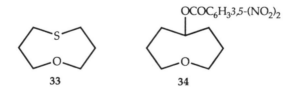

33   34

than oxidation of thiocane under the same conditions.[16] In contrast to this apparent lack of participation by ether oxygen in the rate determining step for thioether oxidation, the rate constant for hydrolysis of **34** is 48,500 times that for the model cyclooctyl derivative.[21] This result provides experimental demonstration that neighboring group participation in solvolysis reactions at carbon cannot be simply extrapolated to neighboring group participation in oxidation of thioethers.

Neighboring group participation by carboxylate but not carboxylic acid or ester groups in oxidation of thioethers has been reported. Aqueous iodine oxidation of O-methyl-thiobenzoate **35** occurs with general base catalysis by the carboxylate group.[22] However,

221

intermolecular nucleophilic catalysis by carboxylate and especially dicarboxylate ions in the aqueous iodine oxidation of thioethers has been reported.[23] The mechanism for this catalysis invokes the formation of acyloxysulfonium ions, $[R_2SOC(=O)R']^+$, via nucleophilic attack by the carboxylate ion on the intermediary iodosulfonium salt although an important role for sulfuranes as intermediates was also suggested.[24] The intramolecular version of this catalysis has been reported in the aqueous iodine oxidation of N-acetyl methionine and N-acetyl-S-methyl cysteine. The rate constants for these oxidations are 1400[25] and 9000[26] times greater, respectively, than that for such oxidation of the corresponding methyl esters. Surprisingly, an "effective" molarity of only 2.3 M was calculated for aqueous iodine oxidation of N-acetylmethionine versus intermolecular carboxylate catalyzed oxidation of the corresponding methyl ester. This result was rationalized in terms of a rate-limiting process involving a sulfurane intermediate. Even larger rate accelerations have been reported for neighboring carboxylate participation in the aqueous iodine oxidation of $S(CH_2CH_2CO_2^-)_2$ and $MeSCH_2CH_2CO_2^-$. The rate constants for oxidation of these thioethers are 10^6 times that for oxidation of simple thioethers.[26]

Neighboring carboxylate participation was also reported in the bromide-catalyzed electrochemical oxidation of thioether 36.[20] In addition to the peak potential at 0.6 V, iso-topic labeling studies provided evidence for the formation of acyloxysulfonium salt 37. The

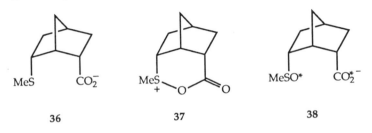

| 36 | 37 | 38 |

use of this indirect method for detecting 37 is important because acyloxysulfonium are ther-mally unstable and cannot be isolated at room temperature. Controlled potential electrolysis of 36 in the presence of a small amount of O–18 labeled water and 2,6-di-t-butylpyridine as base, afforded sulfoxide-carboxylate 38 in which both the sulfoxide and the carboxylate moieties incorporated the O–18 label (O*). This result is accounted for by competitive nuc-leophilic attack through the labeled water on sulfur of 37, resulting in O–18 incorporation in the sulfoxide moiety, and nucleophilic attack at the acyl carbon of 37, resulting in O–18 in-corporation into the carboxylate group. The result provides convincing evidence for the inter-mediary of 37 in the electrochemical oxidation of 36.

Participation by Sulfur

Aqueous iodine oxidation of 1,5-dithiocane 39 is 10^7 to 5×10^8 times faster than that for thiolane under comparable conditions.[15] Dication 40 which is formed in the rate-determining

| 39 | 40 |

step in this reaction has been prepared by other means and isolated as a crystalline salt. The interesting structure of this dication bis(trifluoromethane)sulfonate has been elucidated by x-ray crystallographic methods.[27]

Note that neighboring thioether, amine, and carboxylate participation in the reactions of sulfoxides and related compounds have been extensively studied. Many of these reactions are mechanistically related to those discussed here. Indeed, some are the reverse of the reactions reviewed here. Nevertheless, there is no room to be discussed in this review.[14a,15,28]

NEIGHBORING GROUP PARTICIPATION IN RADICALS

The preceding discussion was concerned with neighboring group participation in the formal two-electron oxidation of thioethers coupled with atom transfer. That is, category (1) oxidations of thioethers. This last section is concerned with neighboring group participation in thioether oxidations by general reactions (2) or (3). The evidence for such participation is direct detection of the intramolecularly bonded, one-electron oxidation products, namely, radicals and increased stability of such species compared with simple dialkyl sulfur radical cations.

Sulfur-Nitrogen Interaction

Oxidation of 3-(methylthio)propylamine **41**, n = 3 by hydroxyl radicals via reaction (3)

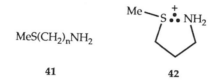

MeS(CH$_2$)$_n$NH$_2$

41   **42**

with X$^{\bullet}$ = $^{\bullet}$OH, in water between pH 3 and 9 using the techniques of pulse radiolysis produced a transient with λ_{max} = 385 nm and a half-life of approximately 600 μs.[29] Simple unassociated radical cations derived by such one-electron oxidation of dialkyl thioethers would have half-lives of less than 1 μs under these conditions. Structure **42** with a two-centered, three-electron bond between S and N resulting in a five-membered ring was assigned to this transient. Such oxidation of methionine afforded a comparable species with λ_{max} = 400 nm but, owing to rapid decarboxylation, this species had an exceedingly short half-life of 220 ns.[29] Decarboxylation is precluded in the corresponding ethyl ester and oxidation of ethyl methioninate gave a relatively long-lived transient with λ_{max} = 385 nm. No comparable species, which would have a four-membered ring on intramolecular cyclization, was detectable on such oxidation of 2–(methylthio)ethylamine **41**, n = 2. Oxidation of 5-methyl-5,1-azathiocane, **23**, with Br$_2^{\bullet-}$ in aqueous solution at pH 10, again via reaction (3) and an R$_2$SBr intermediate, using the technique of pulse radiolysis afforded the corresponding cation radical with a transannular S–N, 2c,3e-bond and with λ_{max} 500 nm and a half-life of 30 μs.[30] This species had previously been prepared by the reaction of **23** with its corresponding dication in acetonitrile and verified by its ESR spectrum.[31] Interestingly, this radical was not formed on treatment of **23** with one-equivalent of the one-electron oxidant NOPF$_6$. However, with two equivalents of oxidant, the corresponding dication with an N–S bond was produced.

Sulfur–Oxygen Interaction

Oxidation of bicyclic tertiary alcohol **31**, R = Me by hydroxyl radicals in aqueous solution yielded a neutral transient with λ_{max} = 400 nm and a half-life of 200 μs at pH 8 and another, positively charged, species with λ_{max} = 420 nm and a half-life of 30 μs at pH 4.[32] Since such species were not observed in the oxidation of the isomer of **31**, R = Me in which the –CMe$_2$OH group is exo and intramolecular S–O interaction is geometrically precluded, they were assigned to structures **43** and **44**, respectively. Comparable oxidation of bicyclic acid **36** yielded transient **45** with λ_{max} = 390 nm and a half-life of 30, 60, and 50 μs at pH 3, 7, and 10, respectively. No comparable species formed on oxidation of the methyl ester of **36** nor of the isomer of **36** with the carboxyl group exo. Oxidation of 4-(methylthio)butanoicacid with hydroxyl radicals in aqueous solution yielded a transient analogous to **45** with

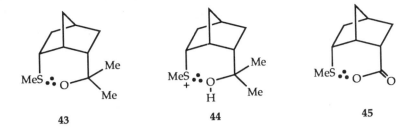

43
44
45

$\lambda_{max} = 410$ nm, but a much shorter half-life of 3 μs.[29] The additional geometric constraints and perhaps gem-dialkyl effect[1] render neighboring group participation more favorable on one-electron oxidation of **36** than 4-(methylthio)butanoic acid despite the formation of six-membered rings in both cases. Owing to especially rapid decarboxylation a transient resulting from neighboring group participation on one-electron oxidation of S-methylcysteine is not observed.[33] However, if such decarboxylation is prevented by protonation on oxygen, then a transient with an S–O bond is observed ($\lambda_{max} = 385$ nm, $\tau_{1/2} = 27$ μs at pH 0).[29] Similar five-membered ring transients with S–O bonds are formed on oxidation of 3-(methylthio)propanoic acid and S(CH$_2$CH$_2$CO$_2$H)$_2$. These transients, devoid of an α-amino group which promotes decarboxylation, have half-lives of 100 and 270 μs, respectively, at pH 3.[32b]

Sulfur-Sulfur Interaction

One-electron oxidation of mesocyclic dithioethers by hydroxyl radical in water using the technique of pulse radiolysis affords the corresponding cation radicals with S–S 2c, 3e-bonds.[34] This is exemplified by oxidation of 1,5-dithiocane **39** to its cation radical. This

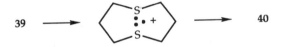

species is unusually stable and can be prepared from 1,5-dithiocane in organic solvents by NOBF$_4$ or Cu(ClO$_4$)$_2$ in acetonitrile, by reaction with its corresponding dication,[31b] or by pulse radiolysis.[34] The dication is formed by addition of two-equivalents of nitrosonium salts to 1,5-dithiocane. The ESR spectrum for the cation radical of 1,5-dithiocane provides excellent evidence for its assigned structure.[35] Electrochemical oxidation of 1,5-dithiocane provides further evidence for neighboring group participation in the oxidation of 1,5-dithiocane.[36] The measured peak potential of 0.34 V for this oxidation reveals that it occurs far more easily than that of ordinary thioethers. Multicenter thioether participation is indicated in the oxidation of trithioether **46** with NOPF$_6$.[37] With two equivalents of this oxidant the sulfurane dication **47** is formed.

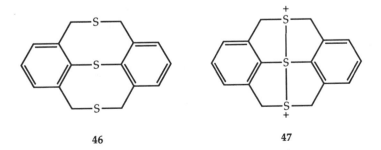

46
47

REFERENCES

1. B. Capon and S.P. McManus, *in:* "Neighboring Group Participation", Vol. 1, Plenum, New York (1976).
2. M.I. Page, *Chem. Soc. Rev.* 2:295 (1973).
3. F.M. Menger, *Acc. Chem. Res.* 18:128 (1985).
4 a. A.E. Dorigo and K.N. Houk, *J. Am. Chem. Soc.* 109:3698 (1987).
   b. F.M. Menger and M.J. Sherrod, *J. Chem. Soc. Perkin Trans.* 2 1509 (1988).
   c. A.E. Dorigo and K.N. Houk, *J. Org. Chem.* 53:1650 (1988).
5. J.E. Baldwin, *J. Chem. Soc. Chem. Commun.* 734 (1976).
6. E. Block, *in:* "Reactions of Organosulfur Compounds", Chp. 4, Academic Press, New York (1978).
7. E.N. Rengevich, V.I. Staninets, and E.A. Shilov, *Dokl. Akad. Nauk SSSR (Engl. Trans.)* 146:787 (1962).
8. R.G. Bernett, J.T. Doi, and W.K. Musker, *J. Org. Chem.* 50:2048 (1985).
9. J.C. Martin, *ACS Symp. Ser.* 69:71 (1978).
10. C.W. Perkins, J.C. Martin, A.J. Arduengo, W. Lau, A. Alegria, and J.K. Kochi, *J. Am. Chem. Soc.* 102:7753 (1980).
11. T. Lavine, *J. Biol. Chem.* 151:281 (1943); S. Mann, Z. *Anal. Chem.* 173:112 (1960); K.-H. Gensch and T. Higuchi, *J. Pharm. Sci.* 56:177 (1967).
12. R.S. Glass and J.R. Duchek, *J. Am. Chem. Soc.* 98:965 (1979); D.O. Lambeth and D.W. Swank, *J. Org. Chem.* 44:2632 (1976).
13. P.R. Young and L.-S. Hsieh, *J. Am. Chem. Soc.* 100:7121 (1978).
14 a. J.T. Doi, W.K. Musker, D.L. deLeeuw, and A.S. Hirschon, *J. Org. Chem.* 46:1239 (1981).
   b. A.S. Hirschon, M.M. Olmstead, J.T. Doi, and W.K. Musker, *Tetrahedron Lett.* 23:317 (1982).
   c. D.L. deLeeuw, M.H. Goodrow, M.M. Olmstead, W.K. Musker, and J.T. Doi, *J. Org. Chem.* 48:2371 (1983).
15. J.T. Doi and W.K. Musker, *J.Am.Chem.Soc.*103:1159(1981).
16. K.A. Williams, J.T. Doi, and W.K. Musker, *J. Org. Chem.* 50:4 (1985).
17. J.T. Doi, P.K. Bharadwaj, and W.K. Musker, *J. Org. Chem.* 52:2581 (1987).
18. J. Klein and H. Stoller, *Tetrahedron* 30:2541 (1974).
19. A.S. Hirschon, J.D. Beller, M.M. Olmstead, J.T. Doi, and W.K. Musker, *Tetrahedron Lett.*22:1195 (1981).
20. R.S. Glass, A. Petsom, M. Hojjatie, B.R. Coleman, J.R. Duchek, J. Klug, and G.S. Wilson, *J. Am. Chem. Soc.* 110:4772 (1988).
21. L.A. Paquette and M.K. Scott, *J. Am. Chem. Soc.* 94:6760 (1972).
22. W. Tagaki, M. Ochiai, and S. Oae, *Tetrahedron Lett.* 6131 (1968).
23. T. Higuchi and K.-H. Gensch, *J. Am. Chem. Soc.* 88:5486 (1966); K.-H. Gensch, I.H. Petman, and T. Higuchi, *ibid* 90:2096 (1968); P.R. Young and L.-S. Hsieh, *ibid* 104:1612 (1982).
24. P.R. Young and M. Till, *J. Org. Chem.* 47:1416 (1982).
25. P.R. Young and L.-S. Hsieh, *J. Org. Chem.* 47:1419 (1982).
26. J.T. Doi, M.H. Goodrow, and W.K. Musker, *J. Org. Chem.* 51:1026 (1986).
27. H. Fujihara, R. Akaishi, and N. Furukawa, *J. Chem. Soc. Chem. Commun.* 930 (1987); F. Iwasaki, N. Toyoda, R. Akaischi, H. Fujihara, and N. Furukawa, *Bull. Chem. Soc. Jpn.* 61:2563 (1988).
28. J.T. Doi and W.K. Musker, *J. Am. Chem. Soc.* 100:3533 (1978).
29. K.-D. Asmus, M. Göbl, K.-O. Hiller, S. Mahling, and J. Mönig, *J. Chem. Soc. Perkin Trans.* 2 641 (1985).
30. W.K. Musker, P.S. Surdhar, R. Ahmad, and D.A. Armstrong, *Can. J. Chem.* 62:1874 (1984).
31 a. W.K. Musker, A.S. Hirschon, and J.T. Doi, *J. Am. Chem. Soc.* 100:7754 (1978).
   b. W.K. Musker, *Acc. Chem. Res.* 13:200 (1980).

32 a. R.S. Glass, M. Hojjatie, G.S. Wilson, S. Mahling, M. Göbl, and K.-D. Asmus, *J. Am. Chem. Soc.* 106:5382 (1984).

  b. S. Mahling, K.-D. Asmus, R.S. Glass, M. Hojjatie, and G.S. Wilson, *J. Org. Chem.* 52:3717 (1987).

33. M.J. Davies, B.C. Gilbert, and R.O.C. Norman, *J. Chem. Soc. Perkin Trans.* 2 731 (1983).

34. K.-D. Asmus, *Acc. Chem. Res.* 12:436 (1979).

35. T.G. Brown, A.S. Hirschon, and W.K. Musker, *J. Phys. Chem.* 85:3767 (1981).

36. G.S. Wilson, D.D. Swanson, J.T. Klug, R.S. Glass, M.D. Ryan, and W.K. Musker, *J. Am. Chem. Soc.* 101:1040 (1979); M.D. Ryan, D.D. Swanson, R.S. Glass, and G.S. Wilson, *J. Phys. Chem.* 85:1069 (1981).

37. H. Fujihara, J.-J. Chiu, and N. Furukawa, *J. Am. Chem. Soc.* 110:1280 (1988).

DESIGN, SYNTHESIS, AND CONFORMATIONAL ANALYSIS OF COMPOUNDS

TAILORED FOR THE STUDY OF SULFUR-CENTERED REACTIVE INTERMEDIATES

Richard S. Glass

Department of Chemistry
The University of Arizona
Tucson, AZ 85721, USA

Neighboring group participation in the oxidation of thioethers provides anchimeric assistance, renders the oxidation potential of thioethers less positive, and stabilizes novelly-bonded reactive intermediates. The nature of the neighboring group has been varied and the participating atom has been nitrogen, oxygen, or sulfur. The effect of the size of the ring formed on participation has been ascertained and formation of five-membered rings is preferable to six-membered rings which, in turn, are favored over other sized rings, in general. Additional geometric constraints have been built into the thioethers studied. Two additional constrained systems that have yielded especially interesting results on oxidation are 2-endo-substituted-6-endo-methylthiobicyclo[2.2.1]heptanes and medium-sized ring thioethers. This review will survey design features in such systems and the synthesis and conformations of such compounds. As will be elaborated upon in this review S–S lone pair-lone pair interactions will also be highlighted.

BICYCLO[2.2.1]HEPTANE DERIVATIVES

Design and Geometry

Owing to the rigidity of the bicyclo[2.2.1]heptyl ring, substituents in the 2-endo and 6-endo positions are held relatively close together. Such preorganization entropically favors

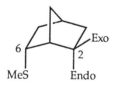

participation, as does relief of steric strain and dipole-dipole repulsions between such substituents, on oxidation. Furthermore, excellent model compounds, needed to assess the consequences of participation if it occurs, are readily at hand. If either the 2- or 6-methyl-thio-substituents or both are exo, then neighboring group participation is geometrically prohibited but inductive effects in the model compounds would be the same as in the 2-endo-substituted-6-endo-methylthio compounds. Although the geometries are relatively fixed in these bicyclo[2.2.1]heptyl derivatives there is rotation about the C(2)-substituent and C(6)-

Sulfur-Centered Reactive Intermediates in Chemistry and Biology, Edited by
C. Chatgilialoglu and K.-D. Asmus, Plenum Press, New York, 1990

227

sulfur single bonds. In addition, the bicyclic framework distorts to a limited extent as well.[1] The parent alkane bicyclo[2.2.1]heptane has C_{2v} symmetry and is comprised of a six-membered ring, i.e. C(1), C(2), C(3), C(4), C(5), C(6), in a boat conformation and two-five-membered rings, i.e. C(1), C(2), C(3), C(4), C(7) and C(1), C(7), C(4), C(5), C(6), in envelope conformations. However, substituted derivatives are distorted. Such distortions have been systematically investigated. They are conceptualized as partial pseudorotations of the five-membered rings and six-membered ring from symmetrical envelope and boat conformations, respectively. A convenient, simplified picture of such distortions has been suggested which focuses on the overall bicyclic system rather than its component rings. Viewing down the C(1)–C(4) vector, if C(3) and C(5) are *both* displaced clockwise (denoted +) or counter-clockwise (denoted –) relative to C(2) and C(6), respectively, then such distortion is called a

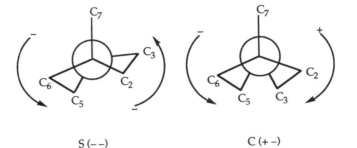

$$S\,(--) \qquad\qquad C\,(+\,-)$$

synchro twist and labeled S. If these displacements are in opposite directions, then this distortion is called a *contra* twist and labeled C. With 2-endo-substituted-6-endo-methyl-thiobicyclo[2.2.1]heptane derivatives C(+–) distortions are found[2-4] and C(1)–C(2)–C(3)–C(4) dihedral angles of up to 13° have been reported (for the parent envelope conformation this angle is 0°). Such twists move the substituents away from each other and are achieved by opening the C(6)–C(1)–C(2) bond angle. Increasing the S–C(6)–C(1) and C(1)–C(2)–(first substituent atom) bond angles also moves the substituents away from each other and is also observed. Despite these distortions, 2-endo-substituted-6-endo-methylthiobicyclo[2.2.1]hep-tane derivatives are valuable systems for studying neigh-boring group participation.

Synthesis

The bicyclo[2.2.1]heptane framework is readily synthesized by Diels-Alder ($_\pi 4_s + _\pi 2_s$) cycloadditions using 1,3-cyclopentadiene as the 4π-diene component. For example, 1,3-cyclo-pentadiene and acrylic acid yield cycloadducts **1a** and **1b** each with the bicyclo[2.2.1]heptane

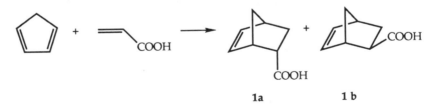

1a   **1 b**

framework. A bonus in the Diels-Alder reaction is that the desired endo-isomers, such as **1a**, are preferentially formed in the reaction ("Alder" rule) owing to secondary orbital interactions, even though the corresponding exo isomers are thermodynamically more stable. Endo isomer **1a** can be separated on a large scale from exo isomer **1b** by the procedure outlined below. Endo acid **1a** but not exo acid **1b** undergoes iodolactonization as its salt inaqueous sodium bicarbonate on treatment with iodine. Attack on the double bond of endo acid **1a** but not **1b** by iodine from the less-hindered exo direction can be accompanied by neighboring carboxylate participation. The endo acid **1a** can be regenerated from iodolacotone **2** by treatment with zinc in acetic acid. The carboxylic acid moiety in 1a is then converted to the

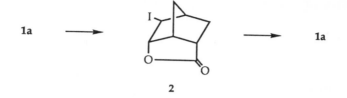

1a ⟶   ⟶ 1a

2

corresponding monothiocarboxylic acid, –COSH, via the corresponding acid chloride. On subjection to acid, cyclization of the monothiocarboxylic acid to thiolactone **3** occurs probably by an electrophilic addition mechanism although a freeradical mechanism is not ruled out. This thiolactonization necessarily results in introducing the sulfur atom in the desired C(6) endo position. Hydrolysis and alkylation of thiolactone **3** provides acid **4a**. The carboxylic acid moiety in **4a** can readily be converted to other functional groups of interest. In addition, exposure of the methyl ester of **4a** to sodium methoxide in methanol results in isomerization to the thermodynamically more stable exo isomer i.e. the methyl ester of **4b**. At equilibrium

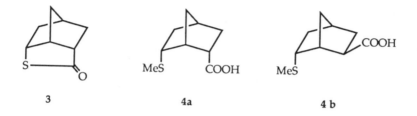

3   4a   4 b

the ratio of the methyl ester of **4b** to the methyl ester of **4a** is approximately 99:1. Saponification of the methyl ester of **4b** so produced containing the small amount of the methyl ester of **4a** yielded the corresponding acids which on recrystallization provided pure **4b**. The carboxylic acid moiety in **4b** could be converted to other functional groups of interest and these compounds served as model compounds in which intramolecular interaction between sulfur and the substituent is precluded geometrically.

    As already pointed out, attack on the double bond of **1a** occurs from the sterically less-hindered direction. This is a point of general importance in designing stereocontrolled syntheses of bicyclo[2.2.1]heptane derivatives. There are many examples of this preferential mode of attack. One especially well-studied example is hydride reduction of bicyclo[2.2.1]-heptane-2-one **5**. Hydride attack on this ketone occurs preferentially from the exo direction

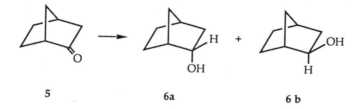

5   6a   6 b

to afford endo alcohol **6a** selectively over exo alcohol **6b**. Reaction of ketone **5** with lithium aluminum hydride gives endo alcohol **6a** and exo alcohol **6b** in a ratio of 89:11. Even greater stereoselectivity results from reduction with lithium trimethoxyaluminum hydride in which the ratio of alcohols **6a**:**6b** is 98:2. Reduction with lithium tri(sec-isoamyl)borohydride, LiAl[(CH$_3$)$_2$CHCH(CH$_3$)]$_3$H at 0° provides 99.6% stereoselective production of endo alcohol **6a**.

MEDIUM-SIZED HETEROCYCLES

Design and Conformational Analysis

The conformations of medium-sized rings, eight- to eleven-membered rings, have unique aspects which render transannular interactions important. Such interactions result in favoring neighboring group participation in oxidation of medium-sized ring thioethers. This review will summarize the conformational analysis of such compounds and focus on transannular interactions.[6]

General Features

There is substantial strain in medium-sized rings owing to contributions from angle, torsional, and steric strain. A characteristic feature of medium-sized cycloalkanes is transannular non-bonded strain. The conformations of such rings results in close juxtaposition of transannular hydrogen atoms. Similar juxtapositions of transannular heteroatoms, in the case in which the sulfur atom is a ring member and the other heteroatom is attached to the ring and the case in which both heteroatoms are ring members, are key in determining the neighboring group participation properties of such systems. Conformational analysis of medium-sized ring cycloalkanes provided a very different picture from that for such analysis of cyclohexane. The latter is dominated by two interconverting rigid chair forms. Flexible twist and boat forms are known but, generally, are of higher energy than the chair forms. In contrast, there are many conformers of comparable energy of medium-sized ring cycloalkanes which are separated by low barriers i.e. medium-sized ring cycloalkanes are "floppy". Consequently, conformational analysis of such cycloalkanes initially proved of limited value unlike the invaluable such studies of cyclohexanes. This picture has changed recently because substituents have been found to remarkably limit the conformations of medium-sized ring cycloalkanes. In addition, conformational constraints such as endocyclic cis double bonds and aromatic ring fusions also dramatically change this picture. Of particular relevance to this review, substituting heteroatoms for ring carbon atoms often restricts the conformational possibilities for medium-sized rings and heteroatom-heteroatom bonds are especially effective conformational constraints. Of course, it was long known that cyclooctasulfur has a well-defined "crown" conformation.

Bond Lengths and Bond Angles

Replacement of methylene groups in cycloalkanes with sulfur has important conformational consequences because geometrical parameters for sulfur are very different from those of carbon and two hydrogen atoms are removed. Such consequences include changing the degree of ring puckering, the extent of transannular interactions, and barriers to interconversion of conformers. These geometric parameters are illustrated in the following examples. The C–S bond length is longer than the C–C bond length. The C–S bond length in simple dialkyl thioethers is approximately 1.82 Å. The C–S–C bond angle is less than the C–C–C bond angle. In dimethylsulfide the C–S–C bond angle is 98.9° and such angles are in the range of 100 - 105° for other thioethers. An important result of these differences in molecular parameters is that the repulsive interaction in the twist forms of thianes is less than in cyclohexane. The S–S bond length in di- and tri-sulfides varies from 2.03 to 2.08 Å and the C–S–S bond angles range from 101 to 106°. The gauche conformation about the S–S bond is preferred in disulfides and the C–S–S–C torsion angle averages 85°. In addition, the rotational barrier about the S–S bond is greater than that about a C–C bond. The van der Waals radius for sulfur is much larger than that for carbon and is taken as 1.80 - 1.85 Å. However, the van der Waals radius for sulfur is probably not spherically uniform about the sulfur atom in thioethers. In contrast to the preference for the anti conformer about the C–C bond e.g. in n-butane, there is a slight preference for the gauche conformer about the C–S bond e.g. in ethyl methyl sulfide. In comparing the conformations of sulfur heterocycles with their corresponding oxygen compounds there are special effects that merit attention because they

230

often result in major differences between these conformers. The gauche conformation about the C–C bond is favored for O–C–C–O fragments, but disfavored by approximately 1.5 kcal/mol with respect to the anti conformation in S–C–C–S fragments. As already mentioned, there is a slight preference for the gauche conformation about the C–S bond in C–C–S–C fragments, but this conformer is strongly disfavored about the C–O bond in C–C–O–C fragments.

Lone Pair - Lone Pair Interaction

The conformations of medium-sized ring dithioethers, e.g. 1,5-dithiocane, often result in transannular juxtapositions of heteroatoms. In such species there is repulsive lone pair-lone pair interaction. Such interactions are not well-understood but are believed to be important in the redox chemistry of these systems. Consequently, it is appropriate to briefly discuss such interactions and photoelectron spectroscopy which provides a quantitative measure of such interactions and can also be used in concert with force-field and molecular orbital calculations to determine the conformation of these molecules in the gas phase.

The two nonbonding pairs of electrons around a thioether sulfur atom are believed to be in two nonequivalent orbitals rather than two equivalent sp^3-hybrid atomic orbitals.[7] One nonbonding pair is in a 3p-type orbital perpendicular to the C–S–C plane which is the highest occupied molecular orbital and the other is in an in-plane orbital such as an sp^2-hybrid orbital which is of lower energy than the $3p_z$ orbital. Theoretical calculations, photoelectron spectra, and x-ray crystallographic studies are all consistent with this suggestion. For example, examination of close nonbonded interactions with sulfur in crystal structures of divalent sulfur compounds revealed that electrophiles preferentially approached the sulfur atom approximately 20° from the perpendicular to the Y–S–Z plane.[8] Nucleophiles preferred approaches in the Y–S–Z plane along extensions to one of the covalent bonds to sulfur but avoiding approach along the extension of the bisector of the Y–S–Z bond angle. These observations support assigning the 3p orbital as the highest occupied molecular orbital and in an in-plane sp^2 orbital the other nonbonding pair of electrons with its major extension along the bisector of the Y–S–Z bond angle. Such interactions may not be reliable electron-density probes because they may represent "incipient" reactions. However, low temperature x-ray determinations of the deformation density on the sulfur atoms of cyclooctasulfur (S_8) support the orbital assignment for the nonbonding electron pairs.[9] These studies reveal that the lone pair density is ellipsoidal in the S–S–S bisecting plane with the longest axis of the ellipsoid perpendicular to the S–S–S plane.

The preferred geometries of close approach of the Y–S–Z moieties of two separate molecules has been ascertained by examination of close nonbonded S···S interactions in crystal structures.[10] Close approach of two nonbonded sulfur atoms is defined as not greater than 3.7 Å which is twice the van der Waals radius for sulfur. Typically, one divalent sulfur atom assumes the role of a nucleophile and the other sulfur that of an electrophile. The nucleophilic sulfur approaches the electrophilic sulfur along the extension of a sulfur bond as seen before for other nucleophiles. The electrophilic sulfur, in turn, approaches roughly perpendicular (at least > 75°) to the Y–S–Z plane of the nucleophilic sulfur in a similar way as other electrophiles. However, at very close S···S distances, exemplified by the S···S nonbonded contact of 3.25 Å in meso-lanthionine, $S[CH_2CH(NH_2)CO_2H]_2$, a different geometric pattern is encountered. This geometry is antiparallel as depicted in **7**.[7] This preferred

7   8

approach is further evidence for a nonspherical distribution of the nonbonding electrons around the sulfur atoms and is best understood in the following way. The approach minimizes lone pair-lone pair repulsions as shown by EH semiempirical calculations. The major extensions of the nonbonding 3p orbitals on each of the sulfur atoms are parallel to each other resulting in little overlap even at short S···S distances, and the in-plane sp^2-hybrid orbitals avoid each other as shown in 7. As a consequence of the modest overlap between lone pair orbitals there are no large lone pair orbital splittings even at very small separations between sulfur atoms. In contrast, other approach geometries result in large orbital splittings. For example, in the 3p···3p approach 8 the 3p orbitals perpendicular to the Y–S–Z planes approach each other directly along their maximum extensions. This results in extensive overlap and lone pair orbital splitting. The antisymmetric combination of nonbonding 3p orbitals rises to high energy at short separations in this geometry.

The discussion on close approach of two divalent sulfur atoms in separate molecules is relevant to the close approach of sulfur atoms in the same molecules e.g. medium-sized ring dithioethers. However, there will be an interplay between the preferred geometries for S···S approach and the conformational preferences imposed by the rest of the molecule. The following two examples illustrate this point. The transannular sulfur atoms in 1,5-dithiocane 9 approach each other such that there is substantial overlap between the nonbonding 3p-orbitals on each of the sulfur atoms. This results from the conformational constraints of the eight-membered ring which thrust the sulfur atoms on the same side of the ring and close to

9   10

each other(approximately 3.4 Å apart) with their $3p_z$-orbitals directed somewhat, but not directly, at each other. On the other hand, the approach geometry of the two sulfur atoms in dithioether 10 contrasts with that in 1,5-dithiocane 9. The two sulfur atoms in 10 are connected by three carbon atoms as are those in 1,5-dithiocane, but are constrained to be even closer to each other by the rigid bicyclo[2.2.1]heptyl ring than those in 9. In the solid state, the intramolecular, nonbonded S···S distance in 10 is 3.179(1) Å.[3] However, rotation is possible about the C(2)–S and C(6)–S single bonds in 10; whereas, comparable rotation in 9 is precluded by ring constraints. The two C–S–C planes in 10 are almost the same so that the $3p_z$ orbitals on the sulfur atoms, which are perpendicular to the C–S–C plane, are parallel to each other as in "antiparallel" 7 and there is less 3p overlap in 10 than in 9.

As is clear from the preceding discussion, it is important to determine the extent of lone pair-lone pair overlap in dithioethers. Both from the point of view of conformational analysis of these compounds and the lone pair-lone pair repulsions that may be important in the redox chemistry of these systems. Photoelectron spectroscopy is an invaluable method for measuring lone pair-lone pair interactions and the basis for this methodology will be briefly reviewed.[11]

Photoelectron Spectroscopy Data

The ionization potentials for a compound are obtained from its photoelectron spectrum. Assuming Koopmans' theorem, the ionization potentials correspond to the negative of the energies of the molecular orbitals from which the electrons are removed. This theorem is valid for the orbitals of highest energies which are the ones of interest here. For a dialkyl thioether the highest occupied molecular orbital is the nonbonding 3p orbital localized on the sulfur atom. If there are two thioether groups in a molecule, then the 3p lone pair orbitals on each of the sulfur atoms can interact with each other by through-bond, through-space, or both effects. As a result of this interaction, the lone pair orbitals must be treated together as linear combinations rather than separately. This combination results in molecular orbitals of different energy, i.e. a splitting of 3p lone pair orbitals, and corresponding ionization potentials measured by photoelectron spectroscopy. The greater the interaction, the greater the orbital splitting, the greater the separation of ionization potentials. For medium-sized ring dithioethers, such as 1,5-dithiocane 9, the interaction between the 3p-type lone pair orbitals on the two-sulfur atoms occurs predominantly through-space and not through-bond. This through-space interation depends on two geometric parameters. These are the S···S nonbonded distance and the dihedral angle between the two C–S–C planes. This dihedral angle reflects the orientation of the $3p_z$-orbitals, which are perpendicular to the C–S–C planes, to each other. For a given separation of sulfur atoms, the extent of orbital overlap and consequent orbital splitting will be greater if the orbitals are pointing directly at each other as in 3p···3p 8 than if they are parallel to each other as in "antiparallel" 7. This orientation effect is illustrated by comparing the 3p-lone pair orbital splitting in 1,5-dithiocane 9 with that in 10. The nonbonded S···S distance in 10 is less than that in 9, but the 3p-lone pair splitting is greater in 9 (0.43 eV as measured by photoelectron spectroscopy)[12] than in 10 (0.28 eV)[3]. The reason for this is that the 3p-lone pair orbitals in 10 are parallel to one another resulting in modest overlap, but point somewhat at each other in 9 resulting in more substantial overlap. Thus, the observed orbital splittings measured by photoelectron spectroscopy provide insight into possible S···S distances and C–S–C dihedral angles for dithioethers. Given the geometric constraints imposed by medium-sized rings, the geometric information obtained by photoelectron spectroscopy allows a determination of the conformation of the medium-sized ring dithioether in the gas phase. In practice, the photoelectron spectrum is fitted with the orbital energies obtained by computations (force field and molecular orbital calculations).

Survey of the Conformations of Medium-sized Heterocycles

Cyclooctane typifies the medium-sized cycloalkanes in being flexible geometrically.[6] Its predominant conformation is a boat-chair 11 which pseudo-rotates rapidly. The chair-

11   12

chair 12 or twist-chair-chair conformation which also pseudo-rotates is also present. As already pointed out, the all sulfur eight-membered ring is in a crown conformation (chair-chair family). 1,5-Dithiocane is in a boat-chair or twist-chair-chair conformation in which the transannular sulfur atoms are on the same side of the ring and relatively close to each other as shown by the large lone pair-lone pair splitting in its photoelectron spectrum and computational analysis. This dithioether is not crystalline at room temperature, but x-ray crystallographic structure studies on several of its derivatives have been reported. In some of these, the 1,5-dithiocane ring is in a boat-chair conformation, in others a chair-chair conformation, and in the bis-iodine complex of 1,5-dithiocane, both conformations are present in the asymmetric unit. 1,3,5,7-Tetra-thiocane adopts a boat-chair conformation but N,N'-diphenyl-3,7-diaza-1,5-dithiocane and the corresponding N,N'-dimethyl analogue have a chair-chair conformation. The eight-membered ring in 2-phenyl-2-chloro-1,3,6-trithia-2-stannocane 13 is in a boat-chair conformation such that there is 1,5-transannular tin-sulfur interaction. This interaction is shown by the short Sn···S distance of 2.806 Å and the trigonal bipyramidal geometry about the tin atom. Replacing the chloro substituent on the tin atom by a phenyl group results in decreased transannular Sn···S interaction. The Sn···S distance is 3.246 Å and the tin atom is intermediate in geometry between tetrahedral and trigonal bipyramidal in this case. There is no evidence for transannular P⋯S interaction in the eight-membered ring of 2-thio-2-tert-butyl-1,3,6-2-trithiaphosphocane. Both a chair and boat-type conformer are present in solutions of 1,5-dithiocane **14** with two conformationally constraining

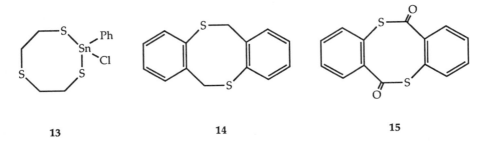

| 13 | 14 | 15 |

benzene rings fused to the eight-membered ring. The x-ray crystal structure of dithio-salicylide **15** reveals the eight-membered ring to be a boat. NMR studies in solution on a derivative of **15** substituted in the aromatic ring show a boat conformation which undergoes ring inversion with $\Delta G^{\neq}$ = 24.6 kcal/mol.

Cyclononane adopts three conformations: the symmetrical D_3 [333] form (twist-boat-chair) **16**, the C_2 [12222] form **17**, and to a much lesser extent the twist-chair-chair [441]. X-ray

| 16 | 17 | 18 |

crystallographic structural analysis and photoelectron spectroscopic analysis of 1,5-dithio-nane and 1,4,7-trithionane show that the nine-membered rings in these compounds have twist-boat-chair [333] conformations. The sulfur atoms in 1,5-dithionane are on opposite sides of the ring and consequently the lone pair orbitals on sulfur do not interact and there is very little splitting of the 3p-lone pair orbitals. However, the three sulfur atoms in 1,4,7-trithionane **18** are on the same side of the ring and close to each other.[13] The molecule has C_3 symmetry and the nonbonded S⋯S distances are 3.451(2) Å. Consequently, the 3p sulfur lone-pair orbitals in 1,4,7-trithionane overlap and are split by 0.6 eV as shown in the photoelectron spectrum of this compound.[12] It is worth noting that owing to the conformational differences between oxygen and sulfur, as outlined above, crown ethers adopt conformations in which the ether oxygen atoms line the center of the crown, i.e. they point toward the center of the ring (endodentate), but in thiacrowns the sulfur generally is directed away from the center of the ring (exodentate). Although, there are both endo- and exo- dentate sulfur atoms in hexathia-18-crown-6.[14] However, all three sulfur atoms in 1,4,7-trithionane are endodentate. This unique conformation renders this molecule a valuable ligand for complexation of transition metal ions. Owing to its preorganization, this compound can coordinate metal ions more strongly than other polythioethers and due to its ability to act as a facially coordinating tridentate ligand confer unusual coordination geometries around metal ions. Such complexes show distinctive properties.

There are several possible conformations for cyclodecane, but the most stable conformer appears to be the boat-chair-boat [2323] conformer **19**. Cyclodecasulfur crystallizes in a conformation of D_2 symmetry and not as the crown conformer of D_5 symmetry suggested by calculations. The conformation of the ten-membered ring in 1,6-dithiecane is the boat-chair-boat [2323] conformation but, curiously, the sulfur atoms occupy conformationally different

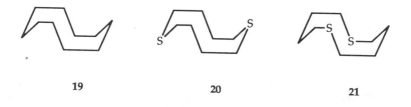

19 20 21

sites in the solid state as compared with the gas phase or solution. In the solid state conformer **20** is preferred as shown by x-ray studies. This conformation would give a much smaller lone pair-lone pair splitting than the 0.15 eV measured in the gas phase by photoelectron spectroscopy. The best fit of the photoelectron spectrum and calculations is conformer **21** in the gas phase.[12]

To learn more about lone pair interactions between divalent sulfur atoms, dithioether **22** was prepared and studied.[15] The idea is as follows. Substituents in the 1,8-"peri" positions of naphthalene, as examples, the two thioether sulfur atoms in **22** and **23**, are held relatively rigidly, relatively close together. In dithioether **23**, the C–S–C planes should be

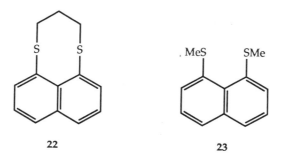

22   23

approximately coplanar with the naphthalene ring. The $3p_z$-orbitals on both sulfur atoms in this compound will be perpendicular to this plane and parallel to each other as in "antiparallel" **7** and dithioether **10**. In this geometry, the $3p_z$-orbitals can overlap in a π-fashion with the π-orbitals of the naphthalene ring. As is clear from the x-ray crystal structure of **23**, this picture is slightly modified in reality. There is rotation about the C(1)–S and C(8)–S bonds in the same direction (conrotatory) by about 28.5° in **23** to alleviate the steric strain between the methyl substituents on each sulfur and the hydrogen atoms ortho to the methylthio groups. Nevertheless, the $3p_z$-orbitals on the sulfur atoms are still approximately parallel to each other and overlap extensively with the π-system of the naphthalene ring. If the rotation about the C(1)–S and C(8)–S bonds is in opposite directions (disrotatory) and continued until the S–Me bonds are perpendicular to the naphthalene plane then the $3p_z$-orbitals on each of the sulfur atoms are orthogonal to the naphthalene π-system and point directly at each other as in $3p\cdots3p$ **8**. This is approximately the geometry in **22** constrained in this orientation by its eight-membered ring. The x-ray crystal structure of **22** shows that the S–CH$_2$ bonds make an angle of approximately 82° with the average plane of the naphthalene ring and the nonbonded S$\cdots$S distance is 3.227(1) Å. Consequently, there should be an enormous splitting of the 3p nonbonding orbitals and intense lone pair-lone pair repulsion in dithioether **22**. This compound is a derivative of 1,5-dithiocane, but owing to the constraints of the naphtho fusion dithioether **22** has a geometry which enforces greater lone pair-lone pair interaction than that in 1,5-dithiocane. (Note that the C(1)–S and C(8)–S bonds are constrained to be almost coplanar in **22**.) The sulfur atoms in **22** are constrained to be closer to each other than those in 1,5-dithiocane (3.227 Å versus approximately 3.4 Å) and the 3p-type orbitals on the sulfur atoms point almost directly at each other in dithioether **22**, but somewhat askew in 1,5-dithiocane.

To determine the lone pair-lone pair splitting in dithioether **22** its photoelectron spectrum was measured and analyzed. This analysis proved very difficult because both π-ionizations, lone pair ionizations, and their combinations had to be distinguished. Semiempirical molecular orbital analysis using the AM–1 (Austin Model 1) method with an updated version of sulfur parameters proved to be the most useful computational method. A new experimental method for distinguishing sulfur lone pair ionization from carbon π ionizations of general applicability was developed to provide convincing evidence for ionization assignments.[16] This experimental method is based on comparison of He I and He II photoelectron spectra. Because the cross-sections for photoionization for sulfur 3p and carbon 2p orbitals are comparable for He I radiation, but much lower for sulfur 3p than carbon 2p orbitals with higher energy He II radiation, the intensity of ionizations from a sulfur 3p-nonbonding orbital decreases 60 - 70% relative to ionizations from a carbon π-molecular orbital on changing the ionizing source from He I to He II. The lone pair-lone pair splitting in dithioether **22** was thereby determined to be the unprecedentedly large value of 2 eV.

The molecular orbital analysis of dithioether **23** as a function of rotation about the C(1)–S and C(8)–S bonds yielded a novel result. The energy of the highest occupied molecular orbital in this system was found to be nearly independent of this rotation. This is unexpected because the overlap between the highly directional $3p_z$ orbitals of the two sulfur atoms changes as a function of this rotation in an analogous way to that already pointed out in the discussion on the close approach of two dimethylsulfide molecules. In the 3p⋯3p approach **8**, which is analogous to 90° conrotation about the C(1)–S and C(8)–S bonds away from the geometry in which the S–CH₃ bonds are in the plane of the naphthalene ring, the antisymmetric combination of nonbonding 3p-orbitals rises to high energy at small S⋯S separations. This is in contrast to the antiparallel approach **7**, which is analogous to 0° rotation about the C(1)–S and C(8)–S bonds, i.e. the S-methyl bonds are in the plane of the naphthalene ring. However, the highest occupied molecular orbital of **23** consists of the antisymmetric combination of sulfur 3p-orbitals and antibonding sulfur 3p - naphthalene π-orbitals. As the rotation about C(1)–S and C(8)–S occurs from the geometry in which the S-methyl bonds are in the plane of the naphthalene ring there are two effects on the energy of the highest occupied molecular orbital. The antisymmetric sulfur 3p-orbital combination will raise the energy of the highest occupied molecular orbital, but this will be counterbalanced by the lowering of the energy of this orbital due to decreased sulfur 3p - naphthalene π antibonding interactions.

Synthesis

The synthesis of medium-sized rings is difficult in general. This is due to the strain in such rings as already discussed. One source of such strain in medium-sized ring dithioethers is due to the close juxtaposition of transannular heteroatoms. This proximity of heteroatoms is key in determining the neighboring group participation properties of such systems. Thus, the very effect that results in novel redox properties for these systems contributes to the difficulty in their synthesis.

A standard method for synthesizing thioethers is the reaction of alkyl halides with thiolates. This method using dithiolates and dihalides can be used with varying success for the preparation of medium-sized ring dithioethers. The problem in such reactions is that there is competition in the intermediary halide-thiolate **24** between cyclization to the unfavorable medium-sized ring and intermolecular reaction to ultimately give a larger ring or polymer as shown below. The conditions used in such reactions e.g. high-dilution techniques, and the counterions of the dithiolate are especially important in obtaining higher yields of the medium-sized ring dithioethers. Use of quaternary ammonium or cesium salts of dithiolates have proven to be of value. An especially recommended method is the reaction of dithiol, cesium carbonate and dihalide in N,N-dimethylformamide as solvent.[14,17] These methods proved to be of limited value though in the synthesis of dithioether **22**.[15] Modest yields were obtained irreproducibly with any of these methods. However, this compound

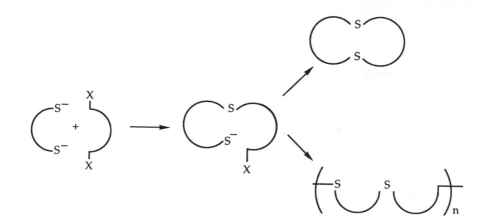

was prepared reproducibly in 85% isolated yield by reduction of disulfide **25** and alkylation with 1,3-dibromopropane under phase transfer conditions. 1,4,7-Trithionane **18** was synthesized in 60% yield by a very elegant template method.[18]

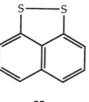

25

Alkylation of the molybdenum dithiolate **26**, obtained by the reaction of the dithiolate salt, $(Me_4N)_2S(CH_2CH_2S)_2$, with $Mo(CO)_3(CH_3CN)_3$, with 1,2-dibromoethane afforded $Mo(CO)_3(1,4,7\text{-TTCN})$ **27**. That is, the 1,4,7-trithionane complex of Mo(0) from

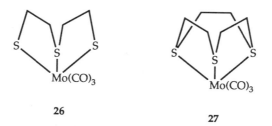

26

27

which the desired 1,4,7-trithionane ligand was liberated by treatment of the complex with dithiolate salt $(Me_4N)_2S(CH_2CH_2S)_2$. This concomitantly produced molybdenum dithiolate **26** which could be recycled.

REFERENCES

1. C. Altona and M. Sundaralingam, *J. Am. Chem. Soc.* 92:1995 (1970).
2. R.S. Glass, J.R. Duchek, U.D.G. Prabhu, W.N. Setzer, and G.S. Wilson, *J. Org. Chem.* 45:3640 (1980).
3. R.S. Glass, B.R. Coleman, U.D.G. Prabhu, W.N. Setzer, and G.S. Wilson, *J. Org. Chem.* 47:2761 (1982).

4. R.S. Glass, M. Hojjatie, W.N. Setzer, and G.S. Wilson, *J. Org. Chem.* 51:1815 (1986).

5. H.C. Brown and S. Krishnamurthy, *Tetrahedron* 35:567 (1979).

6. W.N. Setzer and R.S. Glass, *in*: "Conformational Analysis of Medium-Sized Heterocycles", R.S. Glass, ed., VCH, New York, pp. 151 (1988).

7. D.B. Boyd, *J. Phys. Chem.* 82:1407 (1978).

8. R.E. Rosenfield, Jr., R. Parthasarathy, and J.D. Dunitz, *J. Am. Chem. Soc.* 99:4860 (1977).

9. P. Coppens, Y.W. Yang, R.H. Blessing, W.F. Cooper, and F.K. Larsen, *J. Am. Chem. Soc.* 99:760 (1977).

10. T.N. Guru Row and R. Parthasarathy, *J. Am. Chem. Soc.* 103:477 (1981).

11. R.S. Glass, S.W. Andruski, and J.L. Broeker, *in*: "Reviews on Heteroatom Chemistry", Vol. 1, S. Oae, ed., MYU, Tokyo, pp. 31 (1988).

12. W.N. Setzer, B.R. Coleman, G.S. Wilson, and R.S. Glass, *Tetrahedron* 37:2743 (1981).

13. R.S. Glass, G.S. Wilson, and W.N. Setzer, *J. Am. Chem. Soc.* 102:5068 (1980).

14. S.R. Cooper, *Acc. Chem. Res.* 21:141 (1988).

15. R.S. Glass, S.W. Andruski, J.L. Broeker, H. Firouzabadi, L.K. Steffen, and G.S. Wilson, *J. Am. Chem. Soc.* 111:4036 (1989).

16. R.S. Glass, J.L. Broeker, and M.E. Jatcko, *Tetrahedron* 45:1263 (1989).

17. J. Buter and R.M. Kellog, *J. Org. Chem.* 46:4481 (1981).

18. D. Sellman and L. Zapf, *J. Organomet. Chem.* 289:57 (1985).

THE STEREOCHEMICAL COURSE OF SULPHATE ACTIVATION AND TRANSFER

Gordon Lowe

Dyson Perrins Laboratory
Oxford University
Oxford QX1 3QY, U.K.

The major source of sulphur in all sulphur-containing compounds in the biosphere, such as the amino acids cysteine and methionine, the coenzymes biotin, lipoic acid, thiamine pyrophosphate and coenzyme A, the redox reagent glutathione, etc., is inorganic sulphate. This inert, water soluble and widely available divalent anion is actively transported into cells but before it can participate in biochemical transformations it requires activation.

Robbins and Lipmann[1] were the first to isolate activated sulphate from a living system and showed it to be 3'-phosphoadenosine 5'-phosphosulphate (PAPS) **1** (R = PO_3).

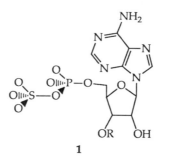

1

This has subsequently been shown to be the active form of sulphate used in all biological systems investigated. It is formed from inorganic sulphate and ATP in the presence of Mg^{2+} by the following route (reactions 1 - 3):

$$ATP^{4+} + SO_4^{2+} + H^+ \rightleftharpoons APS^{2+} + PP_i^{3-} \tag{1}$$

The first step is catalysed by the enzyme ATP sulphurylase; adenosine 5'-triphosphate (ATP) is the activating agent giving rise to the phosphosulphate anhydride bond of adenosine 5'-phosphosulphate (APS) **1** (R = H) and inorganic pyrophosphate. The equilibrium constant for the formation of APS **1** (R = H) however is very unfavourable, $\Delta G_0'$ being about +9.5 kcal/mole at 5 mM Mg^{2+}, 25 °C and pH 7.[2,3] Since $\Delta G_0'$ for hydrolysis of ATP to AMP and PP_i under the same conditions is –10 kcal/mole[4], $\Delta G_0'$ for hydrolysis of the phosphosulphate anhydride bond must be approximately –19.5 kcal/mol. This high group potential can obviously be used to drive other reactions. However, because of the unfavourable $\Delta G_0'$

Sulfur-Centered Reactive Intermediates in Chemistry and Biology, Edited by
C. Chatgilialoglu and K.-D. Asmus, Plenum Press, New York, 1990

for the formation of APS, if no other reactions were involved the concentration of the activated sulphate would be very low.

$$PP_i \quad \rightleftarrows \quad 2P_i \tag{2}$$

As with nucleotidyl transferases generally the liberated inorganic pyrophosphate is hydrolysed by the ubiquitous inorganic pyrophosphatase, and the free energy locked in this phosphoanhydride bond released. $\Delta G_0'$ for this reaction is about -4.5 kcal/mol in the presence of excess Mg^{2+} at pH 7.

$$APS^{2-} + ATP^{4-} \quad \rightleftarrows \quad PAPS^{4-} + ADP^{3-} + H^+ \tag{3}$$

The third enzymic reaction is catalysed by APS kinase which uses another molecule of ATP to phosphorylating the 3'-hydroxy group of APS. $\Delta G_0'$ for breaking the phospho-anhydride bond is between -4 and -5 kcal/mol bringing the overall equilibrium constant for the formation of PAPS from inorganic sulphate close to unity.

In bacteria and fungi the activated sulphate group in PAPS can be reduced by NADPH first to inorganic sulphite with the enzyme PAPS reductase and then to hydrogen sulphide with sulphite reductase.[5,6]

$$PAPS^{4-} + NADPH \quad \rightleftarrows \quad PAP^{4-} + NADP^+ + SO_3^{2-} + H^+ \tag{4}$$

$$SO_3^{2-} + 5\,H^+ + 3\,NADPH \quad \rightleftarrows \quad H_2S + 3\,H_2O + 3\,NADP^+ \tag{5}$$

A number of dithiols can substitute for NADPH as a reductant in the PAPS reductase reaction, e.g. dihydrolipoate 2 which is oxidised to lipoate presumably by the following mechanism,[7]

$$PAPS^{4-} + HSCH_2CH_2CH(SH)[CH_2]_4COOH \qquad \textbf{2}$$

$$\downarrow \tag{6}$$

$$PAP^{4-} + O_3SSCH_2CH_2CH(SH)[CH_2]_4COOH$$

$$\downarrow \tag{7}$$

so that it may be that the function of the NADPH is to reduce a natural disulphide cofactor.

The mechanisms of sulphite reductases have not yet been elucidated but many contain a heme prosthetic group known as siroheme.[8a,b] The available evidence indicates that the siroheme prosthetic group is essential for sulphite reduction and moreover the sulphite interacts directly with it.

Although bacteria and some other organisms use APS and PAPS for the generation of sulphite, hydrogen sulphide and then a multitude of compounds containing divalent sulphur, in plants, fungi and mammals, PAPS is used much more commonly as a sulphating agent. Indeed Baumann as long ago as 1876 isolated phenyl sulphate as its potassium salt from the urine of dogs fed on phenol.[9] The enzymes which catalyse the transfer of the sulphuryl group to suitable acceptors are known as sulphotransferases.

$$PAPS + ROH \quad \rightleftarrows \quad ROSO_3^- + PAP \tag{8}$$

240

ATP sulphurylase has been extensively studied. Initial velocity, product inhibition, dead-end inhibition, and isotope exchange studies have indicated that ATP sulphurylase follows an ordered sequential kinetic mechanism, MgATP adding before sulphate and $MgPP_i$ dissociating before APS.[10-16] MgATP analogues (ATP, CaATP, CrATP, or Mg $\alpha\beta$-methylene ATP), however, were shown to be incapable of inducing the binding of sulphate (or nitrate, a competitive analogue of sulphate), and so it was concluded that either the structural requirements for the metal ion-nucleotide complex are so stringent that only MgATP can induce the conformational change that forms the sulphate binding site, or the mere binding of a metal ion-nucleotide complex is insufficient to promote the binding of sulphate. The latter explanation was favoured, with the implication that the MgATP must be catalytically cleaved to $E\text{-}AMP\cdot MgPP_i$ before the sulphate binding site formed since ATP sulphurylase catalyzes the hydrolysis of MgATP to AMP and $MgPP_i$ in the absence of sulphate,[16] the hydrolysis of APS to AMP and sulphate in the absence of $MgPP_i$,[16,17] and the isotope exchange between $Mg^{32}PP_i$ and MgATP in the absence of sulphate;[16] it is always possible to invoke the concept of substrate synergism[18] where the rates fall below the rate of the overall reaction. In addition, it was considered that the greater binding of APS (dissociation constant about 1 μM) compared with MgATP (dissociation constant about 1 mM) was consistent with the APS being bound predominantly as the $E\text{-}AMP\cdot SO_4$ complex, whereas MgATP was bound predominantly as the E-MgATP complex.[19] It was further proposed that a tyrosine hydroxyl group was the probable site of adenylation by APS, and by MgATP in the presence of NO_3^- to give $E\text{-}AMP\cdot SO_4$ and $E\text{-}AMP\cdot MgPP_i NO_3^-$, respectively, since chemical modification of a tyrosine residue with tetranitromethane led to loss of enzymic activity that could be prevented by APS or MgATP plus NO_3^-, but not by ATP or MgATP alone.[19] (E = enzyme).

Several phosphokinases possess ATPase and/or MgADP-MgATP exchange activity in the absence of their co-substrate, but this has proved to be unreliable evidence for the involvement of a phosphoryl-enzyme intermediate. What has emerged from extensive stereochemical studies is that all phosphokinases which obey sequential kinetics catalyze phosphoryl transfer with inversion of configuration, whereas those which obey ping-pong kinetics catalyze phosphoryl transfer with retention of configuration; there are no known exceptions.[20,21] Since there is good independent evidence for a phosphoryl-enzyme intermediate where ping-pong kinetics are observed but not where sequential kinetics are followed, the simplest interpretation of these results is that inversion of configuration implies a single "in line" displacement reaction, whereas retention of configuration implies a double displacement reaction.

Support for the involvement of an adenylyl-enzyme intermediate in the ATP sulphurylase-catalysed reaction would be provided if the stereochemical course of nucleotidyl transfer was found to proceed with retention of configuration at phosphorus. If, however, inversion of configuration at phosphorus occurs, this would imply an odd number (the simplest being one) of "in line" nucleotidyl displacement reactions. In order to determine this fundamental mechanistic question we synthesized APS made chiral at phosphorus so that the ATP sulphurylase reaction could be run in the thermodynamically favourable direction.

We have developed a general method for the synthesis of phosphate monoesters made chiral at phosphorus by virtue of isotopic substitution.[22] The route is outlined in Scheme 1. The absolute configuration of the $[^{16}O^{17}O^{18}O]$phosphate monoester follows from the absolute configuration of (S)-mandelic acid 3 and the relative configuration of the 2-substituted 2-$[^{17}O]$oxo-$[1-^{18}O]$1,3,2-dioxaphospholan 8. Only the trans-diastereoisomer 8 is formed under thermodynamic control in pyridine solution, hence the route is stereospecific.

It had been reported that AMP can be converted into APS by treatment with sulphur trioxide-triethylamine complex (triethylamine-N-sulphonic acid) over 16 h without significant reaction at the secondary hydroxyl groups or the adenine ring,[23] but this proved

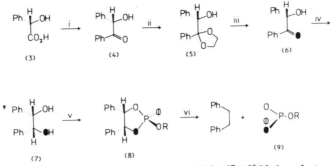

(3)   (4)   (5)   (6)

(7)   (8)   (9)

Scheme 1. A general synthetic route to chiral [^{16}O,^{17}O,^{18}O]phosphate monoesters of known absolute configuration.

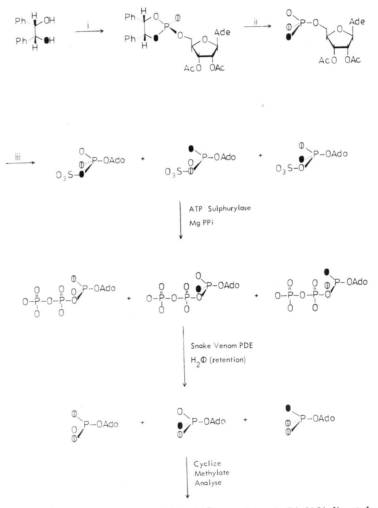

ATP Sulphurylase
Mg PPi

Snake Venom PDE
H$_2$⦶ (retention)

Cyclize
Methylate
Analyse

Scheme 2. ⦶, ^{17}O; ●, ^{18}O. Reagents (i) (a) P^{17}OCl$_3$, C$_5$H$_5$N; (b) 2′,3′-diacetyladenosine, C$_5$H$_5$N; (ii) H$_2$Pd–C, (iii) (a) SO$_3$.NEt$_3$; (b) acetyl esterase.

impossible in our experience even when shorter reaction times were used. Conditions were found, however, in which 2',3'-diacetyladenosine 5'-phosphate was converted exclusively into 2',3'-diacetyladenosine 5'-phosphosulphate with the sulphur trioxide-triethylamine complex in as little as 5 min at room temperature. The acetyl groups could be removed enzymically with acetyl esterase to give APS in good yield. This route was used for the synthesis of the isotopomers of adenosine 5'-[(S)-$^{16}O,^{17}O,^{18}O$]phospho-sulphate as outlined in the upper part of Scheme 2. 2',3'-Diacetyl adenosine 5'-[(S)-$^{16}O,^{17}O,^{18}O$]phosphate was prepared by our general method of synthesis of chiral-[$^{16}O,^{17}O,^{18}O$]phosphate esters[22] and then converted into the isotopomers of adenosine 5'-[(S)-$^{16}O,^{17}O,^{18}O$]phosphosulphate by reaction with the sulphur-trioxide-triethylamine complex followed by enzymic deacetylation.

Incubation of the isotopomers of adenosine 5'-[(S)-$^{16}O,^{17}O,^{18}O$]phosphosulphate and magnesium pyrophosphate with ATP sulphurylase gave isotopically labeled ATP. We have shown that snake venom 5'-nucleotide phosphodiesterase catalyzes the hydrolysis of ATP to AMP with retention of configuration at P_α [24] and have developed a method based on ^{31}P n.m.r. spectroscopy for the analysis of the chirality at phosphorus of adenosine 5'[$^{16}O,^{17}O,^{18}O$]phosphate.[25] By hydrolyzing the isotopomers of ATP with snake venom 5'-nucleotide phosphodiesterase and analyzing the chirality of the adenosine 5'[$^{16}O,^{17}O,^{18}O$]phosphate and its isotopomers formed, it was possible to establish the stereochemical course of the nucleotidyl transfer catalyzed by ATP sulphurylase (Scheme 2). It can easily be shown that greater discrimination between the alternative stereochemical courses can be obtained in the ^{31}P n.m.r. analysis if [^{17}O]water is used for the hydrolysis of the isotopomers of ATP to those of AMP. The [$^{16}O,^{17}O,^{18}O$]AMP and its isotopomers obtained by snake venom 5'-nucleotide phosphodiesterase-catalyzed hydrolysis in [^{17}O]water were cyclized and methylated[25] and the ^{31}P n.m.r. spectrum obtained (Figure. 1). If all the isotopically labeled sites were fully enriched as shown in the Scheme, only the chiral [$^{16}O,^{17}O,^{18}O$]AMP would give rise to species that would appear in the ^{31}P n.m.r. spectrum after cyclization, since the achiral [$^{16}O,^{17}O_2$]- and [$^{17}O_2,^{18}O$]AMP would inevitably give rise to species containing at least one ^{17}O bonded to phosphorus, which would not be observed.[26-28] In practice, the sites are not fully enriched, but since the isotopic composition of the (1R,2S)-1,2-[1-^{18}O]dihydroxy-1,2-diphenylethane and the phosphorus [^{17}O]oxychloride used in the synthesis of 2',3'-diacetyl adenosine 5'[(S)-$^{16}O,^{17}O,^{18}O$]phosphate are known, and the isotopic composition of the [^{17}O]water used in the hydrolysis of the isotopomers of ATP by snake venom 5'-nucleotide phosphodiesterase is known, it is possible to calculate the expected ratios of the isotopomers

Table 1. The observed relative peak intensities of the ^{31}P n.m.r. resonances (from Figure 1) of the diastereoisomeric triesters derived by cyclization followed by methylation[25] of 5'-[$^{16}O,^{17}O,^{18}O$]AMP and its isotopomers, and the calculated values expected for ATP sulphurylase-catalyzed nucleotidyl transfer with retention and inversion of configuration.
The stereoselectivity is determined by the ratio of the intensities of the mono-[^{18}O]triesters compared with the calculated values. ● = ^{18}O.

	Equatorial triester			Axial triester		
	Observed	Calculated		Observed	Calculated	
		Retention	Inversion		Retention	Inversion
MeO–P=O	0.39	0.39	0.39	0.42	0.39	0.39
Me●–P=O	0.89	1.00	0.89	1.00	0.89	1.00
MeO–P=●	1.00	0.89	1.00	0.88	1.00	0.89
Me●–P=●	1.10	1.07	1.07	1.05	1.07	1.07

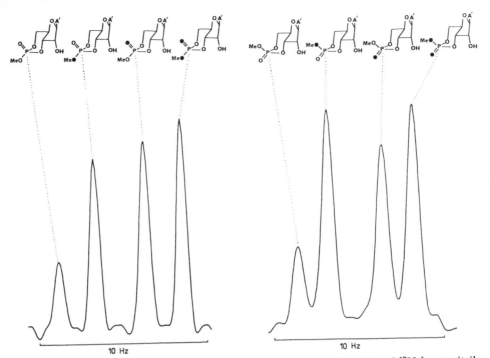

Fig. 1. ^{31}P nmr spectrum (121.5 MHz) in dimethyl sulfoxide (1.2 ml) and [$^{2}H_3$]acetonitrile (1.0 ml) of the equatorial and axial triesters derived by cyclization followed by methylation of 5'-[$^{16}O,^{17}O,^{18}O$]AMP and its isotopomers obtained by hydrolyzing the isotopomers of ATP with snake venom 5'-nucleotide phosphodiesterase in [^{17}O]water. The isotopomers of ATP were derived by incubating the isotopomers of adenosine 5'[(S)-$^{16}O,^{17}O,^{18}O$]phosphosulphate with ATP-sulphurylase and MgPP$_i$. The ^{31}P nmr parameters are: offset 2200 Hz, sweep width 2000 Hertz, acquisition time 2.05 s, pulse width (angle) 15 ms (75) gaussian multiplication (line broadening - 1.0 Hz, gaussian broadening 0.3) in 8K and Fourier transform in 32K. A', N-1-methyladenine; ●, ^{18}O.

of AMP that would be formed if the ATP sulphurylase-catalyzed reaction proceeded with retention or inversion of configuration at phosphorus. The calculated ratios of the isotopomers of the axial and equatorial triesters of 3',5'-cAMP derived from the isotopomers of AMP [the cyclization occurs with inversion of configuration[25]] are compared with the observed relative intensities from Figure 1 in Table 1. From this it is clear that ATP sulphurylase catalyzes nucleotidyl transfer with inversion of configuration at phosphorus.[29]

The finding that ATP sulphurylase catalyzes nucleotidyl transfer with inversion of configuration at phosphorus effectively eliminates the adjacent mechanism followed by a pseudorotation and the double displacement mechanism with an adenylyl-enzyme intermediate as postulated by Segel and co-workers.[16,19] The mechanism must be either an associative or dissociative "in line" mechanism; that is, the adenylyl group must either be transferred directly between substrates in the ternary complex or by way of an adenosine metaphosphate. Although the latter mechanism could conveniently account for some of the observations of Segel and collaborators,[16,19] the lack of precedent for the intermediacy of a metaphosphate ester in the hydrolysis of phosphate diesters[30] and the lifetime necessary to account for MgATP-Mg^{32}PP$_i$ exchange makes the dissociative mechanism unlikely. In fact, we have shown that ATP sulphurylase does not scramble the ^{18}O labels in adenosine $5'[\beta-^{18}O_2]$triphosphate in the presence of nitrate ions (competitive inhibitors of sulphate ions), which effectively eliminates the dissociative "in line" mechanism. We therefore conclude that the sulphate activation by ATP sulphurylase occurs by a direct "in line" nucleophilic substitution by sulphate ion at P_α of ATP.

Since we reported this work[29] Segel *et al.* have shown that if ATP sulphurylase is rigorously purified it does not catalyse ATP hydrolysis or isotope exchange between $Mg^{32}PP_i$ and ATP in the absence of sulphate,[31a-c] which is of course in accord with the direct "in line" nucleophilic substitution mechanism.

SYNTHESIS AND STEREOCHEMICAL ANALYSIS OF CHIRAL [$^{16}O,^{17}O,^{18}O$]SULPHATE MONOESTERS

In order to study the stereochemical course of sulphuryl transfer reactions we needed to develop a general strategy for the synthesis of chiral [$^{16}O,^{17}O,^{18}O$]sulphate monoesters of known absolute configuration and a method for their stereochemical analysis.

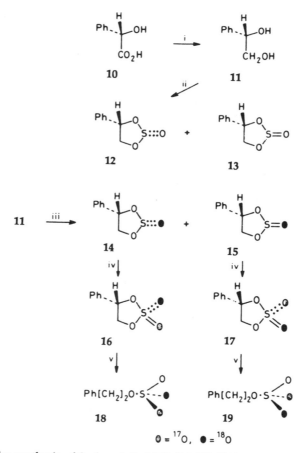

Scheme 3. The synthesis of 2-phenylethyl [(S)-^{16}O-$^{17}O,^{18}O$]sulphate **18** and 2-phenylethyl [(R)-$^{16}O,^{17}O,^{18}O$]sulphate **19**. Reagents: (i) LiAlH$_4$; (ii) SOCl$_2$; (iii) S^{18}OCl$_2$; (iv) RuO$_2$, NaIO$_4$, H$_2$^{17}O; (v) Bu$^n_4$N$^+$BH$_4^-$.

245

The general strategy for synthesis was to take a chiral 1,2- or 1,3-diol, convert it into the diastereoisomeric cyclic sulphites with [^{18}O]thionyl chloride and then oxidise the sulphite esters to the cyclic sulphate diesters with ruthenium [^{17}O]tetroxide. Regioselective ring opening of the cyclic [^{17}O,^{18}O]sulphate diesters should give a chiral [^{16}O,^{17}O,^{18}O]sulphate monoester of known absolute configuration. The first chiral [^{16}O,^{17}O,^{18}O]sulphate monoesters were prepared in this way as outlined in Scheme 3.[32]

(S)-Mandelic acid 10 was reduced by lithium aluminum hydride to 2(S)-phenyl-ethane-1,2-diol 11, which on treatment with thionyl chloride gave the cis- and trans-2-oxo-4(S)-phenyl-1,3,2-dioxathiolanes 12 and 13. Although their configurations could be assigned from spectroscopic evidence[33] they were rigorously established by an X-ray crystallographic analysis of the trans-isomer.[34] [^{18}O]Thionyl chloride was prepared from sulphur [^{18}O$_2$]diox-ide (99 atom% ^{18}O) and phosphorus pentachloride. Reaction of [^{18}O]thionyl chloride with 2(S)-phenylethane-1,2-diol gave cis-2-[^{18}O]oxo-4(S)-phenyl-1,3,2-dioxathiolane 14 and trans-2-[^{18}O]oxo-4(S)-phenyl-1,3,2-dioxathiolane 15 which were separated chromatographically. Their IR spectra had v_{max} (CCl$_4$) (S = ^{18}O) 1173 and 1177 cm^{-1}, respectively, whereas the un-labeled cis- and trans-diastereoisomers had v_{max} (CCl$_4$) (S = ^{16}O) 1215 and 1222 cm^{-1}, respec-tively. Oxidation of the separated diastereoisomers 14 and 15 with ruthenium [^{17}O]tetroxide generated in situ from ruthenium (IV) oxide, sodium periodate, and [^{17}O]water (52.8 atom% ^{17}O, 9.4 atom% ^{16}O, 37.8 atom% ^{18}O) in the presence of (ethanol free) chloroform, gave the diastereotopic 2(R)-[^{17}O,^{18}O]- and 2(S)[^{17}O,^{18}O]dioxa-4(S)-phenyl-1,3,2-dioxathiolanes, 16 and 17, respectively; we have established that the oxidation with ruthenium tetroxide occurs with retention of configuration at sulphur.[35] Finally, reductive cleavage of the benzylic oxygen bond was achieved with tetrabutylammonium borohydride in dimethylformamide. Since this occurs without perturbing any of the sulphur to oxygen bonds the absolute configuration of the chiral [^{16}O,^{17}O,^{18}O]sulphate esters follows from the method of synthesis and the absolute configuration of (S)-mandelic acid. Thus, 2(R)-[^{17}O,^{18}O]dioxa-4(S)-phenyl-1,3,2-dioxathiolane 16 gives 2-phenylethyl (S)-[^{16}O,^{17}O,^{18}O]sulphate 18 and 2(S)-[^{17}O,^{18}O]-dioxa-4-(S)-phenyl-1,3,2-dioxathiolane 17 gives 2-phenylethyl (R)-[^{16}O,^{17}O,^{18}O]sulphate 19 which were isolated as their crystalline tetrabutylammonium salts.

N.m.r. spectroscopy, which had been successfully used for the analysis of chiral [^{16}O,^{17}O,^{18}O]phosphate monoesters[21] did not commend itself for the analysis of chiral [^{16}O,^{17}O,^{18}O]sulphate monoesters since none of the isotopes of sulphur possesses a nuclear spin quantum number of 1/2. Moreover the exocyclic oxygen atoms in a cyclic sulphate diester cannot be alkylated to render them chemically distinguishable as was possible with cyclic phosphate diesters. The analysis of chiral [^{16}O,^{17}O,^{18}O]sulphate monoesters presents, there-fore, a new conceptual problem.

The frequency of an IR vibrational mode is markedly affected if isotopic substitution occurs in the functional group responsible. However, the magnitude of the isotope shift is difficult to predict except in the simplest of molecules, but the shift caused by replacing ^{16}O with ^{18}O in a functional group could theoretically be up to 40 cm^{-1}.[36] We expected that the frequencies of the symmetric and antisymmetric $>$SO$_2$ vibrational modes would be dependent on the location of the heavy oxygen isotope. If so, the isotope effect could form the basis of a method for the analysis of chiral [^{16}O,^{17}O,^{18}O]sulphate monoesters.

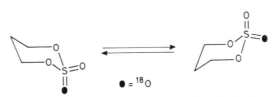

In order to explore the influence of oxygen isotopic substitution at the axial and the equatorial sites of six-membered cyclic sulphate esters, 2,2-[^{18}O]dioxo-1,3,2-dioxathiane was prepared from 2-oxo-1,3,2-dioxathiane by oxidation with ruthenium [^{18}O]tetroxide,[35] since this should exist as an equimolar mixture of chair conformations (neglecting the equilibrium isotope effect) with ^{18}O in the axial and equatorial sites.

The FTIR spectrum is shown in Figure 2(a) (the ^{18}O site is only about 60% enriched). It is evident from this spectrum that the ^{18}O shift in both the symmetric and antisymmetric stretching modes of the $>SO_2$ group is conformationally dependent. It is also apparent that the isotope shift is greater and the intrinsic line-width smaller in the symmetric stretching mode; consequently this vibrational mode should be the most useful for analytical purposes. Separate deconvolution of these two spectral regions gave the resolution enhanced spectrum shown in Figure 2(b).

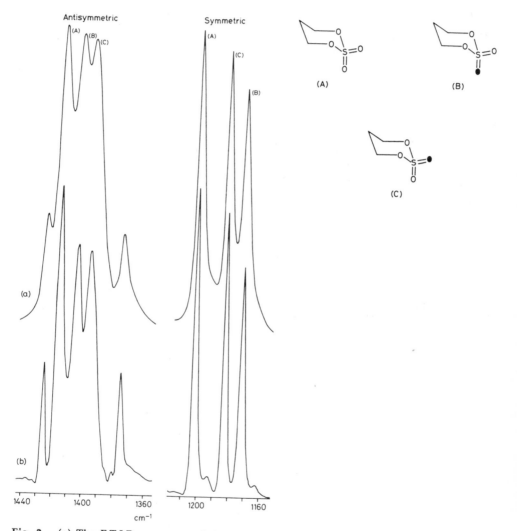

Fig. 2. (a) The F.T.I.R. spectrum of the antisymmetric and symmetric $>SO_2$ stretching region of 2,2-[^{18}O]dioxo-1,3,2-dioxathiane and (b) the same spectrum after deconvolution with an enhancement factor of 1.68 and line-width at half height of 5 cm-1 for the antisymmetric $>SO_2$ stretching region and an enhancement factor of 1.50 and line-width at half height of 4 cm-1 for the symmetric $>SO_2$ stretching region. The spectra were measured on a Perkin-Elmer 1750 F.T.I.R. spectrometer at a resolution of 1 cm^{-1}.

In order to assign the conformations responsible for the two symmetric and two antisymmetric $\verb|>|$S[^{16}O,^{18}O] absorption bands in 2,2-[^{18}O]dioxo-1,3,2-dioxathiane, and to confirm this observation, the *cis*- and *trans*-cyclic sulphite esters obtained by treating (3R)-butane-1,3-diol with thionyl chloride were oxidised with ruthenium [^{18}O]tetroxide to give the isotopomeric cyclic [^{18}O]sulphate esters. Since this oxidation is known to proceed with retention of configuration at sulphur,[35] the *cis*-sulphite must give the cyclic (R_s)-[^{18}O]sulphate **20** and the *trans*-sulphite must give the cyclic (S_s)-[^{18}O]sulphate **21**. As expected, the (R_s)-isotopomer **20** possesses only one symmetric (1172 cm^{-1}) and one antisymmetric (1401 cm^{-1}) S[^{16}O,^{18}O] stretching vibration and likewise the (S_s)-isotopomer **21** possesses only one symmetric (1183 cm^{-1}) and one antisymmetric (1392 cm^{-1}) S[^{16}O,^{18}O] stretching vibration, since the conformation with the methyl group equatorial should be preferred. Although these values differ slightly from those observed for 2,2-[^{18}O]dioxo-1,3,2-dioxathiane, the assignments shown in Figure 2 seem unambiguous.

The three stable oxygen isotopes, ^{16}O, ^{17}O, and ^{18}O, can be arranged to give nine exocyclic isotopomers of (4R)-4-methyl-2,2-dioxo-1,3,2-dioxathiane. Eight isotopomers have been prepared and their FTIR spectra determined and resolution-enhanced by deconvolution. The frequencies of the symmetric and antisymmetric stretching modes for each isotopomer are shown in Table 2.

Table 2. The effect of oxygen isotopic substitution on the symmetric and antisymmetric $\verb|>|$SO$_2$ stretching frequencies (cm^{-1}) of (4R)-4-methyl-2,2-dioxo-1,3,2-dioxathianes. Δ is the isotope shift from the $\verb|>|$SO$_2$ frequency

Isotope		Symmetric stretching frequency	Δ for symmetric stretching mode	Antisymmetric stretching frequency	Δ for antisymmetric stretching mode
Axial	Equatorial				
^{16}O	^{16}O	1201	—	1414	—
^{16}O	^{17}O	1192	9	1401	13
^{17}O	^{16}O	1186	15	1407	7
^{16}O	^{18}O	1183	18	1392	22
^{18}O	^{16}O	1172	29	1401	13
^{17}O	^{18}O	1170	31	1384	30
^{18}O	^{17}O	1163	38	1389	25
^{18}O	^{18}O	1157	44	1378	36

There are a number of features about these data worthy of comment. First, the overall isotope shift is greater for the symmetric stretching mode (e.g. $\verb|>|$SO$_2$ − $\verb|>|$S^{18}O$_2$, 44 cm^{-1}) than for the antisymmetric stretching mode ($\verb|>|$SO$_2$ − $\verb|>|$S^{18}O$_2$, 36 cm^{-1}). Secondly, the shift caused by a heavy oxygen isotope in an axial position is greater than in an equatorial position for the symmetric stretching mode, but the reverse is observed in the antisymmetric stretching mode. Thirdly, the shift caused by ^{18}O is about double that caused by ^{17}O in the same site.

Fourthly, the isotope shifts are approximately additive. Finally, since each of the diastereoisotopomeric pairs are distinguishable, especially in their symmetric stretching mode, IR spectroscopy should provide a means for analyzing chiral [^{16}O,^{17}O,^{18}O]sulphate monoesters after stereospecific cyclization to a chirally substituted six-membered cyclic sulphate.

(S_s)- and (R_s)-(1R)-3-hydroxy-1-methylpropyl [^{16}O,^{17}O,^{18}O]sulphates **29** and **30** were prepared as outlined in Scheme 4. [^{18}O]Thionyl chloride, prepared from sulphur [^{18}O$_2$]dioxide (99 atom% ^{18}O) and phosphorus pentachloride, was used to prepare the cis- and trans-(4R)-4-methyl-2-[^{18}O]oxo-1,3,2-dioxathianes **23** and **24** from (3R)-butane-1,3-diol **22**.[38] The separated diastereoisomers were oxidised with ruthenium [^{17}O]tetraoxide (prepared in situ from ruthenium dioxide, sodium periodate, and [^{17}O]water). Since this oxidation is known to proceed with retention of configuration at sulphur,[35] the cis-[^{18}O]sulphite **23** gives the (2S)-compound **25** and the trans-[^{18}O]sulphite **24** gives the (2R)-compound **26**.

The hydrolytic cleavage of 4-methyl-2,2-dioxo-1,3,2-dioxathiane has been extensively studied, but no conditions were found which gave exclusive cleavage of the primary C-O bond.[39a,b] Ammonia in methanol, however, gave the desired mode of ring cleavage, the primary amines **27** and **28** being isolated virtually quantitatively. The corresponding primary alcohols **29** and **30** were obtained by treatment with nitrous acid in 83 % yield.

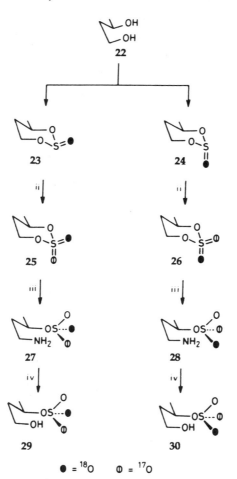

Scheme 4. The synthesis of the (S_S)- and (R_S)-sulphates **29** and **30**. Reagents: (i) S^{18}OCl$_2$, C$_5$H$_5$N; (ii) Ru^{17}O$_4$ (from RuO$_2$, NaIO$_4$, and H$_2$^{17}O); (iii) NH$_3$, MeOH; (iv) NaNO$_2$, aq. AcOH.

249

It was now necessary to develop a stereospecific method for the cyclization of the enantiomeric [^{16}O,^{17}O,^{18}O]sulphate monoesters 29 and 30. Lack of precedent for the formation of cyclic sulphate esters from acyclic sulphate monoesters led to the exploration of several possible reagents. Only two were found, namely trifluoromethanesulphonic anhydride and sulphuryl chloride, the latter giving slightly better yields (*ca.* 40 %).

In order to investigate whether there was any isotope exchange during cyclization, (1R)-3-hydroxy-1-methylpropyl[^{18}O]sulphate was prepared (by the route outlined in Scheme 4, except that in steps i and ii, $SOCl_2$ and $Ru^{18}O_4$, respectively, were used) and cyclized with sulphuryl chloride. The chemical ionization mass spectrum (NH_3) of the cyclic sulphate obtained revealed a molecular ion at *m/z* 172 only (M_r for $C_4H_8SO_4.NH_4^+$ is 172), suggesting that cyclization had occurred by activating the primary alcohol followed by intramolecular displacement by the sulphate monoester (Scheme 5). This mode of cyclization was confirmed by the natural abundance ^{13}C n.m.r. spectrum of the cyclic sulphate which showed C-1 to be split into two resonances at δ 71.784 and 71.749, the endocyclic ^{18}O causing an upfield shift of 0.035 p.p.m. as expected[40a-d] and in a 2:1 ratio of intensity after correcting for the ^{18}O enrichment of the sulphate monoester; thus no loss of isotope had occurred. It was now of interest to investigate the FTIR spectrum of the mixture of isotopomeric cyclic sulphate esters. As expected three absorption bands were observed in both the symmetric and antisymmetric $>SO_2$ stretching regions (Figure 3). For the isotopomer containing ^{18}O in the C–O–S bridge the symmetric and antisymmetric $>SO_2$ absorption bands were at 1201 and 1414 cm^{-1} respectively, i.e. identical (at 1 cm^{-1} resolution) with those for (4R)-4-methyl-2,2-dioxo-1,3,2-dioxathiane (and consequently not resolved from a small amount of unlabeled material). Thus a heavy oxygen isotope in the C–O–S bridge of the cyclic sulphate ester leaves both the symmetric and antisymmetric $>SO_2$ stretching frequencies unperturbed.

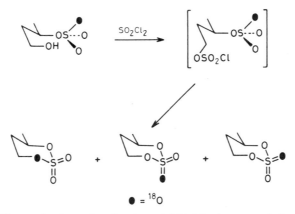

Scheme 5. The mechanism of cyclisation of (1R)-3-hydroxy-methylpropyl sulphate.

Since none of the S-O bonds is broken in the cyclization of (1R)-3-hydroxy-1-methylpropyl sulphate with sulphuryl chloride the cyclization should proceed stereospecifically for a chiral [^{16}O,^{17}O,^{18}O]sulphate with retention of configuration. In order to confirm this prediction the [(S)-^{16}O,^{17}O,^{18}O]sulphate ester 29 and the [(R)-^{16}O,^{17}O,^{18}O]-sulphate ester 30 were cyclized with sulphuryl chloride and the FTIR spectra of the isotopomeric mixture of cyclic sulphate esters measured. The spectra of the symmetric and antisymmetric $>SO_2$ stretching frequencies are shown in Figure 4.

Scheme 6 shows the mixture of isotopomeric (4R)-4-methyl-2,2-dioxo-1,3,2-dioxathianes that should be formed by cyclizing the (S_s)- and (R_s)-chiral [^{16}O,^{17}O,^{18}O]sulphate esters 29 and 30 with retention of configuration at sulphur by the mechanism outlined in Scheme 5. If all three isotopes were fully enriched only the three isotopomers shown on the top row of

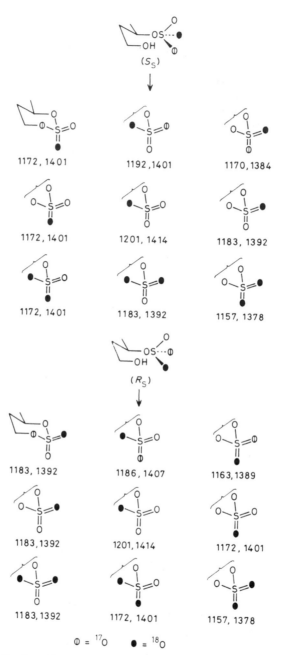

$ \textcircled{0} = {}^{17}O \qquad \bullet = {}^{18}O $

Scheme 6. The cyclisation of the (S_S)- and (R_S)-sulphates **29** and **30** with retention of configuration at sulphur. If the three isotopes were fully enriched only the first three isotopomers of each set would be formed, but in practice the '^{17}O site' contains substantial amounts of ^{16}O and ^{18}O and consequently nine isotopomers should be formed for each chiral [$^{16}O,^{17}O,^{18}O$]sulphate. The frequency (cm^{-1}) of the symmetric and antisymmetric SO_2 stretching bands for each isotopomer are shown below each formula.

251

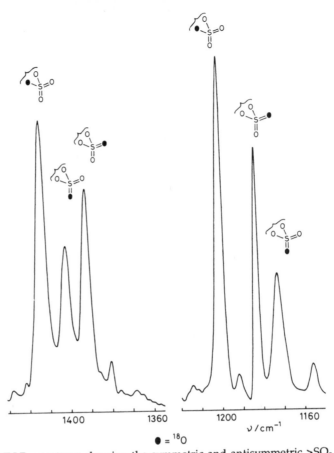

● = ^{18}O

Fig. 3. The F.T.I.R. spectrum showing the symmetric and antisymmetric >SO$_2$ stretching fequencies of the isotopomeric mixture obtained by cyclising (1R)-3-hydroxyl-1-methylpropyl [^{18}O]sulphate with sulphuryl chloride. The spectrum was determined with a Perkin-Elmer 1750 F.T.I.R. spectrometer and a Perkin-Elmer 7300 Professional computer. The spectral resolution was enhanced by Fourier deconvolution: for the symmetric stretching region a line-width of 12 cm^{-1} and an enhancement factor of 1.5 were used whereas for the anti-symmetric stretching region a line width of 20 cm^{-1} and an enhancement factor of 2.0 were used. The symmetric and antisymmetric >SO$_2$ stretching frequencies at 1201 and 1414 cm^{-1}, respectively, coincide with those for unlabelled (4R)-4-methyl-2,2-dioxo-1,3,2-dioxathiane. Only partial structures, showing the isotopic arrangement around sulphur, are illustrated.

each set would be obtained, but in practice the "^{17}O-site" consists of a substantial amount of ^{16}O and ^{18}O, and therefore nine isotopomeric species should be formed; the ^{18}O site is 99 atom% ^{18}O. The symmetric and antisymmetric >SO$_2$ stretching frequencies are shown for each isotopomer.

The spectra shown in Figure 4(a) and 4(b) are easily distinguishable, and therefore provide a method for the stereochemical analysis of chiral [^{16}O,^{17}O,^{18}O]sulphate esters.[41]

It should be noted, however, that the line-width and extinction coefficient for different isotopomers are significantly different. This is well illustrated in Figure 3 where the [$^{18}O_{ax}$,$^{16}O_{eq}$]- and [$^{16}O_{ax}$,$^{18}O_{eq}$]-isotopomers must be present in equimolar amounts. If peak

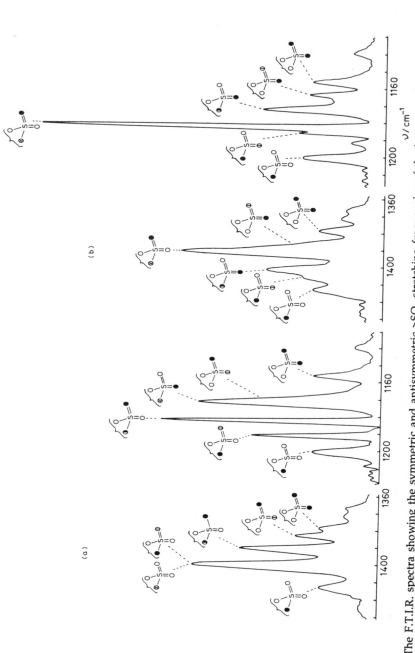

Fig. 4. The F.T.I.R. spectra showing the symmetric and antisymmetric >SO₂ stretching frequencies of the isotopomeric mixture of 4-methyl-2,2-dioxo-1,3,2-dioxathianes obtained by cyclising with sulphuryl chloride: (a) the [(S)-¹⁶O,¹⁷O,¹⁸O] sulphate **29** in which the '¹⁷O-site' consists of 37.4 atom % ¹⁶O, 36.4 atom % ¹⁷O, and 26.2 atom % ¹⁸O; (b) the [(R)-¹⁶O,¹⁷O,¹⁸O]sulphate **30** in which the '¹⁷O-site' consists of 36.0 atom % ¹⁶O, 37.1 atom % ¹⁷O, and 26.9 atom % ¹⁸O.   The spectra were determined as described in the legend to Fig. 3.   Only partial structures, showing the isotopic arrangement around sulphur, are illustrated.   ◐ = ¹⁷O; ● = ¹⁸O; ◑ = ¹⁶O + ¹⁸O; ⊗ = ¹⁶O + ¹⁷O + ¹⁸O.

253

intensities were to be used to quantify the stereochemical analysis this fact must be taken into account. It is much simpler and more accurate to look for the presence and absence of the isotopomers containing ^{17}O in the symmetric $>SO_2$ stretching frequency region of the spectrum. In the set of isotopomers derived from the chiral $[(S)\text{-}^{16}O,^{17}O,^{18}O]$sulphate ester **29** the $[^{16}O_{ax},^{17}O_{eq}]$-isotopomer (1192 cm^{-1}) should be well resolved whereas the $[^{17}O_{ax},^{18}O_{eq}]$ will not. In the set of isotopomers derived from the chiral $[(R)\text{-}^{16}O,^{17}O,^{18}O]$sulphate ester **30**, the $[^{18}O_{ax},^{17}O_{eq}]$-isotopomer (1163 cm^{-1}) should be well resolved whereas the $[^{17}O_{ax},^{16}O_{eq}]$-isotopomer (1186 cm^{-1}) should be only partially resolved. Thus the enantiomeric excess can be calculated from the absorbance at 1192 and 1163 cm^{-1} after taking into account their relative extinction coefficients. In practice the analysis becomes more accurate by subtracting out the spectra of other isotopomers that contain weak bands at these frequencies. The shoulder at 1163 cm^{-1} in Figure 4 (a) and the small peak at 1192 cm^{-1} in Figure 4 (b) arise because the (3R)-butane-1,3-diol used for the synthesis of the chiral $[^{16}O,^{17}O,^{18}O]$sulphate esters **29** and **30** contains 15 % of the (3S)-enantiomer. When this is taken into consideration, the cyclization is seen to occur stereospecifically (within experimental error).

In order to study both chemical and enzyme catalyzed sulphuryl transfer reactions we required phenyl $[^{16}O,^{17}O,^{18}O]$sulphate of known absolute configuration. The route developed for the synthesis of enantiomeric phenyl $[^{16}O,^{17}O,^{18}O]$sulphates and the method for their stereochemical analysis is outlined in Scheme 7.

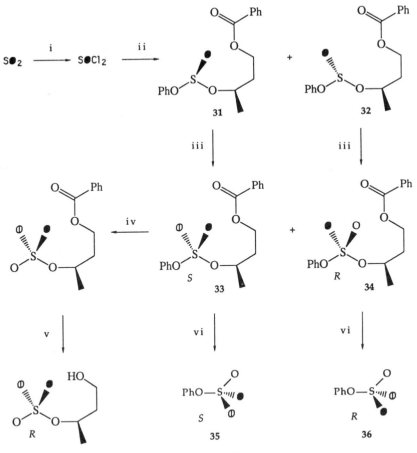

Scheme 7

[^{18}O]Thionyl chloride, prepared by the reaction of [$^{18}O_2$]sulphur dioxide with 1,4-bis(trichloromethyl)benzene in the presence of a catalytic amount of ferric chloride,[42] was reacted first with phenol (one equivalent) and then with (3R)-butan-3-ol-1-benzoate (one equivalent) to give the diastereoisomeric [^{18}O]sulphite esters 31 and 32. Each diastereoisomeric ester was oxidized separately with ruthenium [^{17}O]tetroxide. One of the [$^{17}O,^{18}O$]sulphate diesters was then catalytically hydrogenolysed over Adam's catalyst followed by debenzoylation to give (1R)-3-hydroxy-1-methylpropyl [$^{16}O,^{17}O,^{18}O$]sulphate which was shown to have the R_S configuration by cyclization with sulphuryl chloride and FTIR analysis. The [$^{17}O,^{18}O$]sulphate diester from which it was derived must have the S_S configuration, which on cleavage with ammonia in methanol must give phenyl [(S)-$^{16}O,^{17}O,^{18}O$] sulphate 35. The other diastereoisomeric [$^{17}O,^{18}O$]sulphate diester must have the R_S configuration and so on cleavage with ammonia in methanol must give phenyl [(R)-$^{16}O,^{17}O,^{18}O$] sulphate 36. These enantiomeric phenyl [$^{16}O,^{17}O,^{18}O$]sulphates are now being used to investigate both chemical and enzyme catalyzed sulphuryl transfer reactions.[43]

ACKNOWLEDGMENT

I would like to acknowledge the contribution of Roy Bicknell, Paul M. Cullis, Richard L. Jarvest, Salvatore J. Salamone, Martin J. Parratt and Timothy W. Hepburn whose published work is cited below.

REFERENCES

1.   P.W. Robbins and F. Lipmann, *J. Am. Chem. Soc.* 78:2652 (1956) and *J. Biol. Chem.* 229:837 (1957).
2.   P.W. Robbins and F. Lipman, *J. Biol. Chem.* 233:686 (1958).
3.   J.M. Akagi and L.L. Campbell, *J. Bacteriol.* 84:1194 (1962).
4.   R.W. Gwynn, L.T. Webster Jr., and R.L. Veech, *J. Biol. Chem.* 249:3248 (1974).
5.   N.M. Kredich, *J. Biol. Chem.* 246:3474 (1971).
6.   J. Dreyfuss and K.J. Monty, *J. Biol. Chem.* 238:1019 (1963).
7.   M. Tsang, E. Goldschmidt, and J.A. Schiff, *Plant. Physiol.* 47:20 (1971).
8 a.   M.J. Murphy, L.G. Wilson, L.M. Siegel, H. Kamin, and D. Rosenthal, *J. Biol. Chem.* 248:2801 (1973).
  b.   M.J. Murphy, and L.M. Siegel, *J. Biol. Chem.* 248:6911 (1973).
9.   E. Baumann, *Arch. Gesamte Physiol. Menschen Tiere* 12:69 (1876); *Ber. Deut. Chem. Ges.* 9:54 (1876).
10.   R.S. Bandurski and L.G. Wilson, *Proc. Int. Symp. Enzyme Chem. I.U.B. Symp. Ser.* 2:92 (1958).
11.   L.G. Wilson and R.S. Bandurski, *J. Biol. Chem.* 233:975 (1958)
12.   E.R. Stadtman, "The Enzymes", P.D. Boyer, ed., 3rd Ed., Vol. 8, pp. 35, Acadmic Press, New York (1973)
13.   J.W. Tweedie and I.H. Segel, *J. Biol. Chem.* 233:975 (1958)
14.   M. Shoyab, L.Y. Su, and W. Marx, *Biochem. Biophys. Acta* 258:113 (1972)
15.   J.R. Farley, D. F. Cryns, Y.H.J. Yang, and I.H. Segel, *J. Biol. Chem.* 251:4389 (1976)
16.   J.R. Farley, G. Nakayama, D. Cryns, and I.H. Segel, *Arch. Biochem. Biophys.* 185:376 (1978).
17.   Z. Reuveny and P. Filner, *Anal. Biochem.* 75:410 (1976).
18.   W.A. Bridger, W.A. Millen, and P.D. Boyer, *Biochemistry* 7:3608 (1968).
19.   P.A. Seubert, P.A. Grant, E.A. Christie, J.R. Farley, and I.H. Segel, *Ciba Found. Symp.* 72:19 (1980).
20.   J.R. Knowles, *Annu. Rev. Biochem.* 49:877 (1980).
21.   G. Lowe, *Acc. Chem. Res.* 16:244 (1983).
22.   P.M. Cullis and G. Lowe, *J. Chem. Soc. Perkin Trans.* 1 2317 (1981).

23. R. Cherniak and E.A. Davidson, *J. Biol. Chem.* 239:2986 (1964).

24. R.L. Jarvest and G. Lowe, *Biochem. J.* 199:447 (1981).

25. R.L. Jarvest, G. Lowe, and B.V.L. Potter, *J. Chem. Soc. Perkin Trans.* 1 3186 (1981).

26. G.Lowe, B.V.L. Potter, B.S. Sproat, and W.E. Hull, *J. Chem. Soc. Chem. Commun.* 733 (1979).

27. M.-D. Tsai, *Biochemistry* 18:1468 (1979).

28. M.-D. Tsai, S.L. Huang, J.F. Kozlowski, and C.C. Chang, *Biochemistry* 19:3531 (1980).

29. R. Bicknell, P.M. Cullis, R.L. Jarvest, and G. Lowe, *J. Biol. Chem.* 257:8922 (1982).

30. C.A. Bunton and S.J. Farber, *J. Org. Chem.* 34:767 (1969).

31a. P.A. Seubert, L. Hoang, F. Renosto, and I.H. Segel, *Arch. Biochem. Biophys.* 225:679 (1983).

  b. P.A. Seubert, F. Renosto, P. Knudson, and I.H. Segel, *Arch. Biochem. Biophys.* 240:509 (1985).

  c. I.H. Segel, personal communication.

32. G. Lowe and S.J. Salamone, *J. Chem. Soc. Chem. Commun.* 466 (1984).

33. C.H. Green and D.G. Hellier, *J. Chem. Soc. Perkin Trans.* 2 243:1966 (1973).

34. G. Lowe, S.J. Salamone, and R.H. Jones, *J. Chem. Soc. Chem. Commun.* 262 (1984).

35. G. Lowe, and S.J. Salamone, *J. Chem. Soc. Chem. Commun.* 1392 (1983).

36. S. Pinchas and I. Laulicht, "Infrared Spectra of Labelled Compounds", Academic Press, London and New York, 238 (1971).

37. A.J. Kirby, "The Anomeric Effect and Related Stereoelectronic Effects at Oxygen", Springer-Verlag, Berlin, Heidelberg and New York (1983).

38. (3R)-Butane-1,3-diol from Aldrich, $[\alpha]^{20}$ - 22.05° (c 1, EtOH) contains 15% of the (3S)-enantiomer as determined by the method of R.C. Anderson and M.J. Shapiro, *J. Org. Chem.* 49:1304 (1984). The highest recorded optical rotation for (3R)-butane-1,3-diol is $[\alpha]^{20}$ - 31.6° (c 1, EtOH) by S. Murakami, T. Harada, and A. Tai, *Bull. Chem. Soc. Jpn.* 53:1356 (1980).

39a. J. Lichtenberger, *Bull. Soc. Chim. Fr.* 1002 (1948).

  b. J. Lichtenberger and L. Durr, *J. Am. Chem. Soc.* 664 (1956).

40a. J.S. Risley and R.L. Van Etten, *J. Am. Chem. Soc.* 101:252 (1970).

  b. J.C. Vederas, *J. Am. Chem. Soc.* 102:374 (1980).

  c. J.E. King, S. Skonieczny, K.C. Khemani, and J.B. Stothers, *J. Am. Chem. Soc.* 105:6514 (1983).

  d. P.E. Hansen, *Annu. Rep. NMR Spectrosc.* 15:105 (1983).

41. G. Lowe and M.J. Parratt, *Bioorganic Chemistry* 16:283 (1988).

42. T.W. Hepburn and G. Lowe, *J. Labelled Comps. Radiopharm.*, in press (1990).

43. T.W. Hepburn and G. Lowe, *J. C. S. Chem. Commun.*, submitted for publication

ANIONIC ORGANOSULFUR COMPOUNDS IN SYNTHESIS:

NEW APPLICATIONS OF THE RAMBERG-BÄCKLUND REACTION

Eric Block, David Putman, and Adrian Schwan

Department of Chemistry
State University of New York at Albany
Albany, N.Y. 12222, USA

ORGANOSULFUR CARBANIONS IN SYNTHESIS

Organosulfur carbanions derived from saturated and unsaturated sulfides, sulfoxides and sulfones, from dithioacetals (e.g. 1,3-dithiane) and trithioformates ($(RS)_3CH$), from sulfinate and sulfonate esters, as well as from other sulfur-containing systems may undergo the entire gamut of reactions characteristic of organometallic compounds.[1] A partial list of such reactions would include protonation/deuteration, carbonation, inter- and intramolecular alkylation, hydroxyalkylation with epoxides (nucleophilic ring opening) and with aldehydes and ketones (carbonyl addition), acylation with carboxylic acid derivatives, addition to carbon-carbon and carbon-nitrogen multiple bonds (including conjugate addition), elimination and fragmentation, oxidative coupling, and rearrangement. In addition to their ease of preparation and versatile chemistry, organosulfur carbanions have been widely and artfully used in organic synthesis because of the diverse possibilities for further functionalization of the primary products as well as removal of sulfur through elimination, reduction, or hydrolysis.[1]

Representative examples of applications of organosulfur carbanions in synthesis are shown in Schemes 1-4. Scheme 1 illustrates a direct synthesis of biotin via alkylation of an α-sulfinyl carbanion (THF = tetrahydrofuran; HMPA = hexamethylphosphoramide).[2] In this example methyllithium rather than n-butyllithium is used to generate the carbanion since n-butyllithium attacks sulfoxide sulfur. Scheme 2 employs a variation[3a] of the sulfoximine carbanion procedure developed by Johnson[3b] as an alternative to the Wittig reaction (MCPBA = meta-chloroperbenzoic acid; NBS = N-bromosuccinimide; LDA = lithium diisopropylamide). The sulfur function is removed by aluminum amalgam reduction. Scheme 3 combines Diels-Alder addition of a thioketone with fragmentation of an α-sulfenyl carbanion.[4] Scheme 4 illustrates nucleophilic attack of an α-sulfonyl carbanion on a tris-epoxide followed by 1,3-elimination of benzenesulfonate induced by dimsyl potassium (the potassium salt derived from deprotonation of dimethyl sulfoxide).[5] The sulfonyl function is then removed by sodium/liquid ammonia reduction.

RAMBERG-BÄCKLUND REACTION

General

Among the diverse synthetic procedures involving anionic organosulfur compounds, one of the most useful is the Ramberg-Bäcklund reaction,[1,6] illustrated in Scheme 5 in Nicolaou's

Sulfur-Centered Reactive Intermediates in Chemistry and Biology, Edited by
C. Chatgilialoglu and K.-D. Asmus, Plenum Press, New York, 1990

257

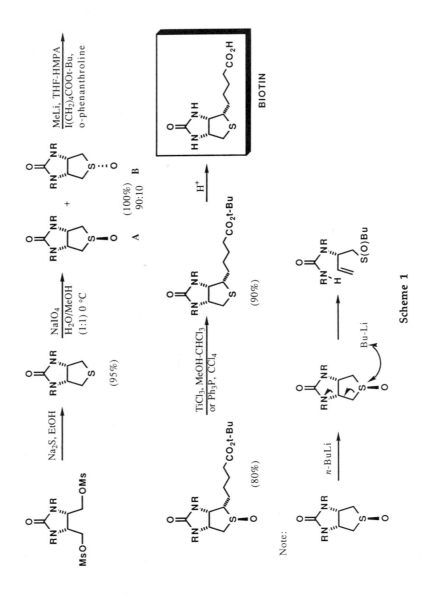

Scheme 1

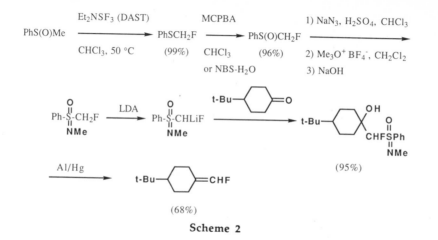

Scheme 2

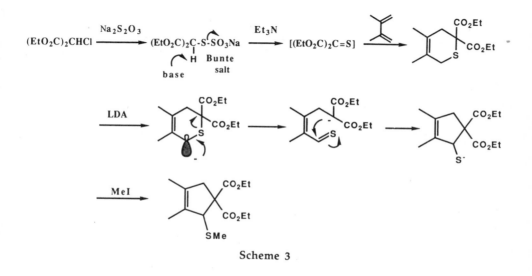

Scheme 3

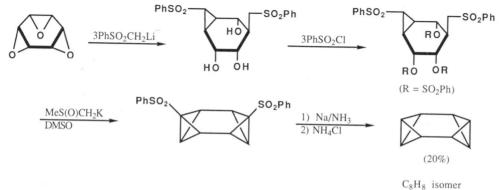

Scheme 4

259

model system for esperamycin action.[7] As with the example in Scheme 4, the Ramberg-Bäcklund reaction involves a 1,3-elimination but *across* rather than adjacent to the sulfonyl group. The Ramberg-Bäcklund reaction is general for molecules containing the structural elements of a sulfonyl group, an α-halogen (or other suitable leaving group), and at least one α'-hydrogen atom. With few exceptions the Ramberg-Bäcklund reaction allows the clean replacement of a sulfonyl group by a double bond. The required α-halogen atom may be introduced by treatment of the corresponding α-sulfonyl carbanion with a source of X^+ (BrCN, I_2 and Cl_3CSO_2Cl are convenient sources of Br^+, I^+, and Cl^+, respectively). A particularly

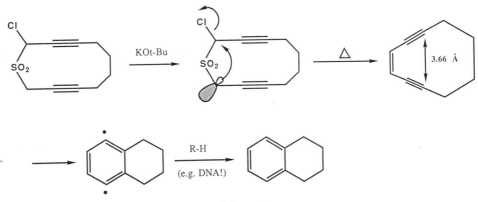

Scheme 5

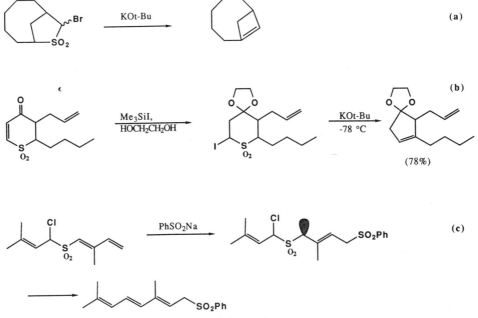

Scheme 6

useful modification of the Ramberg-Bäcklund reaction has been developed by Meyers et al. whereby sulfones may be taken directly to olefin without isolation of α-halosulfones. Carbon tetrachloride serves as the halogen source.

A variety of synthetic applications of the Ramberg-Bäcklund reaction are shown in Schemes 6 and 7. Scheme 6a shows the preparation of an unusual, hindered bicyclic alkene.[8a] In Scheme 6b iodine is introduced using iodotrimethylsilane.[8b] Scheme 6c illustrates a novel Michael addition induced Ramberg-Bäcklund reaction where the benzenesulfinate is the attacking nucleophile.[8c] In contrast Scheme 7a represents a case where benzenesulfinate acts as a leaving group.[9a] Scheme 7b[9b] illustrates the Meyers modification of the Ramberg-Bäcklund reaction while Scheme 7c illustrates a typical case with introduction of halogen at the sulfide stage.[9c] (TMS = trimethylsilyl).

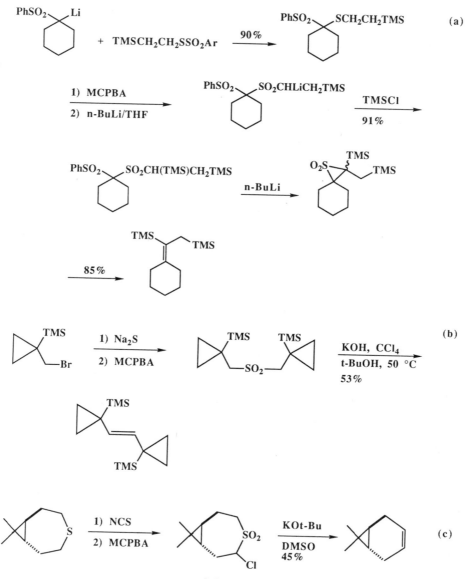

Scheme 7

Several mechanistic features of the Ramberg-Bäcklund reaction are of interest. Step "a" in Scheme 8 is reversible as shown by deuterium exchange studies. Indeed by conducting the Ramberg-Bäcklund reaction in D_2O, good yields of deuterated olefins can be easily obtained. A double inversion displacement mechanism (W geometry) seems likely for step "b" in Scheme 8, based on studies by Bordwell as well as the finding by Corey and Block that a bridgehead α-halosulfone readily undergoes rearrangement (Scheme 9). Recently a thiirane S,S-dioxide has been isolated from a Ramberg-Bäcklund reaction conducted under particularly mild conditions[9d] (Scheme 10).

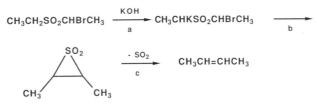

Scheme 8

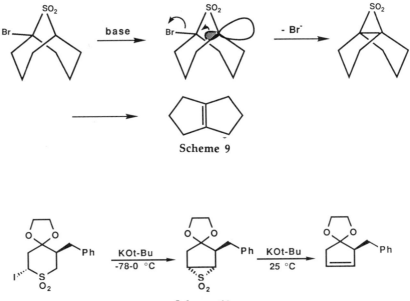

Scheme 9

Scheme 10

Because the preparation of α-haloalkyl sulfones entails multistep procedures, e.g. preparation of sulfides followed by α-halogenation, applications of the Ramberg-Bäcklund reaction have been limited. In the course of seeking new applications of the Ramberg-Bäcklund reaction, particularly those in which the necessary reaction components, sulfonyl group and α-halogen, are already present in the same compound, we have developed two new series of readily prepared reagents, α-haloalkanesulfonyl bromides $RCHXSO_2Br$ (1), and α-chloromethyl 1,2-propadienyl sulfone $CH_2=C=CHSO_2CH_2Cl$ (2). Reagents 1 and 2 package all of the components required for the Ramberg-Bäcklund reaction into one reactive unit, requiring only an olefinic substrate and base. Reagent 1 offers a convenient stereo- and regioselective approach to the synthesis of dienes and higher conjugated polyenes and enones, among other products, while reagent 2 is the basis for a two-step iterative procedure for cyclo-homologation of 1,3-dienes.

Vinylogous Ramberg-Bäcklund reaction

Bromomethanesulfonyl bromide (3) is readily prepared on a large scale in 42-48% yield by bromination of an aqueous suspension of 1,3,5-trithiane.[10,11] Light initiated addition of 3 to an alkene such as 1-octene gives a single 1:1 adduct in 94% yield. Treatment of this adduct with triethylamine affords in 97% yield a 10:1 mixture of (E)- and (Z)-bromomethyl 1-octenyl sulfone, (E)- and (Z)-$C_6H_{13}CH=CHSO_2CH_2Br$ (4E and 4Z), separable by recrystallization. Treatment of pure 4E with KOt-Bu produces in 59% distilled yield a 83:17 mixture of (Z)- and (E)-1,3-nonadiene.[10] In a similar manner pure 4Z gives in 61% yield a 6:94 mixture of the same dienes.

The stereoselectivity of the reaction of 4E with base, which may be termed a "vinylogous Ramberg-Bäcklund reaction", can be attributed to a stabilizing, attractive interaction between the developing negative charge at the α-position and the CH_2 group at the δ-position (a "syn effect") favoring the transition state leading to the (Z)-diene over that leading to the (E)-diene (see Scheme 11).[10] In the case of 4Z the possibility of a stabilizing syn interaction between the α- and δ-position is precluded for steric reasons (see Scheme 11). A particularly useful application of reagent 3 is in the preparation of 1,2-bismethylene-cycloalkanes, as illustrated in Scheme 12.

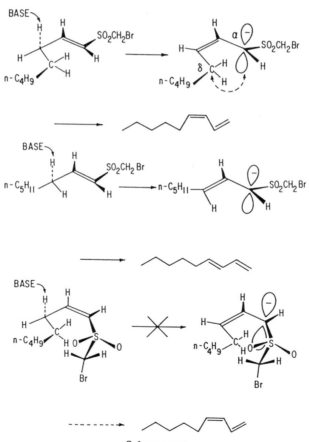

Scheme 11

Scheme 12

The two- or three-step olefin to conjugated diene transformation using **3**/base can be extended to the conversion of conjugated dienes to conjugated trienes, representing an extension of the "vinylogous Ramberg-Bäcklund reaction" to include the case with *two* intervening double bonds.[10] Thus light-initiated addition of **3** to (E)-1,3-heptadiene gave an adduct which was treated with triethylamine followed by potassium *tert*-butoxide to afford a 1.7:1 mixture of (E,Z)- and (E,E)-1,3,5-octatriene in 16% overall yield. The former octatriene is known as fucoserratene, the female sex attractant from the ova of the seaweed *Fucus serratus L.* The stereoselectivity seen in the formation of linear trienes can be rationalized along lines similar to those used above to explain the stereoselectivity seen in diene synthesis. Thus the modest preference of the triene 5,6-double bond for Z geometry can be explained in terms of an energetically favorable "syn effect" in the transition state leading to the anionic intermediate (see Scheme 13). In as much as the dienes used in our methods may be prepared from olefins as already described, our triene synthesis involves a repetitive procedure in which *both* the number of carbon atoms in the chain and the number of double bonds in conjugation increase by one in each cycle, e.g. Scheme 14. Reagent **3** can also be used to convert 2-methyl-1-alkenes into 2-alkyl-1,3-butadienes, methylenecycloalkenes into 1-vinyl-1-cycloalkenes, cycloalkenes into 3-methylene-1-cycloalkenes, alkynes into enynes, and trimethylsilyl enol ethers into α-alkylidene ketones and 1,3-oxathiole 3,3-dioxides (Scheme 15).

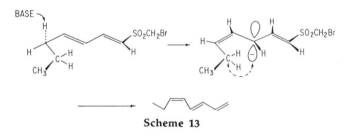

Scheme 13

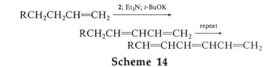

$$RCH_2CH_2CH{=}CH_2 \xrightarrow{\text{2; Et}_3\text{N; } t\text{-BuOK}}$$
$$RCH_2CH{=}CHCH{=}CH_2 \xrightarrow{\text{repeat}}$$
$$RCH{=}CHCH{=}CHCH{=}CH_2$$

Scheme 14

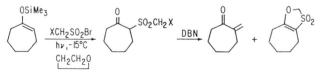

Scheme 15

264

Allenyl Chloromethylsulfone: New Dienophile-Diene Synthon

While reagent **3** can be used in an iterative process for increasing the number of double bonds in an unsaturated compound without altering the number of rings present, our new organosulfur reagent, chloromethyl 1,2-propadienyl sulfone (**2**), $H_2C=C=CHSO_2CH_2Cl$, can be used in an iterative two-step process for "cyclohomologation" of dienes.[12] Compound **2** functions as a potent dienophile whose Diels-Alder adducts give 1,3-dienes with base. The choice of **2** was suggested by the known high reactivity of sulfonyl allenes as dienophiles (at the double bond next to the sulfone function) due to their low LUMO,[13] the anticipated susceptibility of the allylic sulfone Diels-Alder adduct toward base-induced elimination, and a simple projected synthesis of **2** via coupling of readily available chloromethylsulfenyl chloride, $ClCH_2SCl$,[14] with propargyl alcohol giving S-chloromethyl propargyl sulfenate, $ClCH_2S$-O-$CH_2C\equiv CH$, [2,3]-sigmatropic rearrangement[15] of the latter to chloromethyl 1,2-propadienyl sulfoxide, $H_2C=C=CS(O)CH_2Cl$, and oxidation of this compound in turn to **2** (Scheme 16). Thus, crude $ClCH_2SCl$ was added to an ethereal solution of the lithium salt of propargyl alcohol at -78°C, filtered to remove LiCl after warming to ca. 0°C, and directly oxidized with an equivalent of MCPBA in CH_2Cl_2 overnight at room temperature affording after a simple workup **2** in 56% yield as a colorless crystalline solid, mp 39.0-39.5°C, ^{1}H NMR ($CDCl_3$) δ 6.26 (t, 1 H, J = 6.3 Hz), 5.62 (d, 2 H, J = 6.3 Hz), 4.53 (s, 2 H); ^{13}C NMR δ 212.3 (C), 95.2 (CH), 84.3 (CH₂), 58.3 (CH₂), IR (neat) 3000 (m), 1965 (m), 1328 (s), 1247 (m), 1120 (s), 872 (m) cm⁻¹. A 2:1 mixture of 1,2-bismethylenecyclohexane and **2** was warmed to 60°C for 3 h, the product diluted with THF and treated with an equivalent of KO-*t*-Bu at 0°C, and worked up to directly afford the bicyclic triene shown in Scheme 17 in 57% overall yield. Repetition of the process using a 1.2:1 ratio of triene:**2** gave the corresponding tetraene (see Scheme 17) in 85% overall yield. The above procedure can be extended to furan, cyclo-pentadiene, 1,3-cyclohexadiene and 2,3-dimethyl-1,3-butadiene giving homologated dienes in overall yields of 85%, 82%, 60%, and 57%, respectively (see Schemes 17, 18).

Compounds related to the 1,2-bismethylenecyclohexane derivatives described herein, but prepared by lengthier procedures, have been employed in syntheses of pentacene.[16] *cis*-1,4-Dichloro-2-butene has also been employed as a dienophile-diene synthon but requires "severe and carefully controlled reaction conditions [typically several days at 190-200°C], was somewhat erratic" and gave only moderate yields.[17] *cis*-1,4-Dichloro-2-butene does not react with either furan or 1,3-cyclohexadiene.[18]

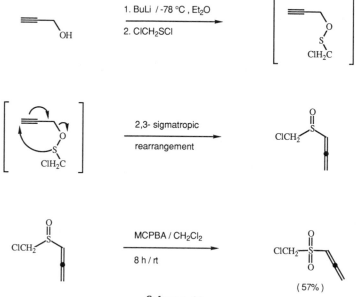

Scheme 16

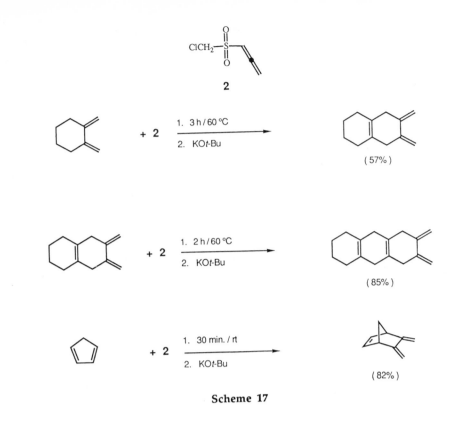

2

1. 3 h / 60 °C
2. KO*t*-Bu

+ **2**

(57%)

1. 2 h / 60 °C
2. KO*t*-Bu

+ **2**

(85%)

1. 30 min. / rt
2. KO*t*-Bu

+ **2**

(82%)

Scheme 17

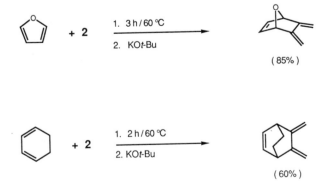

1. 5 h / 80 °C
2. KO*t*-Bu

+ **2**

(57%)

1. 3 h / 60 °C
2. KO*t*-Bu

+ **2**

(85%)

1. 2 h / 60 °C
2. KO*t*-Bu

+ **2**

(60%)

Scheme 18

ACKNOWLEDGMENT

We thank David M. Block for assistance in synthesis. We gratefully acknowledge support for our research from the donors of the Petroleum Research Fund, administered by the American Chemical Society, the Herman Frasch Foundation, the Société Nationale Elf Aquitaine, and the National Science Foundation.

REFERENCES

1. E. Block, "Reactions of Organosulfur Compounds", Academic Press, New York (1978).
2. S. Lavielee, S. Bory, B. Moreau, M.J. Luche, and A. Marquet, *J. Am. Chem. Soc.* 100:1558 (1978).
3 a. M.L. Boys, E.W. Collington, H. Finch, S. Swanson, and J.F. Whitehead, *Tetrahedron Lett.* 29:3365 (1988).
   b. C.R. Johnson, J.R. Shanklin, and R.A. Kirchoff, *J. Am. Chem. Soc.* 95:6462 (1973).
4. S.D. Larsen, *J. Am. Chem. Soc.* 110:5932 (1988).
5. C. Rücker and B. Trupp, *J. Am. Chem. Soc.* 110:4828 (1988).
6. L.A. Paquette, "Organic Reactions", John Wiley, New York, Vol. 25, p. 1 (1977).
7. K.C. Nicolaou, G. Zuccarello, Y. Ogawa, E.J. Schweiger, and T. Kumazawa, *J. Am. Chem. Soc.* 110:4866 (1988).
8 a. R. Neidlein and H. Dörr, *Ann.* 1540 (1980).
   b. G. Casy and R.J.K. Taylor, *J. Chem. Soc., Chem. Commun.* 454 (1988).
   c. J.J. Burger, T.B.R.A. Chen, E.R. deWaard, and H.O. Huisman, *Tetrahedron* 37:417 (1981).
9 a. M.G. Ranasinghe and P.L. Fuchs, *Am. Chem. Soc.* 111:779 (1989).
   b. L.A. Paquette, G.J. Wells, and G. Wickham, *J. Org. Chem.* 49:3618 (1984).
   c. P.G. Gassman and K. Mlinaric-Majerski, *J. Org. Chem.* 51:2398 (1986).
   d. A.G. Sutherland and R.J.K. Taylor, *Tetrahedron Lett.* 30:3267 (1989).
10. E. Block, M. Aslam, V. Eswarakrishnan, K. Gebreyes, J. Hutchinson, R. Iyer, J.-A. Laffitte, and A. Wall, *J. Am. Chem. Soc.* 108:4568 (1986).
11. E. Block and M. Aslam, *Org. Syn.* 65:90 (1987).
12. E. Block and D. Putman, *J. Am. Chem. Soc.* 112:0000 (1990).
13. K. Hayakawa, H. Nishiyama, and K. Kanematsu, *J. Org. Chem.* 50:512 (1985).
14. I.B. Douglass, R.V. Norton, R.L. Weichman, and R.B. Clarkson, *J. Org. Chem.* 34:1803 (1969).
15. S. Braverman and Y. Stabinsky, *Israel J. Chem.* 5:125 (1967).
16. W.J. Bailey and C.-W. Liao, *J. Am. Chem. Soc.* 77:992 (1955).
17. Y.-S. Chen and H. Hart, *J. Org. Chem.* 54:2612 (1989).
18. M.A.P. Bowe, R.G.J. Miller, J.B. Rose, and D.G.M. Wood, *J. Chem. Soc.* 1541 (1960).

THE SYNTHESIS AND METAL COMPLEXES OF SULFUR-CONTAINING MODELS

FOR NITROGENASE ENZYMES

Eric Block, Michael Gernon, and Jon Zubieta

Department of Chemistry
State University of New York at Albany
Albany, N.Y. 12222, USA

NITROGENASE ENZYME ACTIVE SITE

Recently there has been a great deal of interest in synthetically modelling the active site of the enzyme nitrogenase.[1] Nitrogenase is a member of a class of metalloenzymes which bind molybdenum at their active sites. The molybdoenzymes are responsible for many important naturally occurring redox processes. In the case of nitrogenase the redox process is the reductive fixation of atmospheric dinitrogen, N_2. In nitrogen fixation, N_2 is first bound to the active site of nitrogenase and then through a series of unknown intermediates reduced to ammonia. The exact structure of the active site of nitrogenase is not known, but EXAFS data shows it to consist of a cluster of approximate composition $MoFe_{(6-8)}S_{(4-8)}$.[2] The molybdenum atom is believed to be the site where N_2 is bound, but the precise nature of the coordination is not known.[1] It is hoped that a good synthetic model of the active site will show the mode of nitrogen binding.

Furthermore, during the nitrogen fixation process molybdenum is believed to cycle between two or more oxidation states, and the coordinating environment around molybdenum must be stable to the changes which occur.[3] Thus, a good model of the active site of nitrogenase might also elucidate how sulfur ligation stabilizes the various states of the active site during the nitrogen fixation process. It is remarkable that nature has found a coordination environment consisting only of iron and sulfur to be suitable for the problem of nitrogen fixation. In fact, all the molybdoenzymes contain active sites with mostly sulfur ligation, whether as sulfide (nitrogenase) or thiolate (all other molybdoenzymes) donors, and it is probable that certain arrays of sulfur ligands offer unique properties for the stabilization of iron-molybdenum clusters.[4]

Efforts to model the active site of nitrogenase have centered on molybdenum-iron clusters in a coordination environment of sulfido and thiolato sulfur. It is assumed that a complex which resembles the active site of nitrogenase will reversibly bind dinitrogen. However, the only success to date in binding N_2 has been with low valent molybdenum complexes or with complexes having transition metals other than molybdenum and iron.[5,6,7] Clearly, no cluster which reversibly binds N_2 and resembles the actual active site in nitrogenase has been synthesized so far.

Sulfur-Centered Reactive Intermediates in Chemistry and Biology, Edited by
C. Chatgilialoglu and K.-D. Asmus, Plenum Press, New York, 1990

One prerequisite of a nitrogenase active site model is coordinative unsaturation, as the binding of dinitrogen requires at least one open coordination site on the molybdenum. The preparation of coordinatively unsaturated molybdenum-thiolate complexes is complicated by the tendency of thiolates to act as bridging ligands and thus block potentially reactive sites.[8,9] In addition, the reaction of thiols with molybdenum compounds can be complicated by redox processes and by the relatively facile dealkylation of certain alkanethiols at metal centers.[8,9] It has been found that arenethiols having sterically bulky substituents around the sulfhydryl group are ideal for circumventing many of these problems.[10,11] Arenethiols with sterically bulky groups in the vicinity of the mercapto function are often referred to as hindered arenethiols, and henceforth in this chapter the term hindered arenethiol will be used to describe them. Steric bulk prevents the bridging of thiolate ligands by creating steric strain when too many thiolates try to crowd into the same area. Also, steric bulk hinders redox chemistry by preventing the approach of the mercapto group to the metal, and the use of arenethiols eliminates the possibility of dealkylation. The 2,4,6-trialkylbenzenethiols and 2,6-dialkylbenzenethiols are very commonly used when sterically blocked arenethiols are desired. 2,4,6-Triisopropylbenzenethiol (HTIPT) and 2,6-diisopropylbenzenethiol (HDIPT) are probably the most popular choices among these for hindered thiolate ligands.[12]

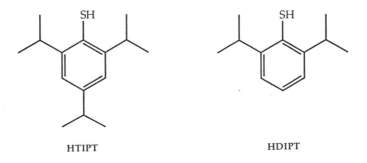

HTIPT   HDIPT

Hindered arenethiols with noncoordinating groups can be considered the simplest ligands to use in potential nitrogenase active site models, and the majority of the hindered arenethiols that have been used to date are benzenethiols with alkyl groups ortho to the sulfhydryl group. Other than HTIPT and HDIPT, such arenethiols as 2,6-dimethyl-benzenethiol, 2,4,6-trimethylbenzenethiol, 2,3,5,6-tetramethylbenzenethiol, and 2,4,6-tri(t-butyl)benzenethiol have been used.[8] An interesting alternative to the simple hindered arenethiols discussed here would be polyarenethiols, particularly homologues of the class of specifically chelating polydentate macrocyclic polyphenols ligands known as calixarenes.[13,14] If macrocyclic polyarenethiols analogous to the calixarenes could be produced, then they might display strong ligation of molybdenum. Such macrocyclic poly-arenethiols might well be the best ligands for potential nitrogenase active site models, especially if a synthetic procedure which allows for variation in the positioning of the mercapto groups in such compounds could be developed. With such a synthetic method the cavity of polyarenethiols could be progressively modified to produce a functional model of the active site of nitrogenase.

We have initiated a program to synthesize calixarene-like cyclic tetramers of thiophenol, e.g. 1, which should be capable of binding metals such as molybdenum and providing a cavity accessible to small linear molecules such as nitrogen. The use of silicon to bridge the rings in 1 should provide a larger cavity size than is possible with carbon and could offer opportunities for bridge functionalization, e.g. by having Si-H or Si-Si bonds for construction of an end-capping group. It was our hope that synthesis of a system such as 1 could be achieved by repetitive *ortho*-lithiation of a thiophenol equivalent and coupling

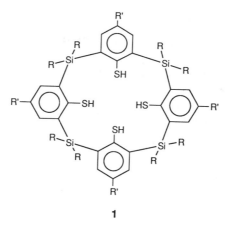

1

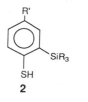

2

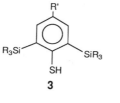

3

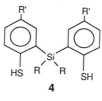

4

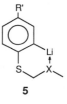

5

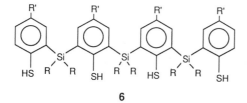

6

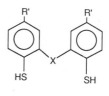

7 (X = S, RP)

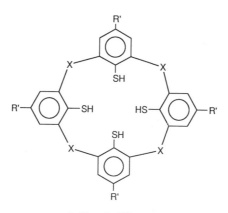

8 (X = S, RP or R$_2$Si)

with a dichlorosilane under dilute conditions. Since satisfactory syntheses of 2-(triorgano-silyl)thiophenols (2), 2,6-bis(triorganosilyl)thiophenols (3) and bis(2-mercaptoaryl)dialkyl-silanes (4) were not available, our immediate goals were to prepare 2-4 by way of a suitable *ortho*-lithiothiophenol equivalent such as 5. In the longer term we hope to extend the procedure to prepare acyclic and cyclic polythiophenols such as 6 and 1. Related types of polythiophenol ligands can be envisioned with the rings linked by other atoms or groups in place of, or in addition to, silicon, e.g. sulfide sulfur or phosphorus (see 7, 8). Synthesis of 7 and 8 would require reaction of *ortho*-lithiothiophenol equivalents with appropriate sulfur and phosphorus electrophiles.

SILYLATED METHANETHIOL DERIVATIVES AS LIGANDS

Our research on nitrogenase models involved development of the α-lithiomethanethiol (LiCH₂SH) and 2-lithiobenzenethiol (o-LiC₆H₄SH) synthons for synthesis of hindered thiols and polythiols. With respect to the α-lithiomethanethiol (LiCH₂SH) synthon, a reagent containing carbon geminally substituted with a thiomethyl group and an oxygen substituent seemed appropriate since the latter group could assist deprotonation of the thiomethyl group by metal coordination[15] and could subsequently facilitate hydrolytic release of thiol along with water-soluble carbonyl byproducts. Of the several reagents examined 2-(methylthio)-tetrahydropyran (9) seemed ideal. Compound 9 can be easily prepared on a large scale in 84% yield by pyridinium *p*-toluenesulfonate catalyzed addition of methanethiol to 2,3-dihydropyran. Deprotonation of 9 with *tert*-butyllithium in 10:1 THF/HMPA at -90°C gave (2-tetrahydropyranl)(thiomethyl)lithium (10). (HMPA = hexamethylphosphoramide). The latter could be converted in good yield to (trimethylsilyl)methanethiol or bis(mercapto-methyl)dimethylsilane by treatment with chlorotrimethylsilane or dichlorodimethylsilane, respectively (eq. 1).[16]

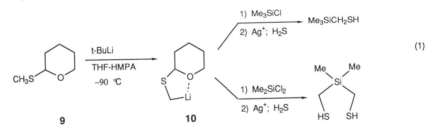

(1)

Highly hindered bis- and tris-(triorganosilyl)methanethiols, such as (Me₃Si)₃CSH and (Me₂PhSi)₃CSH, and (Me₃Si)₂CHSH can be prepared by repetitive silylation of 9 or from the silanes such as (Me₃Si)₃CH by lithiation followed by treatment with elemental sulfur.[17] A series of silver and nickel complexes prepared from these ligands, e.g [AgSCH(SiMe₃)₂]₈, [AgSC(SiPhMe₂)₃]₃, [AgSC(SiMe₃)₃]₄, [Ag₄(SCH₂SiMe₃)₃]ₙ(OMe)ₙ, and [Ni(SCH₂SiMe₃)₂]₅, characterized by X-ray crystallography, reveals the steric control of aggregation that the changing bulk of the ligands introduces.[17,18]

Monolithiation α to a protected alkanethiol group (α-lithiation) has been achieved in alkyl dimethylthiolcarbamates (Scheme 1; TMEDA = N,N,N',N'-tetramethylethylenedi-amine).[19] Also, several procedures exist for the α-lithiation of alkylthio-heterocyclic compounds such as 2-methylthiobenzothiazoles,[20] 2-methylthiooxazolines[21] and 2-methylthio-thiazolines,[22] all of which could presumably be converted into thiols (Scheme 2; LDA = lithium diisopropylamide).

$$(CH_3)_2N-\overset{\overset{\displaystyle O}{\|}}{C}-S-CH_2CH_3 \quad \xrightarrow{s\text{-BuLi, TMEDA}} \quad (CH_3)_2N-\overset{\overset{\displaystyle O}{\|}}{C}-S-\overset{\overset{\displaystyle Li}{|}}{C}HCH_3 \quad \xrightarrow{E^+} \text{products}$$

Scheme 1

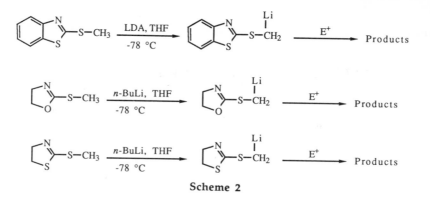

Scheme 2

The deprotonation of THP-protected benzhydrylthiol has been developed by Holm *et al.* as a means of producing tridentate 2,6-bis[(mercaptodiphenyl)-methyl]pyridine compounds (Scheme 3; pyH+Tos− = pyridinium tosylate).[23] These tridentate compounds were used as ligands in the preparation of potential active site models for the oxidase/reductase class of molybdoenzymes.[23] The oxidases and reductases are distinct from nitrogenase, differing mainly in the nature of the transformations which they catalyze. The oxidases and reductases almost always take part in reactions which involve oxygen transfer, and the active site in all of the oxidase/reductase molybdoenzymes contains oxo-bound molybdenum. Oxo-molybdenum compounds, like molybdenum-thiolate complexes, are prone to bridging reactions which can lead to polymerization. Thus, the preparation of active site models of the oxidases and reductases requires ligands which will tend to favor monomeric complexes. The properties which such ligands must have are similar to those which are needed for nitrogenase active site models, and the incorporation of sterically bulky groups near all coordinating functions is usually necessary for success.

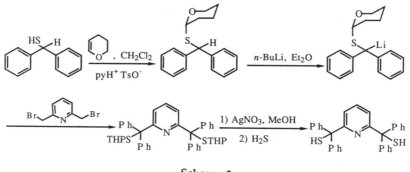

Scheme 3

o-SUBSTITUTED BENZENETHIOL DERIVATIVES AS LIGANDS

Our success in lithiating 2-(methylthio)tetrahydropyran (9) encouraged us to attempt *ortho*-deprotonation of 2-(phenylthio)tetrahydropyran (11) under the conditions used with 9. We find that under carefully controlled conditions (1.2 eq *tert*-butyllithium, 9:1 THF:HMPA, −90°C) 11 could be deprotonated in the *ortho*-position and the resultant lithio-compound trapped with chlorotrimethylsilane giving adduct 12. The THP group of 12 could be removed with mercuric chloride followed by hydrogen sulfide to give 2-(trimethylsilyl)benzenethiol (13) in 63% overall yield (eq. 2). In a similar manner 2-(triethylsilyl)benzenethiol (14) (not

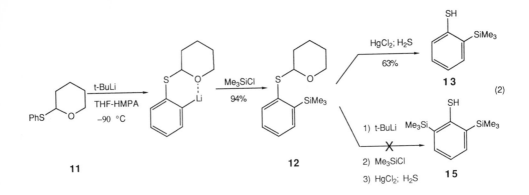

shown as structure) could be prepared in 42% overall isolated yield. We were unable to convert **12** into 2,6-bis(trimethylsilyl)benzenethiol (**15**) by repetition of the above procedure (see eq. 2). We surmise that the presence of the bulky *ortho*-trimethylsilyl group in **12** prevents the 2-tetrahydropyranylthio group from adopting a conformation suitable for chelating the *ortho*-lithio group.

Encouraged by the report of Posner[24] on the *ortho*-lithiation of phenol, we examined the reaction of thiophenol **13** with *tert*-butyllithium in tetrahydropyran. It was hoped that steric effects might be less pronounced with **13** than with **12**. In the event **13** could be dilithiated with *tert*-butyllithium; silylation gave **15** together with a mixture of other bis- and tris-silylated thiophenols (see below). Following our discovery of the lithiation of **13** we learned that Martin and Figuly had independently discovered that thiophenol itself can readily be converted into an *ortho*-dilithio compound **16** which can be condensed with various electrophiles.[25] Information was exchanged with Professor Martin and it was agreed that our two groups would work on different applications of *ortho*-lithiothiophenol equivalents. After submission of our work for publication we learned that Professor K. Smith and co-workers at the University College of Swansea had also independently discovered the ortho-lithiation of thiophenol. Three papers published back to back in the Journal of the American Chemical Society report the independent findings of the three research groups.[26]

2-(Trialkylsilyl)- or 2-(triarysilyl)-thiophenols can be prepared by adding excess chlorosilane to the cyclohexane suspension of dilithio salt; workup requires refluxing in methanol to remove the silyl group from sulfur (eq. 3). Yields decreased as the bulk of the silylating agent increased. Thus the isolated yields of 2-(trimethylsilyl)-, 2-(triethylsilyl)-, 2-(triphenylsilyl)- and 2-(*tert*-butyldimethylsilyl)benzenethiols from reaction of dilithio salt **16** using the Martin and Figuly procedure decreased in the order 92%, 79%, 43% and 28%, respectively. The Martin and Figuly procedure can also be applied to substituted thio-phenols. We have been able to prepare 4-(*tert*-butyl)-2-(trimethylsilyl)benzenethiol (**18**) and 4-(*tert*-butyl)-2-(triphenylsilyl)benzenethiol (**19**) from 4-(*tert*-butyl)benzenethiol via lithium 4-(*tert*-butyl)-2-lithiobenzenethiolate (**17**).

Efforts to silylate the 6-position of 2-(trimethylsilyl)benzenethiol (**13**) are complicated by the fact that **13** undergoes deprotonation and silylation at *both* the 6-position and the methyl groups attached to silicon giving rise to difficultly separable mixtures of 2,6-bis(trimethylsilyl)benzenethiol (**15**), 2-[(dimethyl trimethylsilylmethyl)silyl]benzenethiol (**20**) and 2-(trimethylsilyl)-6-[(dimethyl trimethylsilylmethyl)silyl]benzenethiol (**21**) (eq. 3). Fortunately we find that if the lithiation of **13** is conducted in hexane pure **15** can be isolated by fractional distillation in 60% yield. Silylation α to silicon could also be prevented through use of 2-(triethylsilyl)benzenethiol (**14**), although conversion of **14** to 2,6-bis-(triethylsilyl)-benzenethiol (**22**) is poor even after several days. From our results it is

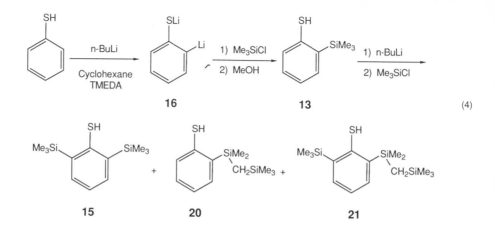

16   **13**   (4)

15   **20**   **21**

clear that the Martin and Figuly procedure is simpler and gives superior yields compared to procedures involving a THP group, primarily because of difficulties in removing the THP group from sulfur. In particular, strongly acidic conditions cannot be used to deprotect silylated THP-thiophenol derivatives because of possible protiodesilylation.

The Martin and Figuly procedure employing the cyclohexane suspension of dilithio salt **16** and **17** cannot be used to prepare bis(thiophenol) **4** since both the thiolate and carbanionic centers compete for silicon, leading to polymer. However if solid **16** and **17** are allowed to settle, the solvent removed via syringe, the solid washed several times with pentane and then dissolved in the minimal amount of cold tetrahydrofuran, reactions involving **16** and **17** can now be conducted at -78°C with a resultant increase in selectivity ("modified-Martin and Figuly" procedure). We were delighted to discover that, using this modified procedure, **16** reacted with dichlorodimethylsilane, dichlorodiethylsilane and 1,2-dichloro-1,1,2,2-tetramethyldisilane exclusively at the carbanionic center affording good yields of the corresponding bis(2-mercaptophenyl) derivatives (**23**, **24** and **25**) as products which could be directly crystallized from pentane giving colorless solids (eq. 4). Bisthiophenol **23** could be oxidized in high yield to 5,5-dimethyldibenzo[b,e]-5-sila-1,2-dithiepin (**26**) (eq. 4). In a similar manner, reaction of the dilithio salt **17** of 4-(*tert*-butyl)benzenethiol with dichloro-diethylsilane gave diethyl bis(2-mercapto-4-*tert*-butylphenyl)silane (**27**) (not shown as

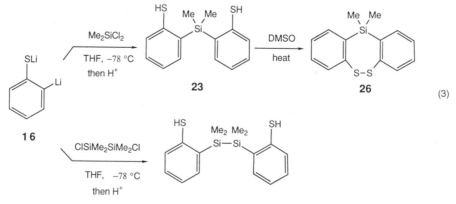

16

23   **26**   (3)

25

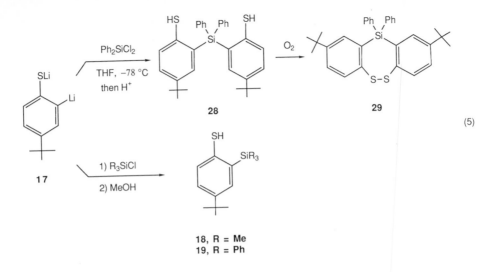

(5)

structure) while dichlorodiphenyl-silane directly gave **29**, a bis-*tert*-butyl derivative of 5,5-diphenyldibenzo[b,e]-5-sila-1,2-dithiepin (eq. 5). In this latter case dithiol **28** undergoes rapid oxidation by oxygen to **29**; **28** can be isolated if care is taken to exclude oxygen during workup. The cyclic disulfides contain a new type of heterocyclic ring system and offer possibilities for preparation of metal complexes by insertion of metals into the S-S bonds. Using the modified Martin and Figuly procedure we could also directly prepare *ortho*-(trialkylsilyl)thiophenols in high yields without the methanol reflux step.

Illustrative of the considerable utility of the Martin and Figuly procedure is our observation that benzene-1,2-dithiol (**32a** [eq. 6]) can be prepared in 84% distilled yield in a one pot process from thiophenol. In a similar manner 4-(*tert*-butyl)-benzenethiol could be converted to 4-(*tert*-butyl)benzene-1,2-dithiol (**32b**) in 71% isolated yield. Our preparation of **32** compares favorably with earlier procedures in terms of ease of reaction as well as yield and may offer advantages due to the mild conditions. Furthermore, treatment of dithiol **32a** with excess *n*-butyllithium followed by sulfur directly gave benzene-1,2-3-trithiol (**34**) in 35% isolated yield. This reaction presumably involves trilithio species **33**.

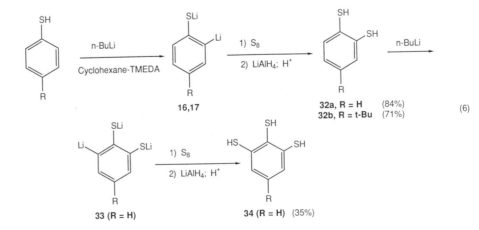

(6)

We have prepared and characterized a series of metal complexes from 2-(trimethylsilyl)-benzenethiol including $(Et_4N)_2[Cd(SC_6H_4-o-SiMe_3)_4]$, $[Cu(SC_6H_4-o-SiMe_3)]_{12}$, $[\{Ag(SC_6H_4-o-SiMe_3)\}_4]_2$,[27a] and a related rhenium complex.[27b]

The reaction of **1** with chlorodiphenylphosphine, dichlorophenylphosphine, phosphorus trichloride, chlorodiphenylphosphine oxide, dichlorophenylphosphine oxide, and phosphorus oxychloride affords the corresponding arenethiols in 50% to 80% yield (Scheme 4).[28]

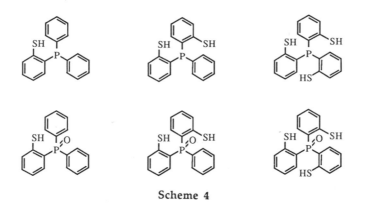

Scheme 4

Such (2-mercaptophenyl)phosphine compounds are potentially very useful as ligands for many transition metals. These compounds are hindered arenethiols with an additional coordinating phosphorus atom. Because of the proximate positioning of the phosphorus and sulfhydryl functions the chances of polymerization in reactions of (2-mercaptophenyl)-phosphine compounds with transition metals is reduced. The arrangement of coordinating groups should improve the ability of these compounds to accommodate the coordination sphere of a single metal ion. Thus, the tendency for polymerization via one ligand bonding to two or more metal ions will be reduced. However, additional steric hindrance ortho to the mercapto groups would retard potential polymerization reactions even more, and efforts were made to prepare (2-mercaptophenyl)phosphines with bulky 6-substituents.

Reaction of lithium 2-lithio-6-trimethylsilylbenzenethiolate (**34**) with the chlorophosphines and chlorophosphine oxides mentioned above affords the expected [2-mercapto-(6-trimethylsilyl)phenyl]phosphines (Scheme 5).

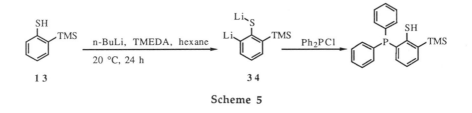

Scheme 5

These compounds combine a proximate positioning of coordinating groups with steric hindrance, and they should make excellent ligands for many transition metals. With these ligands we have thus far prepared the molybdenum complexes $[Mo_2(SC_6H_4-o-PPh_2)_3Cl_3]$, $[Mo(SC_6H_3-2-SiMe_3-6-P(O)Ph_2)_2Cl_2]$, and $[Mo(\{SC_6H_4\}_2-o-PPh)_2NN(Me)Ph]$ (see Figure 1).[29] The first is a binuclear complex with a Mo-Mo triple bond which exhibits nonequivalent metal coordination geometries, $[MoS_3P_2Cl]$ and $[MoS_3PCl_2]$. The second and third complexes show $[MoO_2S_2Cl_2]$ and $[MoS_4P_2N]$ coordination geometry, respectively.[29]

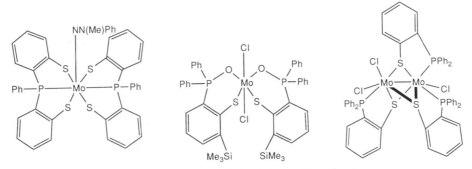

Fig. 1   Structures of Molybdenum Complexes

2-PYRIDINETHIOL DERIVATIVES AS LIGANDS

Our investigations have centered on the utility of the mercapto group in directing *ortho*-lithiations in carbocyclic aromatic substrates. We have extended our study to 2-pyridinethiol (**35**), as this represents the 2-pyridyl analogue of benzenethiol.[30] Using THF as solvent and lithium diisopropylamide (LDA) as base a slow equilibrium-type deprotonation of the 3-position of 2-pyridinethiolate occurred. A mixture of LDA, 2-pyridinethiol, and chlorotrimethylsilane (TMSCl) after 8 h of stirring in THF at room temperature gave a 55% yield of the previously unknown 3-trimethylsilyl-2-pyridinethiol (**36**) (Scheme 6).

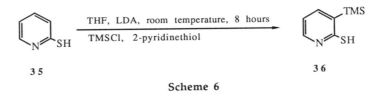

Scheme 6

Similar reactions yield the previously unknown compounds 3-triethylsilyl-2-pyridinethiol (**37**) and 3-phenyldimethylsilyl-2-pyridinethiol (**38**) in 45% and 36% yield, respectively. Surprisingly, when more hindered chlorosilanes such as chloro-*tert*-butyldimethylsilane, chlorotriisopropylsilane, and chloro-*tert*-butyldiphenylsilane were used the 3,6-disubstituted-2-pyridinethiols were obtained (Scheme 7).

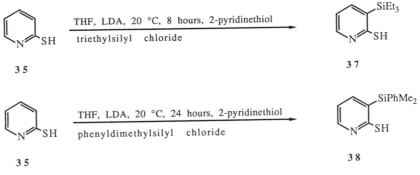

Scheme 7

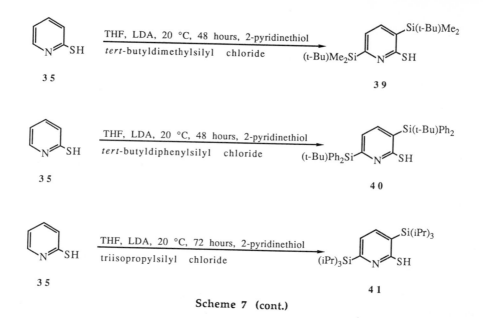

Scheme 7 (cont.)

Thus, the previously unknown 3,6-bis(*tert*-butyldimethylsilyl)-2-pyridinethiol (**39**), 3,6-bis-(*tert*-butyldiphenylsilyl)-2-pyridinethiol (**40**), and 3,6-bis(triisopropylsilyl)-2-pyridine-thiol (**41**) were obtained in 60%, 40% and 30% yield, respectively. There appears to be a tendency for the more hindered chlorosilanes to substitute twice in the 3- and 6-positions while the less hindered chlorosilanes stop after one substitution in the 3-position.

When 2-pyridinethiol (**35**) is reacted with limiting LDA and a hindered chlorosilane such as chloro-*tert*-butyldimethylsilane then the 6-substituted product is obtained. Thus, the reaction of **35** with 2.5 equivalents of LDA and chloro-*tert*-butyldimethylsilane (*t*-BDMSCl) gives a 60% yield of a ca. 1:1 mixture of **39** and 6-*tert*-butyldimethylsilyl-2-pyridinethiol (**42**) (Scheme 8).

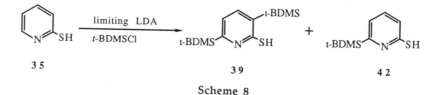

Scheme 8

A reduction in the amount of LDA to below 2.5 equivalents resulted in a loss of yield without a concurrent increase in the purity of the product. The lithiation of the 6-position in compound **35** must be competitive with lithiation of the 3-position in **42**. It is possible to obtain pure **39** only when excess base is used. However, the compounds **39** and **42** are easily separable by column chromatography and this method can be used to prepare **43** on a large scale.

When the compound 3-trimethylsilyl-2-pyridinethiolate (**35**) was treated with LDA and triethylsilyl chloride the unsymmetric compound 3-(trimethylsilyl)-6-(triethylsilyl)-2-pyri-dinethiol (**43**) was isolated in 28% yield (Scheme 9).The range of possible steric environments which can be placed around the mercapto group and the pyridine nitrogen is quite large. It should be possible by adjusting the 3- and 6-substituent of a 2-pyridinethiol to finely control

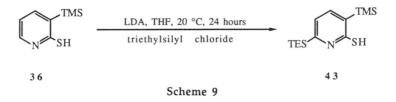

36

43

Scheme 9

its reactivity in such metal complex formations. We have prepared and characterized by X-ray crystallography two similar complexes of ligand **36** with copper and silver, $[Cu_6(SC_5H_3NSiMe_3)_6]$ and $[Ag_6(SC_5H_3NSiMe_3)_6]$, whose structures are best described as molecular paddlewheels (reminiscent of the cyclophanes) as well as a series of molybdenum complexes including two which have N,N-methylphenylhydrazine ligands, namely, $[Mo(SC_5H_3N-3-SiMe_3)_2(NNMePh)_2]$ and $[MoCl_2(SC_5H_3N-3-SiMe_3)_2(NNMePh)].$[31]

ACKNOWLEDGMENT

We gratefully acknowledge support from the National Science Foundation, the National Institutes of Health, the donors of the Petroleum Research Fund, administered by the American Chemical Society, the Herman Frasch Fund, and Société National Elf Aquitaine.

REFERENCES

1.   S.J.N. Burgmayer and E.I. Stiefel, *J. Chem. Ed.* 62:943 (1985).
2.   S.P. Cramer, K.O. Hodgson, W.O. Gillum, and L.E. Mortenson, *J. Am. Chem. Soc.* 100:3398 (1978).
3.   R.H. Holm, *Chem. Soc. Rev.* 10:455 (1981).
4 a. A. Muller, E. Diemann, R. Jostes, and E. Bogge, *Angew. Chem. Int. Ed. Engl.* 20:934 (1981).
   b. T.D.P. Stack, M.J. Carney, and R.H. Holm, *J. Am. Chem. Soc.* 111:1670 (1989).
5.   J. Chatt, J.R. Dilworth, and R.L. Richards, *Chem. Rev.* 78:589 (1978).
6.   T.A. George and R.C. Tisdale, *J. Am. Chem. Soc.* 107:5157 (1985).
7.   D. Sellman and W. Weiss, *Angew. Chem. Int. Ed. Eng.* 17:269 (1978).
8.   P.J. Blower and J.R. Dilworth, *Coord. Chem. Rev.* 76:121 (1987).
9.   I.G. Dance, *Polyhedron* 5:1037 (1986).
10. J.R. Dilworth, J. Hutchinson, and J.A. Zubieta, *J. Chem. Soc. Chem. Commun.* 1034 (1983).
11. N. Ueyana, H. Zaima, and A. Nakamura, *Chem. Lett.* 1481 (1985).
12. P.J. Blower, P.T. Bishop, J.R. Dilworth, T.C. Hsieh, J. Hutchinson, T. Nicolson, and J. Zubieta, *Inorg. Chim. Acta* 101:63 (1985).
13. C.D. Gutsche, *Acc. Chem. Res.* 16:161 (1983); C.D. Gutsche, *Top. Curr. Chem.* 51:742 (1986) and references therein.
14. M.M. Olmstead, G. Sigel, H. Hope, X. Xu, and P.P. Power, *J. Am. Chem. Soc.* 107:8087 (1985).
15. H.W. Gschwend and H.R. Rodriguez, *Org. React.* 26:1 (1979).
16a. E. Block and M. Aslam, *J. Am. Chem. Soc.* 107:6729 (1985).
   b. E. Block and M. Aslam, *Tetrahedron* 44:281 (1988).
17. K. Tang, M. Aslam, E. Block, T. Nicholson, and J. Zubieta, *J. Inorg. Chem.* 26:1488 (1987).
18. B.-K. Boo, E. Block, H. Kang, S. Liu, and J. Zubieta, *Polyhedron* 7:1397 (1988).
19. P. Beak and P.D. Becker, *J. Org. Chem.* 47:3855 (1982).
20. A.R. Katritzky, J.M. Aurrecoechea, and L.M. Vazquez de Miguel, *J. Chem. Soc. Perkin Trans. I* 769 (1987).

21. A.I. Meyers and M.E. Ford, *J. Org. Chem.* 41:1735 (1976).
22. C.R. Johnson and K. Tanaka, *Synthesis* 413 (1976).
23. J.M. Berg and R.H. Holm, *J. Am. Chem. Soc.* 107:917,925 (1985).
24. G.H. Posner and K.A. Canella, *J. Am. Chem. Soc.* 107:2571 (1985).
25. G.D. Figuly, *Ph.D. Thesis*, University of Illinois-Urbana (1981).
26a. G.D. Figuly, C.K. Loop, and J.C. Martin, *J. Am. Chem. Soc.* 111:654 (1989).
   b. E. Block, V. Eswarakrishnan, M. Gernon, G. Ofori-Okai, C. Saha, K. Tang, and J. Zubieta, *J. Am. Chem. Soc.* 111:658 (1989).
   c. K. Smith, C.M. Lindsay, and G.J. Pritchard, *J. Am. Chem. Soc.* 111:665 (1989).
27a. E. Block, M. Gernon, H. Kang, G. Ofori-Okai, and J. Zubieta, *Inorg. Chem.* 28:1263 (1989).
   b. E. Block, H. Kang, G. Ofori-Okai, and J. Zubieta, *Inorg. Chim. Acta* 156:27 (1989).
28. E. Block, G. Ofori-Okai, and J. Zubieta, *J. Am. Chem. Soc.* 111:2327 (1989).
29. E. Block, H. Kang, G. Ofori-Okai, and J. Zubieta, *Inorg. Chim. Acta* 166:155 (1989).
30. E. Block, M. Gernon, H. Kang, and J. Zubieta, *Angew. Chem. Int. Edn. Engl.* 27:1342 (1988).
31. E. Block, M. Gernon, H. Kang, S. Liu, and J. Zubieta, *Inorg. Chim. Acta* 167:143 (1990).

BIOLOGICALLY ACTIVE ORGANOSULFUR COMPOUNDS FROM GARLIC AND ONIONS:

THE SEARCH FOR NEW DRUGS

Eric Block

Department of Chemistry
State University of New York at Albany
Albany, N.Y. 12222, USA

MEDICINAL PROPERTIES OF GARLIC AND ONION: BACKGROUND

Garlic (*Allium sativum*) and onion (*Allium cepa*), members of the well known and widely appreciated genus allium were valued by the early Egyptians and Romans both as

The Genus Allium

A. Ascalonicum	Shallot
A. Cepa	Onion
A. Moly	
A. Porrum	Leek
A. Sativum	Garlic
A. Schoenoprasum	Chives

important dietary constituents and as medicinals for the treatment of many disorders, for example as evidenced by the discovery of garlic cloves and wooden models of onions among the relics found in the burial chambers of the pharaohs and as reported in the writings of the 1st century Roman naturalist Pliny the Elder.[1] Pliny wrote: "Garlic has very powerful properties, and is of great utility to persons on changes of water or locality. The very smell of it drives away serpents and scorpions, and, according to what some persons say is a cure for wounds made by every kind of wild beast, whether taken with the drink or food, or applied topically. ... Some persons have pre-scribed boiled garlic for asthmatic patients; while others, again have given it raw. ... Three heads of garlic, beaten up in vinegar, give relief in toothache; and a similar result is obtained by rinsing the mouth with a decoction of garlic, and inserting pieces of it in the hollow teeth. ..." Pliny also claimed onion to be effective against 28 different diseases.

Garlic and onion have their detractors, e.g. William Shakespeare ("... eat no onions or garlic for we are to utter sweet breath." [Midsummer Night's Dream, Act IV, Scene 2]; "The duke would mouth with a beggar, though she smelt brown bread and garlic." [Measure for Measure, Act III, Scene 2]) and of course vampires, as well as its celebrants, who each year gather at garlic festivals in Gilroy, California, Arleux, France and elsewhere around the world to sing its praise. The Jews who fled Egypt to wander the Sinai wilderness for forty years fondly remembered "... the fish which we did eat in Egypt so freely, and the pumpkins and melons, and the leeks, onions and garlic." (The Bible, Numbers 11.5)

Sulfur-Centered Reactive Intermediates in Chemistry and Biology, Edited by
C. Chatgilialoglu and K.-D. Asmus, Plenum Press, New York, 1990

Both the popularity of garlic and onion as foodstuff as well as their reputation as "cure-alls" stimulated scientific investigations, such as the early work by Pasteur in France (ca. 1858) into garlic's antibacterial activity and by Wertheim (1844) and later Semmler (1882) in Germany into the composition of onion and garlic oils (the former is primarily di-n-propyl disulfide while the latter is mainly diallyl disulfide, $CH_2=CHCH_2SSCH_2CH=CH_2$, (1)). Only with the advent of modern spectroscopic and chromatographic techniques has it become possible to determine the molecular basis for the odor, taste and biological activity of fresh or processed garlic and onion. Key discoveries on this subject were made by Cavallito (1944) in the United States, by Stoll and Seebeck in Switzerland (1948), by Virtanen in Finland (1962) and by Wilkens, Brodnitz and Pascale in the United States (1964-1971). Cavallito discovered an unstable, odoriferous liquid substance in extracts of fresh garlic he termed allicin, $CH_2=CHCH_2S(O)SCH_2CH=CH_2$ (2), which possessed antibacterial properties while Stoll and Seebeck identified the immediate precursor of allicin as alliin, $CH_2=CHCH_2S(O)$-$CH_2CH(NH_2)COOH$ (3) (Scheme 1). Virtanen, Wilkens, Brodnitz and Pascale all did seminal work on the formation and identity of the onion lachrymatory factor.

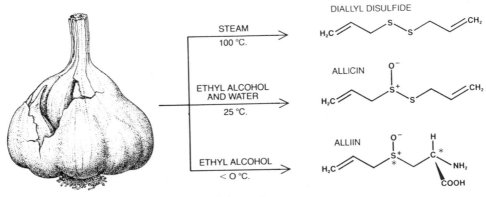

Scheme 1

COMPOUNDS IN GARLIC EXTRACTS

In the early 1970's my co-workers and I began an investigation of the reactions of the simpler homologue of allicin, methyl methanethiosulfinate, $CH_3S(O)SCH_3$, and discovered among other reactions a facile elimination process affording thioformaldehyde (CH_2S) and methanesulfenic acid (CH_3SOH) and a novel rearrangement reaction yielding 2,3,5-trithiahexane 5-oxide, $CH_3S(O)CH_2SSCH_3$, (Scheme 2).[2] These observations proved useful when some years later we turned our attention to the chemistry of allicin itself. With Rafael Apitz-Castro and Mahendra K. Jain it was discovered that extracts of fresh garlic

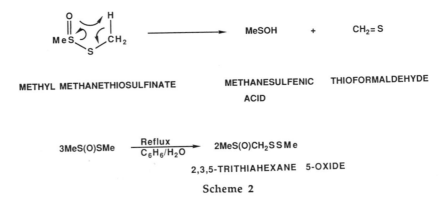

Scheme 2

284

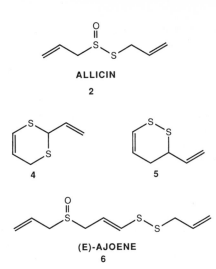

ALLICIN

2

4   **5**

(E)-AJOENE

6

Scheme 3. Components of Extract of Garlic.

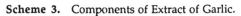

A   3   Allicin $\xrightarrow{\text{heat}}$ *trans*-Ajoene   +

H₂O-acetone

cis-Ajoene   +   H₂O

Mechanism:

2    $\xrightarrow[\text{H}^+]{-\nearrow\text{SOH}}$ $\xrightarrow{-\nearrow\text{SOH}}$

$\xrightarrow{-\text{H}^+}$ *cis*- and *trans*-Ajoene

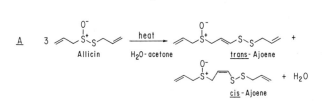

B $\xrightarrow{\text{H}_2\text{O}}$ SH   +   SO₂H

2-propenethiol   2-propenesulfinic
acid

then   SH   + $\longrightarrow$ S-S   +   SOH

and $\longrightarrow$ propylene   +   SO₂

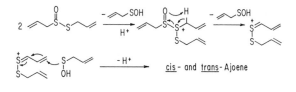

C $\longrightarrow$ SOH   +   thioacrolein

2-propenesulfenic

then $\longrightarrow$ +   H₂O

and   + $\xrightarrow{\text{major}}$ 2-vinyl-[4H]-1,3 dithiin

+ $\xdashrightarrow{\text{minor}}$ 3-vinyl-[4H]-1,2 dithiin

Diels-Alder
Reaction

Scheme 4

which show antithrombotic activity contain 2-vinyl-4*H*-1,3-dithiin (**4**), 3-vinyl-4*H*-1,2-dithiin (**5**) and (E,Z)-4,5,9-trithiadodeca-1,6,11-triene 9-oxide (**6**) (Scheme 3).[3] The latter compound, which we gave the trivial name ajoene (from the Spanish word "ajo", pronounced "aho", meaning garlic), showed significant antithrombotic, antifungal, and lipoxygenase (LO) inhibitory activity. The formation of compounds **4-6** as well as other compounds present in garlic extracts and distillates can be rationalized in terms of various remarkable transformations of allicin as shown in Scheme 4, leading one colleague to describe this chemistry as "Allicin Wonderland".

COMPOUNDS IN ONION EXTRACTS

It is interesting that in the onion, which is a close botanical relative of garlic, the allicin isomer 1-propenyl 1-propenethiosulfinate, $CH_3CH=CH-S(O)S-CH=CHCH_3$ (**7**), also undergoes remarkable transformations. Compound **7** is presumably formed by bimolecular condensation of 1-propenesulfenic acid, $CH_3CH=CHSOH$ (**8**), formed by enzymatic cleavage of $CH_3CH=CHS(O)CH_2CH(NH_2)COOH$ (**9**), (Schemes 5, 6). Compound **8** is also known to rearrange to (Z)-propanethial S-oxide, $CH_3CH_2CH=S^+-O^-$, the lachrymatory factor of the onion, which in turn can dimerize to *trans*-3,4-diethyl-1,2-dithietane 1,1-dioxide.[1,4] Recent collaborative research with Professor H. Wagner and his colleagues in Munich has led to the characterization and a biosynthetically-modelled one-step stereospecific synthesis of a pair

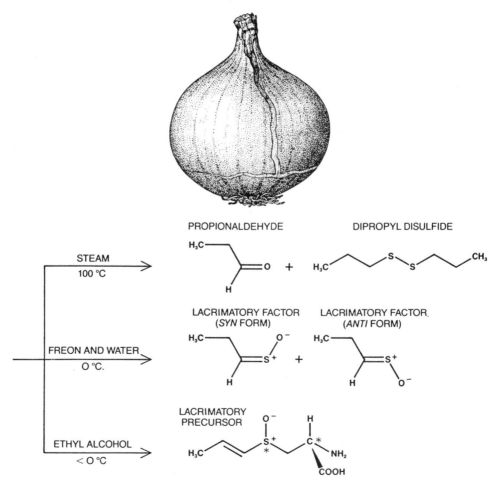

PROPIONALDEHYDE   DIPROPYL DISULFIDE

STEAM
100 °C

LACRIMATORY FACTOR   LACRIMATORY FACTOR
(*SYN* FORM)   (*ANTI* FORM)

FREON AND WATER
0 °C.

LACRIMATORY
PRECURSOR

ETHYL ALCOHOL
< 0 °C

Scheme 5

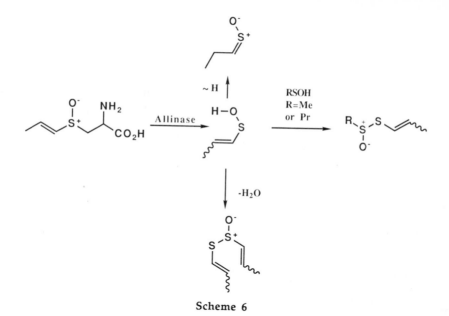

Scheme 6

of biologically active heterocycles termed "zwiebelanes" which are formed by stereospecific rearrangement of 7 (Scheme 7).[5] The zwiebelanes were discovered in the course of research on antiasthmatic agents from onions.

Other compounds recently discovered by Wagner include the unsaturated thiosulfinates $CH_3CH=CHSS(O)Pr-n$ and a series of potent compounds termed "cepaenes" (Scheme 8), among them $CH_3CH=CHS(O)CHEtSSPr-n$ and $CH_3CH=CHS(O)CHEtSSCH=CHCH_3$ which inhibit the cyclooxygenase of sheep seminal microsomes and the 5-lipoxygenase of porcine leucocytes.[6] Cepaenes are presumably formed through interaction of 1-propenesulfenic acid and (Z)-propanethial S-oxide. Related compounds isolated from onions inhibit thrombocyte aggregation.[7]

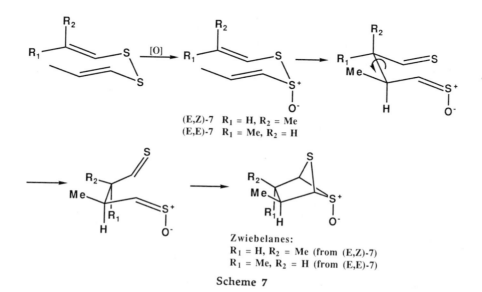

(E,Z)-7 R_1 = H, R_2 = Me
(E,E)-7 R_1 = Me, R_2 = H

Zwiebelanes:
R_1 = H, R_2 = Me (from (E,Z)-7)
R_1 = Me, R_2 = H (from (E,E)-7)

Scheme 7

287

Cepaenes

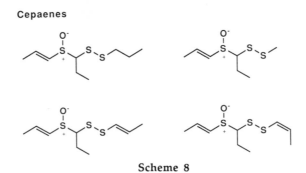

Scheme 8

COMPOUNDS IN GARLIC ESSENTIAL OIL

A fascinating epidemiological study conducted in the People's Republic of China reveals that in a region of China where gastric cancer rates are high, a significant reduction in gastric cancer risk parallels increasing consumption of garlic and onions and other allium foods.[8] Persons in the highest quartile of intake of these plants experienced only 40% of the risk of those in the lowest.[8] Following up on collaborative studies with Sidney Belman and earlier work by Belman and others on tumor inhibition by onion and garlic oils,[9] we sought to determine the molecular basis for the anticancer/antitumor and lipoxygenase (LO) inhibitory activity of garlic essential oil (EOG) preparations (garlic constituents that are steam volatile). Careful GC-MS analysis of EOG samples as well as samples prepared by heating diallyl disulfide, (1), the major component of EOG, led us to the discovery of a group of cyclic and acyclic compounds with two to five sulfur atoms (Schemes 9 and 10).[10] The formation of these unusual compounds, which show LO inhibitory activity, can be accommodated with a mechanism involving cycloaddition reactions of thioacrolein in a key step (Schemes 11 and 12).[10] The viability of such processes could be independently established using known thermal sources of thioacrolein (including allicin!).

While considerable progress has been made on the structures, mechanisms of formation, and syntheses of the numerous unusual organosulfur compounds present in garlic and onion extracts and distillates, there are still significant discoveries to be made and important questions for the chemist to answer.

$$All_2S_2 \longrightarrow All_2S + All_2S_3 + All_2S_4$$

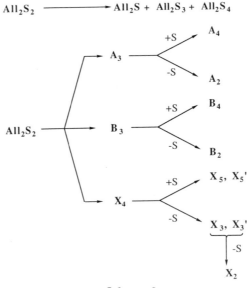

Scheme 9

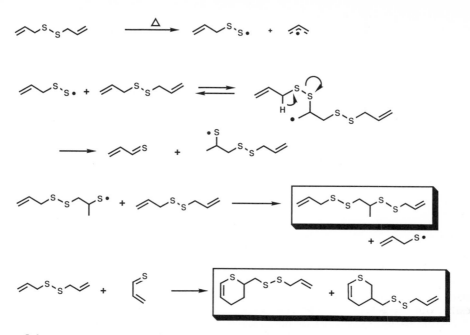

Scheme 10. Lipoxygenase-inhibiting Components of the Essential Oil of Garlic.

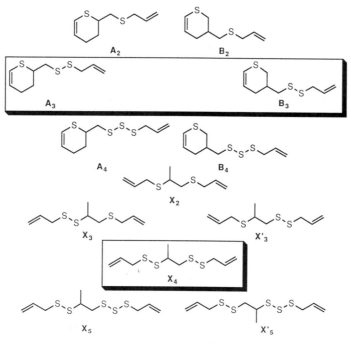

Scheme 11

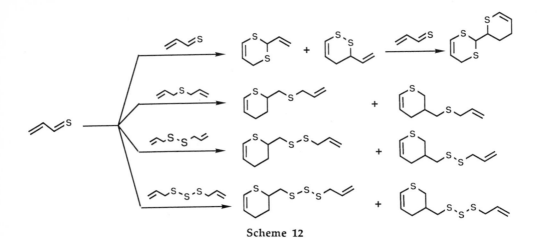

Scheme 12

ACKNOWLEDGMENT

I gratefully acknowledge support for our research from the donors of the Petroleum Research Fund, administered by the American Chemical Society, the Herman Frasch Foundation, the Société Nationale Elf Aquitaine, the National Science Foundation, NATO, and the McCormick Company.

REFERENCES

1.   E. Block, *Sci. Amer.* 252:114 (1985).
2.   E. Block and J. O'Connor, *J. Am. Chem. Soc.* 96:3921,3929 (1974).
3 a. E. Block, S. Ahmad, M.K. Jain, R.W. Crecely, R. Apitz-Castro, and M.R. Cruz, *J. Am. Chem. Soc.* 106:8295 (1984).
   b. E. Block, S. Ahmad, J. Catalfamo, M.K. Jain, and R. Apitz-Castro, *J. Am. Chem. Soc.* 108:7045 (1986).
4 a. E. Block, R.E. Penn, and L.K. Revelle, *J. Am. Chem. Soc.* 101:2200 (1979)
   b. E. Block, A.A. Bazzi, and L.K. Revelle, *Tetrahedron Lett.* 21:1277 (1980)
   c. E. Block, A.A. Bazzi, and L.K. Revelle, *J. Am. Chem. Soc.* 102:2490 (1980).
5.   T. Bayer, H. Wagner, E. Block, S. Grisoni, S.H. Zhao, and A. Neszmelyi, *J. Am. Chem. Soc.* 111:3085 (1989).
6 a. T. Bayer, H. Wagner, V. Wray, and W. Dorsch, *Lancet* 906 (1988).
   b. T. Bayer, W. Breu, O. Seligmann, V. Wray, and H. Wagner, *Phytochem.* 28:2373 (1989).
7.   S. Kawakishi and Y. Morimitsu, *Lancet* 330 (1988).
8.   W.-C. You, W.J. Blot, Y.-S. Chang, A. Ershow, Z.T. Yang, Q. An, B.E. Henderson, J.F. Fraumeni, Jr., and T.-G. Wang, *J. Natl. Cancer Inst.* 81:162 (1989).
9 a. S. Belman, *Carcinogenesis* 4:1063 (1983)
   b. M.J. Wargovich, *Carcinogenesis* 8:487 (1987).
   c. S. Belman, J. Solomon, A. Segal, E. Block, and G. Barany, *J. Biochem. Toxicol.* 4:151 (1989).
10. E. Block, R. Iyer, S. Grisoni, C. Saha, S. Belman, and F.P. Lossing, *J. Am. Chem. Soc.* 110:7813 (1988).

FREE RADICAL PROCESSES IN THE REACTIONS OF ORGANOMETALLICS

WITH ORGANIC SULFUR COMPOUNDS

Glen A. Russell and Preecha Ngoviwatchai

Department of Chemistry
Iowa State University
Ames, Iowa 50011, USA

Sulfur-centered radicals, $RSO_n{}^\bullet$ with n = 0 or 2 and presumably 1, react readily with certain organometallic reagents to generate alkyl radicals. These reactions can involve a formal S_H2 substitution at a metal atom such as Li or Hg,[1-3] e.g., reaction 1, or electron transfer with an easily oxidized organometallic species, reaction 2.[3] The rates of reactions 1 and 2

$$RSO_n{}^\bullet + R^1HgCl \rightarrow RSO_nHgX + R^{1\bullet} \tag{1}$$

$$RSO_n{}^\bullet + (R^1)_2Cu^- \rightarrow RSO_n{}^- + R^{1\bullet} + R^1Cu \tag{2}$$

increase at least qualitatively with the electrophilicity of the attacking radical and with the stability of the alkyl radical formed. Thus, reaction 1 occurs slowly when the attacking radical is a nucleophilic alkyl radical, such as t-Bu$^\bullet$ or t-BuCH$_2$ĊHSPh, but readily with electrophilic species such as t-BuCH$_2$ĊHSO$_2$Ph or heteroatom-centered radicals including $PhSO_n{}^\bullet$ or halogen atoms.

Reactions 1 and 2 provide a route to recycle heteroatom-centered radical species to simple alkyl radicals such as $R^1 = t$-Bu. Sulfur-centered radicals can also react with organometallic reagents to produce metal-centered species by an addition-elimination process. For example, reactions 3 and 4 occur readily with X = halogen, Sn(IV), Hg(II), Pb(IV), Ge(IV), or Zr(IV).[4] Analogous reactions occur with 1-alkynyl and propargyl derivatives.[5,6]

$$RSO_n{}^\bullet + PhCH=CHX \rightarrow Ph\dot{C}H–CH(X)(SO_nR) \rightarrow PhCH=CHSO_nR + X^\bullet \tag{3}$$

$$RSO_n{}^\bullet + {}^\bullet CH_2=CHCH_2X \rightarrow RSO_n\dot{C}H_2-\dot{C}HCH_2X \rightarrow RSO_nCH_2CH=CH_2 + X^\bullet \tag{4}$$

Chain reactions can occur when the alkyl radicals (from reactions 1 or 2) or the heteroatom-centered radicals X$^\bullet$ (from reactions 3 or 4) react to regenerate $RSO_n{}^\bullet$ by reactions such as S_H2 atom transfers (reaction 5),[1,4] addition-elimination with alkenyl, alkynyl, allylic or propargylic derivatives (reaction 6),[2] by electron transfer with $RSO_n{}^-$ (reaction 7),[4] by displacement from a metal salt such as $(RSO_n)_2Hg$, reaction 8,[7] or decomposition of eliminated Hg(I) species such as PhSHg$^\bullet$ (reaction 9).[4]

Sulfur-Centered Reactive Intermediates in Chemistry and Biology, Edited by
C. Chatgilialoglu and K.-D. Asmus, Plenum Press, New York, 1990

291

$$R^{1\bullet} \text{ (or } X^\bullet) + YSO_nR \rightarrow R^1\text{-}Y \text{ (or } X\text{-}Y) + RSO_n^\bullet \tag{5}$$

$$R^{1\bullet} + PhCH=CHSO_nR \rightarrow Ph\dot{C}HCH(R^1)SO_nR \rightarrow PhCH=CHR^1 + RSO_n^\bullet \tag{6}$$

$$ClHg^\bullet + RSO_n^- \rightarrow Cl^- + Hg^\circ + RSO_n^\bullet \tag{7}$$

$$X^\bullet + (RSO_n)_2Hg \rightarrow RSO_nHgX + RSO_n^\bullet \tag{8}$$

$$PhSHg^\bullet \rightarrow PhS^\bullet + Hg^\circ \tag{9}$$

Combination of reactions 1 - 4 which consume $RSO_n^\bullet$ with reactions 5 - 9 which regenerate $RSO_n^\bullet$ leads to a variety of free radical chain processes which for organomercurial substrates can be readily initiated by fluorescent light. Other chain propagating reactions can, of course, be added to the sequence such as unimolecular reactions of $R^{1\bullet}$ (e.g., cyclization) or the addition of $R^{1\bullet}$ or $X^\bullet$ to an unsaturated system to generate chain-carrying adduct radicals.

REACTIONS OF ALKYLMERCURIALS WITH $RSO_n^\bullet$

The chain sequence of Scheme 1 is readily observed with photochemical or thermal initiation.[1,8] Among the substrates RSO_nY which readily participate in Scheme 1 are disulfides, mercpatans and sulfonyl derivatives with Y = halogen, PhS or PhSe. Table 1 presents the relative reactivities of a series of RSO_nY derivatives towards $R^{1\bullet} = t\text{-}Bu^\bullet$ measured in competitive reactions with $t\text{-}BuHgCl$ in PhH or Me_2SO. Additional data on the rate constant for reaction 5 has been obtained using the cyclizable probe, $R^{1\bullet} = 5\text{-hexenyl}^\bullet$ generated from 5-hexenylmercury chloride in PhH.[1]

$$R^1HgX \xrightarrow{h\nu} R^{1\bullet} + XHg^\bullet$$

$$XHg^\bullet + R^1HgX \rightarrow HgX_2 + Hg^\circ + R^{1\bullet}$$

$$R^{1\bullet} + RSO_nY \rightarrow R^1Y + RSO_n^\bullet$$

$$RSO_n^\bullet + R^1HgX \rightarrow RSO_nHgX + R^{1\bullet}$$

Scheme 1. (R^1 = alkyl, X = halogen, carboxylate, alkyl).

The relative reactivities of $t\text{-}BuHgCl : i\text{-}PrHgCl : n\text{-}BuHgCl$ in Me_2SO at 35°C toward $PhS^\bullet$ have been measured by competitive photostimulated reactions with PhSSPh as 1.0 : 0.08 : <0.003.[8] This suggests that reaction 1 with $PhS^\bullet$ occurs in a concerted fashion with a rate constant determined by the stability of the incipient alkyl radical. However, it is difficult to exclude a fast reversible formation of the intermediate $PhSHg^\bullet(R^1)X$ followed by decomposition to yield $R^{1\bullet}$. Alkyl radical exchange with R^1HgX can be excluded because the observed relative reactivities are independent of the concentration of PhSSPh or of the more reactive PhSeSePh with which $k(t\text{-}Bu) : k(n\text{-}Bu) = 1 : <0.004$ in PhH.

Another route to $RSO_n^\bullet$ is β-elimination from radicals, e.g. $>\!C\text{-}\dot{C}SO_nR$ or $-\dot{C}=C\text{-}SO_nR$ formed by radical addition to an alkenyl or alkynyl substrate[2] or by radical abstraction reactions. With $CH_2=CHSO_nPh$ addition of alkyl radicals occurs to form $R^1CH_2\dot{C}HSO_nPh$ which in the case of n = 1 or 2 reacts with R^1HgCl to form $R^1CH_2CH(HgCl)SO_nPh$ and regenerate $R^{1\bullet}$ (Scheme 2a).[9] Only electrophilic radicals will rapidly displace an alkyl radical

Table 1[1,8]   Reactivities of PhSO$_n$-Y towards *tert*-Butyl and 5-Hexenyl Radicals at 35 °C

Substrate	Rel. React., *t*-Bu$^\bullet$	React., 5-Hexenyl$^\bullet$ (M^{-1} s^{-1})[a]
CH$_2$=CHP(O)(OEt)$_2$	1.0[b]	—
PhSH	—	≈ 8 x 10^7
PhSO$_2$Cl	11	—
PhTeTePh	—	5 x 10^7
PhSeSePh	—	1 x 10^7
p-MeC$_6$H$_4$SO$_2$SePh	—	3 x 10^6
PhSSPh	4.3	8 x 10^4
MeSSMe	0.3	—
BuSSBu	0.08	—
i-PrSSPr-*i*	0.0035	—
t-BuSSBu-*t*	0.00045	—

[a]   Based upon the unimolecular rate constant for cyclization of 1 x 10^5 s^{-1} [D. Griller and K.U. Ingold, *Acc. Chem. Res.* 13:317 (1980)].

[b]   The rate constant for *t*-Bu$^\bullet$ addition to CH$_2$=CHP(O)(OEt)$_2$ [J.A. Babau and B.P. Roberts, *J. Chem. Soc. Perkin Trans 2* 161 (1981)] is calculated to be 5 x 10^5 M^{-1} s^{-1} at 35 °C.

(a)   R$^1{}^\bullet$ + CH$_2$=CHSO$_n$Ph → R^{1}CH$_2\overset{\bullet}{\text{C}}HSO_n$Ph

R^{1}CH$_2\overset{\bullet}{\text{C}}HSO_n$Ph + R^{1}HgCl → R^{1}CH$_2$CH(HgCl)SO$_n$Ph + R$^1{}^\bullet$

(b)   R^{1}CH$_2\overset{\bullet}{\text{C}}HSO_n$Ph + R^{1}HgH → R^{1}CH$_2CH_2SO_n$Ph + R^{1}Hg$^\bullet$

R^{1}Hg$^\bullet$ → R$^1{}^\bullet$ + Hg°

Scheme 2.   (R^1 = alkyl).

from an alkylmercurial, but a chain reaction can be sustained with nucleophilic radicals such as R^{1}CH$_2\overset{\bullet}{\text{C}}$HSPh by the use of R^{1}HgCl/BH$_4^-$ (Scheme 2b).[10] Relative reactivities towards *t*-Bu$^\bullet$ at 25°C in CH$_2$Cl$_2$ in competitive reactions following Scheme 2b for CH$_2$=CHSO$_2$Ph : CH$_2$=CHSOPh : CH$_2$=CHSPh are 74:2:1.[11] The reactivity of sterically unhindered alkenes towards nucleophilic radicals such as *t*-Bu$^\bullet$ are determined mainly by polar effects and the high reactivity of CH$_2$=CHSO$_2$Ph ($k_{add.}$ ≈ 3.7 x 10^5 M^{-1} s^{-1})[11] relative to the sulfide is not surprising. However, the low reactivity of the sulfoxide is surprising since both the sulfoxide and sulfone should be able to provide stabilization to the transition state for radical addition, i.e., R$^+$CH$_2\overset{\bullet\,-}{=\!\!=}$CHSO$_n$Ph.

With β-styrenyl derivatives addition of *t*-Bu$^\bullet$ radicals occurs to yield mainly Ph$\overset{\bullet}{\text{C}}$HCH(*t*-Bu)SO$_n$Ph while with the phenylethynyl derivatives only products derived from Ph$\overset{\bullet}{\text{C}}$=C(*t*-Bu)SO$_n$Ph are observed (Scheme 3).[3] Table 2 lists the yields of products

Table 2.[3,11] Photostimulated Reactions of Unsaturated Phenyl Sulfones, Sulfoxides and Sulfides with RHgCl at 35°C in Me_2SO or PhH (E= trans, Z = cis).

Substrate	Products [% (E/Z)][a] $R^1 = t\text{-Bu}, R^2 = c\text{-}C_6H_{11}$	Rel. Reactivity $t\text{-Bu}^\bullet \quad (c\text{-}C_6H_{11})^\bullet$	
$Ph_2C=CHI$	$Ph_2C=CHR$ [86, 95]	1.0[b]	1.0
$CH_2=CHSO_2Ph$	$RCH_2CH_2SO_2Ph$ [97, 69][c]	74[d]	—
$CH_2=CHSOPh$	$R^1CH_2CH_2SOPh$ [5][c]	2.0[d]	—
$CH_2=CHSPh$	$R^1CH_2CH_2SPh$ [34][c,e]	1.0[d]	—
$PhC\equiv CSO_2Ph$	$PhC\equiv CR$ [57,[f] 66]	2.1	12
$PhC\equiv CSPh$	$PhC\equiv CR$ [44, 46]	0.4	0.8
$Ph_2C=CHSO_2Ph$	$Ph_2C=CHR$ [88, 91]	—	6.5
$Ph_2C=CHSPh$	$Ph_2C=CHR^2$ [58]	—	4.2
$(E)\text{-PhCH}=CHSO_2Ph$	$PhCH=CHR$ [43 (50), 74 (43)]	1.3(β)	3.3
	$PhCH(R^1)CH_2SO_2Ph$ [16][c]	0.50(α)	—
$(E)\text{-PhCH}=CHSOPh$	$PhCH=CHR^1$ [32 (21)]	—	—
$(E)\text{-PhCH}=CHSPh$	$PhCH=CHR$ [36 (5), 43 (4.9)]	0.40(β)	1.1
	$PhCH(R^1)CH_2SPh$ and		
	$PhC(R^1)=CHSPh$ [12]	0.13(α)	—
$CH_2=CHCH_2SO_2Ph$	$RCH_2CH=CH_2$ [88, 42]	0.73	—
$CH_2=CHCH_2SPh$	$RCH_2CH=CH_2$ [63, 57]	0.45	—
$CH_2=CHCMe_2SO_2Ph$	$PhSO_2CH_2CH=CMe_2$ [86]	—	—
$CH_2=CHCMe_2SPh$	$R^1CH_2CH=CMe_2$ [45]	—	—
	$PhSCH_2CH=CMe_2$ [22]	—	—
$CH_2=CHCH(CH_2Ph)SO_2Ph$	$R^1CH_2CH=CHCH_2Ph$ [49 (5.2)]	—	—
	$PhSO_2CH_2CH=CHCH_2Ph$ [23 (5.6)]	—	—
$HC\equiv CCH_2SPh$	$R^1CH=C=CH_2$ [<10, 26[f]]	—	—
$HC\equiv CCH_2SO_2Ph$	$R^1CH=C=CH_2$ [<10]	—	—

a Products observed with 2-5 equiv of RHgCl with photolysis from a 275 W sunlamp for 5 - 25 h; when two yields are given, the first refers to R = t-Bu and the second to R = c-C_6H_{11}.

b Competitive reaction with $CH_2=CHP(O)(OEt)_2$ (Table 1) yields $k = 4.8 \times 10^4$ M^{-1} s^{-1} at 35°C.

c Workup with $NaBH_4$.

d Relative reactivity in CH_2Cl_2 (Scheme 2b).

e Also observed, $R^1CH=CHSPh$ (25%), $R^1CH_2CH(R^1)SPh$ (8.5%) and two isomers of $R^1CH_2CH(SPh)CH(SPh)CH_2R^1$.

f Yield in the presence of 1 - 2 equiv of I$^-$ per equiv of RHgCl.

$$t\text{-Bu}^\bullet \;+\; \text{PhCH=CHSO}_2\text{Ph} \;\rightarrow\; \text{Ph}\overset{\bullet}{\text{C}}\text{HCH}(t\text{-Bu})\text{SO}_2\text{Ph} \;+\; \text{PhCH}(t\text{-Bu})\overset{\bullet}{\text{C}}\text{HSO}_2\text{Ph}$$

$$\text{Ph}\overset{\bullet}{\text{C}}\text{HCH}(t\text{-Bu})\text{SO}_2\text{Ph} \;\rightarrow\; (E)\text{-PhCH=CHBu-}t \;+\; \text{PhSO}_2{}^\bullet$$

$$\text{PhSO}_2{}^\bullet \;+\; t\text{-BuHgCl} \;\rightarrow\; t\text{-Bu}^\bullet \;+\; \text{PhSO}_2\text{HgCl}$$

$$\text{PhCH}(t\text{-Bu})\overset{\bullet}{\text{C}}\text{HSO}_n\text{Ph} \;+\; t\text{-BuHgCl} \;\rightarrow\; \text{PhCH}(t\text{-Bu})\text{CH}(\text{HgCl})\text{SO}_2\text{Ph} \;+\; t\text{-Bu}^\bullet$$

$$\text{PhCH}(t\text{-Bu})\text{CH}(\text{HgCl})\text{SO}_2\text{Ph} \;\xrightarrow{\;\text{H}^+\text{ or BH}_4{}^-\;}\; \text{PhCH}(t\text{-Bu})\text{CH}_2\text{SO}_2\text{Ph}$$

Scheme 3

observed with excess t-BuHgCl and the relative reactivities observed with a deficiency of t-BuHgCl for a series of unsaturated sulfones and sulfides which react by free radical chain reactions of the type described in Scheme 3.

Table 2 contains some examples of allylic S_H2' substitutions resulting from terminal attack of t-Bu$^\bullet$ upon an allylic sulfide or sulfone.[12] With α-substituted allylic derivatives rearrangement to the γ-substituted derivative is a serious problem. The eliminated PhSO$_n{}^\bullet$ ($n = 0, 2$) can generate t-Bu$^\bullet$ by reaction (1) or add to the allylic system to yield the rearranged product after elimination. Propargyl derivatives show a similar rearrangement to propadienyl derivatives. From the products observed, it appears that PhS$^\bullet$ is trapped more effectively by t-BuHgCl than is PhSO$_2{}^\bullet$ although the rates of addition of PhSO$_n{}^\bullet$ to the allylic systems also need to be considered.

Addition of t-Bu$^\bullet$ to the substrates of Table 2 is considered to be irreversible. The relative reactivities measured thus are for the radical addition step and do not reflect the ease of β-elimination of PhSO$_n{}^\bullet$. The sulfones are more reactive than the sulfides in trapping t-Bu$^\bullet$ not only for CH$_2$=CHSO$_n$Ph but also for (E)-PhCH=CHSO$_n$Ph, Ph$_2$C=CHSO$_n$Ph and PhC≡CSO$_n$Ph. The sulfone group activates relative to a thiyl group for t-Bu$^\bullet$ attack whether the attack is β (CH$_2$=CHSO$_n$Ph) or α (Ph$_2$C=CHSO$_n$Ph) to the substituent.

Towards t-Bu$^\bullet$ or c-C$_6$H$_{11}{}^\bullet$, 1-alkenyl and 1-alkynyl sulfides have very similar reactivities. However, for the sulfones the acetylenic derivatives are considerably more reactive, possibly reflecting a steric effect. Steric effects may also be responsible for the regiochemistry observed, i. e., only β-attack upon PhC≡CSO$_2$Ph but $k_\beta/k_\alpha = 2.7$ for (E)-PhCH=CHSO$_2$Ph (α and β are now relative to the phenyl group).[3]

PARTICIPATION OF ALKYL RADICALS IN REACTIONS OF OTHER ORGANOMETALLIC REAGENTS WITH UNSATURATED SULFONES AND SULFIDES

The relative reactivity data of Table 2 for t-Bu$^\bullet$ attack has been compared with the classical carbanionic alkylating agents t-BuLi, t-BuMgCl and the $tert$-butyl cuprates.[3] Completely different reactivities or regioselectivities would exclude free radical attack for these $tert$-butylating agents. On the other hand, similar chemo and regioselectivities are at least suggestive of radical attack. In particular, radical and anionic attack upon PhCH=CHSO$_2$Ph would be expected to show greatly different regioselectivities with α-attack for the carbanion and preferred β-attack for the radical ($k_\beta/k_\alpha = 2.7$ for t-Bu$^\bullet$, Table 2).

Table 3 compares the products observed in competitive reactions of an excess of a mixture of 1-alkenyl and 1-alkynyl derivatives with $tert$-butylating agents. The data for $tert$-butyllithium at 0 or 45 °C is particularly definitive for radical attack. Apparently a

$R^\bullet + PhCH=CHSO_nPh \rightarrow Ph\overset{\bullet}{C}HCH(R)SO_nPh + PhCH(R)\overset{\bullet}{C}HSO_nPh$

$R^\bullet + PhC\equiv CSO_nPh \rightarrow Ph\overset{\bullet}{C}=C(R)SO_nPh$

$Ph\overset{\bullet}{C}HCH(R)SO_nPh \rightarrow PhCH=CHR + PhSO_n^\bullet$

$Ph\overset{\bullet}{C}=C(R)SO_nPh \rightarrow PhC\equiv CR + PhSO_n^\bullet$

$PhSO_n^\bullet + t\text{-BuLi} \rightarrow PhSO_nLi + t\text{-Bu}^\bullet$

$PhCH(R)\overset{\bullet}{C}HSO_nPh + t\text{-BuLi} \rightarrow PhCH(R)CH(Li)SO_nPh + t\text{-Bu}^\bullet$

Scheme 4. (R = t-Bu, n = 0 or 2).

Table 3. Relative Reactivities of PhCH=CHSPh (**1**), PhC≡CSPh (**2**), PhCH=CHSO₂Ph (**3**) and PhC≡CSO₂Ph (**4**) in Substitution Reactions

Organometallic	Conditions	k (rel), **1 : 2 : 3 : 4**[a]	k_α/k_β for **3**[b]
t-BuHgCl	Me₂SO, 45°C, hv	1.0 : 1.0 : 3.3 : 5.3	1 : 2.7
t-BuLi	THF, 45°C	1.0 : 1.0 : 2.9 : 5.3	
t-BuLi	THF, 0°C	1.0 : 1.3 : 3.4 : 6.8	1 : 2.8
t-BuMgCl	THF, 25°C	1.0 : 0.7 : 2.2 : 92	1 : 3.7
t-BuLi	THF, –78°C	1.0 : 0.7 : 29 : 50	1 : 32
(t-Bu)₂CuLi	THF, 0°C	1.0 : 0.9 : 3.2 : 7.6	1 : 3.5
(t-Bu)₂CuLi	THF, –78°C	1.0 : 0.8 : 18 : 37	1 : 13
(t-Bu)₂Cu(CN)Li₂	THF, 0°C		1 : 4
(t-Bu)₂Cu(PBu₃)Li	THF, 0°C		1 : 4

[a] Competition involved two substrates each 0.5 M reacting over a 2 - 4 h period with the organometallic reagent at an initial concentration of 0.1 M.

[b] α-Attack leads to PhCH(t-Bu)CH₂SO₂Ph; β-attack leads to PhCH = CHBu-t.

radical chain mechanism (Scheme 4) involving attack of PhSO$_n^\bullet$ upon t-BuLi occurs readily and is spontaneously initiated. With t-BuMgCl at 25 °C, PhC≡CSPh, PhCH=CHSPh and PhCH=CHSO₂Ph appear to react by the radical mechanism, but PhC≡CSO₂Ph is much more reactive although it gives only PhC≡CBu-t. An ionic reaction involving complex formation with the sulfone has been previously suggested for acetylenic sulfones.[13] At 0 °C the reagent 2 t-BuLi/CuI forms products similar to those observed from t-BuLi or t-BuHgCl/hv. The more stable (t-Bu)₂Cu(CN)Li₂ and (t-Bu)₂Cu(PBu₃)Li give only β-substitution products for PhC≡CSPh or PhCH=CHSPh (0 or –78 °C), and although they fail to react with PhCH=CHSO₂Ph at –78 °C, at 0 °C the regiochemistry is similar to that expected for radical attack (k_β/k_α = 4). With PhC≡CSO₂Ph the stable cuprates give mainly or exclusively α-addition at 0 °C (to yield the product expected from anionic addition, PhC(t–Bu)=CHSO₂Ph), but at –78 °C approximately equal parts of α-addition and β-substitution are observed. The enhanced β-substitution for PhC≡CSO₂Ph observed for t-BuMgCl

(25°C), $(t\text{-Bu})_2\text{Cu(CN)Li}_2$ (–78 °C) or $(t\text{-Bu})_2\text{Cu(PBu}_3)\text{Li}_2$ (–78 °C) seems to be connected with complex formation involving the sulfone moiety. This complex formation may also account for the enhanced reactivity in β-substitution observed for $\text{PhCH=CHSO}_2\text{Ph}$ and $\text{PhC}\equiv\text{CSO}_2\text{Ph}$ at –78 °C with $t\text{-BuLi}$ or $(t\text{-Bu})_2\text{CuLi}$. With these reagents the relative reactivities observed for $\text{PhC}\equiv\text{CSPh}$ (β-substitution only), PhCH=CHSPh (β-substitution) and $\text{PhCH=CHSO}_2\text{Ph}$ (α-addition, the minor reaction channel) appear to be consistent with the free radical reactivities observed at higher temperatures with $t\text{-BuHgCl/h}\nu$.

At least three reaction pathways must be involved for the substrates of Table 3. The classical carbanionic α-addition seems to be involved only for $\text{PhC}\equiv\text{CSO}_2\text{Ph}$ reacting with $(t\text{-Bu})_2\text{Cu(CN)Li}_2$, $(t\text{-Bu})_2\text{C(PBu}_3)\text{Li}$ or $t\text{-BuCu(I)Li}$ at 0 °C. With $\text{PhC}\equiv\text{CSO}_2\text{Ph}$ a presumably ionic process, possibly involving complex formation and leading to β-substitution, is observed for $t\text{-BuMgCl}$ at 25 °C and this process becomes dominant for $t\text{-BuLi}$ and the various cuprates at –78 °C with either $\text{PhC}\equiv\text{CSO}_2\text{Ph}$ or $\text{PhCH=CHSO}_2\text{Ph}$. Attack of $t\text{-Bu}^{\bullet}$ seems to be the major reaction pathway for $\text{PhC}\equiv\text{CSPh}$, PhCH=CHSPh and $\text{PhCH=CHSO}_2\text{Ph}$ at 0 °C with $t\text{-BuLi}$, $t\text{-BuMgCl}$ or the t-butyl cuprates while at –78 °C radical attack on $\text{PhC}\equiv\text{CSPh}$ (β-substitution), PhCH=CHSPh (β-substitution) or $\text{PhCH=CHSO}_2\text{Ph}$ (α-addition and β-substitution) is still important, but for sulfones the radical pathway is often overshadowed by β-substitution involving complex formation.

β-ELIMINATION OF $\text{RSO}_n{}^{\bullet}$

The chain reaction described in the previous section involved the β-scission reaction of radicals of the type $>\dot{\text{C}}\text{–CH(R)SO}_n\text{Ph}$ or $-\dot{\text{C}}\text{=C(R)SO}_n\text{Ph}$. The radicals $\text{PhSO}_n{}^{\bullet}$ with n = 0 - 2 are all eliminated readily. Some knowledge about the rates of elimination can be deduced from the stereospecificity of radical elimination processes. Initial work by Shevlin et al. involved the measurement of (E)/(Z) ratios of 2-butene formed from the various diastereomers

$$\text{BrCH(CH}_3)\text{CH(CH}_3)\text{SO}_n\text{Ph}$$

4

of **4** upon reaction with Bu_3SnH.[14] With n = 0 or 2 the reactions were not stereospecific indicating an equilibrium between intermediates **4a** and **4b** (Figure 1).

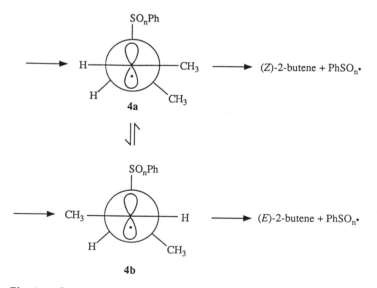

Fig. 1.   Conformation Equilibration in the Elimination Reactions of **4**.

However, with n = 1 the two erythro and the two threo isomers of **4** (the sulfoxide is a third stereogenic center) gave somewhat different (E)/(Z) ratios of 2-butene although the configuration at the sulfur atom was about as important as the erythro/threo stereochemistry of the C_2 fragment. Although the stereospecificity was low [(E)/(Z) = 0.4 to 2.4], it appears that PhSO is eliminated more readily than $PhS^\bullet$ or $PhSO_2{}^\bullet$ and that the elimination of PhSO can compete with rotation about the single bond in the radical (Figure 1) although other conformation effects including bridging could be involved. (The relative reactivities of **4** with n = 0 and 1 towards $Bu_3Sn^\bullet$ differ by less than a factor of 2). More recently Ono et al. have reported stereospecificity in the elimination reaction of β-nitro sulfones with Bu_3SnH.[15] Ono found that the EtC(Me)=C(Me)(CN) formed varied form (E)/(Z) = 99/1 for **5a** to 1/99 for **5b**. For the diastereomers of **6** the diastereoselectivity of the elimination reaction forming Bu_3SnONO and $PhSO_nH$ increased from PhS to PhSO to $PhSO_2$. Ono's explanation of the observed stereospecificity was that elimination occurred in a concerted fashion from an intermediate nitroxide radical formed by addition of $Bu_3Sn^\bullet$ to the nitro group. The sulfone, having the highest preference for the anti periplanar conformation, leads selectively to the (E) and (Z) alkenes from the two diastereomers.

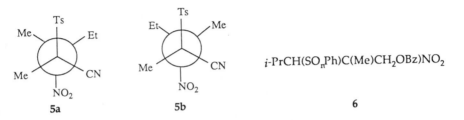

$i\text{-}PrCH(SO_nPh)C(Me)CH_2OBz)NO_2$

5a   **5b**   **6**

Rates of β-elimination of the radicals $X^\bullet$ have been reported by Wagner et al. for photochemically generated 1,4-diradicals, $Ph^\bullet C(OH)CH_2CH_2{}^\bullet CHX$.[16] The results confirm Shevlin's stereochemical results since relative k's of 1 : 456 : 2.9 : 660 : 1380 : 29 are observed for X = BuS, BuSO, $BuSO_2$, PhSO, PhS and $PhSO_2$, respectively, with $k_{elim.}$ for BuS ≈ 2.7 × 10^5 M^{-1} s^{-1} (values for Cl, Br and I are 4 × 10^6, > 3 × 10^8, > 5 × 10^9).

GENERATION OF $PhSO_n{}^\bullet$ BY REACTIONS OF HETEROATOM-CENTERED RADICALS

Reaction 5 with $X^\bullet$ = $ClHg^\bullet$, $Bu_3Sn^\bullet$, $R_3Pb^\bullet$, $(Cp)_2Zr(Cl)^\bullet$, reaction 7 with $X^\bullet$ = $ClHg^\bullet$ or reaction 8 with $X^\bullet$ = $I^\bullet$, $ClHg^\bullet$ or $Bu_3Sn^\bullet$ occur readily.[17] Chain reactions with the appropriate vinyl derivatives of X will occur according to Scheme 5. With X = Bu_3Sn or ClHg the regioselectivity of $PhSO_n{}^\bullet$ attack is very high, even when R' is phenyl. Similar reactions are observed with allyl, propargyl or 1-alkynyl derivatives of X, but in general the 1-alkynyl-metals are much less reactive than the 1-alkenylmetals towards $PhSO_n{}^\bullet$. Some typical photosensitized chain reactions are summarized in Table 4.

$$R'CH=CHX + RSO_n{}^\bullet \rightarrow R'\overset{\bullet}{C}H\text{--}CH(X)SO_nR$$

$$R'\overset{\bullet}{C}H\text{--}CH(X)SO_nR \rightarrow R'CH=CHSO_nR + X^\bullet \quad [5]$$

$$X^\bullet + Y\text{--}SO_nR \rightarrow XY + RSO_n{}^\bullet \quad [7]$$

$$XHg^\bullet + RSO_n{}^- \rightarrow X^- + Hg^\circ + RSO_n{}^\bullet \quad [8]$$

$$X^\bullet + (PhSO_n)_2Hg \rightarrow PhSO_nHgX + PhSO_n{}^\bullet$$

Scheme 5

Table 4.   Substitutions Occurring by the Addition of $PhSO_n \cdot$

Reactants (equivalents)	Conditions[a]	Product	% Yield (E/Z)
(E)-t-BuCH=C(H)HgSR, R = Ph; Bu	PhH, S, 3 h; 12 h	t-BuCH=CHSR	100 (>50); 100 (>50)
Ph_2C=C(R)HgSPh, R = H; Me	PhH, S 6 h	Ph_2C=C(R)SPh	100; 100
(E)-t-BuCH=C(H)HgCl and PhYYPh (1), Y = S; Se; Te	PhH, S, 6 h; 2 h; 18 h	t-BuCH=CHYPh	100 (>50); 95 (>50); 89 (>50)
Ph_2C=C(H)HgBr and RSSR, R = Me (1); i-Pr (10)	PhH, S, 2 h; 20 h	Ph_2C=CHSR	100; 98
Ph_2C=C(H)HgBr and HgA_2, A = PhS (1); $PhSO_2$ (5); PhSe (1)	PhH, R, 20 h	Ph_2C=CHA	100; 100; 80
Ph_2C=C(H)I and A_2Hg (1), A = PhS; $PhSO_2$	Me_2SO, R, 12 h	Ph_2C=CHSO$_2$Ph	100; 93
CH_2=C(H)HgCl and A_2Hg, A = PhS (1); PhSe (1); $PhSO_2$ (5)	Me_2SO, R, 20 h; 20 h; 12 h	CH_2=CHA	76; 39; 43
(Ph_2C=CH)$_2$Hg and PhSSPh (1)	PhH, S, 3 h	Ph_2C=CHSPh	99[b]
(E)-t-BuCH=C(H)HgCl and $PhSO_2$Cl (1); ($PhSO_2$)$_2$Hg (5); $PhSO_2$Na (1.2)	Me_2SO, S, 3 h; 12 h; t-BuOH/Me_2SO, S, 30 h	t-BuCH=CHSO$_2$Ph	99 (>50); 42 (>50); 81 (>50)
(E)-t-BuCH=C(H)HgCl and RSO_2Cl (1.0), R = Me; p-tolyl	PhH, S, 22 h	t-BuCH=CHSO$_2$R	32 (>50); 75 (>50)
(E)-t-BuCH=C(H)HgCl and $PrSO_2$Na (1.2)	t-BuOH/Me_2SO, S, 30 h	t-BuCH=CHSO$_2$Pr	75 (>50)
(E)-t-BuCH=C(H)HgOAc and RSNa (1.2), R = Ph; $PhCH_2$	PhH, S, 17 h	t-BuCH=CHSR	100 (>50); 97 (>50)
(E)-EtC(OAc)=C(Et)HgCl and PhSNa (1.2)	PhH, S, 11 h	EtC(OAc)=C(Et)SPh	92
(Z)-HOCH$_2$C(Cl)=C(H)HgCl and PhSNa (1.2)	PhH, S, 5 h	HOCH$_2$C(Cl)=CHSPh	61
R_2C=C(H)SnBu$_3$ and PhSSPh (1.2), R = H; Me; Ph	PhH, S, 4 h; 2 h; 2 h	R_2C=CHSPh	91; 97; 93
R_2C=C(H)SnBu$_3$ and PhSeSePh (1.2), R = H; CH_3	PhH, S, 24 h	R_2C=CHSePh	0; 0
(E)-PhCH=C(H)SnBu$_3$ and RSSR (1.2), R = Ph; $PhCH_2$	PhH, S, 4 h; 10 h	PhCH=CHSR	86; 85
(E)-PhCH=C(H)SnBu$_3$ and RSO_2Y (1.2), R, Y = Ph, Cl; p-tolyl, SePh	PhH, S, 4 h	PhCH=CHSO$_2$R	88 (>50); 84 (>50)
R_2C=C(H)SnBu$_3$ and $PhSO_2$Cl (1.2), R = H; Me; Ph	PhH, S, 4 h	R_2C=CHSO$_2$Ph	89; 90; 76
(a)-MeO$_2$CCH=C(H)SnBu$_3$ and PhSSPh (1.6), a = E; Z	PhH, R, 8 h	MeO$_2$CCH=CHSPh	79 (3.8); 91 (3.7)
(a)-MeO$_2$CCH=C(H)SnBu$_3$ and $PhSO_2$Cl (1.6), a = E; Z	PhH, R, 10 h	MeO$_2$CCH=CHSO$_2$P	68 (>50); 76 (>50)
(E)-PhCH=C(H)SnBu$_3$ and Cl$_3$CSO$_2$Cl (1.5)	PhH, S, 24 h	PhCH=CHCCl$_3$	48 (>50)
(a)-ClCH=C(H)HgCl and PhSSPh (0.5), a = E; Z	Me_2SO, R, 6 h	ClCH=CHSPh PhSCH=CHSPh	20 (2.5); 16 (0.9) 40 (0.6); 32 (0.7)
MeC≡CCH$_2$SnPh$_3$ and RSO_2Cl (1), R = Ph; Pr	PhH, S, 35 h	CH_2=C=C(Me)SO$_2$R	42; 31
HC≡CCH$_2$SnPh$_3$ and RSO_2Cl (1), R = Ph; Pr	PhH, 70°C, AIBN, 8 h	CH_2=C=CHSO$_2$R CH_2=C=CHSnPh$_3$	36; 31 36; 43
CH_2=CHCH$_2$SnBu$_3$ and RSSR (1), R = Ph; $PhCH_2$	PhH, S, 6 h	CH_2=CHCH$_2$SR	85; 77
MeCH=CHCH$_2$SnBu$_3$ and RSO_2Cl (1), R = Ph; Pr	PhH, S, 9 min; 3 min	CH_2=CHCH(Me)SO$_2$R	72; 46

[a]   3-10 mmol of substrate in 10 ml of solvent, S = 275 W fluorescent sunlamp ca. 20 cm from Pyrex reaction vessel; R = 350 nm Rayonet Photoreactor.

[b]   Based on 2 mol of product/mol of mercurial.

In the reaction between 1-alkenylmercury halides and RSO_2^- or RS^-, the chain reaction involves three steps, (a) electron transfer to $ClHg^\bullet$ to generate $RSO_2^\bullet$ or $RS^\bullet$, (b) addition of $RSO_2^\bullet$ or $RS^\bullet$ α to the mercury atom in the vinylmercurial and (c) β-elimination of $ClHg^\bullet$. These same types of reactions are involved in the $S_{RN}1$ reactions of aliphatic or aromatic substrates with appropriate anions. However, in $S_{RN}1$ processes, the sequence of the three reactions is different in that radical addition to the anion is followed by electron transfer to the substrate followed by a decomposition (elimination) reaction of the substrate radical anion.

Additions of $PhSO_n^\bullet$ to double and triple bonds are reversible. However, competitive reactions according to Scheme 5 with PhSSPh or $(PhS)_2Hg$ as the source of $PhS^\bullet$ gave a consistent series of relative reactivities for a variety of tin and mercury derivatives.[5,17] Possibly in the adduct radical $>\overset{\bullet}{C}-CH(SPh)(MX_n)$, the elimination of $^\bullet MX_n = Bu_3Sn^\bullet$ or $^\bullet HgCl$ occurs more rapidly than the elimination of $PhS^\bullet$. In fact with $(E)/(Z)$ isomers of $ClC(H)=C(H)HgCl$ stereospecificity (retention) is observed in the free radical substitution of HgCl by $t-Bu^\bullet$, $c-C_6H_{11}^\bullet$ or $PhS^\bullet$. With (E) and (Z) $MeO_2CCH=CHSnBu_3$ stereospecificity is observed with $c-C_6H_{11}^\bullet$ but not with $t-Bu^\bullet$, $PhS^\bullet$ or $PhSO_2^\bullet$.[7] Relative reactivities in $PhS^\bullet$ substitution for a metal atom in PhH or Me_2SO at 35°C are measured to be: $PhC\equiv CSnBu_3$ (0.008), $CH_2=C(H)SnBu_3$ (0.05), $CH_2=CHCH_2SnBu_3$ (0.13), $Me_2C=C(H)SnBu_3$ (0.60), $(E)-PhCH=C(H)SnBu_3$ (2.4), $(E)-PhCH=C(H)HgCl$ (5.2), $Ph_2C=C(H)SnBu_3$ (6.0), $Ph_2C=C(H)HgCl$ (7.0). Towards $PhS^\bullet$ or $c-C_6H_{11}^\bullet$ the reactivities of 1-alkynyl derivatives of tin or mercury are much less reactive than the 1-alkenyl derivatives while the 1-alkynyl iodides or sulfones are more reactive or of comparable reactivity with the corresponding 1-alkenyl derivatives (Table 5).

Table 5.[5,8,17] Relative Reactivity Data towards $PhS^\bullet$ and $c-C_6H_{11}^\bullet$ in Me_2SO or PhH at 35°C, Addition to C_β

Substrate	Relative Reactivity $PhS^\bullet$	$(c-C_6H_{11})^\bullet$
$Ph_2C=CHI$	1.0	1.0
$(E)-PhCH=CHI$	—	0.7
$PhC\equiv CI$	0.23	3.8
$Ph_2C=CHSO_2Ph$	0.4	6.4
$(E)-PhCH=CHSO_2Ph$	—	3.3
$PhC\equiv CSO_2Ph$	1.2	13
$Ph_2C=CHSnBu_3$	6	0.8
$(E)-PhCH=CHSnBu_3$	2.4	0.7
$PhC\equiv CSnBu_3$	0.008	0.2
$Ph_2C=C(H)HgCl$	7.0	1.9
$(E)-PhCH=C(H)HgCl$	5.2	1.5
$(PhC\equiv C)_2Hg$	—	0.2[a]
$Ph_2C=CH_2$	—	3.5[b]
$PhCH=CH_2$	$2 \times 10^7 \ M^{-1} s^{-1}$[c]	4.5[b]
$PhC\equiv CH$	$8 \times 10^5 \ M^{-1} s^{-1}$[c]	0.45[b]

[a] Per molecule.

[b] $t-Bu^\bullet$, relative to $Ph_2C=CHI$, ref. 11.

[c] absolute rate constants; O. Ito, R. Omori, and M. Matsuda, *J. Am. Chem. Soc.* 104:3934 (1982).

Towards the 1-alkynyl derivatives PhC≡CX the reactivity towards either c-C_6H_{11}• or PhS• decreases from X = SO_2Ph to I to SnR_3 or HgR. This indicates that towards alkynes both c-C_6H_{11}• and PhS• behave as nucleophilic radicals, $R^+X\ C{\overset{\bullet}{=}}C$-Ph (R = c-C_6H_{11} or PhS). However, towards the 1-alkenyl derivatives, PhCH=CHX or Ph_2C=CHX the relative reactivities towards c-C_6H_{11}• decrease from X = $PhSO_2$ to I ≈ Bu_3Sn ≈ HgCl, but towards PhS• the relative reactivities increase from X = I ≈ $PhSO_2$ to Bu_3Sn ≈ HgCl. Towards 1-alkenyl derivatives c-C_6H_{11}• behaves as a nucleophilic radical with possible perturbations from steric effects. However, PhS• now appears to be an electrophilic radical with a low reactivity for X = I or SO_2Ph and a high reactivity for X = Bu_3Sn or HgX, i.e., the transition state is stabilized by PhS⁻ XCH $\overset{\bullet}{\underset{}{\pm}}$ CHPh. The net effect of c-C_6H_{11}• reacting as an electron donor, but PhS• reacting as either a donor (alkynes) or acceptor (alkenes) results in a complex relationship when the relative reactivities of substituted alkenes and alkynes are compared for the two radicals, Table 6, although in general electron-withdrawing groups increase the reactivity of the alkyne relative to the alkene and electron-supplying groups have the reverse effect for both radicals.

Table 6. Relative Reactivities of
PhC≡CX/Ph_2C=CHX at 35°C

X	Relative Reactivity	
	PhS•	c-C_6H_{11}•
$PhSO_2$	8	4
I	0.6	5
H	0.04	0.1
Bu_3Sn	0.003	0.3
Hg(II)	—	0.07

ACKNOWLEDGMENT

Financial support has been provided by the National Science Foundation and the donors to the Petroleum Research Fund.

REFERENCES

1.   G.A. Russell, H. Tashtoush, *J. Am. Chem. Soc.* 105:1398 (1983).
2.   G.A. Russell, H. Tashtoush, and P. Ngoviwatchai, *J. Am. Chem. Soc.* 106:4622 (1984).
3.   G.A. Russell and P. Ngoviwatchai, *P. Org. Chem.* 54:1836 (1989).
4.   G.A. Russell and J. Hershberger, *J. Am. Chem. Soc.* 102:7603 (1980).
5.   G.A. Russell and P. Ngoviwatchai, *Tetrahedron Lett.* 27:3479 (1986).
6.   G.A. Russell and L.L. Harold, *J. Org. Chem.* 50:1037 (1985).
7.   G.A. Russell and P. Ngoviwatchai, *Tetrahedron Lett.* 26:4975 (1985).
8.   G.A. Russell, P. Ngoviwatchai, H.I. Tashtoush, A. Pla-Dalmau, and R.K. Khanna, *J. Am. Chem. Soc.* 110:3530 (1988).
9.   G.A. Russell, W. Jiang, S.S. Hu, and R.K. Khanna, *J. Org. Chem.* 51:5498 (1986).
10. B. Giese, *Angew. Chem. Int. Ed. Engl.* 24:553 (1985).
11. G.A. Russell, *NATO ASC Series C* 257:13 (1989).
12. G.A. Russell, P. Ngoviwatchai, and Y.W. Wu, *J. Am. Chem. Soc.* 111:4921 (1989).
13. R.L. Smorada and W.E. Truce, *J. Org. Chem.* 44:3444 (1979).

14a. T.E. Boothe, J.L. Greene, Jr., and P.B. Shevlin, *J. Am. Chem. Soc.* 98:951 (1976).

  b. T.E. Boothe, J.L. Greene, Jr., P.B. Shevlin, M.R. Willcott, III, R.R. Inners, and A. Cornelius, *J. Am. Chem. Soc.* 100:3874 (1978).

  c. T.E. Boothe, J.L. Green, Jr., and P.R. Shevlin, *J.Org. Chem.* 45:794 (1980).

15. N. Ono, A. Kamimura, and A. Kaji, *J. Org. Chem.* 52:5111 (1987).

16a. P.J. Wagner, J.H. Sedon, and M.J. Lindstrom, *J. Am. Chem. Soc.* 100:2579 (1978);

  b. P.J. Wagner, M.J. Lindstrom, J.H. Sedon, and D.R. Ward, *J. Am. Chem. Soc.* 103:3842 (1981).

17. G.A. Russell, P. Ngoviwatchai, H. Tashtoush, and J. Hershberger, *Organometallics* 6:1414 (1987).

THE INVENTION OF RADICAL CHAIN REACTIONS

OF VALUE IN ORGANIC SYNTHESIS

Francesco Minisci

Dipartimento di Chimica
Politecnico di Milano
20133 Milano, Italy

The title of this lecture was suggested by Professor Barton, who has brought a very relevant contribution to the organic synthesis by this approach, particularly with sulphur derivatives and, in fact, should have been the lecturer on this subject. Since Professor Barton, unfortunately, was unable to participate at this meeting, I have the ungrateful task to substitute for such a scientific master. Maintaining just the title is, however, easier since it covers also my own research activities.

This synthetic approach, actually, involves most of the aspects of the free-radical chemistry. However, considering that the main objective of this NATO Institute is the state-of-the-art in research on sulphur derivatives and sulphur-centered reactive intermediates, the lecture will be limited to free-radical reactions in this field.

Particular emphasis will be focussed on synthesis involving some recent free-radical reactions with the S=O and C=S functional groups, neglecting more classical reactions, such as those of thiyl and sulphonyl radicals, well documented in the literature.

Selective syntheses by free-radical reactions become important when effective chain processes can be achieved. This effectiveness is mainly determined by the kinetic features of the steps involved (rate constants and kinetic length of the chain). Two main chain processes are generally of value in organic synthesis: i) redox chains, in which a catalyst is continuously regenerated in redox processes with the reagents and the intermediate radicals, ii) free-radical chains in which free radicals are involved in all the steps of the chain. The invention of new radical chain reactions can, therefore, be based on the knowledge of the rate constants of the elementary steps involved. This knowledge is much more important in free-radical than in ionic-reactions, since rate constants for free-radical reactions are rarely affected macroscopically by the reaction medium, contrary to the behavior of ionic reactions. The invention can be also based on more qualitative concepts, such as the enthalpic factor (mainly the influence of the strength of the bonds involved), the polar factor, the redox features of the reagents, the intermediate radicals, the nature of the catalyst etc.

Since the free radicals generally are very reactive species and the rate constants for all the possible competitive reactions are often not available, it is usually the combination of both quantitative kinetic data and qualitative concepts of free-radical reactivity which contribute to the invention of effective chain processes of synthetic value.

Sulfur-Centered Reactive Intermediates in Chemistry and Biology, Edited by
C. Chatgilialoglu and K.-D. Asmus, Plenum Press, New York, 1990

Free-radical reactions of synthetic value involving the functional groups C=S and S=O are known for very long time already, although their mechanistic nature often was incorrectly or not at all recognized. Thus the syntheses of aryl xanthates[1] (eq. 1) and aryl-sulphonic acids[2] (eq. 2) from diazonium salts with potassium xanthate and SO_2, respectively, are known for almost a century.

$$ArN_2^+ \; + \; {}^-SCSOEt \quad \rightarrow \quad Ar\text{–}SCSOEt \; + \; N_2 \tag{1}$$

$$ArN_2^+ \; + \; SO_2 \; + \; H_2O \quad \xrightarrow{Cu^+} \quad ArSO_3H \; + \; N_2 \; + \; H^+ \tag{2}$$

These reactions can now be well explained by a free-radical and redox chain process. The synthesis of xanthate (eq. 1) may, in fact, turn into an explosion unless suitable experimental conditions are maintained. This can be related to a fast free-radical chain (eqs. 3 and 4) due to the high nucleophilicity (high reducing character) of the intermediate radical anion adduct

$$Ar^\bullet \; + \; \underset{\underset{S^-}{|}}{S{=}C}\text{–}OEt \quad \rightarrow \quad Ar\text{–}S\text{–}\underset{\underset{S^-}{|}}{\overset{\bullet}{C}}\text{–}OEt \tag{3}$$

$$Ar\text{–}S\text{–}\underset{\underset{S^-}{|}}{\overset{\bullet}{C}}\text{–}OEt \; + \; ArN_2^+ \quad \rightarrow \quad Ar\text{–}S\text{–}\underset{\underset{S}{\|}}{C}\text{–}OEt \; + \; Ar\text{–}N{=}N^\bullet \tag{4}$$

$$\downarrow$$

$$Ar^\bullet \; + \; N_2$$

Initiation can be achieved via formation and homolysis of the diazoxanthate, Ar–N=N–SCSOEt. Very effective are also redox processes initiated by traces of, e.g., I^-, which support the chain character. The synthesis of the sulphonic acids (eq. 2), for example, is explained by an effective redox chain (Scheme 1)

$$ArN_2^+ \; + \; Cu^+ \quad \rightarrow \quad Ar^\bullet \; + \; N_2 \; + \; Cu^{2+}$$

$$Ar^\bullet \; + \; SO_2 \quad \rightarrow \quad ArSO_2^\bullet$$

$$ArSO_2^\bullet \; + \; Cu^{2+} \; + \; H_2O \quad \rightarrow \quad ArSO_3H \; + \; Cu^+ \; + \; H^+$$

Scheme 1

Obviously the authors of these syntheses could not yet speculate about radical mechanisms at the time being.

Many years later, but still considerable time ago, I have shown[3] the radical nature of reactions similar to those of eqs. 1 and 2. Using alkyl and aryl radicals generated from per-oxides the general character of the reactions of carbon-centered radicals with xanthates or SO_2 could be demonstrated.

The redox decomposition of cyclohexanone peroxide in the presence of SO_2 and potassium xanthate leads, in fact, to sulphonic acids and alkylxanthates according to the eqs. 5 and 6.

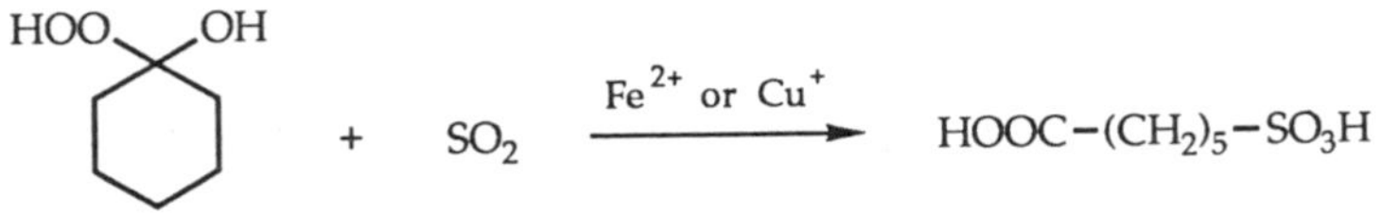

$$\text{(cyclohexane ring with HOO and OH substituents)} \; + \; SO_2 \quad \xrightarrow{Fe^{2+} \; \text{or} \; Cu^+} \quad HOOC\text{–}(CH_2)_5\text{–}SO_3H \tag{5}$$

HOO_OH (cyclohexane structure) + ⁻SCSOEt $\xrightarrow{Fe^{2+} \text{ or } Cu^+}$ $HOOC-(CH_2)_5-SCSOEt$ + OH^- (6)

I have explained the formation of the sulphonic acid by a redox chain (eqs. 7, 8 and 9) quite similar to that of Scheme 1.

HOO_OH (cyclohexane) + Fe^{2+} → Fe^{3+} + OH^- + •O_OH (cyclohexane)   (7)

↓

$$HOOC-(CH_2)_4-CH_2^\bullet$$

$HOOC-(CH_2)_4-CH_2^\bullet$ + SO_2 → $HOOC-(CH_2)_5-SO_2^\bullet$ (8)

$HOOC-(CH_2)_5-SO_2^\bullet$ + Fe^{3+} + H_2O → $HOOC-(CH_2)_5-SO_3H$ + Fe^{2+} + H^+ (9)

The addition of carbon-centered radicals, generated e.g. by hydrogen abstraction, onto sulphurdioxide has extensively been employed in the chemical industry for many years now. Recently carbon-centered radicals generated photochemically,[4] from organocobalt precursors,[5] or N-hydroxy-2-thiopyridone esters[6] (Scheme 2), have been successfully utilized.

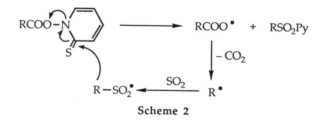

Scheme 2

The relatively high rate constant (4×10^5 M^{-1} s^{-1} at 25°C)[7] for the addition of ethyl radicals to SO_2 is consistent with these results.

I have related the formation of alkyl xanthates (eq. 6) to our finding,[3,8] that the Sandmeyer reaction of diazonium salts, whose radical mechanism was first suggested by Waters,[9] was a particular case of quite general redox chains of carbon-centered radicals (alkyl and aryl) with halogen or pseudo-halogen ligands of Fe(III) or Cu(II) salts (eq. 10).

HOO_OH (cyclohexane) + X^- $\xrightarrow{Fe^{2+} \text{ or } Cu^+}$ $HOOC-(CH_2)_5-X$ (10)

X = Cl, Br, I, N_3, SCN, SCSOEt etc.

Accordingly, I have explained[3,8] also reaction (6) by a ligand-transfer mechanism in a redox chain involving eqs. 7 and 11.

$HOOC-(CH_2)_4-CH_2^\bullet$ + FeX^{2+} → $HOOC-(CH_2)_5-X$ + Fe^{2+} (11)

Recent kinetic studies of Ingold and coworkers which indicate the fast addition of carbon-centered radicals to the C=S bond, and the numerous synthetic results of Barton and coworkers involving the radical addition to the C=S bond now suggest that the most probable mechanism involves the redox chain characterized by eqs. 7, 12, and 13.

$$HOOC-(CH_2)_4-CH_2{}^\bullet \;+\; S=C-OEt \;\rightarrow\; HOOC-(CH_2)_5-S-\overset{\bullet}{C}-OEt \tag{12}$$
$$\underset{S^-}{|} \qquad\qquad\qquad\qquad \underset{S^-}{|}$$

$$HOOC-(CH_2)_5-S-\overset{\bullet}{\underset{|}{C}}-OEt \;+\; Fe^{3+} \;\rightarrow\; HOOC-(CH_2)_5-S-\overset{\parallel}{\underset{S}{C}}-OEt \;+\; Fe^{2+} \tag{13}$$
$$\qquad\qquad\quad S^-$$

I have obtained quite similar results for chain processes involving the addition of aryl radicals, generated from arylhydrazone peroxides (eq. 14), to SO_2 and potassium xanthate[8]

$$Ar-N=N-\overset{R}{\underset{\underset{OOH}{|}}{C{\sim}R}} \;+\; Fe^{2+} \;\rightarrow\; Fe^{3+} \;+\; OH^- \;+\; Ar-N=N-\overset{R}{\underset{\underset{O^\bullet}{|}}{C{\sim}R}} \tag{14}$$

$$\downarrow$$
$$Ar^\bullet \;+\; N_2 \;+\; R_2CO$$

Radical addition to S=O and C=S bonds is therefore quite an old story.

In the following I will discuss the most recent synthetic developments, in which the S=O and C=S functional groups act as mediators in generating free radicals useful for selective syntheses. The recent years have brought a rapid development in the use of alkyl radicals for the formation of C–C bonds and in the synthesis of target molecules.[10] The availability of general, simple, cheap and selective sources of alkyl radicals is therefore of great synthetic interest.

DMSO AS MEDIATOR FOR THE GENERATION OF FREE RADICALS

We have developed a general and simple source of alkyl radicals, useful for selective syntheses, from alkyl iodides by using DMSO as mediator.[11] This involves hydroxyl radicals and, in some way paradoxally, these very reactive and unselective radical species could be utilized for developing highly selective syntheses even of complex molecules.

The invention of effective redox chains was based, in this case, on the knowledge of rate constants for the various steps involved in the chain and of equilibrium constants for a system descrbing iodine abstraction by the methyl radical from alkyl iodides. Most rate constants for the reaction of hydroxyl radicals with organic and inorganic compounds (hydrogen abstraction, addition to unsaturated groups, oxidation etc.) are in the range of 10^7 - 10^{10} $M^{-1}s^{-1}$ and it could appear unwise to utilize such a reactive radical for selective syntheses. However, applying the kinetic data of Asmus,[12] and Norman and Gilbert[13] we could show that the problem of the high reactivity, and consequently low selectivity, of the hydroxyl radical can be overcome by using DMSO as solvent. Hydroxyl radicals are easily obtained, for example, through the well-known redox decomposition of H_2O_2 by Fe(II) salt (eq. 15). They react very fast with DMSO (eq. 16) and thus the yield of competitive, and possibly unselective reactions with other substrates is minimized owing to the excess of the solvent. This includes $^\bullet OH$ reaction with alkyl iodides[14] (eq. 17) utilized as source of alkyl radicals.

$$H_2O_2 \;+\; Fe^{2+} \;\rightarrow\; {}^\bullet OH \;+\; OH^- \;+\; Fe^{3+} \tag{15}$$

306

$$\bullet OH \ + \ MeSOMe \ \rightarrow \ \underset{Me-S-Me}{\overset{HO \diagdown \diagup O\bullet}{}} \qquad\qquad k = 7 \times 10^9\,M^{-1}s^{-1} \qquad\qquad (16)$$

$$R-I \ + \ \bullet OH \ \rightarrow \ R-\overset{\bullet}{I}-OH \qquad\qquad k > 10^9\,M^{-1}s^{-1} \qquad\qquad (17)$$

The $\bullet OH$ radical adduct to DMSO undergoes a fast β-fission acting as selective source of methyl radicals (eq. 18).

$$\underset{Me-S-Me}{\overset{HO \diagdown \diagup O\bullet}{}} \ \rightarrow \ Me^\bullet \ + \ MeSO_2H \qquad\qquad k = 1.5 \times 10^7\,s^{-1} \qquad\qquad (18)$$

Now we have utilized the rate and equilibrium constants investigated by Griller[15] for the iodine abstraction from alkyl iodides by the methyl radical (eq. 19) in order to develop a general source of alkyl radicals.

$$R-I \ + \ Me^\bullet \ \rightleftarrows \ R^\bullet \ + \ MeI \qquad\qquad \overrightarrow{k} > 10^6\,M^{-1}s^{-1} \qquad\qquad (19)$$

The iodine abstraction is fast and successfully competes with most of other possible reactions of the methyl radical: it occurs according to the equilibria reported in Table 1.

Table 1. Equilibrium constants[15] for
$$Me^\bullet + R-I \ \rightleftarrows \ Me-I + R^\bullet$$

R	K
Et	20.1
i-Pr	468
t-Bu	1.7×10^4

Based on these rate and equilibrium data we have developed new effective and general methods of homolytic alkylation via Fe^{2+} catalyzed redox chains in $RI/H_2O_2/DMSO$ systems.[11]

A general method of alkylation of heteroaromatic bases is shown by eq. 20.

$$+ \ RI \ + \ MeSOMe \ + \ H_2O_2 \ \xrightarrow{Fe^{2+}} \qquad + \ MeI \ + \ MeSO_2H \ + \ H_2O \qquad (20)$$

Some results for a variety of heteroaromatic bases and alkyl iodides are summarized in Table 2. This type of reaction has successfully been applied also to complex molecules, such as iodosugars,[11] which lead to C-nucleosides, interesting for their biological activity.

A complex, but selective redox chain is working in reaction 20. Eqs. 15, 16, 18, and 19 are involved in the initial steps of the chain. The knowledge of the rate constants for the addition of the alkyl radicals to protonated heteroaromatic bases (eq. 21) and the high oxidizability of the pyridinyl type radical intermediates (eq. 22) have contributed to the invention of effective redox chains.[16]

Table 2. Alkylation of Heteroaromatic Bases by Alkyl Iodides, H_2O_2 and DMSO[11c]

Heteroaromatic base	Alkyl iodide	Orientation (Substitution site)	Conversion[a] (%)	Yields[b] (%)
Lepidine	i-propyl	2	90	88
"	i-butyl	2	93	91
"	c-hexyl	2	88	92
"	n-propyl	2	75	82
"	n-butyl	2	73	78
"	t-butyl	2	98	86
Quinaldine	i-propyl	4	95	96
"	c-hexyl	4	94	97
"	n-butyl	4	72	81
Quinoline	i-propyl	2(25%), 4(36%) 2,4(39%)	97	94
"	i-butyl	2(27%), 4(38%) 2,4(35%)	95	92
"	c-hexyl	2(23%), 4(33%) 2,4(44%)	98	91
"	n-propyl	2(36%), 4(39%) 2,4(25%)	78	82
"	n-butyl	2(40%), 4(43%) 2,4(17%)	75	77
"	t-butyl	2	96	87
Isoquinoline	i-propyl	1	86	88
"	c-hexyl	1	79	84
Acridine	i-propyl	9	78	92
"	c-hexyl	9	83	96
4-Cyanopyridine	i-propyl	2(67%), 2,6(33%)	86	95
"	c-hexyl	2(62%), 2,6(38%)	92	93
"	t-butyl	2(58%), 2,6(42%)	96	94
4-Acetylpyridine	i-propyl	2(68%), 2,6(32%)	85	95
4-Methylpyridine	c-hexyl	2	35	99
Pyrazine	c-hexyl	2	45	78
Quinoxaline	c-hexyl	2(68%), 2,3(32%)	86	81
Benzothiazole	c-hexyl	2	38	87

[a] % of converted heteroaromatic base
[b] Yield based on the converted base

$$\text{(structure: N-protonated pyridine)} + R^\bullet \xrightarrow{k} \text{(radical adduct)} \rightarrow \text{(product)} + H^+ \qquad k \approx 10^6 - 10^8 \, M^{-1} s^{-1} \qquad (21)$$

$$\text{(dihydropyridyl radical)} + Fe^{3+} \rightarrow \text{(substituted pyridine)} + Fe^{2+} + H^+ \qquad (22)$$

The fact that the equilibria of eq. 19 are shifted to the right is not in itself a sufficient condition to achieve a high selectivity because the reaction rates of the Me$^\bullet$ and R$^\bullet$ radicals can be quite different. When the enthalpic factor governs the reactivity, as in the iodine abstraction (eq. 19), the methyl radical is more reactive than primary, secondary, tertiary and in general α-substituted alkyl radicals; this may counterbalance unfavorable equilibria. The radical source becomes selective in the heteroaromatic substitution because the polar effects are dominant. The rates for alkyl radical addition to protonated heterocyclic rings are strongly affected by the nucleophilic character of the radicals, and generally all alkyl radicals without electron-withdrawing groups in α-position are more nucleophilic than the methyl radical. Thus, in the combined system of eqs. 19 and 21 equilibrium 19 is governed by the enthalpic factor, whereas the addition of the alkyl radicals to the protonated heterocyclic ring (eq. 21) is controlled by the polar factor. Both factors work in the same direction rendering R$^\bullet$ the more favoured radical compared with Me$^\bullet$ in equilibrium 19 (for enthalpic reasons) and also more reactive in its reaction with the heterocyclic compounds (for polar reasons). The overall result is a highly selective substitution.

For the same reasons discussed with the heteroaromatic bases this radical source is equally effective for the alkylation of quinones (eq. 23) in a similar redox chain.[11b]

$$\text{(benzoquinone)} + RI + MeSOMe + H_2O_2 \xrightarrow{Fe^{2+}} \text{(alkylquinone)} + MeI + MeSO_2H + H_2O \qquad (23)$$

Another class of electron-deficient organic compounds, diazonium salts, are effectively alkylated by nucleophilic alkyl radicals, which add to diazonium group in a fast reaction leading to a diazo coupling reaction (eq. 24)[11a]

$$\overset{+}{Ar-N{\equiv}N} + RI + MeSOMe + H_2O_2 + 2\,Fe^{2+} \rightarrow$$

$$\rightarrow Ar-N{=}N{-}R + MeI + MeSO_2H + 2\,Fe^{3+} + OH^- \qquad (24)$$

In this case a redox chain does not occur. Owing to its high electrophilic character the radical adduct (eq. 25) is quantitatively reduced to the azoderivative (eq. 26) by the Fe(II) salt in a fast stoichiometric process.

$$\overset{+}{Ar-N{\equiv}N} + R^\bullet \rightarrow \overset{\bullet+}{Ar-N{=}N{-}R} \qquad\qquad k \approx 10^6 - 10^8 \, M^{-1}s^{-1} \qquad (25)$$

$$\overset{\bullet+}{Ar-N{=}N{-}R} + Fe^{2+} \rightarrow Ar-N{=}N{-}R + Fe^{3+} \qquad (26)$$

When the alkyl iodide bears electron-withdrawing groups in α-position, iodine abstraction by methyl radical (eq. 27) is still favored for polar as well as enthalpic reasons. However, the resulting radicals now exhibit electrophilic character and do not react with electron-deficient substrates, such as protonated heteroaromatic bases or diazonium salts. If the polar character of the radical is reversed, it is necessary to also reverse the polarity of the substrate, i.e. to use electron-rich substrates (aromatics or olefins) in order to reestablish favorable thermodynamic and kinetic conditions for selective syntheses. Thus, in systems containing iodoacetic acid favorable conditions are achieved by using, for example, anisole (eq. 28) or olefins (eq. 29). In this case the radical $^\bullet CH_2COOH$ is present in large excess in its equilibrium with the methyl radical (eq. 27), and it is more reactive than the methyl radical with respect to addition to the aromatic ring or to the olefinic bond.[11]

$$Me^\bullet + I\text{-}CH_2COOH \rightleftarrows MeI + {}^\bullet CH_2COOH \tag{27}$$

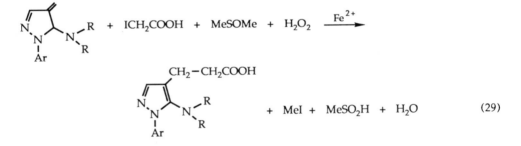

$$\text{o } 77\ \%,\ \text{p } 18.5\ \%,\ \text{m } 4.5\ \%$$

A question, which arises for this general source of alkyl radicals is why a small difference in the C–I bond energies (56.5 kcal/mol for Me–I and 52.4 kcal/mol for t-Bu–I) goes along with such a big difference in selectivity. This fact is striking if we compare the iodine abstraction with the hydrogen abstraction from C–H bonds, for which larger differences of energy of the C–H bonds are reflected in smaller selectivity changes.

Under complete thermodynamic control this behavior would be explained by the fact that small variations in bond strengths are completely exploited. However, for processes under kinetic control thermochemistry will only partially be reflected in the transition states, as shown by the Evans-Polanyi ($E_a \cong \Delta H^0 + C$) and similar relationships. A reasonable explanation therefore is that thermochemistry is reflected in the transition states of the iodine abstraction to a greater extent than in the transition states of the hydrogen abstraction.

An alternative explanation, based on an addition-elimination process, (eq. 31), would rely on only circumstantial evidence, namely that hydroxyl radicals, for example, undergo fast addition to the alkyl iodides (eq. 17); however, the hydroxyl radical is quite a particular species and it can be dangerous to extrapolate its behavior to carbon-centered radicals.

310

$$R-I + R'^{\bullet} \rightarrow [R \cdots I \cdots R']^{\ddagger} \rightarrow R^{\bullet} + I-R' \tag{30}$$

$$R-I + R'^{\bullet} \underset{k_{-1}}{\overset{k_1}{\rightleftarrows}} R-\overset{\bullet}{I}-R' \underset{k_{-2}}{\overset{k_2}{\rightleftarrows}} R^{\bullet} + I-R' \tag{31}$$

Whenever the rate of iodine abstraction is in the diffusion controlled range, as with aryl radical (R' = aryl), it becomes less meaningful to distinguish between the mechanism of eqs. 30 and 31 because the fast addition of the aryl radical to the iodine atom has to occur practically simultaneously to the breaking of the C–I bond in the alkyl iodide. The overall rate for the iodine transfer is comparatively lower, however, for R' being methyl or, in general, an alkyl radical. Considering this and the other kinetic constants involved in eq. 31, the selectivity could be determined by the homolysis of the C–I bonds of the intermediateradical adduct, $R-\overset{\bullet}{I}-R'$.

N-HYDROXY-2-THIOPYRIDONE AND ANALOGOUS DERIVATIVES AS MEDIATORS FOR THE GENERATION OF FREE RADICALS

A quite general approach to the generation and synthetic application of carbon and other radicals is based on the decarboxylation of carboxylic acids by using N-hydroxy-2-thio-pyridone as mediator, through the formation of the corresponding thiohydroxamic esters (Barton reaction).[17]

Actually, a variety of even more simple and cheap methods of oxidative decarboxylation of carboxylic acids are available for the generation of carbon radicals. However, the invention of the N-hydroxy-2-thiopyridone method is of particular synthetic use for two reasons:

i) Given the mildness of the reaction conditions the method can find wide use in manipulation of complex and often fragile natural products.

ii) Effective free-radical chains take place.

Two factors contribute to the effectiveness of the free-radical chains: the fast addition of the radicals to the C=S bond (eq. 32) and the low energy of the N–O bond, which suffers fast β-fission of the radical adduct (eq. 33).

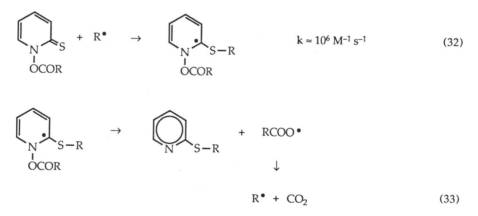

$$k \approx 10^6 \ M^{-1} \ s^{-1} \tag{32}$$

$$R^{\bullet} + CO_2 \tag{33}$$

Ingold and coworkers[18] have shown that the addition of free radicals to the C=S bond is generally a fast reaction. More recently, Newcomb[19] has reported a value of $1.9 \times 10^6 \ M^{-1}s^{-1}$ at 40°C for the rate of addition of the n-octyl radical to the esters of N-hydroxy-2-thiopyridone. The β-fission of the radical adduct (eq. 33) is fast, not only for the intrinsic low energy that, in general, characterizes N–O bonds, but also because the homolysis of the N–O bond in this case leads to the rearomatization of the pyridinyl radical.

The synthetic utilization of this general radical source is related to the availability of radical traps suitable for the interception of carbon radical intermediates (Schemes 3 and 4) by chain processes in competition with the radical chain of eqs. 32 and 33.

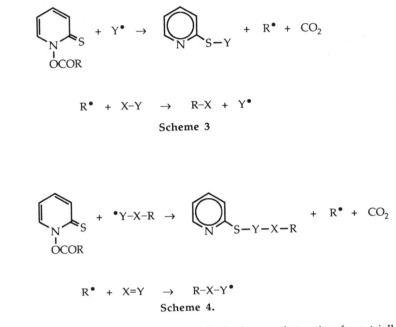

$$R^{\bullet} + X\text{-}Y \quad \rightarrow \quad R\text{-}X + Y^{\bullet}$$

Scheme 3

$$R^{\bullet} + X{=}Y \quad \rightarrow \quad R\text{-}X\text{-}Y^{\bullet}$$

Scheme 4.

A variety of radical traps proved to be successful: hydrogen abstraction from trialkyl tin hydride[17] (X–Y = H–SnR$_3$), halogen abstraction from polyhalomethanes[17] (X–Y = Br–CCl$_3$, Cl–CCl$_3$, I–CHI$_2$), reaction with oxygen[17] (R–X–Y$^{\bullet}$ = R–O–O$^{\bullet}$), addition to unsaturated derivatives[17] (olefins, aromatics, quinones, R–X–Y$^{\bullet}$ = R–Ċ-Ċ$^{\bullet}$), to SO$_2$, (Scheme 2),[6] to isocyanides[20] (R–X–Y$^{\bullet}$ = R–C$^{\bullet}$=NR).

Two main synthetic limitations concern the general procedures of Schemes 3 and 4. To be successful the carbon radical must react faster with the radical trap than with the thiopyridone in order to minimize the chain process of eqs. 32 and 33. This is the case, for example, with oxygen as radical trap. Since the rate of the radical addition to the C=S bond of the thiopyridone is relatively high (10^6 M^{-1}s^{-1}) the reactivity of several of the other frequently utilized radical traps (R$_3$SnH, CCl$_4$, olefins, aromatics, SO$_2$ etc.) with carbon radicals occur with rates of the same order of magnitude. In these cases it is necessary to apply an excess of the radical trap in order to minimize the processes of eqs. 32 and 33 and also to increase the selectivity of the processes of Schemes 3 and 4.

The second synthetic limitation is related to the fact that, according to Scheme 4, the reaction product is always a thiopyridinyl derivative, in which the pyridinyl group still needs to be removed. Usually this can be achieved by a reductive process.

The stereochemical aspects of these reactions are of particular interest. A high retention of configuration was observed[21] for reactions falling into Scheme 4 and eqs. 32 and 33 by utilizing the monoester of the ketal of tartaric acid (Scheme 5), disproving once again the general assumption that free-radical reactions cannot be stereospecific.

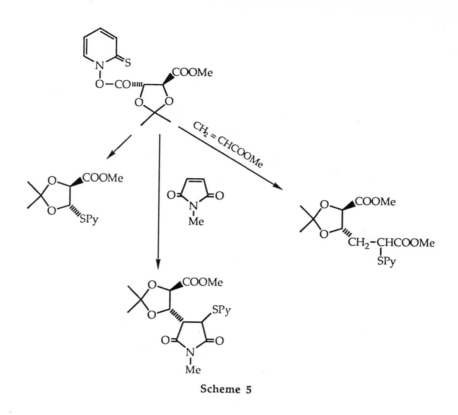

Scheme 5

High stereospecificity is achieved for radicals[22] generated from uronic esters of 2-thio-pyridone which add to electron-poor alkenes to yield carbon-4 functionalized chain-elongated furanosides and D-ribonucleosides. Newcomb has shown that the Barton method can be used to generate nitrogen-centered radicals.[33] Irradiation of carbamates in acidic medium with a tungsten lamp leads to amino radical cations, which can undergo intramolecular processes according to Scheme 6.

The amino radical cations, due to their electrophilic character, are interesting reactive species because of the exceptional selectivity in their reactions with subsitituted alkanes, olefins, acetylenes and aromatics.[24]

Unfortunately, a general competitive process (eq. 34) limits the synthetic applications under conditions where a sufficiently strong acidic medium cannot be used.[25]

$$R\text{--}CH_2\text{--}\overset{+}{\underset{\bullet}{N}}H\text{--}R \rightarrow R\text{--}\underset{\bullet}{C}H\text{--}NH\text{--}R + H^+ \tag{34}$$

Thus, olefins which are not deactivated by electron-withdrawing groups, and aromatic compounds activated by strongly electron-releasing groups give good results for inter- or intra-molecular additions under relatively weak acidic conditions owing to almost diffusion controlled rates.[24] The synthetically useful reactions with the somewhat less reactive alkanes and non-activated aromatics (inter- and intramolecular) require a medium of much higher acidity in order to overcome the competitive eq. 34 and are, therefore, unsuitable for the Barton method.

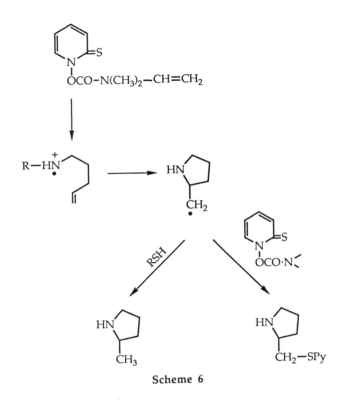

Scheme 6

Alkoxy radicals[26] can also be obtained according to the Barton method by chain processes from N-alkoxypyridine-2-thione. Thus, when the N-cyclopentyloxypyridine-2-thione was heated with $BrCCl_3$ and AIBN (α,α'-azo-bis-isobutyronitrile) in benzene, the sole product (>95%) was 5-bromo-pentanol. Its formation is consistent with a chain mechanism involving the generation of the cyclopentyloxy radical by attack of $^{\bullet}CCl_3$ to pyridinethione (eq. 35), followed by β-fission of the cyclopentyloxy radical (eq. 36), bromine abstraction and regeneration of $^{\bullet}CCl_3$ (eq. 37).

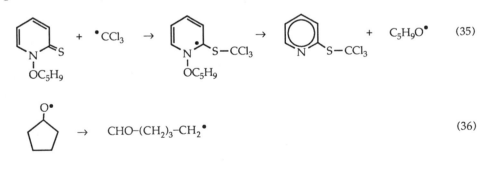

$$CHO\text{-}(CH_2)_3\text{-}CH_2{}^{\bullet} + Br\text{-}CCl_3 \rightarrow CHO\text{-}(CH_2)_4\text{-}Br + {}^{\bullet}CCl_3 \qquad (37)$$

The Barton-McCombie[27] deoxygenation of primary and secondary alcohols by means of the reaction of their dithiocarbonate ester derivatives with trialkyltin hydride (Scheme 7) falls in the general picture of chain processes involving the C=S group. There have been spectacular synthetic applications[17,27,28] of this reaction, which illustrate the exceptional tolerance of other function groups.

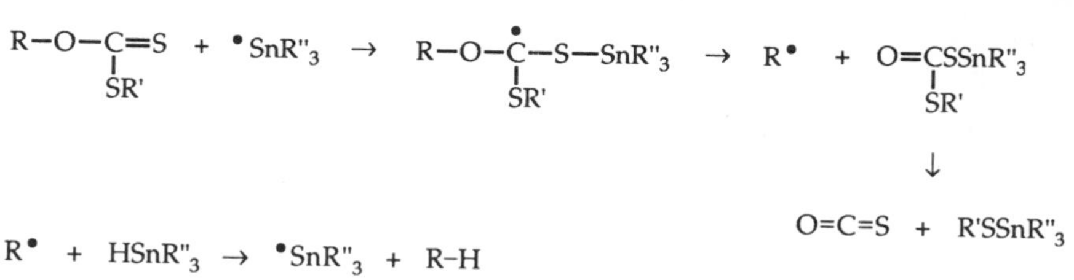

$$R-O-\underset{\underset{SR'}{|}}{C}=S \ + \ {}^\bullet SnR''_3 \ \rightarrow \ R-O-\underset{\underset{SR'}{|}}{\overset{\bullet}{C}}-S-SnR''_3 \ \rightarrow \ R^\bullet \ + \ O=\underset{\underset{SR'}{|}}{C}SnR''_3$$

$$\downarrow$$

$$O=C=S \ + \ R'SSnR''_3$$

$$R^\bullet \ + \ HSnR''_3 \ \rightarrow \ {}^\bullet SnR''_3 \ + \ R-H$$

Scheme 7

An alternative mechanism was suggested by Beckwith[29] (Scheme 8).

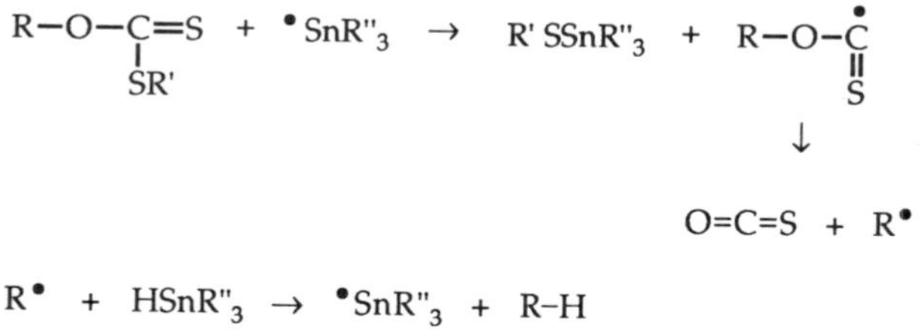

$$R-O-\underset{\underset{SR'}{|}}{C}=S \ + \ {}^\bullet SnR''_3 \ \rightarrow \ R'SSnR''_3 \ + \ R-O-\underset{\underset{S}{\|}}{\overset{\bullet}{C}}$$

$$\downarrow$$

$$O=C=S \ + \ R^\bullet$$

$$R^\bullet \ + \ HSnR''_3 \ \rightarrow \ {}^\bullet SnR''_3 \ + \ R-H$$

Scheme 8.

Bachi[30] has provided an elegant support of the Barton mechanism using an alkenyl-dithio-carbonate, a probe reacting via cyclization (Scheme 9).

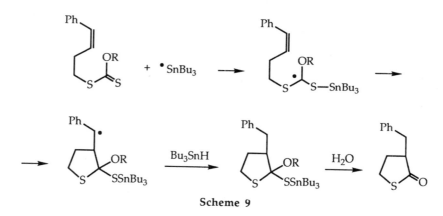

Scheme 9

It has, however, been suggested[31] that the Bachi results could also be explained by the Beck-with mechanism, involving a complex addition-elimination-addition process (synchronous or by discrete steps) (Scheme 10); but this clearly appears to be an "Occam razor", considering in particular the increasing evidence for relatively high rates for the addition of free radicals to the C=S bond and the fact that hydrogen transfer from Bu_3SnH to carbon radicals (very likely also to the alkoxythiocarbonyl radical) is faster than the intermolecular addition of the thiocarbonyl radical to an internal olefin.

Acyl xanthates are also useful precursors of acyl and alkyl radicals in photochemically initiated chain processes[32] (Scheme 11). When the decarbonylation of the acyl radicals (eq. 38) is fast, as with phenylacetyl or pivaloyl radicals, the corresponding alkyl radicals can be trapped in similar chain processes.

$$R-\overset{\bullet}{C}O \ \rightarrow \ R^\bullet \ + \ CO \tag{38}$$

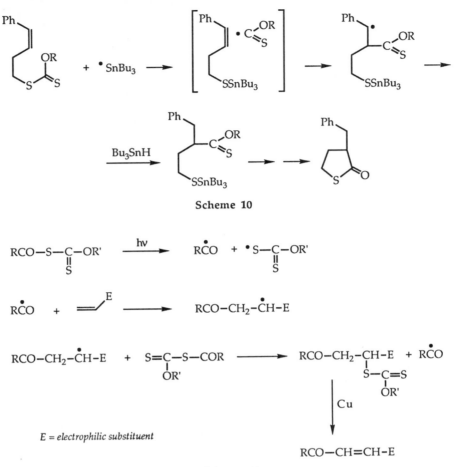

Scheme 10

E = *electrophilic substituent*

Scheme 11

In conclusion, the synthetic success of all the free radical chains discussed above has been related, at least from a qualitative point of view, to the presence of what Barton has called a "disciplinary group", which forces the radical chains into fixed, selective directions. For a long time most of the synthetic developments have preceeded quantitative kinetic investigations which, as we have seen, contribute such much to the rationalization of the results. However, it becomes more and more frequent to utilize quantitative kinetic data determined beforehand for the invention of new selective chain processes of value in organic synthesis.

REFERENCES

1. R. Leuckart, *J. Prakt. Chem.* 41:187 (1890).
2. L. Landsberg, *Chem. Ber.* 23.I:454 (1890).
3. F. Minisci, *Ital. Pat.* 580012 (17/5/1957); *Angew. Chem.* 70:599 (1958); *Gazz. Chim. Ital.* 89:626 (1958), 90:2439 (1959).
4. R.M. Wilson and S.W. Wunderly, *J. Am. Chem. Soc.* 96:7350 (1974).

5 a. P. Bongeard, M.D. Johnson, and G. Lampan, *J. Chem. Soc., Perkin Trans. I.* 849 (1982),

   b. M.R. Ascheroft, P. Bongeard, A. Burry, C.J. Cooksey, M.D. Johnson, J.M. Hungerford, and G.M. Lampman, *J. Org. Chem.* 49:1751 (1984);

   c. V.F. Patel and G. Pattenden, *Tetrahedron Lett.* 28:145 (1987).

6. D.H.R. Barton, B. Lacher, B. Misterkiewiez, and S.Z. Zard, *Tetrahedron* 44:1153 (1988).

7. A. Good and J.C.J. Thynne, *Trans. Farad. Soc.* 63:2708, 2720 (1967).

8. F. Minisci, *Gazz. Chim. Ital.* 89:1910, 1922, 1941, 2428 (1959); 90:1307 (1960); *Rend. Accad. Lincei 8* 24:538 (1958); *Acc. Chem. Res.* 8:165 (1975).

9. W.A. Waters, *J. Chem. Soc.* 266 (1942).

10 a. B. Giese, "Radicals in Organic Synthesis: Formation of Carbon-Carbon Bonds", J.E. Baldwin, ed., Pergamon Press, Oxford (1986).

   b. "Free Radicals in Synthesis and Biology", F. Minisci, ed., Kluwer Acad. Publ., Dordrecht (1989).

11. E. Minisci, E. Vismara, and F. Fontana

   a. *Tetrahedron Letters* 28:6373 (1987), 29:1975 (1988); ref. 10b, p. 53.

   b. "Paramagnetic Organometallic Species in Activation, Selectivity, Catalysis", M. Chanon, ed., Kluwer Acad. Publ., Dordrecht, pp. 29 (1989).

   c. *J. Org. Chem.*, 54:5224 (1989).

12. D. Veltwisch, E. Janata, and K.-D. Asmus, *J. Chem. Soc., Perkin Trans. 2* 146 (1980).

13. B.C. Gilbert, R.O.C. Norman, and R.C. Sealy, *J. Chem. Soc., Perkin Trans. 2* 303 (1975).

14. K.-D. Asmus, E. Anklam, and H. Mohan, *in* "Proceeding of the Fifth International Symposium on Organic Free Radicals", H. Fischer, ed., Springer Verlag, pp. 3 and references therein (1988).

15. J.A. Hawari, J.M. Kanabus-Kaminska, D.D. Wayner and D. Griller, "Substituent Effects in Radical Chemistry", H.G. Viehe, ed., D. Reidel Publ. Co., Dordrecht, pp. 81 (1986).

16. F. Minisci, E. Vismara, and F. Fontana, *Heterocycles* 28:489 (1989).

17 a. D.H.R. Barton and S.Z. Zard, *Pure and Appl. Chem.* 58:675 (1986); ref. 15, p. 443,

   b. D.H.R. Barton and N. Ozbalik, ref. 11b, p. 1.

18 a. J.C. Scaiano and K.U. Ingold, *J. Am. Chem. Soc.* 98:4727 (1976);

   b. J.C. Scaiano, J.P. Tremblay, and K.U. Ingold, *Can. J. Chem.* 54:3407 (1976);

   c. D. Forrest, K.U. Ingold, and D.H.R. Barton, *J. Phys. Chem.* 81:915 (1977);

   d. D. Forrest and K.U. Ingold, *J. Am. Chem. Soc.* 100:3868 (1978).

19. M. Newcomb and J. Kaplan, *Tetrahedron Lett.* 28:1615 (1987).

20. D.H.R. Barton, N. Ozbalik, and B. Vacher, *Tetrahedron* 44:3501 (1988).

21. D.H.R. Barton, A. Gateau-Olesker, S.D. Gero, B. Lacher, C. Tachdjian, and S.Z. Zard, *J. Chem. Soc., Chem. Commun.* 1790 (1987).

22. D.H.R. Barton, S.D. Gero, B. Quillet-Sire, and M. Somadi, *J. Chem. Soc., Chem. Commun.* 1372 (1988).

23. M. Newcomb and T.M. Deeb, *J. Am. Chem. Soc.* 109:3163 (1987).

24. F. Minisci, ref. 15, p. 391.

25. S. Auricchio, M. Bianca, A. Citterio, F. Minisci, and S. Ventura, *Tetrahedron Lett.* 3373 (1984).

26. A.L.J. Beckwith and B.P. Hay, *J. Am. Chem. Soc.* 110:4415 (1988).

27 a. D.H.R. Barton and S.W. McCombie, *J. Chem. Soc., Perkin Trans. 1* 1574 (1975);

   b. D.H.R. Barton, D. Crick, A. Löbberding, and S.Z. Zard, *J. Chem. Soc., Chem. Commun.* 646 (1985); *Tetrahedron* 42:2329 (1986).

28. W. Hartwig, *Tetrahedron* 39:2609 (1983).

29. P.J. Barker and A.L.J. Beckwith, *J. Chem. Soc., Chem. Commun.* 683 (1984).

30. M.D. Bachi and E. Bosch, *J. Chem. Soc., Perkin Trans. 1* 1517 (1988).

31. D. Crich, *Tetrahedron Lett.* 29:5805 (1988).

32. P. Delduc, C. Tailham, and S.Z. Zard, *J. Chem. Soc., Chem. Commun.* 308 (1988), ref. 10b, p. 263.

RECENT ADVANCES IN THE CHEMISTRY OF COMPOUNDS CONTAINING

S–SI MOIETIES

Marco Ballestri, Chryssostomos Chatgilialoglu, Pasquale Dembech,
Andrea Guerrini and Giancarlo Seconi

I. Co. C. E. A.
Consiglio Nazionale delle Ricerche
40064 Ozzano Emilia (Bologna), Italy

The majority of radical reactions of interest to synthetic chemistry are chain processes in which radicals are generated by some initiation process, and which then undergo a series of propagation steps generating fresh radicals and finally disappear by combination or disproportionation. A synthetically useful radical chain reaction should require as little radical initiator as possible and should form few side products.[1] This is possible only if the chain propagating radical meets certain conditions of reactivity and selectivity; that is (i) the selectivities of the radical involved in the chain must differ from one other and (ii) the reaction between radicals and non radicals must be faster than radical combination reactions.[1]

Probably the best known free radical reactions are the reductions of a variety of organic substrates by tributyltin hydride. A review on the use of tributyltin hydride as a reagent in organic chemistry via radicals has recently appeared.[2] The reactions consist of at least two-step chain processes, viz.,

$$R^\bullet \;+\; Bu_3SnH \;\rightarrow\; RH \;+\; Bu_3Sn^\bullet \tag{1}$$

$$Bu_3Sn^\bullet \;+\; RZ \;\rightarrow\; [R\overset{\bullet}{Z}SnBu_3] \;\rightarrow\; Bu_3SnZ \;+\; R^\bullet \tag{2}$$

where $[R\overset{\bullet}{Z}SnBu_3]$ is a reactive intermediate or a transition state. Of great importance are the chain reactions in which the intermediate carbon centered radical, $R^\bullet$, forms a C–C bond inter- or intramolecularly prior to reaction with tin hydride. Indeed, in his recent review on the radical reactions of interest in organic synthesis,[3] Curran dedicated almost half of the several hundred citations to C–C bond formation via the "tin-method".

In certain cases, where the formation of a C–C bond is slow relative to reaction of alkyl radicals with tin hydride, a slightly less active hydrogen donor that can nevertheless fulfil the other requirements of these chain processes could provide a very useful alternative to the usual tin hydride. Trialkylgermanium hydrides for example are less reactive hydrogen donors than tin hydrides, while the corresponding germyl radicals are at least as reactive as the stannyl radicals in reactions with organic substrates.[4] However, due mainly to the elevated cost of germanium hydrides, their properties have been obtained by working with low concentrations of tin hydrides.[1]

Sulfur-Centered Reactive Intermediates in Chemistry and Biology, Edited by
C. Chatgilialoglu and K.-D. Asmus, Plenum Press, New York, 1990

319

Notwithstanding the widespread utility of tin hydrides, there are certain limitations to their use; that is, the disadvantages of the tin hydrides include the high toxicity of triorganotin compounds and the difficulty for complete elimination of the tin compounds from the desired products. Therefore, the replacement of tin and germanium compounds by cheaper and ecologically acceptable hydrogen transfer agents are required. One of us recently showed that tris(trimethylsilyl)silane, $(Me_3Si)_3SiH$, can be used in place of tributyltin hydride as a radical-based reducing agent[5] including the reaction which the $(Me_3Si)_3SiH$ is used as mediator for the formation of C–C bond either intermolecularly or intramolecularly.[6]

We have continued to explore the use of silanes as radical-based reducing agent in order to obtain compounds with increased selectivity and/or reactivity. As the kinetic and thermo-dynamic factors that control propagation steps in chain reactions are fundamental ones for the invention of new free radical reducing agents, the purpose of this article is to report some of the available data concerning radicals which contain a S–Si moiety and rationalize them from the point of view of the reducing ability of their parent compounds. This chapter has been divided into two sections. In the first one we will deal with compounds containing S–Si–H group whereas in the second section the Si–S–H functionality will be considered.

S–Si–H FUNCTIONALITY

In order to form the basis of discussion, the bond dissociation energies (BDE) of some group 14 hydrides with the absolute rate constants of their reactions with *tert*-butoxyl radicals (eq. 3) are collected in Table 1.

$$R_3MH + t\text{-}BuO^\bullet \rightarrow R_3M^\bullet + t\text{-}BuOH \tag{3}$$

The factors influencing the strength of silicon-hydrogen bonds are somehow different from those which control carbon-hydrogen bond dissociation energies. For example, (i) the weakening effect of the C–H bonds by successive substitution of alkyl groups at the C–H function is absent in the silicon congeners and (ii) the effect produced by phenyl substitution at the Si–H function, in analogy with C–H, is small. In other words, a striking feature of a large class of silanes is the almost constant silicon-hydrogen bond strength of ca. 90 kcal mol^{-1}.[7] Furthermore, the reactivity of these silanes towards *tert*-butoxyl radicals is similar, i.e. $(6\pm3) \times 10^6$ M^{-1} s^{-1}. However, it has been recently shown that successive substitution of silyl or thiyl groups at the Si–H function can have a profound effect on the silicon hydrogen bonds and on the reactivity of these silanes towards *tert*-butoxyl radicals (see Table 1).[8,9] Since silyl radicals are generally more reactive than germyl and stannyl radicals with

Table 1. Bond Dissociation Energies and Related Kinetic Data for Some Group 14 Hydrides

R_3MH	BDE, kcal mol^{-1}	k_3, M^{-1} s^{-1}
Et_3Si–H	90.1	4.6×10^6
Cl_3Si–H	91.3	4.0×10^7
$(MeS)_3Si$–H	82.5	4.4×10^7
$(Me_3Si)_3Si$–H	79.0	1.0×10^8
Bu_3Ge–H	82-84	8.0×10^7
Bu_3Sn–H	74.0	2.2×10^8

organic substrates, either in atom abstraction or in addition to multiple bonds, it followed that trialkylthiosilanes would be capable of sustaining radical chain reductions analogous to reactions 1 and 2.

Good to excellent yields of reduction products from a variety of organic substrates were obtained using trialkylthiosilanes.[10] Salient results are presented in Table 2 for $(MeS)_3SiH$. Evidence in favour of a free radical mechanism (Scheme 1) was provided by the following observations: The reactions were catalyzed by thermal and photochemical sources of free radicals and retarded by inhibitors such as 2,6-di-*tert*-butyl-4-methyl-phenol.

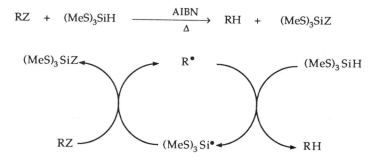

$$RZ \ + \ (MeS)_3SiH \ \xrightarrow[\Delta]{AIBN} \ RH \ + \ (MeS)_3SiZ$$

AIBN = azoisobutyronitrile

Scheme 1

Table 2. Reduction of Organic Substrates by $(MeS)_3SiH$ [a]

RZ	Yield $RH,$[b] %
$CH_3(CH_2)_{14}CH_2-Br$	94
$CH_3(CH_2)_{14}CH_2-I$	97
$c\text{-}C_6H_{11}-N=C$	92
$c\text{-}C_6H_{11}-OC(S)SMe$	97
$c\text{-}C_6H_{11}-SePh$	98

a   General procedure: a solution containing the compound to be reduced, trimethyliosilane (1.2 eq.) and a catalytic amount of AIBN in toluene was heated at 75 - 90 °C for 30 - 60 minutes and then analysed by GC.

b   Yields were quantified by GC using tetradecane as an internal standard and based on formation of RH.

Thus, trialkylthiosilanes are effective reducing agents of organic substrates and it would appear to offer a cheaper alternative to trialkylgermanium hydrides. As we mentioned in the introduction, tributyltin hydride has been frequently employed in systems where the initially formed organic radical must first undergo a unimolecular or intramolecular reaction with the formation of a carbon-carbon bond if the desired product is to be obtained (cf. Scheme 2, where M = Sn). When the latter process is slow relative to reduction by tin hydride, a slightly less active hydrogen donor could provide a very useful alternative. We suggest that trialkylthiosilanes have the necessary chemical properties to meet these requirements. That is, silyl radicals react with rates similar to or even faster than those of $Bu_3Sn^\bullet$ radicals while $(MeS)_3SiH$ is only ca. 1/12 as good as Bu_3SnH as an H-atom donor to primary alkyl radicals.

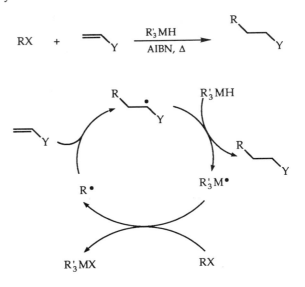

AIBN = azoisobutyronitrile

Scheme 2

Si–S–H FUNCTIONALITY

It is well known that thiols are rather good H-atom donors towards alkyl radicals even though sulfur-hydrogen bonds in the alkanethiols are relatively strong (see Table 3).[11] It is believed that "polar effects" in the transition state play an important role in these reactions. However, alkanethiyl radicals are very poor atom-abstracting agents and therefore do not support chain reactions analogous to reactions 1 and 2. On the contrary, trialkylsilyl radicals are the most reactive species known for halogen atom abstraction and addition to multiple bonds, whereas the corresponding trialkylsilanes are rather poor H atom donors towards alkyl radicals and tend therefore not to support chain reactions except under forcing conditions.

Table 3. Bond Dissociation Energies and Related Kinetic Data from Some Compounds Containing Heteroatom-hydrogen Bonds

Compound	BDE, kcal mol^{-1}	$k_{RCH_2}{}^\bullet$, M^{-1} s^{-1}
$R_3Sn–H$	74.0	7.5×10^6
$R_3Si–H$	90.1	4.0×10^2
$RS–H$	88.4	8.0×10^6

Recently, Roberts and his coworkers[12] reported that trialkylsilane reduces alkyl halides to corresponding hydrocarbons in the presence of alkanethiols, which act as polarity reversal catalysts for hydrogen transfer from the silane to the alkyl radical. The reaction consist in three-step chain process as shown in Scheme 3.

$$RX \ + \ Et_3SiH \ \xrightarrow[\text{TBHN, }\Delta]{\text{R'SH}} \ RH \ + \ Et_3SiX$$

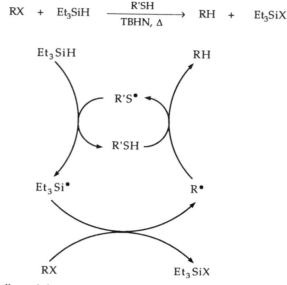

TBHN = di-*tert*-butylhyponitrite

Scheme 3

The above considerations suggest that any class of compounds with an appropriate molecular arrangement that allow the transformation of a thiyl radical to a silyl radical via an intramolecular rearrangement will potentially be a good reducing agent. In 1968, Pitt and Fowler found that trisilanethiols readily undergo a radical-induced skeletal rearrangement, providing the first examples of 1,2 shifts of the silyl and disilanyl group which occur in solution under mild conditions, viz.,[13]

$$\equiv Si \diagdown \underset{/\ |}{Si}-S^\bullet \ \longrightarrow \ ^\bullet \underset{/\ |}{Si}-S \diagup^{Si\equiv} \tag{6}$$

Tris(trimethylsilyl)silylthiol,[14] which we have prepared in an one-pot reaction from the corresponding silane in an 95% overall yield, viz.,

$$(Me_3Si)_3SiH \ \xrightarrow{CCl_4} \ (Me_3Si)_3SiCl \tag{7}$$

$$(Me_3Si)_3SiCl \ \xrightarrow{NH_2} \ (Me_3Si)_3SiNH_2 \tag{8}$$

$$(Me_3Si)_3SiNH_2 \ \xrightarrow{H_2S} \ (Me_3Si)_3SiSH \tag{9}$$

When heated at 80 °C in toluene containing azobisisobutyronitrile (AIBN) the silylthiol rearranges to an isomer in about 30 minutes. The identification of the rearranged product as well as the observation that the reaction was retarded by inhibitors such as duroquinone are in agreement with the Pitt and Fowler findings (cf. Scheme 4):

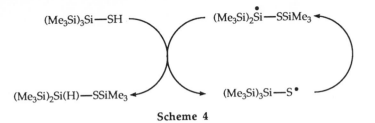

$(Me_3Si)_3Si—SH$

$(Me_3Si)_2\overset{\bullet}{Si}—SSiMe_3$

$(Me_3Si)_2Si(H)—SSiMe_3$

$(Me_3Si)_3Si—S^{\bullet}$

Scheme 4

When alkyl halides were introduced into the reaction mixture the corresponding hydrocarbons were obtained in good to excellent yield.[14] Salient results are presented in Table 4. We also considered that the rearranged product, $(Me_3Si)_2Si(H)SSiMe$, might be a good hydrogen donor and that it would be capable of substaining radical chain reactions in a manner similar to tris(trimethylsilyl)silane. Our expectation turned out to be correct. The reduction yields are also presented in Table 4. There is evidence that the reduction of alkyl bromides and iodides by $(Me_3Si)_3SiSH$ proceeds via a three-step radical chain mechanism, as shown in Scheme 5.[14]

Table 4. Reduction of Organic Halides by $(Me_3Si)_3SiSH$ and $(Me_3Si)_2Si(H)SSiMe$ [a]

RX	$(Me_3Si)_3SiSH$ conversion RH,[b] %	$(Me_3Si)_2Si(H)SSiMe$ conversion RH,[b] %
$CH_3(CH_2)_{14}CH_2I$	99	100
$CH_3(CH_2)_{14}CH_2Br$	96	91
$CH_3(CH_2)_{14}CH_2Cl$	22	19

a   General procedure: a solution, containing the halide, the appropriate reducing agent (1.2 eq.) and a catalytic amount of AIBN in toluene (3%) was heated at 80 °C for 30 minutes and then analysed by GC.

b   Conversions were quantified by GC using tetradecane as an internal standard and based on formation of RH.

$$RX \ + \ (Me_3Si)_3SiSH \ \xrightarrow[\Delta]{AIBN} \ RH \ + \ (Me_3Si)_2Si(X)SSiMe_3$$

$(Me_3Si)_2Si(X)SSiMe_3$

$(Me_3Si)_3SiSH$

$R^{\bullet}$

RH

RX

$(Me_3Si)_2\overset{\bullet}{Si}SSiMe_3$

$(Me_3Si)_3SiS^{\bullet}$

AIBN = azoisobutyronitrile

Scheme 5

324

ACKNOWLEDGEMENT

We thank Progetto Finalizzato Chimica Fine II (CNR, Rome) for the finanical support.

REFERENCES

1. B. Giese, "Radicals in Organic Synthesis: Formation of Carbon-Carbon Bonds",
    Pergamon Press, Oxford (1986).
2. W.P. Newmann, *Synthesis* 665 (1987).
3. D.P. Curran, *Synthesis* 417 and 489 (1988).
4. J. Lusztyk, B. Maillard, S. Deycard, D.A. Lindsay, and K.U. Ingold, *J. Org. Chem.*
    52:3509 (1987) and references therein.
5. C. Chatgilialoglu, D. Griller, and M. Lesage, *J. Org. Chem.* 53:3641 (1988).
6. B. Giese, B. Kopping, and C. Chatgilialoglu, *Tetrahedron Lett.* 30:681 (1988).
7. C. Chatgilialoglu, *in*: "Free Radicals in Synthesis and Biology", F. Minisci, ed.,
    Kluwer, Dordrecht, pp. 115 (1989).
8. J.M. Kanabus-Kaminska, J.A. Hawari, D. Griller, and C. Chatgilialoglu, *J. Am. Chem.*
    Soc. 109:5287 (1987).
9. C. Chatgilialoglu and D. Griller, manuscript in preparation.
10. C. Chatgilialoglu, A. Guerrini, and S. Seconi, manuscript in preparation.
11. M. Newcomb, A.G. Glenn, and M.B. Manek, *J. Org. Chem.* 54:4603 (1989).
12. R.P. Allen, B.P. Roberts, and C.R. Willis, *J. Chem. Soc., Chem. Commun.* 1387 (1989).
13. C.G. Pitt and M.S. Fowler, *J. Am. Chem. Soc.* 90:1928 (1968).
14. M. Ballestri, C. Chatgilialoglu, and G. Seconi, manuscript in preparation.

REACTION KINETICS OF SULFUR-CENTERED RADICALS

David Griller[*] and José A. Martinho Simões [§]

[*]Division of Chemistry, National Research Council of Canada
Ottawa, Ontario, Canada, K1A OR6

[§] Centro de Química
Estrutural, Insituto Superior Técnico
1096 Lisboa Codex, Portugal

Sulfur-centered radicals are involved in coal and oil chemistry and play important roles in atmospheric and biological processes and yet very little kinetic information is available that describes these reactions in any detail. Figure 1 shows a bar chart that describes the state of the literature in 1984 and shows the amount of space dedicated to the kinetics of radical processes in the comprehensive Landolt-Börnstein compendium.[1] It is quite clear that there is very little information on sulfur centered radicals as compared with their carbon and oxygen-centered analogues and one must wonder why this is the case.

A simple answer to this question is that sulfur-centered radicals are not as important as one might expect and that the literature reflects this. We would argue that experimental difficulties are responsible for the paucity of kinetic data surrounding these reactive intermediates. For example, one of the most basic reactions of alkanethiyl radicals - olefin addition - is generally reversible,[2-10] which makes kinetic analysis of the process exceedingly difficult.

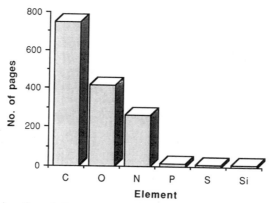

Fig. 1. Plot showing the relative amounts of kinetic data in solution available in 1984 for organic radicals centered on various elements.

Sulfur-Centered Reactive Intermediates in Chemistry and Biology, Edited by
C. Chatgilialoglu and K.-D. Asmus, Plenum Press, New York, 1990

Studies on species as fundamental as alkanethiyl radicals have been particularly difficult because these radicals are hard to detect spectroscopically. As a consequence, time resolved methods cannot be easily applied in the study of their reaction kinetics. For example, electron paramagnetic resonance (EPR) spectra have often been incorrectly assigned to alkanethiyl radicals when they were actually due to perthiyl ($RSS^\bullet$) species.[11] The latter were formed by sulfur-carbon bond cleavage in disulfides or by photolysis of tri- and tetrasulfides that are common impurities in commercially available samples of disulfides. Symons has pointed out[12] that the degeneracy of the p-π orbitals in alkanethiyl radicals and the consequent angular momentum about the molecular z-axis would give rise to extremely broad EPR spectra. In fact, sulfur centered radicals can only be detected by this technique when the degeneracy is lifted as is the case for perthiyl radicals[11] and arenethiyl ($ArS^\bullet$) radicals.[13] Hydrogen bonding to alkanethiyl radicals is too weak to lift the degeneracy to a sufficient extent, although this mechanism does allow the spectra of alkoxyl radicals to be detected.[12]

Again, alkanethiyl radicals illustrate the problems of optical detection by pulse radiolysis or flash photolysis techniques since they have optical spectra[14] in regions where most of their precursors are strongly absorbing. Indeed, the spectra that are normally detected by the use of these techniques are due to perthiyl radicals[11,15] that result from sulfur-carbon bond cleavage in disulfides, eq. 1, or to species that result from subsequent reactions of the thiyl radicals. For example, disulfide radical anions are formed when thiyl radicals are generated in the presence of thiolate anions,[16-21] or sulfuranyl radicals[22,23] may be formed by the reactions of alkanethiyl radicals with disulfides, eq. 2,3. These species have absorption spectra in the range 300 - 500 nm and are therefore readily detected. In some instances, they have been used to probe the lifetimes of alkanethiyl radicals. However, kinetic analysis is complicated by the fact that the reactions are often reversible.

$$RSS{-}R \qquad \rightarrow \qquad RSS^\bullet \ + \ R^\bullet \qquad\qquad (1)$$

$$RS^\bullet \ + \ RS^- \qquad \rightleftharpoons \qquad (RS{\cdot}^\bullet{\cdot}SR)^- \qquad\qquad (2)$$

$$RS^\bullet \ + \ RSSR \qquad \rightleftharpoons \qquad R\overset{\bullet}{S}(SR)_2 \qquad\qquad (3)$$

In contrast to alkanethiyl radicals, arenethiyl radicals have absorption spectra in the 300 to 500 nm range and are therefore easily detected by flash photolysis and pulse radiolysis techniques.[24-31] As a consequence, they have been studied in great detail.

While flash photolysis techniques have been of limited use in the study of simple alkanethiyl radicals in solution, they have been used to advantage in the gas phase. Both $^\bullet SCH_3$ and $^\bullet SCD_3$ have been generated by laser flash photolysis of the corresponding disulfides and have been monitored by following their laser induced fluorescence.[32] While the technique shows considerable promise, complications have already been identified that were due to the fluorescence of reaction products at the monitoring wavelength.[33]

In this review, no attempt will be made to provide a comprehensive compilation of rate constants for reactions involving sulfur-centered radical intermediates since these are often unreliable. Instead we will attempt a critical analysis of some basic reaction kinetics illustrating what is good and why, and what can be improved. In this way, we hope to alert readers to possible pitfalls in the published literature and to establish some directions for future research.

SELF-REACTIONS

Among the most basic processes involving sulfur centered radicals are self-reactions as illustrated in eq. 4 for alkanethiyl radicals. Many of the rate constants for self-reactions

$$RS^{\bullet} + RS^{\bullet} \rightarrow RS\text{--}SR \tag{4}$$

have been measured by kinetic EPR spectroscopy and some typical examples are shown in Table 1. The data indicate that all of the reactions are diffusion controlled processes. While a case could be made that the self reaction of $t\text{-BuS}^{\bullet}\text{O}$ is less than the diffusion controlled limit, experimental uncertainties (vide infra) would tend to undermine this hypothesis.

Measurements of $2k_4$ for sulfonyl radicals, $RS^{\bullet}O_2$, that were made using pulse radiolysis techniques have been criticized for their lack of accuracy.[38,40] However, it is equally likely that the kinetic EPR results are fairly unreliable because the measurements are strongly dependent upon the homogeneity of radical concentration and the accuracy of signal integration, both of which are extremely difficult to optimize. Indeed, values of $2k_4$ for several sulfonyl radicals that were measured by the kinetic EPR approach were found to be independent of the size of the radicals concerned which, at first sight, runs contrary to the notion that the reactions are diffusion controlled. In fact, the results most probably reflect the rather high experimental errors associated with the measurements.

Few product studies have been carried out on self-reactions although it would seem highly likely that for $RS^{\bullet}$ and $RSS^{\bullet}$ the products are simply dimers of the radicals. However, for most substituted arenesulfonyl radicals the reactions are more complex and lead initially to an unstable sulfinyl sulfonate that decomposes to form $ArS^{\bullet}O$ and $ArS^{\bullet}O_3$ radicals. Of these, the $ArS^{\bullet}O$ species have been detected by EPR spectroscopy.

$$Ar\overset{\bullet}{S}O_2 \rightleftarrows Ar\text{-}\overset{O}{\underset{O}{\overset{\|}{S}}}\text{-}O\text{-}\overset{O}{\underset{}{\overset{\|}{S}}}\text{-}Ar \rightarrow Ar\overset{\bullet}{S}O + Ar\overset{\bullet}{S}O_3 \tag{5}$$

$$\downarrow$$

non-radical products

Table 1. Rate Constants for the Self-Reactions of Some Sulfur Centered Radicals in Solution Measured by Kinetic EPR Spectroscopy

Radical	$2k_4$, $M^{-1} s^{-1}$	Temperature, K	Ref.
$BuS^{\bullet}$[a]	6×10^{11}	298	4, 34, 35
$PhS^{\bullet}$[b]	1×10^{9}	296	36
$t\text{-Bu}\overset{\bullet}{S}O$	6×10^{7}	173	37
$CH_3\overset{\bullet}{S}O_2$	5×10^{9}	223	38
$Ph\overset{\bullet}{S}O_2$	5×10^{9}	223	38
$t\text{-BuSS}^{\bullet}$	2×10^{8}	190	39

[a] In the gas phase.
[b] Estimated using diffusion theory.

The mechanism of this reaction has been thoroughly established.[41,42] However, in the kinetic analysis that was used to probe the system no account was taken of the fact that $ArS^\bullet O_2$ must react with the two other radical species in the system. The same criticisms apply to a study[4] of the self-reactions of t-BuSS$^\bullet$ where the radicals were generated via reaction 6. Since a highly reactive chlorine atom is also formed, solvent derived radicals must have participated in the reaction, eq. 7, yet no account was taken of this possibility.

$$t\text{-BuSSCl} \quad \rightarrow \quad t\text{-BuSS}^\bullet \ + \ Cl^\bullet \tag{6}$$

$$Cl^\bullet \ + \ \text{Solvent} \quad \rightarrow \quad HCl \ + \ \text{Solvent(–H)} \tag{7}$$

While most studies of sulfur radical self-reactions have some deficiencies they do establish that the reactions are diffusion controlled. It is therefore likely that the best way to obtain reliable data is to calculate the results by using the "von Smoluchowski" approach,[36] eq. 8. In this equation, D is the diffusion coefficient for an individual radical, r is the reaction diameter, N is Avogadro's number and σ is the spin statistical factor.

$$2k_4 = 8\,\pi \cdot \sigma \cdot \rho \cdot N \cdot D \tag{8}$$

The diffusion coefficient D for the radical is assumed to be equal to that of its parent hydride which can generally be measured using a Taylor diffusion apparatus, and the reaction diameter can be estimated using the molar volumes of the parent hydrides as a model. For most simple radicals, the spin statistical factor that is used in eq. 8 is 1/4. When two doublet radicals meet in a solvent cage the probability that the radical pair will exist in the singlet state that is required for reaction is only 1/4, the remaining 3/4 being triplet encounters that ultimately lead to separation of the partners.

For alkanethiyl radicals, there is a minor difficulty in applying eq. 8 because the extremely anisotropic g-tensor[12] that results from the degeneracy of the π-p orbitals may bring about rapid intersystem crossing so that even triplet radical pairs can attain the reactive singlet state during their encounter in a solvent cage. This rather interesting question has not yet been addressed in the literature so that any attempt to apply the "von Smoluchowski" equation in this context will involve an uncertainty of at least a factor of four in the final estimate.

HYDROGEN ABSTRACTION BY THIYL RADICALS

The sulfur-hydrogen bond dissociation enthalpy in alkanethiols is 88 kcal mol^{-1}, and is 81 kcal mol^{-1} in thiophenol.[43] We would therefore expect alkanethiyl radicals to be able to abstract hydrogen at sites where carbon-hydrogen bonds are weak and that the arylthiyl radical would be a relatively poor hydrogen abstracting agent.

Lunazzi and Pacucci[44] provided an extremely simple demonstration of this possibility in an EPR study of the reaction between the methylthiyl radical and cyclopentene. They found that at low temperatures (173 K) addition of the thiyl radical to the double bond was the dominant mode of attack, eq. 9. However, above 213 K this spectrum disappeared and was

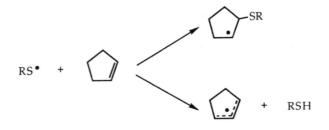

(9)

replaced at 293K by the spectrum of the allylically stabilized cyclopentenyl radical. The authors concluded that addition to the double bond had an activation energy that was considerably less than that for hydrogen abstraction. Interestingly, the almost complete disappearance of spectra between 213 and 293K suggests that the radical species present in highest concentration at these temperatures was the methylthiyl radical itself and the observation supports the concept that alkanethiyl addition to double bonds is readily reversible.

A detailed kinetic analysis was carried out by Pryor, Gojon and Church[45] who looked at the reaction between cyclohexanethiyl radicals and suitable substrates, R–H, eq. 10, 11. To follow the reactions they used the thiol as the solvent and partially tritiated the S–H moiety. The analysis was then carried out by analyzing for tritium in the substrate and by correcting the data for tritium isotope effects. By taking substrates in pairs, the authors established relative rate constants for the respective hydrogen abstractions. However, numerous control experiments were required to ensure that the substrate derived radicals were indeed captured by reaction with the thiol and these were carried out using cumene alone as a test compound. In essence, the kinetic scheme could work well only when there was a fine balance between the efficiencies of reactions 10 and 11.

$$c\text{-}C_6H_{11}S^{\bullet} \quad + \quad R\text{-}H \qquad \rightarrow \qquad c\text{-}C_6H_{11}S\text{-}H \; + \; R^{\bullet} \tag{10}$$

$$c\text{-}C_6H_{11}SH(T) + \; R^{\bullet} \Big\langle \begin{array}{l} c\text{-}C_6H_{11}S^{\bullet} \; + \; R\text{-}H \\[1em] c\text{-}C_6H_{11}S^{\bullet} \; + \; R\text{-}T \end{array} \tag{11}$$

By applying this technique, the authors established the relative reactivities of some 28 compounds with respect to ethylbenzene which was chosen as the standard. For example, dodecane was 500 times less reactive than ethylbenzene while 9,10-dihydroanthracene was 200 times more reactive. An approximate scale of absolute rate constants was formed by reacting ethylbenzene in competition with triphenyl phosphite, reactions 10, 11 (R = PhCHCH$_3$), 12 and 13. The value of k_{12} was known to be ca. 1.5 x 10^7 M^{-1} s^{-1} for the butanethiyl radical (vide infra) and it was assumed that the same value would apply to the cyclohexanethiyl radical. This led to a value of k_{10} for ethylbenzene that was ca. 1 x 10^6 M^{-1} s^{-1}.

$$c\text{-}C_6H_{11}S^{\bullet} \; + \; (PhO)_3P \quad \rightarrow \quad c\text{-}C_6H_{11}^{\bullet} \; + \; (PhO)_3PS \tag{12}$$

$$c\text{-}C_6H_{11}SH \; + \; c\text{-}C_6H_{11}^{\bullet} \quad \rightarrow \quad c\text{-}C_6H_{11}S^{\bullet} \; + \; c\text{-}C_6H_{12} \tag{13}$$

One might wonder how reliable the absolute value of k_{10} actually is. In fact, it is related through three separate competition experiments to the absolute value for thiyl radical addition to a simple olefin in the gas phase. Several assumptions were built into the calculation of the linkage and touch upon the magnitudes of tritium isotope effects, the equivalence of reactivity of thiyl radicals with different structures, the extent to which olefin addition of the alkanethiyl radicals is reversible and the lack of solvent effects. In short, the reported value of k_{10} could easily be in error by an order of magnitude.

HYDROGEN ABSTRACTION FROM THIOLS

Burkhart[46] studied the radical chain reaction between pentanethiol and triethyl-phosphite (cf. eq. 12, 13), in benzene as solvent, and established that reaction 13 was rate determining and therefore that the termination step in the radical chain process involved the self-reaction of two pentyl radicals, eq. 14. Kinetic analysis showed that the ratio

$k_{13}/(2k_{14})^{1/2}$ was equal to 3 $M^{-1/2}$ $s^{-1/2}$. Since $2k_{14}$ is approximately 1 x 10^{10} $M^{-1}s^{-1}$,[47] one can estimate that the rate constant for hydrogen abstraction by the pentyl radical from pentanethiol is ca. 3 x 10^5 $M^{-1}s^{-1}$.

$$2 \; CH_3(CH_2)_4{}^\bullet \quad \rightarrow \quad \text{non-radical products} \tag{14}$$

In subsequent studies, Burkhart and Merrill[48] obtained values for analogous reactions that were in the range 10^5 to 10^6 $M^{-1}s^{-1}$. However, their accompanying measurements of rate constants for the self-reactions of alkyl radicals (c.f. reaction 14) lead to values of k_{14} that were sometimes an order of magnitude less than the diffusion controlled limit, suggesting that the analysis was not entirely accurate. Some of the difficulty may have come from the method of calibrating the apparatus that was used for the measurements. This involved running a polymerization reaction with known kinetics in the apparatus that was used for the thiol experiments. By using this technique the rate of initiation, required for the analysis, was established. Clearly, this approach leads to possibilities for the accumulation of errors. However, these deficiences are exaggerated when translated into values for k_{14} because of the need to square the measured quantity (see above).

The "radical clock" technique[49] has been used to measure rate constants for the reactions of alkyl radicals with alkanethiols. This method is outlined in eq. 15 and 16. In these expressions, $A^\bullet$ is a radical that undergoes a rearrangement with a known rate constant, k_{15}, to form radical $B^\bullet$. Both radicals may then react with the thiol to abstract a hydrogen, reactions 16 and 17. The most convenient way to set up the reaction is to arrange the chemistry so that the thiyl radical reacts with a suitable substrate to regenerate $A^\bullet$, reaction 18. In essence, the rearrangement of $A^\bullet$ is used as a timing device to probe the efficiency of its reaction with the thiol. If the reactions are carried out to low conversion and if all of $B^\bullet$ that is formed is converted to BH, then eq. 19 can be used to solve the reaction kinetics. At high conversions, integrated rate expressions are required.[50]

$$A^\bullet \rightarrow B^\bullet \tag{15}$$

$$A^\bullet + RSH \rightarrow A\text{–}H + RS^\bullet \tag{16}$$

$$B^\bullet + RSH \rightarrow B\text{–}H + RS^\bullet \tag{17}$$

$$RS^\bullet \rightarrow \rightarrow A^\bullet \tag{18}$$

$$[A\text{–}H]/[B\text{–}H] \quad = \quad k_{16}[RSH]/k_{15} \tag{19}$$

The rearrangement shown in eq. 21 was studied by EPR spectroscopy[50] and, as part of this work, the authors used the process as a "clock" reaction to probe the reaction of a primary alkyl radical with t-BuS–H. Although a rate constant of 2.3 x 10^5 $M^{-1}s^{-1}$ was obtained, the authors pointed out that the approach could be in error by an order of magnitude because of the errors associated with the measurements of the Arrhenius parameters for reaction 15 and the need to extrapolate the data to obtain a value for k_{15} at the temperature of the thiol reaction. It was suggested that a more suitable clock might be the rearrangement of the 5-hexenyl radical, eq. 22, since this had been calibrated by two independent techniques and since data were available at ambient temperature. This suggestion was adopted by Newcombe[51] who also used the rearrangement of cyclopropylcarbinyl in his study, reaction 23. The data led to the Arrhenius expression shown in eq. 20 that describes the rate constant for the reaction of a primary alkyl radical with t-BuSH (R in kcal K^{-1} mol^{-1}). The results lead to k_{16} = 8.0 x 10^6 $M^{-1}s^{-1}$ at 298 K.

$$\log (k_{16}/\;M^{-1}s^{-1}) = (8.37 \pm 0.08) - (2.00 \pm 0.09) / 2.3 \; RT \; (R \; in \; kcal \; K^{-1} \; mol^{-1}) \tag{20}$$

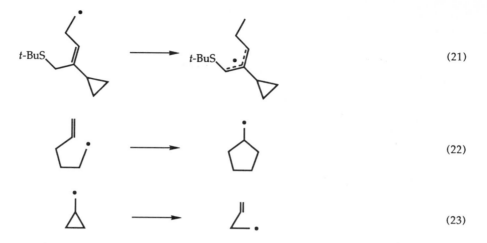

$$\text{(21)}$$

$$\text{(22)}$$

$$\text{(23)}$$

Franz, Bushaw and Alnajjar[52] measured rate constants for hydrogen abstraction from thiophenol by alkyl radicals using a flash photolysis technique. Their approach was to generate alkyl radicals via the reaction of t-butoxyl with trialkylphosphines, eq. 24, 25 and then to monitor the growth of the thiophenoxyl radicals formed in the subsequent abstraction process, eq. 26.

$$t\text{-BuO–OBu-}t \quad \rightarrow \quad 2\ t\text{-BuO}^\bullet \tag{24}$$

$$t\text{-BuO}^\bullet + R_3P \quad \rightarrow \quad t\text{-BuOPR}_2 + R^\bullet \tag{25}$$

$$R^\bullet + PhSH \quad \rightarrow \quad R\text{–H} + PhS^\bullet \tag{26}$$

Although the authors did not discuss the possibility that t-butoxyl would rapidly attack thiophenol to produce the thiophenoxy radical, it would seem that this process would not be significant in light of the high rate constants for the attack at the phosphines ($k_{25} >$ 5×10^8 M^{-1}s^{-1}) and the fact that they always appeared to use a large excess of phosphine over thiophenol.

The values of k_{26} obtained at 298 K for R = butyl, octyl, i-propyl and t-butyl radicals were 1.4×10^8, 9.2×10^7, 1.1×10^8 and 1.5×10^8 M^{-1}s^{-1} respectively. Interestingly, the rate constants were all essentially the same despite changing thermochemistry for the reaction. Although this finding is surprising, the rate constants for reaction with thiophenol are substantially higher than those reported by Newcombe for alkanethiols which is in accord with the rather large difference in the sulfur-hydrogen bond dissociation enthalpies for the two substrates.

Rate constants for the reactions of α-hydroxyalkyl radicals with cysteamine, $H_2NCH_2CH_2SH$ (abbreviated CySH), in water have been measured by a pulse radiolysis technique[53], eq. 27-30, in which the growth of the signal from the disulfide radical anion (reaction 28) was used to probe the lifetime of the thiyl radical.

$$HO^\bullet + CySH \quad \rightarrow \quad H_2O + CyS^\bullet \tag{27}$$

$$CyS^\bullet + CyS^- \quad \rightleftarrows \quad (CyS\cdot^\bullet\cdot SCy)^- \tag{28}$$

$$HO^\bullet + RR'CHOH \quad \rightarrow \quad H_2O + RR'\overset{\bullet}{C}OH \tag{29}$$

$$CySH + RR'\overset{\bullet}{C}OH \quad \rightarrow \quad CyS^\bullet + RR'CHOH \tag{30}$$

The experiment was carried out in two stages. In the first, hydroxyl radicals generated by the radiolysis pulse reacted very rapidly with cysteamine leading to an "instantaneous" growth of the probe signal, eq. 27,28. In the second experiment, alcohol was added to the system. Now, hydroxyl radicals generated by he radiation pulse attacked the alcohol to some extent forming hydroxyalkyl radicals, reaction 29. These, in turn, reacted relatively slowly with cysteamine leading to a correspondingly slow growth of the probe signal, eq. 30 and 28 respectively. Analysis of the slow growth region gave values of k_{30}.

The results are rather confusing. For example, k_{30} = 4.2 x 10^8, 1.4 x 10^8, 4.0 x 10^8 and 1.8 x 10^7 $M^{-1}s^{-1}$ for $(CH_3)_2C^\bullet OH$, $CH_3C^\bullet HOH$, $CH_3C(O)C^\bullet H_2$ and $^\bullet CH_2(CH_3)_2COH$ respectively giving rise to several questions. Why are the values high when compared with data in organic solvents; wouldn't hydrogen bonding in water be expected to slow the reaction down? Since, hydroxyalkyl radicals form relatively weak C–H bonds, why are they more reactive than the primary alkyl radical derived from t-butyl alcohol?

It is easy to point to deficiencies in the scheme. For example, attack by hydroxyl at the alcohols is not regiospecific so that several different alcohol-derived species must contribute to the reaction kinetics. Moreover, there are doubts, in some contexts (vide infra), about the reliability of disulfide radical anions as probe molecules. Nevertheless, the data are supported by an independent study where a different probe molecule was employed.

Wolfenden and Willson[54] used 2,2'-azinobis-(3-ethylbenzthiazoline-6-sulfonic acid), ABTS, as a probe since it rapidly transfers electrons to thiyl radicals producing a highly colored radical cation, eq. 31. They measured the following values for k_{30}: 3 x 10^8 and 2 x 10^8 $M^{-1}s^{-1}$ for $(CH_3)_2C^\bullet OH$ and $CH_3C^\bullet HOH$ respectively in good agreement with the previous study. However, in a note added in proof to their paper the authors remarked upon the fact that reaction 31 was reversible on the timescale of the pulse radiolysis study, which must have implications for the analysis as a whole. Moreover, they did not consider the possibility that hydroxyalkyl radicals, which are powerful reducing agents, could at least partially destroy the radical cation during the timescale of the measurements. Some of these pitfalls could have been explored by an analysis of both the growth kinetics for the probe molecule and the final concentration of radical cation obtained.

$$ABTS + RS^\bullet \quad \rightarrow \quad ABTS^{\bullet +} + RS^- \tag{31}$$

The values of the rate constants in question are rather important since they are relevant to so-called "repair" mechanisms for biological systems in which thiols donate hydrogen to carbon radical centers, eq. 32, that are formed adventitiously. An unrepaired carbon-centered radical, $R^\bullet$, would likely react with oxygen and propagate a radical chain oxidation in the system thus amplifying the damage, reactions 33, 34. Since values of k_{33} are available, the efficiency of say an intracellular repair mechanism can be assessed from a knowledge of k_{32} and the concentrations of intracellular thiol and oxygen.

$$R^\bullet + R'SH \quad \rightarrow \quad R-H + R'S^\bullet \tag{32}$$

$$R^\bullet + O_2 \quad \rightarrow \quad ROO^\bullet \tag{33}$$

$$ROO^\bullet + RH \quad \rightarrow \quad R^\bullet + ROOH \tag{34}$$

In light of the difficulties discussed above, it would certainly be appropriate to measure rate constants for hydrogen abstraction from biologically important thiols, such as glutathione, using a radical clock approach in water or, at least, in highly polar solvents.

Thankfully, the complex question of the reactions between thiyl radicals and oxygen, eq. 35, is a subject that will be discussed by other authors in this publication.[55] It will therefore be given only a brief mention in this chapter on reaction kinetics.

$$RS^\bullet + O_2 \quad \rightarrow \quad RSO_2^\bullet \tag{35}$$

Dimethyl sulfide, methanethiol and dimethyl disulfide are the most abundant naturally occurring organosulfur compounds in the atmosphere.[35] They are particularly abundant above oceans where their tropospheric oxidation is of considerable importance. Thiol radicals are certainly involved and while they can be formed in simple reactions, e.g. hydrogen abstraction from methanethiol, they can also be generated by rather complex reaction mechanisms. An example is shown in eq. 36 - 39.

$$HO^\bullet + CH_3SCH_3 \quad \rightarrow \quad H_2O + CH_3SC^\bullet H_2 \tag{36}$$

$$CH_3SC^\bullet H_2 + O_2 \quad \rightarrow \quad CH_3SCH_2OO^\bullet \tag{37}$$

$$CH_3SCH_2OO^\bullet + NO \quad \rightarrow \quad CH_3SCH_2O^\bullet + NO_2 \tag{38}$$

$$CH_3SCH_2O^\bullet \quad \rightarrow \quad CH_3S^\bullet + H_2CO \tag{39}$$

Methanethiyl radicals do react with oxygen and ultimately yield sulfur dioxide.[33] However, the process must be fairly inefficient since thiol radicals when formed in relatively high concentrations on the photolysis of methanethiol in oxygen predominantly undergo dimerization to give dimethyl disulfide.[56]

Product and kinetic studies[33] indicate that the the methanethiyl radical reacts ca. 10^6 times less readily with oxygen than with nitrogen dioxide for which the rate constant is 8×10^9 $M^{-1}s^{-1}$. It is possible that the thiyl-oxygen reaction, eq. 35, is readily reversible. However, this possibility was considered in that kinetic study[33] and it was concluded that $k_{35} >> k_{-35}$.

The slowness or reversibility of reaction 35 is clearly established in studies of the cooxidation of thiols and olefins,[56,57] where the reaction proceeds via the mechanism shown in reactions 40, 41 and where the reaction between thiyl radicals and oxygen does not play a significant role.

$$RS^\bullet + R'CH=CH_2 \quad \rightarrow \quad R'C^\bullet H-CH_2SR \tag{40}$$

$$R'\overset{\bullet}{C}HCH_2SR + O_2 \quad \rightarrow \quad R'CH(OO^\bullet)CH_2SR \tag{41}$$

In contrast to the above experiments which point to a very slow reaction between thiyl radicals and oxygen, a fairly high rate constant, $k_{35} = 4 \times 10^7$ $M^{-1}s^{-1}$, was reported[58] for the reaction between the thiyl radical derived from penicillamine, $(CH_3)_2C(S^\bullet)CH(NH_2)CO_2H$, and oxygen. In these experiments, the growth of the thiyl radical was monitored directly in pulse radiolysis experiments in the presence and absence of oxygen and under conditions of pH where the disulfide radical anion (reaction 3) was absent.

Much higher rate constants (8×10^9 $M^{-1}s^{-1}$) for reaction 35 have been measured for the cysteine radical in pulse radiolysis experiments using the disulfide radical anion as a probe signal[59] (see above). However, these are almost certainly in error.[58] The measurements were based on the yield of radical anion measured at a fixed time after the pulse. The lifetime of the radical anion was such that it would have undergone decomposition to restore the thiyl radical, before the measurement was made and the thiyl radicals released in this process would have reacted with oxygen to some extent. The net effect of this artefact would have been to reduce the measured concentration of the radical anion signal and this would have been interpreted in terms of a much faster oxygen - thiyl radical reaction than actually took place. Despite this problem, the inference is that thiyl radicals react with oxygen in the microsecond time domain which is very much faster than gas phase experiments would indicate.

The most charitable explanation of all of the data that one can make is that reaction 35 is rapid (k_{35} = ca. 10^7 $M^{-1}s^{-1}$) and reversible but this would imply that some of the gas phase experiments[33] were incorrectly interpreted. Reversibility of reaction 35 has, in fact, been implied in some pulse radiolysis work by Tamba et al. for RSH being glutathione[60] (see also articles by M. Quintiliani and P. Wardman in this book), and is discussed extensively in a paper by Mönig et al.[61] Clearly, this is an area that warrants more detailed investigation.

REACTIONS OF THIYL RADICALS WITH OLEFINS AND ORGANOMETALLIC COMPOUNDS

The reactions of arenethiyl radicals with olefins have been investigated extensively by Itoh, Matsuda and their colleagues[24-31] who used flash photolysis techniques to follow the absorption spectrum of the thiyl radicals. To avoid the problems of reversibility, olefins were chosen that gave stabilized radicals, e.g. styrenes and the reactions were run in the presence of oxygen so as to scavenge the carbon centered radicals produced before elimination took place (see reactions 40 and 41). It is noteworthy that the phenylthiyl radical did not react with oxygen on the microsecond timescale of the experiments. Typical rate constants in solution for styrene, eq. 43, and methylmethacrylate addition, eq. 44, at 298 K were 2.7 x 10^7 and 5.4 x 10^6 $M^{-1}s^{-1}$ respectively. The nuances of these reactions have been explored extensively and the effect of substitution in styrenes and in the arenethiyl radicals have been thoroughly investigated and are described in a recent review.[57]

$$PhS-SPh \quad \rightarrow \quad 2\ PhS^\bullet \tag{42}$$

$$PhS^\bullet + PhCH=CH_2 \quad \rightarrow \quad Ph\overset{\bullet}{C}H-CH_2SPh \tag{43}$$

$$PhS^\bullet + CH_2=C(CH_3)CO_2CH_3 \quad \rightarrow \quad PhSCH_2-\overset{\bullet}{C}(CH_3)CO_2CH_3 \tag{44}$$

Measuring absolute rate contants for the reactions of alkanethiyl radicals has proved to be substantially more difficult because they do not have chromophores that are easily monitored by spectroscopic techniques. However, it was recently shown that 1,1-diphenyl ethylene funtioned as a good "probe" molecule for this purpose.[10] Laser flash photolysis experiments showed that addition of t-BuS$^\bullet$ was rapid (k_{46} = 9.9 x 10^8 $M^{-1}s^{-1}$) and gave rise to a substituted diphenylmethyl radical that had a strong absorption spectrum at 329 nm. Moreover, complementary electron paramagnetic resonance, EPR, experiments showed that the addition reaction was irreversible on the milli- to microsecond timescale.

$$t\text{-BuS–SBu-}t \quad \rightarrow \quad 2\ t\text{-BuS}^\bullet \tag{45}$$

$$t\text{-BuS}^\bullet + Ph_2C=CH_2 \quad \rightarrow \quad Ph_2\overset{\bullet}{C}-CH_2SBu\text{-}t \tag{46}$$

Having established the viability of this probe technique, it was a fairly straightforward matter to measure the rate constants for the reactions of t-BuS$^\bullet$ with other substrates. A typical example is given in eq. 47 that describes the reaction with triethylboron which proceeds by a homolytic displacement, S_H^2, mechanism.

Table 2. Rate constants for the Reactions of t-BuS$^\bullet$ with a Variety of Substrates at 298 K in Isooctane as Solvent

Substrates	$k_s \times 10^{-8}$, M^{-1}s^{-1}
$Ph_2C=CH_2$	9.9
$(c\text{-}C_3H_5)_2C=CH_2$	2.4
Et_3B	1.3
$s\text{-}Bu_3B$	0.8
$(MeO)_3P$	2.7
$(EtO)_3P$	3.1
Bu_3P	9.0
$CH_3(CH_2)_5CH=CH_2$	0.02

$$t\text{-BuS}^\bullet \ + \ Et_3B \quad \rightarrow \quad t\text{-BuSBEt}_2 \ + \ Et^\bullet \tag{47}$$

Rate constants, k_s, for reaction with substrates such as the borane were measured monitoring the rate constants, k_{obs}, for the first order growth of the probe signal as a function of the substrate concentration, eq. 48. In this equation, k_o describes the lifetime for the growth of the probe signal in the absence of substrate. The results[10] are described in Table 2.

$$k_{obs} \ = \ k_o \ + \ k_{obs}[\text{substrate}] \tag{48}$$

Among the substrates used in these experiments[10] was 1,1-dicyclopropylethylene. It was expected that its rapid ring-opening (c.f. reaction 20) would preclude reversible loss of the thiyl radical and that its reactivity would typify that of a normal olefin. While the rearrangement concept worked well, the substrate turned out to be remarkably labile (see Table 2) and was similar in reactivity to trialkyl boranes. The result was surprising since Davies and Roberts[62] had used competition experiments to establish that tributylboron was 110 times more reactive than oct-1-ene in competition for BuS$^\bullet$.

To establish the authenticity of the results, a competition experiment was carried out between 1,1-dicyclopropylethylene and oct-1-ene. Although reversible loss of t-BuS$^\bullet$ from the latter does occur to a significant extent, it can be dealt with if the competition experiments are carried out at different oct-1-ene concentrations. While the method of kinetic analysis is beyond the scope of this review, it was demonstrated that 1,1-dicyclopropylethylene was indeed 130 times more reactive than oct-1-ene. Taking the rate constant for addition to 1,1-dicyclopropylethylene to be 2.4×10^8 M^{-1}s^{-1} (Table 2) led to $k = 2 \times 10^6$ M^{-1}s^{-1}. This rate constant for addition of t-BuS$^\bullet$ with oct-1-ene was in good agreement with the value reported by Sivertz and his colleagues of 7×10^6 M^{-1}s^{-1} for the reaction of butanethiyl with pent-1-ene.[4,34,35]

REACTIONS OF SULFONYL RADICALS

Our understanding of the chemistry of sulfonyl radicals, RS$^\bullet$O$_2$, seems to be somewhat less equivocal than that of thiyl radicals although there are some difficulties with the reaction kinetics. Sulfonyl radicals are readily formed by halogen abstraction from sulfonyl halides using trialkylsilyl radicals, eq. 24, 49, 50, and rate constants for these processes have been studied in some detail. For example, rate constants for chlorine abstraction at $MeSO_2Cl$ and $C_6H_5CH_2SO_2Cl$ were found to be 3.2 x 10^9 and 5.7 x 10^9 $M^{-1}s^{-1}$ respectively demonstrating that the reactions were close to diffusion controlled.[63]

$$t\text{-BuO}^\bullet + Et_3SiH \quad \rightarrow \quad t\text{-BuOH} + Et_3Si^\bullet \qquad (49)$$

$$Et_3Si^\bullet + RSO_2Cl \quad \rightarrow \quad Et_3SiCl + R\overset{\bullet}{S}O_2 \qquad (50)$$

Sulfonyl radicals can also be formed by alkyl radical addition to sulfur dioxide although there is a conflict in this area about the magnitude of the rate constant for the process. For example, two gas phase[64,65] studies give quite different rate constants (5 x 10^6 and 2 x 10^8 $M^{-1}s^{-1}$ at 298 K) for the reactions of methyl radicals with sulfur dioxide, reaction 51. The origins of the difference are not clear although it is possible that the reaction is reversible on the timescale of the measurements (see reactions of thiyl radicals with oxygen). It is noteworthy that for $PhCHS^\bullet O_2$, loss of SO_2 is too fast to be measured by standard laser flash photolysis techniques[63,66] so that $k_{52} > 10^8$ s^{-1} at 295 K. The rapidity of the reaction must reflect the favourable thermochemistry for the formation of the benzyl radical.

$$Me^\bullet + SO_2 \quad \rightarrow \quad Me\overset{\bullet}{S}O_2 \qquad (51)$$

$$PhCH_2\overset{\bullet}{S}O_2 \quad \rightarrow \quad Ph\overset{\bullet}{C}H_2 + SO_2 \qquad (52)$$

Oxygen centered radicals react rapidly with sulfur dioxide. For example, the rate constant for $t\text{-BuO}^\bullet$ addition[67] is ca. 10^7 to 10^8 $M^{-1}s^{-1}$ while that for $MeOO^\bullet$ attack[68] is 6 x 10^6 $M^{-1}s^{-1}$. While sulfonyl radicals seem to be poor at abstracting hydrogen from hydrocarbons,[69] the reverse process - hydrogen abstraction from sulfinic acids - is quite efficient (rate constant = 10^6 $M^{-1}s^{-1}$).[70]

CONCLUSION

Sulfur centered radicals play a vital role in diverse areas of chemistry and yet there is very little reliable kinetic data on their reactions. Many of the experiments that have been attempted have been plagued by the difficulties of monitoring the radicals themselves and by reversibility in many of the reactions. These problems demand fresh initiatives that focus on processes that are of importance in biological and atmospheric chemistry.

REFERENCES

1. Landolt-Börnstein, New Series Group II, vol. 13, H. Fischer, ed., W. de Gruyter, Berlin (1984)
2. For a review see: C. Walling, in "Free Radicals in Solution," John Wiley and Sons Inc., New York, N. Y., 1957, pp. 313-326.
3. R.H. Pullen and C. Sivertz, *Can. J. Chem.* 35:723, (1957).
4. C. Sivertz, *J. Phys. Chem.* 63:34, (1959).
5. C. Walling and W. Helmreich, *J. Am. Chem. Soc.* 81:1144 (1959).
6. C. Walling and M.S. Pearson, *J. Am. Chem. Soc.* 86:2262 (1964).
7. D.M. Graham and J.F. Soltys, *Can. J. Chem.* 48:2173 (1970).
8. R.J. Balla, B.R. Weiner and H.H. Nelson, *J. Am. Chem. Soc.* 109:4804 (1987).
9. A.G. Davies and B.P. Roberts, *J. Chem. Soc. (B)* 1830 (1971).
10. D.J. McPhee, M. Campredon, M. Lesage and D. Griller, *J. Am. Chem. Soc.* 111:7563 (1989).
11. T.J. Burkey, J.A. Hawari, F.P. Lossing, J. Lusztyk, R. Sutcliffe and D. Griller, *J. Org. Chem.* 50:4966 (1985).

12. M.C.R. Symons, *J. Chem. Soc., Perkin Trans.* 2 1618 (1974).
13. Landolt Börnstein, New Series, Group II, vol. 9, part B.
14. A.B. Callear and D.R. Dickson, *Trans. Faraday Soc.* 66:1987 (1970).
15. G.H. Morine and R.R. Kuntz, *Photochem. Photobiol.* 33:1 (1981).
16. For typical examples, see refs. 17-22.
17. G.E Adams, R.C. Armstrong, A. Charlesby, D.E. Michael, and R.I. Wilson, *Trans. Faraday Soc.* 65:1969 (1969).
18. G. Caspari and A. Granzow, *J. Phys. Chem.* 74:836 (1970).
19. M.Z. Hoffman and E. Hayon, *J. Am. Chem. Soc.* 94:7950 (1972).
20. M.Z. Hoffman and E. Hayon, *J. Phys. Chem.* 77:990 (1973).
21. T.-L. Tung and J.A. Stone, *J. Phys. Chem.* 78:1130 (1974)..
22. M. Bonifacic and K.-D. Asmus, *J. Phys. Chem.* 88:6286 (1984).
23. J.R.M. Giles and B.P. Roberts, *J. Chem. Soc., Perkin Trans.* 2 1497 (1981).
24. O. Ito and M. Matsuda, *J. Am. Chem. Soc.* 101:1815 (1979).
25. O. Ito and M. Matsuda, *J. Am. Chem. Soc.* 101:5732 (1979).
26. M. Nakamura, O. Ito and M. Matsuda, *J. Am. Chem. Soc.* 102:698 (1980).
27. O. Ito and M. Matsuda, *Bull Chem. Soc. Jpn.* 57:1745 (1984).
28. O. Ito and M. Matsuda, *Int. J. Chem. Kinet.* 16:909 (1984).
29. O. Ito and M. Matsuda, *J. Org. Chem.* 49:17 (1984).
30. O. Ito, Y. Arito and M. Matsuda, *J. Chem. Soc., Perkin Trans.* 2 869 (1988).
31. O. Ito, S. Tamura, K. Murakami and M. Matsuda, *J. Org. Chem.* 53:4758 (1988).
32. M. Suzuki, G. Inoue and H. Akimoto, *J. Chem. Phys.* 81:5405 (1984).
33. G.S. Tyndall and A.R. Ravishankara, *J. Phys. Chem.* 93:2426 (1989) and references cited therein.
34. R. Back, G. Trick, C. McDonald and C. Sivertz, *Can. J. Chem.* 32:1078 (1954).
35. M. Onyszchuck, C. Sivertz, *Can. J. Chem.* 33:1034 (1955).
36. T.J. Burkey and D. Griller, *J. Am. Chem. Soc.* 107:246 (1985).
37. J.A. Howard and E. Furimsky, *Can. J. Chem.* 55:555 (1974).
38. C. Chatgilialoglu, L. Lunazzi and K.U. Ingold, *J. Org. Chem.* 48:3588 (1983).
39. J.E. Bennett and G. Brunton, *J. Chem. Soc., Chem. Comm.* 62 (1979).
40. J.E. Bennett, G. Brunton, B.C. Gilbert and P.E. Whittal, *J. Chem. Soc., Perkin Trans.* 2 1359 (1988).
41. M.Kobayashi, H. Minato, Y. Miyaji, T. Yoshioka, K. Tanaka and K. Honda, *Bull. Chem. Soc. Jpn.* 45:2817 (1972).
42. C. Chatgilialoglu, B.C. Gilbert, B. Gill and M.D. Sexton, *J. Chem. Soc., Perkin Trans.* 2 1141 (1980).
43. See chapter by D. Griller on "Thermochemistry ..." in this book.
44. L. Lunazzi and G. Placucci, *J. Chem. Soc., Chem. Comm.* 533 (1979).
45. W.A. Pryor, G. Gojon and D.F. Church, *J. Org. Chem.* 43:793 (1978).
46. R.D. Burkhart, *J. Phys. Chem.* 70:605 (1966).
47. D. Griller, Landolt Börnstein, New Series, Group II, Vol. 13a Part 1, pp. 5-1130, 1984.
48. R.D. Burkhart and J.C. Merrill, *J. Phys. Chem.* 73:2699 (1969).
49. D. Griller and K.U. Ingold, *Acc. Chem. Res.* 13:317 (1980).
50. M. Campredon, J.M. Kanabus-Kaninska and D. Griller, *J. Org. Chem.* 53:5393 (1988).
51. M. Newcombe, private communication. We thank Professor Newcombe for making his results available prior to publication.
52. J. A. Franz, B.A. Bushaw and M. Alnajjar, *J. Am. Chem. Soc.* 111:268 (1989).
53. G.E. Adams, G.S. McNaughton and B.D. Michael, *Trans. Faraday Soc.* 64:902 (1968).
54. B.S. Wolfenden and R.L. Willson, *Perkin Trans.* 2 105 (1982).
55. See article by M. Guerra and C. Chatgilialoglu in this book.
56. D.M. Graham and J.F. Soltys, *Can. J. Chem.* 48:2173 (1970).
57. O. Ito and M. Matsuda in "Chemical Kinetics of Small Organic Radicals", Z. B. Alfassi, ed., Vol III, Chap. 15, p. 133.
58. K. Schäfer, M. Bonifacic, D. Bahnemann and K.-D. Asmus, *J. Phys. Chem.* 82:2777 (1978).
59. J.P. Barton and J.E. Packer, *Int. J. Radiat. Phys. Chem.* 2:159 (1970).

60. M. Tamba, G. Simone, and M. Quintiliani, *Int. J. Radiat. Biol.* 50:595 (1986).

61. J. Mönig, K.-D. Asmus, L.G. Forni, and R.L. Willson, *Int. J. Radiat. Biol.* 52:589 (1987).

62. A.G. Davies and B.P. Roberts, *J. Chem. Soc. (B)* 1830 (1971).

63. C. Chatgilialoglu, L. Lunazzi and K.U, Ingold *J. Org. Chem.* 48:3588 (1983).

64. A. Good and J.C.J. Thynne, *Trans. Faraday Soc.* 63:2720 (1967).

65. F.C. James, J.A. Kerr and J. Simons, *J. Chem. Soc., Faraday Trans.* 2124 (1973).

66. I.R. Gould, C. Tung, N.J. Turro, R.S. Givens and B. Matuszewski, *J. Am. Chem. Soc.* 106:1789 (1984).

67. A.G. Davies, B.P. Roberts and B.R. Sanderson, *J. Chem. Soc., Perkin Trans.* 2 626 (1973).

68. C.S. Kan, R.D. McQuigg, M.R. Whitbech and J.G. Calvert, *Int. J. Chem. Kinet.* 11:921 (1971).

69. A. Horowitz and L.A. Rajenbach, *J. Am. Chem. Soc.* 97:10 (1975).

70. B.C. Gilbert, R.O.C. Norman and R.C. Sealy, *J. Chem. Soc., Perkin Trans.* 2 303 (1975).

REDOX SYSTEMS WITH SULPHUR-CENTERED SPECIES

David A. Armstrong

Department of Chemistry
University of Calgary
Calgary, Alberta, Canada T2N 1N4

The reactions discussed below have been chosen to illustrate the potential behaviour of thiyl radicals, disulphide radical anions and other sulphur-containing reactive intermediates, as oxidizing and reducing agents in inorganic, organic and biological systems. The influence of sulphur-containing species on oxidation-reduction reactions has been recognized for many years. Since the first half of this century the chemical and petroleum industries have used mercaptans for the protection of petroleum products (1932) and the stabilization of chlorinated hydrocarbons, such as carbon tetrachloride (1933), against autoxidation.[1] The fundamental role of RSH appears to be that of a scavenger of oxygen and/or an inhibitor of oxidation catalysts.

When studies of radiation chemistry and radiation biology became common in the 1950's, the repair reaction (1):

$$RSH + A^{\bullet} \rightarrow RS^{\bullet} + AH \tag{(1)/(-1)}$$

was proposed as one mechanism of explaining the protective effect of thiol molecules in biological systems.[2] The utility of this reaction depends on the equilibrium lying far to the right and RS$^{\bullet}$ radicals being removed from the system by a mechanism that renders them harmless. During the course of this lecture it will be seen that RS$^{\bullet}$ radicals are capable of a wide variety of different types of reaction. Thus, the mechanisms by which they may be removed from a cell must be considered in some detail.

THE REPAIR REACTION

Usually it is assumed that the A$^{\bullet}$ radical is a DNA fragment species or some other cellular target from which a hydrogen atom has been removed by hydroxyl radical reactions. In principle stable anions, from which electrons have been removed, might also be repaired, viz:

$$RSH + A^{\bullet} \rightarrow RS^{\bullet} + H^+ + A^- \tag{2}$$

or

$$RS^- + A^{\bullet} \rightarrow RS^{\bullet} + A^- \tag{3}$$

Sulfur-Centered Reactive Intermediates in Chemistry and Biology, Edited by
C. Chatgilialoglu and K.-D. Asmus, Plenum Press, New York, 1990

In determining the position of these equilibria at different pH's it is necessary to know the pH dependence of the half cell potential, E_h, of both the sulphydryl couple and the AH or A$^-$ couple. One must be careful to adopt consistent procedures in developing equations for this, and here the reader is referred to the recent work of Wardman.[2] For the general case of an ionizable AH molecule with ionisation constant K_i the following equations apply:

$$A^\bullet \ + \ H^+ \ + \ e^- \ \rightleftarrows \ AH \tag{4}$$

$$E_h \ = \ E^0(A^\bullet/AH) + \frac{RT}{F} \ln \frac{[A^\bullet] \gamma_{A^\bullet}}{[AH]_t \gamma_{AH}} (\{H^+\} + \frac{K_i \gamma_{AH}}{\gamma_{A^-}}) \tag{5}$$

$$A^\bullet \ + \ e^- \ \rightleftarrows \ A^- \tag{6}$$

$$E_h \ = \ E^0(A^\bullet/A^-) + \frac{RT}{F} \ln \frac{[A^\bullet] \gamma_{A^\bullet}}{[A^-] \gamma_{A^-}} \tag{7}$$

Here square brackets refer to concentrations and "$\{\ \}$" to activities. The important relation between $E^0(A^\bullet/AH)$ and $E^0(A^\bullet/A^-)$ is as follows:

$$E^0 (A^\bullet/AH) \ = \ E^0 (A^\bullet/A^-) + \frac{RT}{F} \ln K_i (AH) \tag{8}$$

Equations (4) through (7) can be used to find the pH dependences of E_h for both the AH and the RSH couples. From these the pH dependences of K_1, K_2, and K_3 can be derived.[2] However, in addition to the position of the repair equilibrium, one must also consider the rate of the repair reaction.

The rate of the repair reaction in the forward direction was first studied quantitatively by Adams and co-workers[3a], using the $^\bullet CH_2OH$ and other alcohol radicals as the species to be repaired. As shown in Figure 1, this rate falls off with ionization of the thiol. The "structure" in the region of pH 10 is probably due to different ionic species of 1,2-aminothiol. While the efficiency of repair of an A$^\bullet$ species derived from an AH molecule with high pK (i.e., low acidity) falls off with pH, the reverse is true for a system where the pK is small. This is demonstrated by the second curve in Figure 1, which shows the rate constant for reaction of $I_2^{\bullet-}$ with cysteine.[3b] The main point about the behaviour illustrated is that transfers of H atoms or electrons are generally faster when protons do not have to be acquired from or given to the aqueous medium. A corollary is that, while the equilibrium position for the

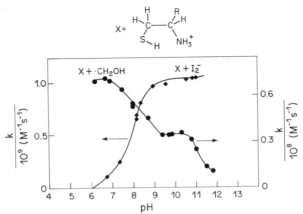

Fig. 1. Effect of pH on second order rate constants for H abstraction from cysteamine by $^\bullet CH_2OH$ and oxidation of cysteine by $^\bullet I_2^-$ from refs. 3a and 3b, respectively. The thiol pK's are 8 to 9.

repair reaction at a given pH may appear extremely favorable, the rate of the process may be trivially slow. In determining the latter it will generally be necessary to calculate the fraction of the sulphydryl species present in the appropriate reactive form. This can be done from the pKs with the aid of conventional equations for acid-base equilibria.

RELATION OF E^0 TO BOND DISSOCIATION ENERGY

At this point it is worth reminding ourselves that the reduction potential of an $A^\bullet/AH$ couple is the cell potential for the reaction defined by equation (9):

$$A^\bullet \ + \ ^1/_2\,H_{2(g)} \ \rightleftarrows \ AH \qquad\qquad \Delta G^0(A^\bullet/AH) \qquad (9)$$

This is the sum of equations (4) and (10):

$$^1/_2\,H_{2(g)} \ \rightleftarrows \ H^+ \ + \ e^- \qquad\qquad (10)$$

The addition of equations (11) and (12) to (9) gives equation (13).

$$A_{(g)}{}^\bullet \ \rightleftarrows \ A^\bullet \qquad\qquad \Delta G^0{}_{soln.}(A^\bullet) \qquad (11)$$

$$AH \ \rightleftarrows \ AH_{(g)} \qquad\qquad -\Delta G^0{}_{soln.}(AH) \qquad (12)$$

$$A_{(g)}{}^\bullet + \ ^1/_2\,H_{2(g)} \ \rightleftarrows \ AH_{(g)} \qquad\qquad \Delta G_{(g)}{}^0(A^\bullet/AH) \qquad (13)$$

When bond dissociation energies are available, the free energy change in reaction (13) can readily be calculated (entropies can be calculated from molecular properties quite easily, see other paper by Armstrong, ref. 11a). Free energies of solvation of AH molecules are usually available and those of the $A^\bullet$ radicals can be found by using suitable models (e.g., for $CH_3S^\bullet$ one may use CH_3Cl, PhF for $PhO^\bullet$, and Ar for $Cl^\bullet$).

The relationship between solution reduction potential and bond dissociation energy is important, not only for checking the reliability of both sets of data, but also for gaining an

Table 1. Comparison of Experimental and Calculated $E^0(R^\bullet/R^-)$, Solvation Energy Discrepancies, and Electron Affinities of Radicals

$R^\bullet$	$E^0(R^\bullet/R^-)^a$ /V	$(\Delta G^0{}_{soln}\{R\} - \Delta G^0{}_{soln}\{Model\})$ /(kJ Mol^{-1})	E.A. of R^b /eV
$RS^\bullet$	0.79 (0.84)	−5	≈2.0
$PhO^\bullet$	0.80 (0.90)	−10	2.4
$I^\bullet$	1.33 (1.41)	−8	3.08
$Br^\bullet$	1.93 (2.07)	−14	3.36
$Cl^\bullet$	2.41 (2.62)	−20	3.61

[a] Data for $I^\bullet$, $Br^\bullet$ and $Cl^\bullet$ from H.A. Schwarz and R.W. Dodson, *J. Phys. Chem.* **88**:3643 (1984), others from ref. 11 of first paper of Armstrong in this book. Values in parenthesis calculated, others are experimental. Uncertaincies ±0.05 V (±0.1 V).

[b] From L.G. Christophorou, "Atomic and Molecular Physics", Wiley, New York (1971).

insight into special interactions which may occur between the free radicals and the solvent water. Table 1 shows the experimental and calculated values of E^0 for a number of electron-affinic radicals. In each case the experimental potential is apparently slightly less than the calculated, which means that the radical species is more stable in solution than the model solute chosen for it. The excess negative free energy of solution of the radicals is shown in the third column of the Table. Only for the last two cases does it actually exceed the currently unavoidably high uncertainties in the calculations and provide a definite reason to suspect that there may be specific interactions with the H_2O molecules. As can be seen from the fourth column of Table 1 these atoms also have the highest electron affinities and thus the best match of orbital energies for possible $\sigma\sigma^*$ interactions with lone pairs of H_2O. For the thiyl radical, if such an interaction does exist, the energy is not very large because the excess free energy of solution ($\Delta G^0_{soln}\{R\} - \Delta G^0_{soln}\{Model\}$) is well within the overall uncertainty of ± 10 kJ Mol^{-1} or ± 0.1 V.

Further evidence relating to interactions between the thiyl radicals and the water molecules may be obtained from an examination of spectra. Treinin and Hayon[4] observed that the absorptions of highly electron affinic radicals in water were far to the red of the gas phase absorptions of lowest energy, and attributed this to a predominance of solvent to solute electron charge transfer transitions. From the electron affinity of thiyl of about 2.0 eV, one would expect a λ_{max} in the region of 220 nm from their data. The observed λ_{max} is at 340 nm. Thus, this appears to be due to an intrinsic transition of the radical rather than charge transfer. The gas phase transitions occur in the region of 370 nm. It appears that in water these are blue shifted. For the perthiyl radical the gas phase and aqueous absorption maxima occur at very similar wavelengths.

RS$^\bullet$ AS AN OXIDANT

Probably the most important reactions of RS$^\bullet$ as an oxidant are back reactions to the repair process, i.e. reaction (–1). As illustrated by the following two examples, the rate constants for these processes are small. However, if the RS$^\bullet$ radicals are sufficiently long lived, the consequences may be quite important[5a] (Me$_2$THF = 1,2-Dimethyltetrahydrofurane).

RS$^\bullet$ + Me$_2$THF $\rightarrow$ RSH + Me$_2$THF(–H)$^\bullet$

$k = 5 \times 10^3$ [5a]

RS$^\bullet$ + HCO$_2^-$ $\rightarrow$ RSH + $^\bullet$CO$_2^-$

$k \geq 3.6 \times 10^4$ M^{-1}s^{-1} [5b]

Examples of additional oxidizing reactions of RS$^\bullet$ are given in Table 2. Here it will be seen that RS$^\bullet$ can act as an oxidant for both organic and inorganic species. Also it can participate in both hydrogen atom and electron transfer processes. Some of the latter involve such large complex systems as the ferrocytochrome c system.

It might be worthwhile to point out an experimental problem for the ferrocyanide system. It results from substitution of the ligands by the thiolate present in the solution to provide a source of RS$^\bullet$ radicals. This process is enhanced by the photolysis of the ferrocyanide in the analyzing light beam of the pulse radiolysis facility. If only fresh ferrocyanide is pulsed one sees the oxidation to ferricyanide by the RS$^\bullet$ species. However, if the ferrocyanide is pre-exposed to light, photoaquation takes place and one of the cyanides appears to be substituted by RS$^-$. The resulting Fe(II)(CN)$_5$(RS)$^{4-}$ species then reacts with ferricyanide to give a thiolate-substituted ferricyanide, Fe(III)(CN)$_5$(RS)$^{3-}$. This has exceedingly strong charge transfer transitions occurring at about 700 nm as shown in Figure 2.

Table 2. Rate Constants for Some Oxidation/Reduction Reactions of RS$^\bullet$ and RS.$^\bullet$.SR$^-$

Reaction [a]			$k/M^{-1}s^{-1}$	Ref.
GS$^\bullet$ + NADH	→	GSH + NAD$^\bullet$	2.3×10^8	6a
RS$^\bullet$ + FlH$^-$	→	RS$^-$ + $^\bullet$FlH	3.2×10^9	7a
GS$^\bullet$ + FerroCyt(II)c	→	GS$^-$ + FerriCyt(III)c	2.5×10^8	6b
RS$^\bullet$ + Fe(CN)$_6^{4-}$	→	RS$^-$ + Fe(CN)$_6^{3-}$	1.0×10^8	7b
RS$^\bullet$ + Cu(I)(RS$^-$)(CN$^-$)$_2$	→	RS$^-$ + Cu(II)(RS$^-$)(CN$^-$)$_2$	3.3×10^9	8a
CysS$^\bullet$ + Cu(I)(CysS$^-$)$_2$	→	CysS$^-$ + Cu(II)(CysS$^-$)$_2$	1.8×10^9	8b
PenS$^\bullet$ + Cu(I)(PenS$^-$)$_2$	→	PenS$^-$ + Cu(II)(PensS$^-$)$_2$	4.5×10^8	8b
CysS.$^\bullet$.SCys$^-$ + Cu(I)(CysS$^-$)$_2$	→	2 CysS$^-$ + Cu(II)(CysS$^-$)$_2$	2.7×10^8	8b
$^\bullet$CO$_2^-$ + Cu(II)(PenS$^-$)$_2$	→	CO$_2$ + Cu(I)(PenS$^-$)$_2$	1.3×10^8	8c
PenS$^\bullet$ + Cu(II)(PenS$^-$)$_2$	→	PenSSPen + Cu(I)(PenS$^-$)^{b}	3.3×10^8	8c
RS.$^\bullet$.SR$^-$ + Fl	→	RSSR + Fl$^-$	4.0×10^8	9b
R'S.$^\bullet$.SR'$^-$ + O$_2$	→	R'SSR' + O$_2$$^{\bullet-}$	1.3×10^9 ^{c}	10a
			0.4×10^9 ^{d}	10b

^{a} RSH = β–mercaptoethanol, GSH = glutathione, CysSH = cysteine, PenSH = penicillamine.

^{b} Proceeds through the complex: $RS \overset{-}{-}-Cu(II)\begin{matrix} S\!-\!R \\ \cdot \; - \\ \cdot S\!-\!R \end{matrix}$

^{c} R'SH = lipoic acid ^{d} R'SH = cysteine

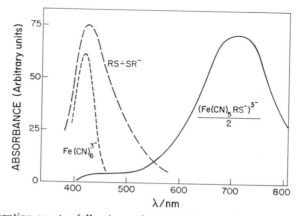

Fig. 2. Absorption spectra following pulse radiolysis of 10^{-4} M ferrocyanide in 10^{-2} M β-mercaptoethanol. Initial spectrum of RS.$^\bullet$.SR$^-$ present with RS$^\bullet$ (— —). Final product obtained with fresh solutions (-----). Strongly absorbing final product obtained, if Fe(CN)$_5$RS^{4-} has formed by photo-aquation prior to pulse; absorbance multiplied x 1/2 (———) (data from ref. 7b).

A systematic study of the oxidation of thiolate-liganded Cu(I)(RS)$_2$ species has shown that the cysteinyl radical oxidation of the cysteine-liganded system proceeds somewhat faster than the penicillamine thiyl radical oxidation of the penicillamine liganded Cu(I) (see Table 2). The "RS" species referred to in Table 2 is β-mercaptoethanol. Where the majority of the Cu(I) species had only one thiolate ligand and two cyanide ligands, the rate constant is somewhat larger than for the bulkier amino-thiol systems. These features suggest that steric factors are fairly important in determining the rates.

For the cysteine liganded Cu(I) species it was demonstrated that the disulphide anion of cysteine as well as the cysteinyl radical oxidized the Cu(I) (Table 2). The rate constant for the radical anion is, however, almost a factor of ten less than for the thiyl radical. This can be attributed to the greater bulk of the anion and/or its lower reduction potential (see Table 2 of first paper by Armstrong in this book). At the same time one must recall that the electron transferred to the anion must be accepted into an antibonding orbital. Thus, dissociation of the anion must occur simultaneously with or precede the electron transfer.[9a]

RS.$^{•}$.SR$^-$ AS A REDUCTANT

The reduction of the Cu(II) penicillamine complex to Cu(I) by $^{•}CO_2^-$ has been shown to proceed with the rate constant of 1.3×10^8 M^{-1}s^{-1}.[8c] An interesting feature is that reaction of this Cu(II) complex with RS$^{•}$ radicals also results in reduction to Cu(I) with a rate constant of a similar magnitude: 3.3×10^8 M^{-1}s^{-1} (Table 2). Mezyk and Armstrong[8c] have concluded that this reduction is caused by attack of the penicillamine thiyl radical on the sulphur atom of one of the penicillamine ligands. The resulting RS.$^{•}$.SR$^-$ can then transfer an electron to Cu(II). Here it must be recalled that while RS.$^{•}$.SR$^-$ has a reduction potential of 0.57 V, which is sufficient to oxidize Cu(I) to Cu(II), it can also act as a strong reducing agent, because the reduction potential of RSSR to RS.$^{•}$.SR$^-$ is –1.57 V (Table 2, first paper by D. A. Armstrong in this book).

In effect the complexation of RS$^{•}$ with RS$^-$ converts an oxidant into a strong reductant, and, as Ahmad and Armstrong pointed out,[9a] this can have a very significant effect on reactions in thiol-containing systems. Some examples of rate constants for reductions by disulphide radical anions are given in Table 2. In this particular connection it is worth noting that recent work has suggested that disulphide radical anions in biological systems may in fact be very important repair agents for target radicals.

When RS$^{•}$ radicals are generated by low intensities of ^{60}Co γ-rays the lifetimes of the radicals are sufficiently long that the equilibrium of equation (14) can be established.

$$RS^{•} + RSH \rightleftharpoons RS.^{•}.SR^- + H^+ \tag{14}$$

$$\tag{15}$$

Under these circumstances it was shown that photo-stationary concentrations of reduced flavin could be obtained.[9] These concentrations depended on the hydrogen ion and RSH concentrations as expected on the basis of equilibrium (14). For the disulphydryl systems, where equilibrium (15) applies, there was, as expected, a dependence only on hydrogen ion concentration.

REDOX REACTIONS OF CYSTEINE RESIDUES IN PROTEINS

A significant number of proteins contain free sulphydryl groups, which do not have a specific catalytic function and in many instances appear to be somewhat inert. Also common is the disulphide linkage where the sulphydryls of two different residues are co-oxidized to form a disulphide. This helps to provide stability in the tertiary structure of the protein. Such disulphides can be reduced by $e^-_{(aq)}$, and sometimes by $^\bullet CO_2^-$. The disulphide radical anions thus formed are extremely long-lived and their decay may result in damage to the protein molecule.[11]

Of somewhat greater interest are the sulphydryl groups which have a specific function. In metal containing-enzymes the metal ions are frequently bound by thiolate anions. This is true in superoxide dismutase and alcohol dehydrogenase. Thiolate binding of copper occurs in several of the "blue" copper proteins. In the sections "RS$^\bullet$ as an Oxidant" and "RS$\bullet$$\bullet$SR$^-$ as a Reductant" we observed the oxidation and reduction of copper bound by such ligands.

Sulphydryl groups in proteins may also have other important catalytic functions, such as substrate binding. In such cases the sulphur atom may be made strongly nucleophilic by the presence of a base. This is illustrated below, where the base-sulphydryl combination is shown to form an ion pair.

Papain and glyceraldehyde phosphate dehydrogenase are examples of enzymes with activated nucleophilic thiol groups (here denoted ES$^-$) involved in the active centres.[12] Papain was used extensively to investigate the effects of radiation-produced entities on such activated thiols.

Trapping of Enzyme Sulphenic Acid and Mixed Disulphide

The following equations show the mechanism for the reaction between hydrogen peroxide and sulphydryl molecules such as cysteamine.[12] Only the ionized form of the sulphydryl appears to react at a measurable rate, and so at pHs below the pK the primary reaction is relatively slow. The second reaction appears to be comparatively swift (see work of Packer et al. cited in ref. 12).

$$CysSH + H_2O_2 \rightarrow CysSOH + H_2O \tag{16}$$
$$(k \leq 1\ M^{-1}s^{-1})$$

$$CysSOH + CysSH \xrightarrow{\text{(fast)}} CysSSCys + H_2O \tag{17}$$

$$ES^- + H_2O_2 \rightarrow ESOH + OH^- \tag{16a}$$

For papain as the enzyme ES$^-$ in equation (16a) the primary reaction with hydrogen peroxide is much faster. Also the sulfenic acid which is formed cannot react further with other papain molecules and is therefore stabilized. When smaller sulphydryl molecules are added to the system the reactions in Table 3 occur. It is worth noting here that the methyl groups of penicillamine prevented it from regenerating the free enzyme from its ESSPen mixed disulphide form. Free enzyme can, however, be obtained from the mixed disulphide of papain and penicillamine by cleaving the disulphide with cyanide or a smaller more active sulphydryl.[12]

Table 3. Rate Constants for H_2O_2-Papain-CysSH System at pH = 6.0 ± 0.2[12]

Reaction		$k/M^{-1}s^{-1}$	
		298K	274K
$ES^- + H_2O_2$	$\rightarrow$ ESOH + OH^-	66	66
ESOH + CysSH	$\rightarrow$ ESSCys + H_2O	$\geq$800	$\geq$800
ESSCys + CysSH	$\rightarrow$ ES^- + H^+ + $(CysS)_2$	11	3
ESOH + PenSH	$\rightarrow$ ESSPen + H_2O	$\approx$800	–
ESSPen + PenSH	$\rightarrow$ *no reaction*	$\approx$0	–
ESOH + $\begin{array}{c}HS \\ HS\end{array}\Big]$ §	$\rightarrow$ $\begin{array}{c}S \\ S\end{array}\Big]$ + ES^- + H^+ + H_2O	$\geq$800	–

§ Dithiothreitol

Trapping of RS$^\bullet$

Reactions of H_2O_2 can be completely suppressed by incorporating micromolar concentrations of catalase into the papain solutions.[12] Under those circumstances mixed disulphides of papain with cysteine, penicillamine and glutathione were still observed when the enzyme was irradiated in the presence of these thiols under a nitrous oxide atmosphere (see Figure 3). From the magnitude of this it was concluded that the following reactions were responsible:

$$\bullet OH + RSH \rightarrow H_2O + RS^\bullet \tag{18}$$

$$ES^- + RS^\bullet \rightarrow ES.^\bullet.SR^- \tag{19}$$

$$ES.^\bullet.SR^- + RS^\bullet \rightarrow ESSR + RS^- \tag{20}$$

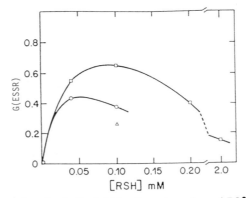

Fig. 3. Yield of papain mixed disulphides from trapping of RS$^\bullet$ radicals of cysteine (○), penicillamine (□) and Glutathione (△) in N_2O-saturated thiol solutions containing 20 μM papain and 4×10^{-8} M catalase (based on data from ref. 12).

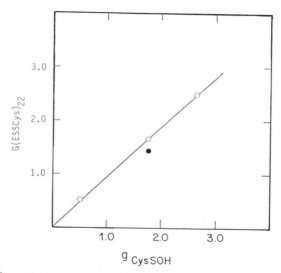

Fig. 4. Yield of papain-cysteine mixed disulphide from radiolysis of N_2O-saturated 20 μM solutions of papain and varying amounts of cystine versus yield of RSOH expected from reaction (21): without (○) and with (●) 4×10^{-8} M catalase (based on data from ref. 13). Data have been corrected for $^{\bullet}OH$ attack on papain and, when no catalase was present, for H_2O_2 reactions.

Trapping of RSOH

When papain is irradiated in the presence of disulphide, the ESOH species from the hydrogen peroxide has no sulphydryl to react with. In any case its formation can be suppressed, if desired, by adding catalase. In the presence of disulphide reaction (21) occurs along with other processes discussed in the first paper by Armstrong in this book.

$$^{\bullet}OH + RSSR \rightarrow RSOH + RS^{\bullet} \tag{21}$$

$$ES^- + RSOH \rightarrow ESSR + OH^- \tag{22}$$

Now with the ES^- of papain reaction (22) can occur, causing the RSOH to be trapped.[13] Figure 4 shows the correlation between the yield of repairable papain mixed disulphide estimated to be due to reaction (22) and the calculated yield (g value) for CysSOH in solutions of increasing cysteine concentration. The correlation is excellent. Here the yields are in μmoles per 10 J of radiation energy absorbed. Note that this also applies in Figures 3 and 5.

Trapping of Products of $RS^{\bullet} + O_2$ Reactions

In aerated solutions containing micromolar concentrations of papain hydrated electrons and hydrogen atoms react with oxygen to form superoxide radicals.[12] In the presence of a sulphydryl at millimolar concentrations the $^{\bullet}OH$ radicals will produce $RS^{\bullet}$. This then will react with oxygen and form other radicals which may react with the papain. A small number of $^{\bullet}OH$ react directly with papain to inactivate it and a correction has been made for this in the results shown in Figure 5. With dithiothreitol the $RS^{\bullet}$ radical rapidly cyclizes and then ionizes to form the cyclic disulphide radical anion (see first paper by Armstrong in this book). This transfers its electron to oxygen (Table 2), which results in an approximate doubling of the yield of $O_2^{\bullet-}$. The species $O_2^{\bullet-}$ has been shown to inactivate papain with an efficiency of 30%.[12] Thus, if all radicals[#] are converted to $O_2^{\bullet-}$, one expects an inactivation

---

[#] Recall that the total yield of radiation-produced radicals is ca. 6 μmole per 10 J.

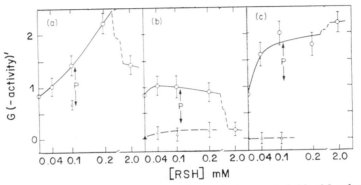

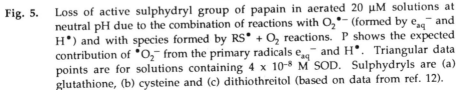

Fig. 5. Loss of active sulphydryl group of papain in aerated 20 μM solutions at neutral pH due to the combination of reactions with $O_2^{\bullet-}$ (formed by e_{aq}^- and $H^\bullet$) and with species formed by $RS^\bullet + O_2$ reactions. P shows the expected contribution of $^\bullet O_2^-$ from the primary radicals e_{aq}^- and $H^\bullet$. Triangular data points are for solutions containing 4×10^{-8} M SOD. Sulphydryls are (a) glutathione, (b) cysteine and (c) dithiothreitol (based on data from ref. 12).

yield of about 2 μmole per 10 J of radiation energy absorbed. The results in Figure 5c demonstrate that this is in fact the case. Furthermore, on addition of superoxide dismutase (SOD), the inactivation is completely suppressed by the removal of the $O_2^{\bullet-}$ (see the triangle data points).

The results with cysteine are significantly different (Figure 5b). Here, there is no significant increase in the inactivation yield. Furthermore, the reduction in inactivation observed on the addition of SOD is seen to correspond reasonably well with the magnitude expected for suppression of the $O_2^{\bullet-}$ formed from $e^-_{(aq)}$ and $H^\bullet$ atoms. This is shown by the vertical arrows marked P. One must conclude, therefore, that reaction of cysteine radicals with oxygen does not lead to the formation of $O_2^{\bullet-}$ or of any other species which reacts with the catalytic sulphydryl group of papain.

The results in Figure 5a with glutathione show that the situation is again completely different. The yield of inactivation is strongly inhanced as the concentration of glutathione rises. The effect of SOD is about what is expected for the removal of the $O_2^{\bullet-}$ produced from $e_{(aq)}^-$ and $H^\bullet$. Thus, the reaction of glutathione derived thiyl radical $GS^\bullet$ and oxygen must produce a further species, which is highly damaging towards the papain activated sulphydryl. It appears from other work that this may be an $RSO^\bullet$ type radical or other oxygen containing species.

CONCLUSION

The results presented above have shown that thiyl radicals can take part in a great variety of oxidation reactions. Removal of $RS^\bullet$ formed in the repair process (equation (1)) is not therefore likely to occur by simple dimerisation, but by a complex non-trivial set of processes. Complexation with RS^- converts $RS^\bullet$ into a disulphide radical anion, which may behave as an oxidant or a reductant. In cells this radical anion may serve as a repair agent or transfer an electron to oxygen to produce $O_2^{\bullet-}$.

The situation is further complicated by the fact that certain biologically active sulphydryls may react preferentially with thiyl radicals or oxygenated products formed from them. For papain this is particularly true of the products of glutathione radical reactions with oxygen.

Finally it must be emphasized that protection by RSH is only complete when the sulphur radicals are removed by conversion to $O_2^{\bullet-}$ or some other species, which can be handled by the normal redox processes of the cell.

ACKNOWLEDGEMENTS

This lecture and the first one have been made possible by the important contributions of many co-workers including M. Lal, A.J. Elliot, K.J. Stevenson, G.M. Gaucher, W. Lin, J.D. Buchanan, G.C. Goyal, J.R. Clement, R. Ahmad, Zhennan Wu, A-D. Leu, S. Mezyk and P.S. Surdhar. The author is deeply indebted to these persons, the late F.C. Adam and other colleagues for inspiration an insight.

REFERENCES

1. E.E. Reid, Organic Chemistry of Bivalent Sulphur, Chemical Publishing Co., New York (1958).
2. P. Wardman, Reduction Potentials of One-electron Couples Involving Free Radicals in Aqueous Solutions, *J. Phys. Chem. Ref. Data* 18:1637 (1989).
3 a. G.E. Adams, G.S. McNaughton, and B.D. Michael, *Trans. Faraday Soc.* 64:902 (1968).
   b. G.E. Adams, J.E. Aldrich, R.H. Bisby, R.B. Cundall, J.L. Redpath, and R.L. Wilson, *Radiation Research* 49:278 (1972).
4. A. Treinin and E. Hayon, *J. Am. Chem. Soc.* 97:1716 (1975).
5 a. M.S. Akhlaq, H.-P. Schuchmann, and C. von Sonntag, *Int. J. Radiat. Biol.* 51:91 (1987).
   b. A.J. Elliot, A.S. Simsons, and F.C. Sopchyshyn, *Radiat. Phys. Chem.* 23:377 (1984).
6 a. L.G. Forni and R.L. Willson, *Biochem. J.* 240:897 (1986).
   b. L.G. Forni and R.L. Willson, *Biochem. J.* 240:905 (1986).
7 a. P.S. Surdhar and D.A. Armstrong, *J. Phys. Chem.* 89:5514 (1985).
   b. Zhennan Wu, R. Ahmad, and D.A. Armstrong, to be published.
8 a. A.-D. Leu and D.A. Armstrong, *J. Phys. Chem.* 90:1449 (1986).
   b. S.P. Mezyk and D.A. Armstrong, *Can. J. Chem.* 67:736 (1989).
   c. S.P. Mezyk and D.A. Armstrong, to be published.
9 a. R. Ahmad and D.A. Armstrong, *Can. J. Chem.* 62:171 (1984).
   b. P.S. Surdhar and D.A. Armstrong, *Can. J. Chem.* 63:3411 (1985).
   c. P.S. Surdhar and D.A. Armstrong, *Int. J. Radiat. Biol.* 52:419 (1987).
10 a. P.C. Chan and B.H.J. Bielski, *J. Am. Chem. Soc.* 95:5504 (1973).
    b. J.P. Barton and J.E. Packer, *Int. J. Radiat. Phys. Chem.* 2:159 (1970).
11. N.N. Lichtin, J. Ogden and G. Stein, *Biochem. Biophys. Acta* 263:14 (1972).
12. D.A. Armstrong and J.D. Buchanan, *Photochem. Photobiol.* 28:743 (1978).
13. M. Lal, W.S. Lin, G.M. Gaucher, and D.A. Armstrong, *Int. J. Radiat. Biol.* 28:549 (1975).

ELECTROCHEMISTRY OF THIOETHERS AS MODELS FOR BIOLOGICAL ELECTRON TRANSFER

George S. Wilson

Department of Chemistry
University of Kansas
Lawrence, KS 66045, USA

Considering the importance of sulfur in biological systems, it is somewhat surprising that thioethers have only recently received attention from the point of view of their redox chemistry. In terms of biological function, there are perhaps three classes of interactions in which they are involved. The first case involves their role as a part of metal centers in redox-active proteins. Examples include cytochrome c where histidine and methionine comprise the two axial ligands in an iron porphyrin, co-factors in methanogenic bacteria F-420 (S–Ni coordination)[1] and in the Type I "Blue Copper" proteins.[2] Examples of this latter group include azurin and plastocyanin. In the latter case the immediate coordination around Cu(II) is one cysteine thiolate (S at 2.1 Å), two histidine imidazoles (N at 2.05 Å and 2.10 Å), and a methionine sulfur (at 2.9 Å) in a distorted trigonal pyramidal array.[3]

The second category of functions involves the possible role of thioethers in the transfer of electrons over long distances (10 - 30 Å) apparently through the protein fabric. Since these reactions are known to occur, it is important to understand how the medium through which the electrons must move can facilitate the efficiencies characteristic of biological reactions. The subject of electron transfer in proteins has recently been reviewed in detail.[4,5] Although the evidence is largely circumstantial, there are a number of indications that thioethers may be good candidates for this role. An analysis of the x-ray structures of a number of redox proteins has indicated the close proximity of methionine residues to π-electron systems of aromatic rings in flavins (isoalloxazine), heme proteins (porphyrin) or side chains of amino acids.[6] The oxidation of yeast cytochrome c peroxidase produces a ferryl species and a non-porphyrin radical ascribed to a reversibly oxidizable amino acid residue. Based on the x-ray structure of this molecule, the involvement of met-172 (methionine moiety at position 172 within molecular sequence of yeast cyotochrome c peroxidase) or trp-51 (trp = tryptophane moiety) which are close to the heme ring seems possible. Using site-directed mutagenesis, the met-172 was replaced with serine, and in a separate experiment, the trp-51 replaced with phenylalanine. The elimination of either of these residues does not cause the EPR signal to disappear, suggesting that perhaps both centers contribute to the observed signal.[7] A recent ENDOR study, however, casts serious doubt on the intermediacy of a sulfur-centered radical.[8]

The third category of interactions involves redox reactions of thioethers such as the P-450 catalyzed conversions of thioethers into their corresponding sulfoxides. It has been shown that the rate of sulfoxidation of a series of thioanisole derivatives (ArSMe) by a reconstituted P-450 complex isolated from rabbit hepatic microsomes correlates well with peak

Sulfur-Centered Reactive Intermediates in Chemistry and Biology, Edited by
C. Chatgilialoglu and K.-D. Asmus, Plenum Press, New York, 1990

potentials from cyclic voltammetry and also with Hammett values.[9] The suggestion that sulfoxidation proceeds through a cation radical has recently been brought into question and an atom transfer mechanism has been suggested as more likely.[10] Other types of reactions in this category include a variety of biosynthetic processes. Ethylene, a plant hormone, is synthesized from L-methionine through an oxidative process.[11] The increasing importance of peptides and proteins as therapeutic agents has made it necessary to understand in detail the conditions under which methionine is converted into its sulfoxide as this process contributes to lowered protein stability and possible loss of biological activity.[12]

There is little question that thioether intermediates in biological systems will not be identified and understood until they can be generated in well-defined model systems. Such systems often must be rather rigid so that the interactions of interest can be properly studied and controlled.

MODELS FOR THIOETHERS IN METAL CENTERS OF PROTEINS

As ligands of transition metals, thioethers have not attracted a great deal of attention. They are weak sigma donors and poor π-acceptors thus yielding low formation constants (log K = 2 - 4) so that ligand displacement occurs very readily. In view of these properties, it is somewhat surprising to find met-80 as a ligand in horse heart cytochrome c. A number of studies have shown that the interaction of met-80 with the heme iron does not define the conformation of the protein.[13] Rather it is the protein folding which places the methionine in the proper orientation with respect to the iron. Methionine coordination does help to establish the rather high redox potential of cytochrome c by preferentially stabilizing the Fe(II) form. It can be shown, however, that this stabilization cannot account for the difference of more than 450 mV between simple model compounds (hemes having histidine and methionine as ligands) and the potential of cytochrome c itself. By using recombinant heme peptides, it is possible to show that this positive shift can also be attributed to the creation of a largely hydrophobic pocket into which the heme is inserted.[14] Thus it is not sufficient to model such interactions by looking only at the iron center. The environmental effects can be quite important. To look at thioether coordination in porphyrins, we studied, in collaboration with Prof. David Dolphin, the basket-handle porphyrin shown in Figure 1. The molecule is so designed to inhibit displacement of the thioether by the more strongly interacting nitrogenous ligands such as imidazole (his) (his = histidine moiety). In addition to studying the effects of ligand binding on the redox potential of the Fe(III)/Fe(II) couple, it was also of interest to see

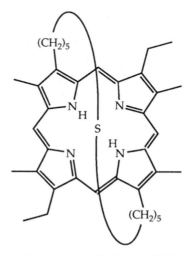

Fig. 1. Structure of "basket handle" porphyrin.

Table 1.   Peak Potentials for the Oxidation of Sulfides

Compound	Peak Potential[a]
Dimethyl sulfide	1.75 V
Di-n-butyl sulfide	1.79 V
S-strap porphyrin[b]	1.51 V
Zn S-strap porphyrin[b]	1.86 V
Fe S-strap porphyrin[c]	2.72 V

[a]   All potentials reported vs. SCE.
[b]   Scan rate 100 mV/s; 0.1 M $NaClO_4$ in acetonitrile.
[c]   Scan rate 100 mV/s; 0.2 M TBATFB in CH_2Cl_2.

if a thioether coordinated to a metal can undergo oxidation. Table 1 summarizes the results of these investigations.[15] The oxidation of "simple" aliphatic sulfides occurs at 1.75 - 1.80 V vs. saturated calomel electrode (SCE) or about 1.4 V vs. Ag/Ag^+. The potential of the S-strap porphyrin is somewhat lower (1.51 V vs. SCE), but as electron density on the sulfur is decreased as result of increasingly stronger coordination with metals, the potential becomes quite high as indicated.

In the course of examining the redox chemistry of small ring polythioethers (thio-crown ethers), we synthesized several transition metal complexes of 1,4,7-trithiacyclononane (TTCN).[16] Detailed discussion of the x-ray structures and spectral properties of the complexes of Cu(II), Fe(II), Co(II) and Cd(II) are beyond the scope of the present discussion. The properties of these thioether complexes can be summarized by saying that they possess greatly enhanced stability (6 - 9 orders of magnitude improvement), they form cofacial complexes consisting of two TTCN molecules with the metal in between, and generally behave as strong field ligands. While they are not models for metal centers in proteins, they do illustrate that the binding properties of thioethers can be substantially altered if the coordination geometry is optimized. Because the sulfur atom is large and polarizable it would seem to be a good candidate for transmitting electrons from the metal core to the electrode surface. A second important consideration, which affects both complex stability and electron transfer rate, is the ability of the ligand(s) to rapidly accommodate the coordination geometry changes necessitated by redox reactions. There are very few examples of Co(II) thioether complexes and still fewer where the Co is low spin.[17]

Figure 2 shows the cyclic voltammogram of $Co(II)(TTCN)_2$. Two reversible reduction peaks are observed which correspond to Co(III) → Co(II) and Co(II) → Co(I), respectively. There is a third irreversible reduction wave at –1.78 V which corresponds to the formation of Co(0). At this point the ligands are lost and their presence in solution can be detected on the basis of the electrochemistry of the free ligand. As noted from Figure 2, it is possible to obtain the heterogeneous electron transfer rates constants for the first two steps and these are reasonably rapid (0.012 cm/s and 0.0075 cm/s, respectively). This result, and similar results obtained for a $Cu(II)(TTCN)_2$ complex, are in contrast with previous studies of thioether-containing macrocycles[18] in which both the stabilities and electron transfer rates are low. In the case of TTCN, all of the ligation sites are endodentate, whereas in the case of many of the larger macrocycles conformational transitions are required to form the complex. Changing coordination geometry is also linked to changes in redox state. Cu(II), for example, prefers octahedral coordination whereas Cu(I) complexes prefer tetrahedral geometry. The geometry

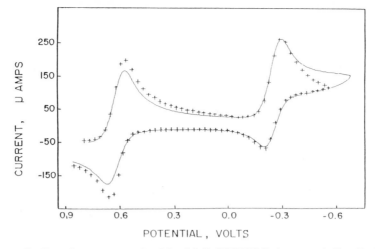

Fig. 2.   Cyclic voltammogram for 0.8 mM Co(II)(TTCN)$_2$ in acetonitrile. Pt electrode, Ag/Ag$^+$ 0.1 M reference. Points: theory; solid line: experimental.

(distorted tetrahedral) of the active sites in blue copper proteins can represent an intermediate geometry between the Cu(II) and Cu(I) state. As noted above, the Cu–S distance in plastocyanin is 2.9 Å leading some workers to conclude that methionine is largely non-interacting.

MODELS FOR "LONG DISTANCE" ELECTRON TRANSFER

The question of the dependence of electron transfer rates on the distance between redox centers has been addressed for a wide range of systems. For proteins, the approach has been to react the redox protein with a small molecule. Alternatively one can examine transfer within multisite proteins or between a native redox site and one which has been inserted.[19] These approaches, however, do not address directly the effects of the intervening medium on the rates of electron transfer. For this reason, well-defined model systems will be needed. Taube[20] has provided some convincing evidence for through-space electronic delocalization in mixed valence complexes of ruthenium ammines when 1,5-dithiocane (DTCO) serves as a bridging ligand. In this case, the interaction is facilitated by a ligand to metal charge transfer between an antibonding orbital from the sulfur lone pairs and the complementary metal orbital. This interaction leads to electron transfer which is effectively adiabatic. It has further been demonstrated that sulfur can facilitate "long range" (11 - 17 Å) tunneling at rates ranging from 8.0 x 10^7 s^{-1} to 3.5 x 10^4 s^{-1}, respectively. The bis-ruthenium systems studied were spiro compounds and were therefore sufficiently rigid to preclude direct ruthenium to ruthenium electron transfer.[21] Such binuclear systems typically gives rise to single broad voltammetric peaks which are the result of two closely-spaced one electron steps. If there are two identical redox centers in the binuclear system then they could be expected to show a peak potential separation of 35.5 mV if they do not interact with each other.[22] This difference is due purely to statistics. If the change in oxidation state of one of the centers is "sensed" by the other then, in general, the separation will be greater than the non-interacting value. A role for thioethers in the facilitation of electron transfer within proteins has been considered.[23] It has been suggested that in c-type cytochromes porphyrin π-orbitals and Fe(III) d-π orbitals can delocalize onto the bridging sulfur atom. In cytochrome c this corresponds to cys-17 (cys = cysteine moiety) where the sulfur is used to form a covalent (thioether) linkage between the porphyrin ring and protein backbone. The thioether group is further exposed to the solvent providing further possibility for interaction.

In summary, the examination of the role of sulfur, especially thioethers, in biological systems is complicated by the low stability of transient intermediates and properties which are neither known nor clearly predictable. Thus the importance of fundamental studies on well-defined model system is clear.

ACKNOWLEDGEMENT

Support of this work by the National Institutes of Health (Grant No. HL 15104) and the National Science Foundation is gratefully acknowledged.

REFERENCES

1. S.L. Tan, J.A. Fox, N. Kojima, C.T. Walsh, and W.H. Orme-Johnson, *J. Am. Chem.Soc.* 106:3064 (1984).
2. D.R. McMillin, *J. Chem. Ed.* 62:997 (1985).
3. K.W. Penfield, A.A. Gewirth, and E.I. Solomon, *J. Am. Chem. Soc.* 107:4519 (1985).
4. R.A. Marcus and N. Sutin, *Biochim. Biophys. Acta* 811:265 (1985).
5. G. McClendon, *Acc. Chem. Res.* 21:160 (1988).
6. B.L. Bodner, L.M. Jackman, and R.S. Morgan, *Biochem. Biophys. Res. Commun.* 94:807 (1980).
7. D.B. Goodin, A.G. Mauk, and M. Smith, *J. Biol. Chem.* 262:7719 (1987).
8. M. Sivaraja, D.B. Goodin, M. Smith, and B.M. Hoffman, *Science* 245:738 (1989).
9. Y. Watanabe, S. Oae, and T. Iyanagi, *Bull. Chem. Soc. Jpn.* 55:188 (1982).
10. A.E. Miller, J.J. Bischoff, C. Bizub, P. Luminoso, and S. Smiley, *J. Am. Chem. Soc.* 108:7773 (1986).
11. S.F. Yang, *in*: "The Chemistry and Biochemistry of Plant Hormones", V.C. Runeckles, E. Sondheimer, and D.C. Walton, eds., Academic Press, New York, pp. 131 (1974).
12. L.C. Teh, L.J. Murphy, N.L. Huq, A.S. Surus, H.G. Friesen, L. Lazarus, and G.E. Chapman, *J. Biol. Chem.* 262:6472 (1987).
13. W.R. Fisher, H. Taniuchi, and C.B. Anfinsen, *J. Biol. Chem.* 248:3188 (1973).
14. H. Wilgus, J.S. Ranweiler, G.S. Wilson, and E. Stellwagen, *J. Biol. Chem.* 253:3265 (1978).
15. D.P. Root, PhD Dissertation, University of Arizona (1984).
16. W.N. Setzer, C.A. Ogle, G.S. Wilson, and R.S. Glass, *Inorg. Chem.* 22:266 (1983).
17. G.S. Wilson, D.D. Swanson, and R.S. Glass, *Inorg. Chem.* 25:3827 (1986).
18. E.R. Dockal, L.L. Diaddario, M. Glick, and D.B. Rorabacher, *J. Am. Chem. Soc.* 99:4530 (1977).
19. H. Elias, M.H. Chou, and J.R. Winkler, *J. Am. Chem. Soc.* 110:429 (1988).
20. C.A. Stein, and H. Taube, *J. Am. Chem. Soc.* 100:1635 (1978).
21. C.A. Stein, N.A. Lewis, and G. Seitz, *J. Am. Chem. Soc.* 104:2596 (1982).
22. F. Ammar and J.M Saveant, *J. Electroanal. Chem.* 47:115 (1973).
23. G. Tollin, L.K. Hanson, M. Caffrey, T.E. Meyer, and M.A. Cusanovich, *Proc. Natl. Acad. Sci. USA* 83:3693 (1986).

FREE-RADICAL REACTIONS INVOLVING THIOLS AND DISULPHIDES

Clemens von Sonntag

Max-Planck-Institut für Strahlenchemie
4330 Mülheim a.d. Ruhr, F.R. Germany

FORMATION OF THIYL RADICALS

The RS–H bond is rather weak compared to the RO–H or the R_3C–H bonds (Table 1; allylic or biallylic C–H bonds are comparable in strength to or weaker than the S–H bond). For this reason, thiols involved in free-radical reactions may act as a sink for the free spin by suffering H-transfer [e.g. reaction (1)].

$$^\bullet CH_2OH + RSH \rightarrow CH_3OH + RS^\bullet \qquad (1)$$

In reaction (1), the original compound which has given rise to the radical (in our case methanol) is restored. Therefore such reactions have been termed *repair reactions* by radiation biologists, or better: *chemical repair reactions*, in order to distinguish them from the enzymatic repair of radiation damage. Some aspects of chemical repair by thiols will be discussed below, and its biological relevance will be dealt with in the article by C. von Sonntag and H.-P. Schuchmann in this book.

Due to the ease of H-transfer reactions such as reaction (1), thiyl radical may be generated from all kinds of other free-radical precursors, in radiation chemistry[1,2] notably by the OH radical (*cf.* first article by D.A. Armstrong in this book).

Photolysis of thiols also leads to S–H bond breakage (reaction (2)).[3] This is not necessarily due to the weakness of the S–H bond, because in the series of aliphatic alcohols the O–H bond is preferentially broken as well, which in this system is the *strongest* bond. In fact, it has been shown that this preferential splitting of the O–H bond is due to the special structure of the excited state which is antibonding with respect to the O–H bond.[4] Another potential source of thiyl radicals may be the photolysis of disulphides [reaction (3)].[3] This reaction is reminiscent of the formation of oxyl radicals from the photolysis of peroxides.

$$RSH + h\nu \rightarrow RS^\bullet + H^\bullet \qquad (2)$$

$$RSSR + h\nu \rightarrow 2 RS^\bullet \qquad (3)$$

Compared to the O–O bond, the S–S bond is rather strong (*cf.* Table 1). While thermolysis of peroxides is a convenient source of oxyl radicals, disulphides do not break up into thiyl radicals as readily at elevated temperatures.

Sulfur-Centered Reactive Intermediates in Chemistry and Biology, Edited by
C. Chatgilialoglu and K.-D. Asmus, Plenum Press, New York, 1990

Table 1. Gas phase bond dissociation energies (approximate), in units of kcal mol^{-1}

Bond	BDE	Bond	BDE
RO–H	104	RCH$_2$–H	98
PhO–H	86	R$_2$CH–H	95
R$_2$N–H	91	R$_3$C–H	93
RS–H	91		
		HOCH$_2$–H	94
RO–OR	38	HOCH(R)–H	93
R$_2$N–NR$_2$	59	HOC(R$_2$)–H	91
RS–SR	72	RSCH$_2$–H	96
		PhCH$_2$–H	88
CH$_2$=O	172	C–H (allylic)	89–81[a]
CH$_2$=S	129	C–H (biallylic)[b]	73

[a] depending on substitution.
[b] 1,3-cyclohexadiene.
Data were taken from: D.F. McMillen and D.M. Golden, *Ann. Rev. Phys. Chem.* 33:493 (1982), and S.W. Benson, *Chem. Rev.* 78:23 (1978).

REACTIVITY OF RS$^{\bullet}$ AS COMPARED TO RO$^{\bullet}$ RADICALS

Radiation biologists have often considered thiyl radicals to be rather unreactive and to undergo barely any other reaction than to combine with one another [reaction (4)].

$$2\ RS^{\bullet} \rightarrow RSSR \tag{4}$$

They are indeed rather inert when compared to their oxygen-centered analogues, the oxyl radicals, which undergo efficient hydrogen abstraction from hydrocarbons [reaction (5)]. Further, in aqueous solutions they undergo a rapid 1,3-H shift if they happen to carry an α-hydrogen [reaction (6)], or readily fragment if they are tertiary oxyl radicals [reaction (7)]. The half-lives for such reactions are reported to be in the microsecond and sub-microsecond range.

$$RO^{\bullet} + R_3CH \rightarrow ROH + R_3C^{\bullet} \tag{5}$$

$$R_2C(H)O^{\bullet} \text{ (in water)} \rightarrow {}^{\bullet}CR_2OH \tag{6}$$

$$R_3CO^{\bullet} \rightarrow R_2C=O + {}^{\bullet}R \tag{7}$$

Reaction (6) is exothermic by about 10 kcal mol^{-1}, and reaction (7) has only a low activation energy in the gas phase; owing to solvation effects it is even faster in aqueous solutions. The analogues of reactions (6) and (7) involving the corresponding sulphur compounds are endothermic (note, e.g., the weakness of the C=S double bond, Table 1). Such reactions do, therefore, not occur with the corresponding thiyl radicals at room temperature.

The H-donation reaction of thiols [*cf.* reaction (1)] may be used to titrate the yield of a given radical among a mixture of radicals by the quantitative analysis of the resulting product. However, this method fails with allylic and biallylic radicals which react only slowly with thiols, or not at all. An example is given for the three radicals derived by OH radical attack on 1,3-cyclohexadiene.[5]

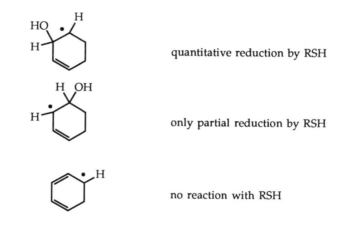

quantitative reduction by RSH

only partial reduction by RSH

no reaction with RSH

The rate constant of the reaction of thiols with carbon-centered radicals is often (but not always, see below) in the order of 10^7 - 10^8 dm^3 mol^{-1} s^{-1} (*cf.* Table 2), i.e. below the limits set by the diffusion. However, according to some preliminary results (Akhlaq, Steenken, and von Sonntag, unpublished) these reactions proceed without any noticeable activation energy (approximately 2 kcal mol^{-1}, i.e. about equivalent to that of the diffusion in water) and the slowness of the reaction is largely due to negative entropic terms.

The rate constant of the H-transfer is not linked to the redox-potential[6] of the radical involved. In fact, in the series of α-hydroxyalkyl radicals (Table 2) it is just the other way round: the most reducing α-hydroxyalkyl radical, i.e. that derived from 2-propanol (E_7^0 = –2.2 V), is reduced by RSH faster than that derived from methanol (E_7^0 = –1.4 V), and the oxidizing formylmethyl radical ($^\bullet$CH$_2$CHO) reacts so slowly with thiols that the H-transfer reaction cannot be measured on the pulse radiolysis time-scale ($k \ll 10^7$ dm^3 mol^{-1} s^{-1}). It is, however, readily reduced via electron transfer by thiolate ions (k = 1.2 x 10^8 dm^3 mol^{-1} s^{-1}[7] for these redox reactions see articles by D.A. Armstrong in this book).

Table 2.   Rate constants of the *repair* of some radicals by thiols (selected from data compiled in ref. 2)

Radical	Thiol	k/dm^3 mol^{-1} s^{-1}
$^\bullet$C(CH$_3$)$_2$OH	2-mercaptoethanol	5.1 x 10^8
$^\bullet$C(CH$_3$)HOH	2-mercaptoethanol	2.3 x 10^8
$^\bullet$CH$_2$OH	2-mercaptoethanol	1.3 x 10^8
$^\bullet$CH$_2$CH$_2$OH	2-mercaptoethanol	4.7 x 10^7
$^\bullet$CH$_3$	methylmercaptane	7.4 x 10^7
$^\bullet$CH$_2$CHO	glutathione	very slow
$^\bullet$SCH$_2$CH$_2$OH	dithiothreitol	1.7 x 10^7

Relevant to the discussion of the *chemical repair* of nucleic acid radicals are the observations that the radicals derived from pyrimidines and purines are also among those radicals which react only slowly with RSH by H-transfer (see also article by C. v. Sonntag and H.-P. Schuchmann in this book).

It can be seen from Table 1 that some C–H bond energies are very close to those of the S–H bond. In particular, an alkoxyl or hydroxyl group in α-position reduces the C–H bond energy by a few kcal mol^{-1} as compared to the unsubstituted alkane. With a C–H bond energy just below that of the S–H bond, the reverse of reaction (8) is slightly endothermic, and reversibility can be observed at room temperature. In fact, if one starts with *trans*-2,5-dimethyltetrahydrofuran, its *cis* isomer is formed *via* a chain reaction as indicated by the set of reactions (8) - (11).[8]

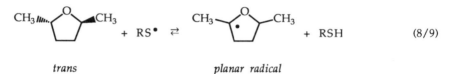

$$\text{trans} \qquad\qquad\qquad \text{planar radical}$$

(8/9)

$$\updownarrow$$

(10/11)

$$\text{cis}$$

Similar reaction sequences have been found with formate, phosphite, and alcohols.[9,10] (For further references see Ref. 2.) The rate constant for H-abstraction from the 2,5-dimethyltetrahydrofurans and from isopropanol are in the order of 10^4 dm^3 mol^{-1} s^{-1},[8,10] while H-abstraction from biallylic sites in polyunsaturated fatty acids are in the order of 10^6 - 10^7 dm^3 mol^{-1} s^{-1}.[11] In this context it is interesting to note that we have been unsuccessful to detect abstraction of any of the allylic H-atoms from the methyl group of 1,3-dimethylthymine. The conditions were such that we should have observed this reaction even if the rate constant was as low as 3×10^3 dm^3 mol^{-1} s^{-1} (Schuchmann and v. Sonntag, unpublished results). The parameters that govern both the H-transfer from the thiols to organic radicals and its reverse reaction are not yet fully understood at present.

ADDITION OF THIYL RADICALS TO C=C DOUBLE BONDS, AND THE REVERSE OF THIS REACTION

Most radicals are capable of adding to C=C double bonds. This is also believed to occur with thiyl radicals [reaction (12)], but usually no products of this reaction are observed. The reverse [reaction (13)] is so fast that the addition radicals do not play a major role owing to their very low concentration.[2]

$$RS^\bullet + R_2C=CR_2 \ \rightleftarrows \ RSC(R_2)-C(R_2)^\bullet \tag{12/13}$$

DECOMPOSITION OF 1,4-DITHIOTHREITOL: A CHAIN REACTION

When free radicals react with 1,4-dithiothreitol in deoxygenated aqueous solutions, considerable amounts of H$_2$S are liberated in a chain reaction which is depicted in reactions

(14) - (19).[12] All the elements of this chain reaction have been discussed previously, thus it is not necessary to present the various steps here in more detail.

$$HSCH_2CH(OH)CH(OH)CH_2S^\bullet \quad \rightleftarrows \quad \overline{SCH_2CH(OH)CH(OH)CH_2S}(H)^\bullet \qquad (14/15)$$

$$HSCH_2CH(OH)CH(OH)CH_2S^\bullet \quad \rightleftarrows \quad HSCH_2\overset{\bullet}{C}(OH)CH(OH)CH_2SH \qquad (16/17)$$

$$HSCH_2\overset{\bullet}{C}(OH)CH(OH)CH_2SH \quad \rightarrow \quad CH_2{=}C(OH)CH(OH)CH_2SH + {}^\bullet SH \qquad (18)$$

$${}^\bullet SH + HSCH_2CH(OH)CH(OH)CH_2SH \quad \rightarrow \quad H_2S + HSCH_2CH(OH)CH(OH)CH_2S^\bullet \qquad (19)$$

(The cyclic radical formed in reaction (14) is a $2\sigma/1\sigma^*$ three-electron bonded species with a relatively weak sulphur-sulphur bond; see also article by K.-D. Asmus in this book.)

REACTION OF RS$^\bullet$ WITH RS$^-$ AND ITS REVERSIBILITY. PROTONATION OF THE DISULPHIDE RADICAL ANIONS

It is known (*cf.* articles by D.A. Armstrong and K.-D. Asmus in this book) that thiyl radicals react very rapidly with thiolate anions [reaction (20)] and that the resulting disulphide radical anions are in equilibrium with these two species [reaction 21)], the equilibrium constant for most systems studied so far being around $K = 10^3$.[2] It is noted that this reaction has its analogies in the reaction of halogen and pseudo-halogen radicals with their corresponding anions (e.g. $Br^\bullet + Br^- \rightleftarrows Br_2{}^{\bullet-}$).[13]

The disulphide radical anions (RS$\bullet$$\bullet$SR)$^-$ react rapidly with protons [reaction (22); $k \approx 10^{10}$ dm^3 mol^{-1} s^{-1}; *cf.* Ref. 14]. These protonated disulphide radical anions are not stable and break up rapidly (k in the order of 10^6 s^{-1}) into thiols and thiyl radicals [reaction (25)].

$$RS^\bullet + RS^- \quad \rightleftarrows \quad (RS\bullet\bullet SR)^- \qquad (20/21)$$

$$(RS\bullet\bullet SR)^- + H^+ \quad \rightleftarrows \quad RS\bullet\bullet SR(H) \quad \rightleftarrows \quad RS^\bullet + RSH \qquad (22/25)$$

Usually, equilibrium (24/25) lies far to the right side, and the RS$\bullet$$\bullet$SR(H) radical is not detectable at low thiol concentrations. However in the case of dithiols, the thiol and the thiyl radical functions are linked together by a carbon chain. This enables stabilization and observation of such intermediates in equilibrium with their open-chain forms [cf. equilibrium (14/15)]. It appears that the ratio of protonated disulphide radical anion and the corresponding open-chain form strongly depends on the ring size. The apparent absorption coefficients of the disulphide radical anions and their protonated forms are given in Table 3. It can be seen from this table that in the case of a 5-membered ring system there is only a slight shift of the absorption maximum, and also the apparent absorption coefficients of the protonated forms and the anions are not very different. This is not so for smaller or larger ring sizes. It is known that thiyl radicals do not absorb significantly in this wavelength region. This rules out any contribution of the open-chain form to the observed absorptions. With the assumption that the ring-closed protonated forms and the radical anions have about the same absorption coefficients, we may conclude that in the 5-membered ring system the cyclic form predominates, while in the other cases the open-chain forms contribute to a greater extent. In Table 3, pK_a values are also listed. These are the *observed* pK_a and include the effect of the equilibrium (24/25). In those cases where the open-chain form plays a role the observed pK_a values are higher than the intrinsic pK_a values given by equilibrium (22/23). In the case of the 7-membered ring system the absorption at 370 nm may, in fact, be largely due to the absorption of the open thiyl radical (*cf.* Ref. 14).

Table 3. Absorption maxima and molar absorption coefficients of cyclic disulphide radical anions, $(\overset{\frown}{X-S-S})^{\bullet-}$ and their protonated forms, $(\overset{\frown}{X-S-S})H^{\bullet}$ Note that for the latter equilibria like (14/15) have to be taken into account and only the apparent absorption coefficient is given

$\overset{\frown}{X-S-S}$	$(\overset{\frown}{X-S-S})^{\bullet-}$		$(\overset{\frown}{X-S-S})H^{\bullet}$		pK_a
	λ_{max}	ε	λ_{max}	ε	
$\overset{\frown}{CH_2-CH_2-S-S}$	500	2990	460	240	6.7
$\overset{\frown}{CH_2-CH_2-CH_2-S-S}$	400	6800	380	4910	6.6
$\overset{\frown}{CH_2-CH_2-((CH_2)_4COOH)-S-S}$ a)	410	9200	385	6900	5.85
$\overset{\frown}{CH_2-CH_2-CH_2-CH_2-S-S}$	400	4540	380	290	7.0
$\overset{\frown}{CH_2-CH(OH)-CH(OH)-CH_2-S-S}$	390	6600	380	450	5.2
$\overset{\frown}{CH_2-CH_2-CH_2-CH_2-CH_2-S-S}$	425	3830	370	155	9.4

a) data from Ref.[21]

It can also be seen from Table 3 that in the case of the 4-membered ring system the absorption maximum is shifted significantly to longer wavelengths than in the other systems. This is in keeping with a weaker three-electron bond due to ring strain.[13]

It is well known from the work of Asmus and his colleagues that the sulphide radical cation readily complexes with another sulphide molecule to give the dimer radical cation [equilibrium (26/27)].[13]

$$R_2S^{+\bullet} + R_2S \rightleftarrows (R_2S\bullet\therefore\bullet SR_2)^+ \qquad\qquad (26/27)$$

Acceptance of another proton by the cyclic forms of the protonated disulphide radical anions would transform them into the dihydrogen analogues of the above-mentioned dimeric sulphide radical cations. These species absorb at different wavelengths, compared to the radical anions or their protonated forms. Thus, on going to lower pH values, it should be possible to observe any further protonation of the cyclic protonated radical anions. We have carried out some preliminary experiments with 1,5-pentanedithiol as the substrate but have *not* observed any change in absorption (neither in λ_{max} nor ε), even down to a pH as low as –1 (5 mol dm^{-3} H$_2$SO$_4$). This is perhaps a surprising result, since in the case of H$_2$S as substrate, the formation of $(H_2S\bullet\therefore\bullet SH_2)^+$ has been reported at pH ≤ 2.[15]

In agreement with expectations, the disulphide radical anions are strongly reducing radicals, while the related dimeric sulphide radical cations are good oxidants.

REDUCTION OF DISULPHIDES YIELDING $(RS\bullet\therefore\bullet SR)^-$. THE COMPLEX KINETICS WITH REDUCING ORGANIC RADICALS

It is well known that the disulphide radical anions are readily formed when solvated electrons react with the disulphide, but other radicals such as the formate radical anion are also capable of undergoing this reduction reaction. In a recent study we have shown that also α-hydroxyalkyl radicals show this reaction, but that it is too slow to be followed by pulse radiolysis.[16] However, the anions of these radicals readily reduce a disulphide such as oxidized dithiothreitol. These reactions appear not to be simple electron transfer as their rate constants show no correlation with the redox potentials of the various α-hydroxyalkyl radical anions. In addition, their Arrhenius plots are considerably curved, indicating that one does not deal with a one-step mechanism. It has been suggested that the α-hydroxyalkyl radical and their anions first form (reversibly) a complex before decomposing into carbonyl compound and disulphide radical anion [reactions (28) - (30)].

$$\bullet CH_2OH + RSSR \ \rightleftharpoons \ RSSR(CH_2OH)^\bullet \qquad\qquad (28/29)$$

$$RSSR(CH_2OH)^\bullet \ \rightarrow \ (RS\bullet\cdot\bullet SR)^- + CH_2O + H^+ \qquad\qquad (30)$$

$$RS^\bullet + R'SSR' \ \rightleftharpoons \ R'SSR'(SR)^\bullet \ \rightleftharpoons \ RSSR' + R'S^\bullet \qquad\qquad (31/33)$$

Such a mechanism is reminiscent of the displacement of $RS^\bullet$ in disulphides by thiyl radicals [reactions (31) - (33)]. The intermediate of this reaction is thought to be a three-electron-bonded species.[17]

The reduction of the disulphide by the formate radical may lead to a chain reaction as has been shown for the case of lipoamide [LipSS; reactions (34) - (36)].[18]

$$LipSS + \bullet CO_2^- \ \rightarrow \ Lip(S\bullet\cdot\bullet S)^- + CO_2 \qquad\qquad (34)$$

$$Lip(S\bullet\cdot\bullet S)^- + H_2O \ \rightarrow \ Lip(S\bullet\cdot\bullet SH) + OH^- \qquad\qquad (35)$$

$$Lip(S\bullet\cdot\bullet SH) + HCO_2^- \ \rightarrow \ Lip(SH)SH + \bullet CO_2^- \qquad\qquad (36)$$

REACTIONS OF THIYL RADICALS AND DISULPHIDE RADICAL ANIONS WITH OXYGEN

The vast majority of radicals studied so far have been carbon-centered radicals. These react readily with oxygen, very often with rate constants which are close to diffusion-controlled.[2] At room temperature, a possible equilibrium between the radical and oxygen, on the one hand, and the peroxyl radical, on the other, lies usually quite far to the side of the peroxyl radical. Recently it has been shown that in the case of some thiyl radicals, observable equilibria are rapidly established (37/38) and that, even in oxygen-saturated solutions, they are not entirely shifted to the peroxyl radical side.[19]

$$RS^\bullet + O_2 \ \rightleftharpoons \ RSO_2^\bullet \qquad\qquad (37/38)$$

Accordingly, the resulting chemistry in such systems must be very complicated and is far from being fully understood at present. There is good evidence from low-temperature ESR studies that the $RSO^\bullet$ radical may be an intermediate.[20] Moreover, there is mounting evidence that at low 2-mercaptoethanol concentrations per initial thiyl radical, half a molecule of disulphide and one molecule of an acid (mainly sulphonic and some sulfuric acid) is formed without any noticeable chain reaction (no effect of dose rate, little effect of thiol concentration).

As the pH of the solution approaches the pK_a value of the thiol, the situation becomes more complicated still since the concentration of thiolate ions is increased and equilibrium (20/21) comes into play. The reaction of the resulting disulphide radical anions with oxygen rapidly leads to the disulphide and a superoxide radical anion (reaction 39).

$$(RS\cdot^{\bullet}\cdot SR)^- + O_2 \;\rightarrow\; RSSR + O_2{}^{\bullet -} \tag{39}$$

The subsequent reactions are still obscure, but is is known that a chain reaction sets in. This is a surprising fact since the superoxide radical anion is known to be a very poor H-abstractor, and is not expected to propagate a chain effectively.

REFERENCES

1. C. von Sonntag and H.-P. Schuchmann, in: "The Chemistry of Functional Groups." The Chemistry of Ethers, Crown Ethers, Hydroxyl Groups, and Their Sulphur Analogues. Part 2. Suppl. E, pp. 971-993, S. Patai, ed., Wiley, New York (1980).

2. C. von Sonntag, "The Chemical Basis of Radiation Biology", Taylor and Francis, London, p. 353 (1987).

3. C. von Sonntag and H.-P. Schuchmann, in: "The Chemistry of Functional Groups." The Chemistry of Ethers, Crown Ethers, Hydroxyl Groups, and Their Sulphur Analogues. Part 2. Suppl. E., pp. 923-934, S. Patai, ed.,Wiley, New York (1980).

4. R.J. Buenker, G. Olbrich, H.-P. Schuchmann, B. Schürmann, and C. von Sonntag, J. Am. Chem. Soc. 106:4362 (1984).

5. X.-M. Pan, E. Bastian, and C. von Sonntag, Z. Naturforsch. 43b:1201 (1988).

6. S. Steenken, in: "Landolt-Börnstein, Neue Serie, Gruppe II", Vol. 13e, p. 147, H. Fischer, ed., Springer Verlag, Heidelberg (1985).

7. M.S. Akhlaq, S. Al-Baghdadi, and C. von Sonntag, Carbohydr. Res. 164:71 (1987).

8. M.S. Akhlaq, H.-P. Schuchmann, and C. von Sonntag, Int. J. Radiat. Biol. 51:91 (1987).

9. P.S. Surdhar, S.P. Mezyk, and D.A. Armstrong, J. Phys. Chem. 93:3360 (1989).

10 C. Schöneich, M. Bonifacic, and K.-D. Asmus, Free Radical Res. Comm. 6:393 (1989).

11. C. Schöneich, K.-D. Asmus, U. Dillinger, and F. v. Bruchhausen, Biochem. Biophys. Res. Commun. 161:113 (1989).

12. M.S. Akhlaq and C. von Sonntag, J. Am. Chem. Soc. 108:3542 (1986).

13. K.-D. Asmus, Acc. Chem. Res. 12:436 (1979).

14. M.S. Akhlaq and C. von Sonntag, Z. Naturforsch. 42c:134 (1987).

15. S.A. Chaudhri and K.-D. Asmus, Angew. Chem. Int. Ed. Engl. 20:672 (1981).

16. M.S. Akhlaq, C.P. Murthy, S. Steenken, and C. von Sonntag, J. Phys. Chem. 93:4331 (1989).

17. M. Bonifacic and K.-D. Asmus, J. Phys. Chem. 88:6286 (1984).

18. Z. Wu, R. Ahmad, and D.A. Armstrong, Radiat. Phys. Chem. 23:251 (1984).

19. M. Tamba, G. Simone, and M. Quintiliani, Int. J. Radiat. Biol. 50:595 (1986).

20. D. Becker, S. Swarts, M. Champagne, and M.D. Sevilla, Int. J. Radiat. Biol. 53:767 (1988).

21. M.Z. Hoffman and E. Hayon, J. Am. Chem. Soc. 94:7950 (1972).

HYDROGEN ABSTRACTION BY THIYL RADICALS FROM ACTIVATED

C–H-BONDS OF ALCOHOLS, ETHERS AND POLYUNSATURATED FATTY ACIDS

Christian Schöneich[#], Marija Bonifačić[§], Uwe Dillinger[¶], and
Klaus-Dieter Asmus[#]

[#] Hahn-Meitner-Institut Berlin, Bereich S, Abteilung Strahlenchemie
Postfach 39 01 28, 1000 Berlin 39, F.R. Germany

[§] Ruder Boskovic Institut, Department of Physical Chemistry
P.O.B. 1016, 41001 Zagreb, Croatia, Yugoslavia

[¶] Universitätsklinikum Rudolf Virchow, Abteilung für Anaesthesiologie
und operative Intensivmedizin, Freie Universität Berlin, 1000 Berlin 19
F.R. Germany

Owing to their reducing properties thiols and thiolates, RSH/RS⁻, participate in many
cellular redox processes serving as an electron donor according to the general reaction (1).[1]

$$RSH \rightarrow RS^{\bullet} + e^- + H^+ \tag{1}$$

Moreover, thiols readily transfer hydrogen atoms onto carbon-centered radicals, particularly
to those of low oxidizing power, as depicted in reaction (2).[1,2]

$$R-CH_2^{\bullet} + RSH \rightarrow R-CH_3 + RS^{\bullet} \tag{2}$$

In radiation biology this type of reaction is commonly known as "repair" reaction.[2] It
may describe, for example, the repair of radicals from DNA or lipids which have been
generated upon interaction of ionizing radiation with living cells. Reactions (1) and (2) both
lead to the formation of thiyl radicals, RS⁺. For quite a long time little attention has been
paid to the chemical properties of these species and their possible role in biological systems.
The only process considered was the dimerization to the corresponding disulfides. In recent
years thiyl radicals have, however, also been identified as reasonably good oxidants,[3,4] and
were found to be frequently involved in reversible addition reactions, e.g. with molecular
oxygen ($RS^{\bullet} + O_2 \rightleftarrows RSOO^{\bullet}$)[5,6] or thiolate ($RS^{\bullet} + RS^- \rightleftarrows RS.^{\bullet}.SR^-$).[7]

The aim of this contribution is to show that thiyl radicals could also abstract hydrogen
atoms from activated C–H bonds as they are present, for example, in alcohols, ethers or polyunsaturated fatty acids, i.e. in biomolecules which form the structure of cell membranes.
Chemically, this process constitutes the reverse of the "repair" process and thus may provide
a rationale for the fact that the protection thiols can offer against radical damage, e.g. in
radiation biology or against lipid peroxidation, never seems to be complete.[8] The mutage-
nicity of thiols, as shown in the Ames test, could perhaps also be related to the generation of
thiyl radicals and their deleterious effects in biological environment.[9-11]

Sulfur-Centered Reactive Intermediates in Chemistry and Biology, Edited by
C. Chatgilialoglu and K.-D. Asmus, Plenum Press, New York, 1990

Alcohols and ethers can be regarded as simple model compounds for the more complex sugar moieties of DNA strands. Hydroxyl radicals which are generated, for example, by interaction of ionizing radiation with aqueous systems or via catalytic decomposition of hydroperoxides easily abstract hydrogen atoms from alcohols, ethers and sugar molecules.[2,12,13] As a consequence they may thus directly initiate DNA damage. The so-formed α-hydroxyl- or α-alkoxyl radicals can be repaired by hydrogen transfer from thiols with rate constants k_{-3} being typically in the order of 10^6-10^8 M^{-1}s^{-1}.[2,12] However, in several studies increasing evidence emerged for the reversibility of the repair process,[14] i.e. for hydrogen atom abstraction by thiyl radicals. Thus the repair mechanism should generally be written as an equlibrium reaction as formulated in equation (3) for thiyl reaction with alcohols and ethers (R^1, R^2, R^3 = H or alkyl).

$$R^1R^2C(OR^3)-H \; + \; RS^\bullet \; \underset{k_{-3}}{\overset{k_3}{\rightleftarrows}} \; R^1R^2C^\bullet(OR^3) \; + \; RSH \tag{3}$$

As of today reliable information on absolute rate constants for the H-atom abstraction by thiyl radicals is still scarce. The few data available suggest rate constants for these reactions generally to be of the order $k_3 < 10^7$ M^{-1} s^{-1}.[14,15]

In case the difference of rate constants is large ($k_{-3} \gg k_3$) proof of equlibrium (3) through direct kinetic measurements is rather difficult if not impossible to obtain. However, using an indirect method it was possible to monitor the adjustment of the equlibrium after or during perturbation through addition of substrates which react selectively with either the thiyl radicals or with the α-hydroxyl/-alkoxyl radicals. The latter are generally good reductants and can be extracted from the equilibrium by, for example, an irreversible reaction with electron acceptors such as CCl_4 or p-nitroacetophenone (PNAP).[16] The oxidizing thiyl radicals, on the other hand, react effectively with antioxidants like vitamin E (α-tocopherol),[16] 2,2′-azinobis-(3-ethylbenzthiazolin-6-sulfonate) (ABTS)[17] or vitamin C (ascorbic acid).[4] All the different routes are outlined in scheme 1 using 2-propanol as a model alcohol.

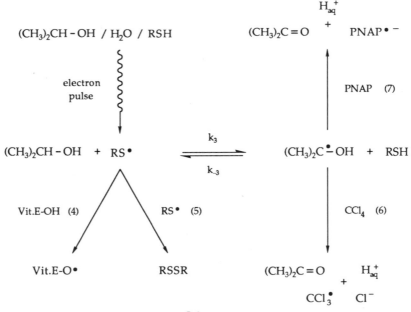

Scheme 1

Thiyl radicals are conveniently produced by pulse irradiation of solutions containing the thiols in the neutral –SH form (i.e. pH<7) in mixtures of water and the respective ethers and/or alcohols.[16] The relevant reactions for the water/2-propanol system are outlined in equations (8–12) with $R'^{\bullet}$ denoting all organic radicals which are formed during the reactions (9) and (11). [Reactions (8) and (9) show the reactive radical species generated from the solvents water and 2-propanol upon radiation with the high energy electrons].

$$H_2O \rightarrow e_{aq}^{-}, \ {}^{\bullet}OH, \ H^{\bullet} \tag{8}$$

$$(CH_3)_2CHOH \rightarrow e_{sol}^{-}, \ (CH_3)_2C^{\bullet}OH, \ {}^{\bullet}CH_3, \ {}^{\bullet}CH_2(CH_3)CHOH \tag{9}$$

$$e_{aq}^{-}, e_{sol}^{-} + N_2O + H_2O \rightarrow {}^{\bullet}OH + {}^{-}OH + N_2 \tag{10}$$

$${}^{\bullet}OH/H^{\bullet} + (CH_3)_2CHOH \rightarrow H_2O/H_2 + (CH_3)_2C^{\bullet}OH, \ {}^{\bullet}CH_2(CH_3)CHOH \tag{11}$$

$$R'^{\bullet} + RSH \rightarrow R'H + RS^{\bullet} \tag{12}$$

The thiyl radicals generally exhibit optical absorption spectra with a maximum in the 300–330 nm region and extinction coefficients around 400–1200 $M^{-1}cm^{-1}$.[18]

The lifetime of the penicillamine thiyl radical (PenS$^{\bullet}$, $^{+}H_3N$–CH(CO$_2^{-}$)–C(CH$_3$)$_2$S$^{\bullet}$) in water/2-propanol (1/1, v/v) as a function of CCl$_4$ concentration is shown in Figure 1. In the absence of CCl$_4$ the decay of PenS$^{\bullet}$ is of pure second order due to the dimerization reaction (5). In presence of CCl$_4$ the decay kinetics change to first order with the half-lives decreasing with increasing CCl$_4$ concentration. As PenS$^{\bullet}$ does not directly react with CCl$_4$ (as shown in an independent set of experiments) this indicats the involvement of equlibrium (3) and reaction (6) in the mechanism of RS$^{\bullet}$ disappearance.

A similiar observation could be made using PNAP instead of CCl$_4$. In this case it is convenient to monitor the build–up kinetics of PNAP$^{\bullet-}$ which is formed according to reaction (7). Both the CCl$_4$ and the PNAP systems, as well as a third system in which the competion between reactions (4) and (6) across equilibrium (3) at different Vit E-OH/CCl$_4$-ratios is analyzed yield the same k_3 as can be seen from Table 1.[16] Good agreement of all methods applied is also obtained for the reaction of different thiyl radicals with 2-propanol.

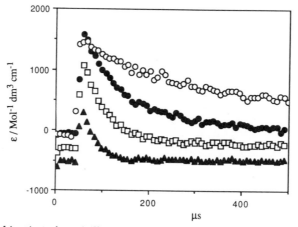

Figure 1. Decay kinetics of penicillamine thiyl radical (PenS$^{\bullet}$) in absence ($\circ$) and presence of CCl$_4$ ($\bullet$ 1.0, $\square$ 2.0 and $\blacktriangle$ 5.2 mM) in pH 5.1 water/2-propanol (1/1, v/v) mixtures containing also 10^{-2} M penicillamine after a 1μs electron pulse (1.6 MeV).

Table 2 summarizes the kinetic data for penicillamine thiyl radicals with different alcohols and ethers. The rate constants correlate well with the bond energies of the broken C–H bond and also reflect the actual number of activated C–H-bonds.

Furthermore, from the equlibrium constants K = k_3/k_{-3} (k_{-3} was obtained independently by measuring the build–up of the RS$^\bullet$ absorption in the reaction of alcohol/ether radicals with RSH at different thiol concentrations) it was possible to calculate the redox potentials E° for some of the α-hydroxyl radicals.[19] They agree very well with previously published data[20,21] which not only supports the existence of equilibrium (3) but also demonstrates the reliability of our kinetic measurements.

The experimental data demonstate that thiyl radicals do indeed react with activated C–H-bonds of alcohols and ethers in an overall reversible process. The actual yield of hydrogen abstraction is dependent on the presence and concentration of electron acceptors which irreversibly extract the produced α-hydroxyl or -alkoxyl radicals from the equlibrium.

Table 1. Rate constants for hydrogen abstraction from 2-propanol by various thiyl radicals in water/2-propanol (1/1, v/v) mixtures, measured via different experimental systems. (PenS$^\bullet$ = penicillamine, GS$^\bullet$ = glutathione, CysS$^\bullet$ = cysteine thiyl radical)

Thiyl radical	System	$k_3/M^{-1}s^{-1}$
PenS$^\bullet$	addition of various concentrations of CCl_4	1.4×10^4
	various ratios of PenSH/PNAP	1.1×10^4
	competition, various ratios of CCl_4/Vit.E-OH	1.0×10^4
GS$^\bullet$	competition, various ratios of CCl_4/Vit.E-OH	1.2×10^4
CysS$^\bullet$	competition, various ratios of CCl_4/Vit.E-OH	2.0×10^4

Table 2. Data for the equlibria between penicillamine thiyl radicals and radicals from various alcohols/ethers (BE = bond energy)

compound	k_3 $M^{-1}s^{-1}$	k_{-3} $M^{-1}s^{-1}$	K (k_3/k_{-3})	BE(C–H) kcal mol^{-1}	E°[–C$^\bullet$(OH)–] V
2-propanol	1.2×10^4	1.2×10^8	1.0×10^{-4}	91	−1.33
1-propanol	6.0×10^3	8.0×10^7	0.8×10^{-4}		
tetrahydrofurane	1.2×10^4	1.0×10^8	1.2×10^{-4}	92	
ethanol	2.3×10^3	9.1×10^7	0.3×10^{-4}	93	−1.15
ethyleneglycol	1.2×10^3	3.9×10^7	0.3×10^{-4}		
methanol	1.3×10^3	3.6×10^7	0.4×10^{-4}	94	−1.07
1,4-dioxane	$<3.0 \times 10^3$	6.7×10^7	$<0.5 \times 10^{-4}$		

Molecular oxygen constitutes a biologically available electron acceptor. It must be recognized, however, that the reaction of C-centered radicals (including α-hydroxyl radicals) generally proceeds via an addition of O_2 to yield the corresponding peroxyl radical, possibly followed by $O_2^{\bullet-}$ elimination, rather than by electron transfer.[2] A quantitative evaluation of the equlibrium kinetics in these systems is further complicated by the presumed reversibility of the oxygen addition to $RS^{\bullet}$ [5] and many other open questions concerning the kinetics and even the identity of the oxygen adduct.[22] Irrespective of what the exact mechanism of the removal of C-centered radicals is, oxygen concentrations around 50-100 μM (i.e. under physiological conditions) would be sufficient to drive equlibrium (3) onto the right hand side.

Analogous considerations could be made for transition metals (e.g. Fe^{III}, Cu^{II}) which are also reducible by α-hydroxyl/alkoxyl radicals (incl. sugar radicals).[23] Processes as shown in scheme 1 may thus serve as an alternative mechanism for an often observed increase in DNA damage in the presence of transition metals or oxygen together with thiols.[24] Also, the observed DNA cleavage in presence of the Cu^{II}/thiol system[25] could possibly be related to thiyl radical attack on the sugar moiety in DNA.

HYDROGEN ABSTRACTION FROM POLYUNSATURATED FATTY ACIDS

Lipids are the major components of biological membranes. They are rich in polyunsaturated fatty acids (PUFAs) which likely undergo lipid peroxidation caused, for example, by the action of free radicals.[26,27] This process is initiated by hydrogen abstraction from the bisallylic methylene groups of the respective fatty acid yielding pentadienyl-type radicals which subsequently react with molecular oxygen to give peroxyl radicals. Since increased lipid peroxidation was observed in biochemical experiments containing thiols, Fe^{III} and O_2,[28,29] a system known for thiyl radical generation, investigations were extended to the measurement of absolute rate constants for hydrogen abstraction from fatty acids by thiyl radicals.[30] The structures of the most common polyunsaturated fatty acids are given below. Comparing bond energies an abstraction from bisallylic centers (BDE_{C-H} = 82 kcal/mol)[27] is expected to occur even faster than from OH/OR-activated C–H bonds of alcohols and ethers [BE(C–H) = 91-94 kcal/mol].

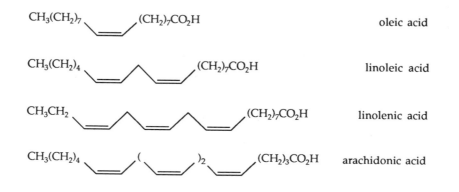

CH$_3$(CH$_2$)$_7$ ‌ (CH$_2$)$_7$CO$_2$H ‌ oleic acid

CH$_3$(CH$_2$)$_4$ ‌ (CH$_2$)$_7$CO$_2$H ‌ linoleic acid

CH$_3$CH$_2$ ‌ (CH$_2$)$_7$CO$_2$H ‌ linolenic acid

CH$_3$(CH$_2$)$_4$ ‌ ( )$_2$ (CH$_2$)$_3$CO$_2$H ‌ arachidonic acid

Pulse irradiation of N_2O-saturated, pH 5.1, ethanol/water mixtures (1/1, v/v) containing 1×10^{-2} M cysteine leads to the formation of cysteinyl radicals, analogous to the production of $PenS^{\bullet}$ in the water/2-propanol/penicillamine system described above. Shortly after the pulse, e.g. at 4 μs, $CysS^{\bullet}$ are the only radical species present in the solution. Optically this is indicated by a weak and rather uncharacteristic absorption in the 260 - 340 nm range. Addition of linolenic acid (18:3), e.g. at 1×10^{-3} M, yields a different and very strong transient UV band shown in Figure 2. It is characterized by a pronounced absorption maximum at 280 nm which matches the known transient spectra obtained in the PUFA

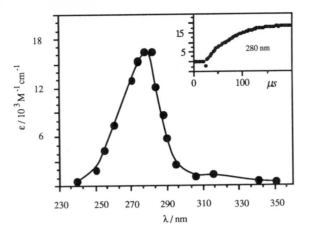

Figure 2. Transient optical absorption spectrum of pentadienyl-type radicals obtained in the reaction of CysS[•] + linolenic acid during pulse radiolysis of N$_2$O-saturated, pH 5.1, water/ethanol (1/1, v/v) mixtures containing 10^{-2} M cysteine and 10^{-3} M linolenic acid (18:3).
Insert: Trace of absorption (280 nm) *vs.* time. Pulse length: 1 μs.

oxidation by other radicals, namely [•]OH,[31] (CH$_3$)$_3$CO[•] [32] or SO$_3$[•−].[33] The spectrum is assigned to a pentadienyl-type radical, formed via hydrogen abstraction from linolenic acid by cysteinyl radicals according to reaction (13).

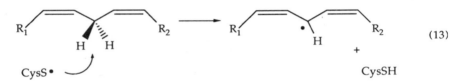

$$(13)$$

Analogous observations were made for different PUFAs or different thiyl radicals as well. Some representative rate constants for hydrogen abstraction by RS[•] are summerized in Table 3.[30,34]

Table 3. Rate constants for hydrogen abstraction from PUFAs by various thiyl radicals in N$_2$O-saturated ethanol/water (1/1,v/v), pH 5.1, [RSH] = 10^{-2} M. (c-18:2 linoleic, c-18:3 linolenic, c-22:3 docosa-trienic, c-20:4 arachidonic acid)

RSH	k_{13} / 10^7 M^{-1}s^{-1} for RS[•] + PUFA			
	c-18:2	c-18:3	c-22:3	c-20:4
CysSH	0.6	0.9	–	1.6
GSH	0.8	1.9	1.9	3.1
CysSH-CO$_2$Et	1.3	1.9	–	2.4
PenSH	0.3	0.4	–	0.5
HO(CH$_2$)$_2$SH	3.1	4.5	–	6.8

The extinction coefficient for such pentadienyl radicals has been reported by Patterson to be $\varepsilon_{280} \approx 30000$ M^{-1} cm^{-1}.[31] The highest possible value obtainable from our measurements at high concentrations of fatty acids (just below the critical micelle concentration) in ethanol/water was $\varepsilon_{280} \approx 25000$ M^{-1}cm^{-1} based on an initial yield for of G(RS$^\bullet$) = 5.5. This difference may, although small within pulse radiolysis error limits, reflect some side reactions of RS$^\bullet$ in our system.

Dependence on fatty acid and thiyl radical structure

As Table 3 shows, the reactivity of thiyl radicals towards the various polyunsaturated fatty acids changes by as much as one order of magnitude. Since the S–H bond energies of the used thiols do not differ to any great extent other reasons must be responsible for this discrepancy. It is noted that the reaction proceeds faster with increasing number of double bonds in the fatty acid and thus is related to the number of bisallylic hydrogen atoms.

Oleic acid (c-18:1) shows no measurable reaction. Therefore, an addition of RS$^\bullet$ to the double bonds seems to be of minor if any importance. This would be in accord with the well known reversibility of thiyl radicals addition to single double bonds, with the equlibrium being located well on the side of the educts.[2] The chain length of the hydrocarbon does not influence the reactivity as can be deduced from a comparison of linolenic (c-18:3) and docosa-trienic acid (c-22:3). Competition experiments conducted in support of the direct measurements fully confirm the kinetc data.

The lowest reactivity is exhibited by the sterically hindered thiyl radical from peni-cillamine. The reaction is accelerated in going from thiyl radicals of strongly hydrophilic thiols (e.g. CysSH) to radicals of more lipophilic derivatives (e.g. mercaptoethanol, HO–(CH$_2$)$_2$–SH). This may be associated with the pronounced lipophilicity of the PUFAs. Other important parameters seem to be the distance of the thiyl radical center to the ionic groups and total number of ionic groups within the molecule. Glutathiyl radicals, for example, in which S$^\bullet$ is located in the middle of the tripeptide and separated from both COO$^-$ and –NH$_3^+$ by a peptide bond, thus react faster than simple cysteine thiyl radicals. Also, radicals from cysteineethylester (CysSH–CO$_2$Et), containing only one charged group (–NH$_3^+$) exceed cysteinyl radicals in reactivity. This effect is easily understood considering that in water/alcohol mixtures the lipophilic fatty acid should be surrounded mainly by the less polar component, i.e. alcohol molecules, establishing an interesting analogy to reversed phase HPLC where the C$_{18}$–phases are coated by, e.g., a methanol double layer in methanol/water mixtures.[35]

In conclusion, the higher the lipophilicity of the thiyl radicals is the better access they should have to the bisallylic methylene groups of the PUFAs. Moreover, thiyl radical centers located at less polar hydrocarbon chains should associate with the fatty acid chain more easily to give a transition state with less activation energy.

Influence of solvent polarity

It is interesting to note that solvent polarity has an influence on the rate of hydrogen transfer. This is exemplified in a series of experiments on the reaction of CysS$^\bullet$ with linolenic acid (18:3) in water/alcohol/acetone mixtures of different composition (alcohols = ethanol, methanol, tert-butanol) to vary the dielectric constants of the solution over a wide range. The relationship between the measured rate constants and the dielectric constants is displayed in Figure 3. No dependence on solvent polarity is observed in the low-polarity region (curve 1) while at higher polarity the reactivity increases significantly and almost linearly with increasing dielectric constant (curve 2).

This result implies that, at least in more polar solvents, the hydrogen transfer proceeds via a polar reaction mechanism including charge separation in the transition state. Such a

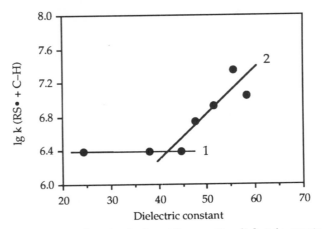

Figure 3: Plot of lg k (CyssS• + linolenic acid) over the dielectric constant of solvent mixture (dielectric constant adjusted with water/alcohol/acetone mixtures of different composition)

behaviour is characteristic for hydrogen transfer by weaker electrophilic radicals like e.g. Br• and RS•.[36] They are likely to be involved in so-called "later" transition states on the reaction coordinate after the actual breakage of the C–H bond has already taken place.[36] A polar mechanism for hydrogen abstraction from aromatic substituted C–H bonds by hexane thiyl radicals was, in fact, proposed by Pryor et al.[15] who found a linear Hammett-Brown correlation for the enhancement of reactivity with increasing electron donating capability of the aromatic ring (e.g. by the p-CH₃O substituent). Both findings suggest the occurance of partial bond breakage in the transition state concomitant with a shift of negative charge onto the attacking thiyl radical. However, comparison of benzenethiyl, cyclohexanethiyl and bromine radical revealed that the determining factors for such charge separation is not the electron affinity of the attacking radical but the polarizability.[15] A comparable solvent dependence was not found for the hydrogen abstraction from alcohols and ethers since stabilization of the transition state by α-hydroxyl/alkoxyl substituents probably exceeds that of the solvent effect.

CONCLUSION

The obtained rate constants clearly demonstrate that thiyl radicals are able to abstract activated hydrogen atoms from biologically relevant molecules. These findings may thus provide an explanation for DNA damage as well as for enhanced lipid peroxidation in presence of thiols or, more precisely, RS• generating systems. More recent results show that hydrogen abstraction takes place even at physiological oxygen concentrations,[34] thereby generating characteristic products of lipid peroxidation. Comparison with oxygen radicals allows to establish the following reactivity scale against polyunsaturated fatty acids:

$$(CH_3)_3CO• > RS• > CCl_3OO• > (CH_3)_2C(OH)OO• > HO_2• > (fatty\ acid)–OO•$$

Thus, regarding reactivity in homogeneous solution, thiyl radicals belong to the group of potentially most damaging radicals and should thus not be considered any longer as biologically harmless. Especially thiyl radicals from more lipid soluble mercaptans, e.g. mercaptoethanol, dithiothreitol, methylmercaptan and ethylmercaptan, are expected to diffuse easily into cell membranes. If and how more hydrophilic thiols may migrate into lipid membranes is still an open question. However, keeping in mind that even the water soluble $O_2^{\bullet-}$ can cause lipid peroxidation within the membrane (probably in its protonated state, $HO_2^{\bullet}$)[37] the same may also be true for water soluble thiyl radicals.

ACKNOWLEDGMENTS

We gratefully acknowledge the financial support provided by the "Association of International Cancer Research" (AICR), and by the "Internationales Büro KFA Jülich" who assisted our work within the terms of an agreement on scientific cooperation between the Federal Republic of Germany and the Socialist Federal Republic of Yugoslavia.

REFERENCES

1. P. Wardman, in: "Glutathione Conjugation", H. Sies and B. Ketterer, eds., Academic Press, New York, p. 43-72 (1988).
2. C von Sonntag, "The Chemical Basis of Radiation Biology", Taylor & Francis, London (1987).
3. L. G. Forni and R. L. Willson, Biochem. J., 240:897 (1986).
4. L. G. Forni, J. Mönig, V O. Mora-Arellano, and R. L. Willson, J. Chem. Soc. Perkin Trans. 2, 961 (1983).
5. J. Mönig, K.-D. Asmus, L. G. Forni, and R. L. Willson, Int. J. Radiat. Biol., 52:589 (1987).
6. M. Tamba, G. Simone, M. Quintiliani, Int. J. Radiat. Biol., 50:595 (1986).
7. K. -D. Asmus, in "Methods Enzymology", vol. 186, L Packer and A.N. Glazer, eds., Academic Press, San Diego, p. 168-180 (1990)
8. D. Schulte-Frohlinde, Free Rad. Res. Comm., 6:181 (1989).
9. H. Glatt, M. Protic-Sabljic, and F. Oesch, Science, 220:961 (1983).
10. M. H. Carter and P. D. Josephy, Biochem. Pharmacol., 35:3847 (1986).
11. A.-A. Stark, A. Arad, S. Siskindovich, D. A. Pagano, and E. Zeiger, Mutation Res., 224:89 (1989).
12. M. Z. Baker, R. Badiello, M. Tamba, M. Quintiliani, and G. Gorin, Int. J. Radiat. Biol., 41:595 (1982).
13. M. Tamba and M. Quintiliani, Radiat. Phys. Chem., 23:259 (1984).
14 M. S. Akhlaq, H. P. Schuchmann, and C. von Sonntag, Int. J. Radiat. Biol., 51:91 (1987).
15. W. A. Pryor, G. Gojon, and D. F. Church, J. Org. Chem., 43:793 (1978).
16. Ch. Schöneich, M. Bonifacic, and K.-D. Asmus, Free Rad. Res. Commun., 6:393 (1989).
17. B. S. Wolfenden and R. L. Willson, J. Chem. Soc Perkin Trans. 2, 805 (1982).
18. M. Z. Hoffman and E. Hayon, J. Phys. Chem., 77:990 (1973).
19. Ch. Schöneich, M. Bonifacic, and K.-D. Asmus, unpublished results.
20. J. Lilie, A. Henglein, and G. Beck, Ber. Bunsenges. Phys. Chem., 75:458 (1971).
21. H. A. Schwarz and R. W. Dodson, J. Phys. Chem., 93:409 (1989).
22. see articles by M. Quintiliani, and Ch. Chatgilialoglu and M. Guerra in this NATO-ASI book.
23. K.-D. Asmus and M. Bonifacic, in: "Landolt-Börnstein. Zahlenwerte und Funktionen", H. Fischer, ed., New Series, Springer Verlag, Berlin, vol. 13b (1984).
24. W. A. Prütz and H. Mönig, Int. J. Radiat. Biol., 52:677 (1987).
25. D. C. A. John and K. T. Douglas, Biochem. Biophys. Res. Commun., 165:1235 (1989).
26 A. A. Horton and S. Fairhurst, CRC Critical. Rev. Toxicol., 18:27 (1987).
27. N. A. Porter and D. G. Wujek, in: "Reactive Species in Chemistry, Biology, and Medicine", A. Quintanilha, ed., NATO ASI Series A: Life Sciences, vol. 146, Plenum Press (1988).
28. A. J. F. Searle and R. L. Willson, Biochem. J., 212:549 (1983).
29. M. Tien, J. R. Bucher, and S. D. Aust, Biochem. Biophys. Res. Commun., 107:279 (1982).

30. Ch. Schöneich, K.-D. Asmus, U. Dillinger, and F. von Bruchhausen, *Biochem. Biophys. Res. Commun.*, 161:113 (1989).

31. K. Hasegawa and L. K. Patterson, *Photochem. Photobiol.*, 28:817 (1978).

32. M. Erben-Russ, Ch. Michel, W. Bors, and M. Saran, *J. Phys. Chem.*, 91:2362 (1987).

33. M. Erben-Russ, W. Bors, R. Winter, and M. Saran, *Radiat. Phys. Chem.*, 27:419 (1986).

34. Ch. Schöneich, K.-D. Asmus, U. Dillinger, and F. von Bruchhausen, unpublished results.

35. R. P. Scott and P. Kucera, *J. Chromatogr.*, 142:213 (1977).

36. G. A. Russell, *in:* "Free Radicals", vol. 1, J. K. Kochi, ed., Wiley & Sons, New York, p. 275-331 (1973).

37. B. H. Bielski, R. L. Arudi, and M. W. Sutherland, *J. Biol. Chem.*, 258:4759 (1983).

THIYL FREE RADICALS: ELECTRON TRANSFER, ADDITION OR

HYDROGEN ABSTRACTION REACTIONS IN CHEMISTRY AND BIOLOGY, AND

THE CATALYTIC ROLE OF SULPHUR COMPOUNDS

Christina Dunster and Robin L. Willson

Biology and Biochemistry Dept.
Brunel University of West London
Uxbridge, Middlesex, UB8 3PH
U.K.

Organic chemists have long been familiar with the variety of addition and hydrogen transfer reactions that thiyl free radicals (RS$^\bullet$) can undergo, and of the chain processes and relatively fast reversible reactions that can often take place.[1-6]

The cis trans isomerization of varoius olefins can be accelerated greatly by the presence of thiols[7]. In the presence of oxygen, thioacetic acid adds to 3,4 benzpyrene[8] and compounds such as styrene and methyl acrylate can be co-oxidised with the ultimate formation of hydroxylated and sulphoxide-containing products.[9-11]

Nearly forty years have passed since Bickel and Kooijman reported how during the thermal decomposition of α,α' azoethyl benzene in inert solvents containing thiols, active hydrogen donors like 9,10-dihydroanthracene can be readily dehydrogenated.[5] Waters and colleagues also reported that only 0.5% of benzyl mercaptan is effective in decomposing 80-90% of various aldehydes[6,12] In the absence of thiol only a trace of carbon monoxide was obtained implicating a mechanism involving both hydrogen abstraction by thiyl radicals (1) and the reverse, hydrogen transfer from thiols, (3).

$$RS^\bullet \;+\; R'CHO \;\rightarrow\; RSH \;+\; R'C^\bullet O \tag{1}$$

$$R'C^\bullet O \;\rightarrow\; R'^\bullet \;+\; CO \tag{2}$$

$$R'^\bullet \;+\; RSH \;\rightarrow\; R'H \;+\; RS^\bullet \tag{3}$$

More recent experiments[13-25] have confirmed the possible reversibility of reactions such as (3). The significance of such reactions and the determination of a number of related absolute rate constants are discussed in the accompanying papers by Armstrong, Schöneich et al., and von Sonntag in this book.

In this paper a variety of extremely rapid *electron transfer* reactions involving thiyl free radicals will be described together with an example of a near *diffusion controlled* reaction of a thiyl free radical with a *conjugated unsaturated compound* in aqueous alcohol solution.

Sulfur-Centered Reactive Intermediates in Chemistry and Biology, Edited by
C. Chatgilialoglu and K.-D. Asmus, Plenum Press, New York, 1990

Such results provide further insights into the possible important protective and catalytic role of thiols in biological systems.[26,27]

ELECTRON TRANSFER REACTIONS: THE VERSATILITY OF THE PULSE RADIOLYSIS TECHNIQUE

Unquestionably, during the last thirty years the technique of pulse radiolysis has provided more information concerning the kinetics and thermodynamics of free radical reactions in aqueous solution than any other technique. Although in a few circles the mistaken view may remain that when a solution is irradiated a host of more or less random esoteric reactions take place, the vast amount of accurate information obtained from radiation chemical studies relevant to chemistry generally, speaks for itself. It cannot be over-stressed that the free radicals studied by pulse radiolysis are of thermal energy and follow the same laws as free radicals formed by conventional chemistry or biology.

Details of the pulse radiolysis technique and the manner in which experiments can be designed and their results interpreted are described elsewhere.[28-30]

In the study of one-electron transfer reactions, the technique which involves simple absorption spectrophotometry could not be easier: unlike flash photolysis no absorbing chromophore is required for free radical formation.

Solvated electrons, and hydroxyl free radicals are the principal free radical species formed when water is irradiated. Although the former is a very strong reductant and the latter a very strong oxidant, systems have been developed by which their individual reactions and the reactions of their free radical products can be studied in isolation.

Where possible, free radical reactions are followed directly by following the overall changes in absorption taking place at different wavelengths due to the loss of a reactant or the formation of a product. Where little, if any, spectral changes occur at a convenient wavelength, indirect competition studies can be made. The concentration of a particular reactant is varied and the effects on a reference reaction that has been previously characterised directly, is assessed.

Electron transfer reactions between a free radical and a molecule can of course be of two types: those in which the free radical is the oxidant and the molecule the reductant, and vice versa. Both types can be readily studied, e.g.:

$A^{\bullet-}$ + B $\rightarrow$ A + $B^{\bullet-}$ radical as reductant

$C^{\bullet+}$ + D $\rightarrow$ C + $D^{\bullet+}$ radical as oxidant

In the case of one electron reductants, the aqueous tert-butanol,[31] 2-propanol-acetone[31] and nitrous oxide-formate[32] systems, have been widely used for intial free radical generation. With the disulphide, lipoate (lipoSS), for example, pulse radiolysis of solutions containing excess tert-butanol, results in the appearence of the spectrum of the lipoate electron adduct (λ_{max} 410 nm) which decays over hundreds of microseconds. In the presence of oxygen the absorption decays exponentially and first order in oxygen concentration in agreement with the corresponding electron transfer reaction and the formation of the superoxide radical.[33]

$e^-_{(aq)}$ + lipoSS $\rightarrow$ lipoSS$^{\bullet-}$ (4)

lipoSS$^{\bullet-}$ + O_2 $\rightarrow$ lipoSS + $O_2^{\bullet-}$ $k = 0.9 \times 10^9$ M^{-1} s^{-1} (5)

Where reactions cannot be conveniently studied directly, compounds such as tetranitro-methane,[34] p-nitroacetophenone[35,36] and various quinones[37,38] and viologens[39,40] have been used as references. Studies such as these, have led to the accumulation of a vast bank of spectral and absolute rate constant data. A large number of oxidation-reduction potentials of free radical-molecule couples have also been obtained from measurements of the relative free radical concentrations at equilibrium or from measurement of the rates of the respective forward and reverse reactions.[41,42]

For the study of oxidising free radicals, systems containing nitrous oxide in the absence or presence of excess bromide, thiocyanate or azide have been widely used.

$$e^-_{(aq)} + N_2O \xrightarrow{H_2O} {}^\bullet OH + OH^- + N_2 \tag{6}$$

$${}^\bullet OH + Br^- \xrightarrow{Br^-} Br_2^{\bullet -} \tag{7}$$

$${}^\bullet OH + SCN^- \xrightarrow{SCN^-} (SCN)_2^{\bullet -} \tag{8}$$

Where reactions are not inhibited by oxygen, solutions containing excess 2-propanol-acetone or tert-butanol and carbon tetrachloride can also be employed.[43-45]

$$e^-_{(aq)} + CH_3COCH_3 \rightarrow (CH_3COCH_3)^{\bullet -} \tag{9}$$

$$(CH_3COCH_3)^{\bullet -} + H_2O \rightarrow CH_3C^\bullet OHCH_3 + OH^- \tag{10}$$

$${}^\bullet OH + CH_3CHOHCH_3 \rightarrow CH_3C^\bullet OHCH_3 + H_2O \tag{11}$$

$$CH_3C^\bullet OHCH_3 + CCl_4 \rightarrow CH_3COCH_3 + CCl_3^\bullet + Cl^- \tag{12}$$

$$CCl_3^\bullet + O_2 \rightarrow CCl_3OO^\bullet \tag{13}$$

Where spectral changes are negligible indirect studies can again be undertaken. Pheno-thiazines such as chlorpromazine,[46] promethazine[47] and metiazininc acid,[48] aminopyrine,[49,50] a water soluble analogue of vitamin E, (Trolox C),[51-53] tetramethyl-phenylenediamine (TMPD)[54] and the peroxidase reagent 2,2'- azino-di-(3-ethyl benzethiazoline-6-sulphonate), (ABTS),[55] can be used as references provided the resulting product free radicals do not themselves react with any of the other species present. Again using these systems a wide selection of one-electron oxidation-reduction potentials, absolute rate constants and spectral data have been obtained.

ELECTRON TRANSFER REACTIONS OF THIYL FREE RADICALS

The possibility that thiyl free radicals might be able to rapidly accept an electron from a wide range of organic compounds including many of biological significance became very apparent with the finding that the thiyl free radicals from cysteamine, cysteine, glutathione and other sulphur compounds could react very rapidly with the salivary peroxidase reagent ABTS.[55]

This finding arrived unexpectedly during competition studies in which ABTS was being used as a reference solute in the determination of the rates of reaction of the hydroxyl free radical with a variety of compounds whose products were only weakly absorbing. In the case of several alcohols the yield of the strongly absorbing cation $ABTS^{\bullet +}$ was reduced in agreement with simple competition. With cysteine however, although the radical-cation absorption was reduced initially a relatively slow formation followed, the magnitude of the absorption after some hundred microseconds approximating that observed in the absence of the amino acid.

Subsequent studies in which the concentration of thiol was greatly increased (so that the reaction of almost all the hydroxyl radicals led to the formation of thiyl free radicals) confirmed that the rates of the slower reactions were first order in ABTS in agreement with the rapid occurence of reaction (15).

$$^\bullet OH \quad + \quad RSH \quad \rightarrow \quad H_2O \quad + \quad RS^\bullet \tag{14}$$

$$RS^\bullet \quad + \quad ABTS \quad \rightarrow \quad RS^- \quad + \quad ABTS^{\bullet +} \tag{15}$$

The reactions of other organic compounds known to readily undergo one electron oxidation were soon examined. In these instances thiyl radicals were also formed in aqueous acetone-isopropanol solutions of the thiol under conditions where the reactions (9)-(11) and (16) predominated initially.

$$CH_3C^\bullet OHCH_3 \quad + \quad RSH \quad \rightarrow \quad CH_3CHOHCH_3 \quad + \quad RS^\bullet \tag{16}$$

Again as in the case of ABTS, with chlorpromazine, (CZ), promethazine, (PZ), or amino-pyrine, (AP), the exponential formation of the characteristic absorptions of the corresponding radical cations were observed, for example, with the glutathione thiyl radical ($GS^\bullet$)

$$GS^\bullet \quad + \quad CZ \quad \rightarrow \quad GS^- \quad + \quad CZ^{\bullet +} \tag{17}$$

$$GS^\bullet \quad + \quad PZ \quad \rightarrow \quad GS^- \quad + \quad PZ^{\bullet +} \tag{18}$$

$$GS^\bullet \quad + \quad AP \quad \rightarrow \quad GS^- \quad + \quad AP^{\bullet +} \tag{19}$$

In the case of ferrocytochrome C spectral changes corresponded to the rapid formation of ferricytochrome C and in the case of NADH the loss of the coenzyme's ground state absorption was readily apparent:[56,57]

$$GS^\bullet \quad + \quad cytIIc \quad \rightarrow \quad GS^- \quad + \quad cytIIIc \tag{20}$$

$$GS^\bullet \quad + \quad NADH \quad \rightarrow \quad GS^- \quad + \quad NAD^\bullet \quad + \quad H^+ \tag{21}$$

Negligible changes in absorption occurred in the absence of the thiol indicating that electron transfer, in contrast to hydrogen transfer from the above compounds to the 2-propanol radical, is relatively slow. Indeed, the fact that a thiol might act as an intermediate catalysing the overall one-electron reduction of carbon centred free radicals was clearly illustrated in the case of glutathione and ascorbate. In the absence of glutathione the slow appearence of the ascorbate free radical was observed in agreement with the relatively slow hydrogen transfer reaction:[58]

$$CH_3C^\bullet OHCH_3 \quad + \quad AH^- \quad \rightarrow \quad CH_3CHOHCH_3 \quad + \quad A^{\bullet -} \tag{22}$$
$$k_{22} = 1.2 \times 10^6 \text{ M}^{-1} \text{ s}^{-1}$$

In the presence of glutathione the characteristic ascorbate free radical absorption appeared but considerably more rapidly in agreement with the following catalytic scheme.[59,60]

hydrogen transfer:

$$CH_3C^\bullet OHCH_3 \quad + \quad GSH \quad \rightarrow \quad CH_3CHOHCH_3 \quad + \quad GS^\bullet \tag{23}$$
$$k_{23} = 1.8 \times 10^8 \text{ M}^{-1}\text{s}^{-1}$$

electron transfer:

$$GS^\bullet \;+\; AH^- \;\;\rightarrow\;\; GSH \;+\; A^{\bullet -} \;+\; H^+ \tag{24}$$

$$k_{24} = 6.0 \times 10^8 \; M^{-1} \, s^{-1}$$

In the case of NADH, the conclusion that the above reactions were taking place by electron transfer rather than hydrogen transfer reactions was supported by studies with the monodeuterated derivative NADD formed enzymatically by the reduction of NAD^+ by deuterated ethanol in the presence of alcohol dehydrogenase.[57] If the reaction takes place by hydrogen transfer it might be anticipated that at least a small isotope effect on the absolute rate constant might be observed, (half that anticipated if both the hydrogen atoms involved in the oxidation-reduction couple were deuterated). No such difference was apparent.

Thiyl free radicals from simple mono thiols can also be generated by one electron reduction of the corresponding disulphide. On pulse radiolysis of aqueous t. butanol solutions containing cystine (CySSCy) and Trolox C (TxOH) for example, the rapid formation of the corresponding phenoxyl free radical was observed.[52] In alkaline solution the reaction occured more rapidly in agreement with the ionisation of the phenol facilitating the electron transfer process. (For "$\bullet\,\therefore\,\bullet$" three-electron bonds see article by K.-D. Asmus in this book).

$$e^-_{(aq)} \;+\; CySSCy \;\;\rightarrow\;\; CyS\bullet\therefore\bullet SCy^- \tag{25}$$

$$CyS\bullet\therefore\bullet SCy^- \;\;\rightleftarrows\;\; CyS^\bullet \;+\; CyS^- \tag{26}$$

$$TxOH \;+\; OH^- \;\;\rightarrow\;\; TxO^- \;+\; H_2O \tag{27}$$

$$CyS^\bullet \;+\; TxOH \;\;\rightarrow\;\; CyS^- \;+\; TxO^\bullet \;+\; H^+ \tag{28}$$

$$k_{28} = 1 \times 10^8 \; M^{-1} \, s^{-1}$$

$$CyS^\bullet \;+\; TxO^- \;\;\rightarrow\;\; CyS^- \;+\; TxO \tag{29}$$

$$k_{29} = 8 \times 10^8 \; M^{-1} \, s^{-1}$$

REACTIONS WITH UNSATURATED COMPOUNDS

Recently thiyl free radicals have been found to react very rapidly with the benzene metabolite and unsaturated conjugated diene, muconic acid, $k = 1.1 \times 10^8 \; M^{-1} \, s^{-1}$.[61,62] It was therefore of interest to see whether essential polyunsaturated fatty acids might behave similarly, and whether vitamin A with its highly conjugated poly-ene side chain might compete effectively and therefore provide protection.[63]

Irradiation of aqueous methanol (60%) solutions of glutathione under these conditions leads to the rapid formation of glutathione thiyl free radicals, $GS^\bullet$, through a series of well documented reactions including:

$$H_2O/CH_3OH \;\;\rightarrow\;\; {}^\bullet OH \;+\; CH_3O^\bullet \;+\; 2e^-_{(sol)} \;+\; 2H^+ \tag{30}$$

$$e^-_{(sol)} \;+\; N_2O \;\;\rightarrow\;\; N_2 \;+\; OH^- \;+\; {}^\bullet OH \tag{6}$$

$${}^\bullet OH \;+\; CH_3OH \;\;\rightarrow\;\; {}^\bullet CH_2OH \;+\; H_2O \tag{31}$$

$${}^\bullet CH_2OH \;+\; GSH \;\;\rightarrow\;\; CH_3OH \;+\; GS^\bullet \tag{32}$$

$$CH_3O^\bullet \;+\; GSH \;\;\rightarrow\;\; CH_3OH \;+\; GS^\bullet \tag{33}$$

On γ-radiolysis of nitrous oxide saturated, aqueous methanol solutions containing glutathione (10mM) and retinol (20μM), the characteristic ground state optical absorption of the vitamin (λ_{max} 325 nm) decreased with increasing radiation dose, in agreement with reactions (6) and (30) - (33) followed by:

$$GS^\bullet \quad + \quad retinol \quad \rightarrow \quad products \tag{34}$$

In the absence of glutathione or in the additional presence of vitamin C (ascorbate, AH^-, 200μM), arachidonate, linolenate or linoleate (1mM), the extent of destruction was markedly reduced. In contrast, no significant protection was afforded by the presence of oleate (1mM). The protection observed was attributed to reaction: (24), (35) - (37), competing with reaction (34):

$$GS^\bullet \quad + \quad arachidonate \quad \rightarrow \quad products \tag{35}$$

$$GS^\bullet \quad + \quad linolenate \quad \rightarrow \quad products \tag{36}$$

$$GS^\bullet \quad + \quad linoleate \quad \rightarrow \quad products \tag{37}$$

To confirm such competition was feasible kinetically, absolute rate constants for the above reactions were determined by pulse radiolysis. On pulse radiolysis of nitrous oxide saturated aqueous methanol solutions containing glutathione (10mM) and retinol (20 - 400μM) a slowly decaying very strong transient absorption (λ_{max} 380nm), was apparent. The absorption formed exponentially and first order in retinol concentration. From the slope of the kinetic plot a value $k_{34} = 1.4 \times 10^9$ M^{-1} s^{-1} was derived.

On pulse radiolysis of aqueous methanol solutions containing glutathione (10mM) and ABTS instead retinol, the exponential formation of the characteristic transient absorption of the related radical cation, $ABTS^{\bullet+}$ (λ_{max} 415nm) was observed in agreement with

$$GS^\bullet \quad + \quad ABTS \quad \rightarrow \quad GS^- \quad + \quad ABTS^{\bullet+} \tag{38}$$

and $k_{38} = 4.3 \times 10^7$ M^{-1} s^{-1}.

With 1mM ABTS, the magnitude of the absorption at 415nm was a factor of 0.90 that observed with 400μM retinol at 380nm. Assuming at these concentrations that the same yield of $GS^\bullet$ reacts with each compound and taking the reported extinction coefficient of $ABTS^{\bullet+}$ ($\varepsilon_{415nm} = 3.6 \times 10^4$ M^{-1} cm^{-1}), a value $\varepsilon_{380nm} = 4.0 \times 10^4$ M^{-1} cm^{-1} for the product retinol free radical was obtained.

On pulse radiolysis of similar solutions of ABTS (200μM) together with arachidonate, linolenate, linoleate or muconate, but not oleate, the magnitude of the $ABTS^{\bullet+}$ absorption decreased progressively with increasing polyunsaturated fatty acid concentration. Kinetic competition plots of the magnitude of the absorptions were in agreement with simple competition and kinetic equations similar to:

$$A_0 / A \quad = \quad 1 \quad + \quad \{k_{35} [arachidonate] / k_{38} [ABTS]\}$$

where A_0 = magnitude of the $ABTS^{\bullet+}$ absorption in the absence of the fatty acid and A in its presence. Taking the above value for k_{38} and the relative rate constants from the slopes of the competition plots, the corresponding absolute rate constants, $k_{35} = 2.2 \times 10^7$ M^{-1} s^{-1}, $k_{36} = 1.9 \times 10^7$ M^{-1} s^{-1} and $k_{37} = 1.3 \times 10^7$ M^{-1} s^{-1} were derived, in good agreement with the values by Schöneich et al.[25] The lack of competition with oleate indicated $k_{39} < 2 \times 10^6$ M^{-1} s^{-1} for

$$GS^\bullet \quad + \quad oleate \quad \rightarrow \quad products \tag{39}$$

The value for the reaction of GS$^\bullet$ with ascorbate, $k_{24} = 6.0 \times 10^8\,M^{-1}\,s^{-1}$ had been determined by directly following the formation of the ascorbyl free radical at 360nm on pulse radiolysis of aqueous solutions of glutathione and vitamin C. A related value $k = 3.6 \times 10^8$ $M^{-1}\,s^{-1}$ has been similarly obtained in solutions containing 60% methanol.

With respect to the mechanisms involved, parallel studies by Asmus and colleagues[25] indicated that thiyl radicals react with essential fatty acids by hydrogen abstraction e.g.:

$$GS^\bullet \quad + \quad arachidonate \quad \rightarrow \quad GSH \quad + \quad arachidonate(\text{-}H)^\bullet \tag{40}$$

The much more rapid reaction of GS$^\bullet$ with retinol and muconate may be attributed to the fact that in these compounds the unsaturated system is conjugated and addition/or electron transfer reactions may also occur:

$$GS^\bullet \quad + \quad retinol \quad \rightarrow \quad GSH \quad + \quad retinol(\text{-}H)^\bullet \tag{41}$$

$$GS^\bullet \quad + \quad retinol \quad \rightarrow \quad (GS\text{-}retinol)^\bullet \tag{42}$$

$$GS^\bullet \quad + \quad retinol \quad \rightarrow \quad GS^- \quad + \quad retinol^{\bullet+} \tag{43}$$

The absorption spectrum of the free radical, supposedly the radical anion, formed by one electron addition to retinol in methanol (λ_{max} 370 nm) has been published along with those of related compounds[63] and further studies are in progress to identify the species formed from the reaction of thiyl radicals.

Prompted by the findings with muconate it has since been found that in aqueous methanol solutions, styrene too reacts at a nearly diffusion controlled rate with thiyl radicals.[64] Clearly there is much scope for further studies.

$$C_6H_5CH=CH_2 \quad + \quad GS^\bullet \quad \rightarrow \quad products \quad (k_{44} = 4.1 \times 10^8\,M^{-1}\,s^{-1}) \tag{44}$$

THIOL CATALYSIS AND PROTECTION IN FREE RADICAL BIOLOGY.

The pulse radiolysis technique and related radiation methods are proving to be extremely useful in improving our understanding of the reactions of thiyl free radicals in aqueous organic solutions. The relevance to chemistry is obvious. With the increasing interest in the role of free radicals in biology and medicine, such information is now also attracting the attention of many biochemists.

Thiyl free radicals can be formed by a variety of chemical and enzymic reactions of biological interest. For example, in addition to being formed by the action of light or radiation, they can be formed during the iron catalysed oxidation of cysteine and other thiols, alone,[65] or in the presence of lipids[66,67] or whole cells, or by the action of horseradish peroxidase[69] or prostaglandin ^{H} synthase.[70]

With respect to free radical biology generally, it is perhaps in the area of protection that the role of thiols will prove to be of greatest interest. Radiobiologists have long been familiar with the way thiols such as cysteine, glutathione and cysteamine can protect against the lethal effects of radiation. The reactions that are thought to be involved may well be relevant to other forms of injury.

Nucleic acid sugar radicals for example may be repaired by hydrogen transfer and base radicals such as those from guanine by electron transfer.[71,72] The effectiveness of such reactions will depend on the subsequent reactions of the resulting thiyl free radicals. If these are with an electron donor such as ascorbate or NADH, then this may facilitate repair.

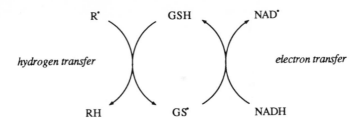

If on the other hand a reaction with a lipid occurs such reactions may exacerbate the situation. To complicate matters further, the possible reactions of oxygen with carbon centered and sulphur centered free radicals must be considered at every stage.[73-78]

addition:

$$GS^{\bullet} + GS^{-} \rightleftarrows GS{\cdot}^{\cdot}{\cdot}SG^{-} \tag{45}$$

$$GS^{\bullet} + O_2 \rightleftarrows GSOO^{\bullet} \tag{46}$$

$$GS^{\bullet} + -CH{=}CH- \rightleftarrows -CH(GS)C^{\bullet}H- \tag{47}$$

hydrogen transfer:

$$GS^{\bullet} + RH \rightleftarrows GSH + R^{\bullet} \tag{48}$$

electron transfer:

$$GS^{\bullet} + X \rightleftarrows GS^{-} + X^{\bullet +} \tag{49}$$

$$GS{\cdot}^{\cdot}{\cdot}SG^{-} + O_2 \rightleftarrows GSSG + O_2^{\bullet -} \tag{50}$$

$$GS{\cdot}^{\cdot}{\cdot}SG^{-} + Y \rightleftarrows GSSG + Y^{\bullet -} \tag{51}$$

$$R^{\bullet} + O_2 \rightleftarrows ROO^{\bullet} \tag{52}$$

Finally it is of interest to go back half a century again, to the time when the possible role of free radicals in enzyme action was first considered. In principle, a free radical scheme involving thiols can be written for the action of a dehydrogenase enzyme along the lines originally considered by Haber, Willstätter, Waters and others.[1,5,79]

electron transfer:

$$ES^{\bullet} + NADH \rightleftarrows ES^{-} + NAD^{\bullet} + H^{+} \tag{53}$$

$$RCHO + NAD^{\bullet} + H^{+} \rightleftarrows RC^{\bullet}HOH + NAD^{+} \tag{54}$$

hydrogen transfer:

$$RC^{\bullet}HOH + ESH \rightleftarrows ES^{\bullet} + RCH_2OH \tag{55}$$

acid dissociation:

$$ES^{-} + H^{+} \rightleftarrows ESH \tag{56}$$

net:

$$NADH + RCHO + H^+ \rightleftarrows RCH_2OH + NAD^+ \qquad (57)$$

The possibilty that such reversible reactions might occur even in a concerted biological environment must, remain extremely speculative. The constraints that would have to be imposed in order that the reactions predominate in a particular direction and are uninfluenced by side reactions with oxygen or other biological molecules, are enormous. On the basis of our present knowledge such a scheme, whilst interesting to contemplate, seems highly improbable.

ACKNOWLEDGEMENT

Financial support from the Cancer Research Campaign and the assistance of Drs. B. D. Michael, B. Voynovich and colleagues from the Gray Laboratory, and of Mrs. K. Kalsi, are gratefully acknowledged.

REFERENCES

1. W. A. Waters, "The Chemistry of Free Radicals", Oxford University Press, London(1946).
2. W. A. Pryor, "Mechanisms of Sulphur Reactions", McGraw-Hill, New York, p. 75-93, (1962).
3. A. Ohno and S. Oae, *in:* "Organic Chemistry of Sulphur", S. Oae, ed., Plenum Press, New York and London, p. 119-186, (1975).
4. M. S. Kharasch, A. T. Read, and F. R. Mayo, *Chem. and Ind. (London)*, 57:792 (1938).
5. A. F. Bickel and E. C. Kooijman, *Nature*, 170:211 (1952).
6. E. F. P. Harris and W. A. Waters, *Nature*, 170:212 (1952).
7. C. Walling and W. Helmreich, *J. Amer. Chem. Soc.*, 81:1144 (1959).
8. A. L. J. Beckwith and L. B. See, *Austr. J. Chem.*, 17:109 (1964).
9. M. S. Kharasch, W. Nudenberg, and G. J. Mantell, *J. Org. Chem.*, 16:524 (1951).
10. H. Bredereck, A. Wagner, and A. Kottenhahn, *Chem. Ber.*, 93:2415 (1960).
11. A. A. Oswald and T. J. Wallace, *in:* "Organic Sulfur Compounds", "Anionic Oxidation of Thiols and Co-oxidation of Thiols with Olefins", N. Kharasch and C. Y. Meyers, eds., Pergamon Press, New York, Vol. 2, pp. 205-232 (1966).
12. K. E. J. Barrett and W. A. Waters, *Discussions Faraday Soc.*, 14:221 (1953).
13. G. E. Adams, G. S. McNaughton, and B. D. Michael, *in:* "In Excitation and Ionisation", G. Scholes and G. R. A. Johnson, eds., Taylor and Francis, London, pp. 281-293, (1967).
14. G. E. Adams, G. S. McNaughton, and B. D. Michael, *Trans. Faraday Soc.*, 64:902 (1968).
15. G. E. Adams, R. C. Armstrong, A. Charlesby, B. D. Michael, and R. L. Willson, *Trans. Faraday Soc.*, 65:732 (1969).
16. J. L. Redpath, *Radiat. Research*, 54:364 (1973).
17. T. Miyashita, M. Iiono, and M. Matsuda, *Bull. Chem. Soc. Jpn.*, 50:317 (1977).
18. W. A. Pryor, G. Gojon, and D. F. Church, *J. Org. Chem.*, 43:793 (1978).
19. M. Morita, K. Sasai, M. Tajima, and M. Fujimaki, *Bull. Chem. Soc. Jpn.*, 44:2257 (1971).
20. A. J. Elliot and F. C. Sopchyshyn, *Radiat. Phys. Chem.*, 19:417 (1982).
21. A. J. Elliot, A. S. Simsons, and F. C. Sopchyshyn, *Radiat. Phys. Chem.*, 23:377 (1984).
22. K. Schäfer and K-D. Asmus, *J. Phys. Chem.*, 85:852 (1981).
23. M. S. Akhlaq, H-P. Schuchmann, and C. von Sonntag, *Int. J. Radiat. Biol.*, 51:91 (1987).
24. C. Schöneich, M. Bonifacic, and K-D. Asmus, *Free Radical Res. Commun.*, 6:393 (1989).

25. C. Schöneich, K-D. Asmus, U. Dillinger, and F. V. Bruchausen, *Biochem. Biophys. Res. Commun.*, 161:113 (1989).

26. R. L. Willson, *in*: "Radioprotecters and Anticarcinogens", O. F. Nygaardand M. G. Simic, eds., Academic Press, New York, p. 1-14 (1983).

27. R. L. Willson, C. A. Dunster, L. G. Forni, C. A. Gee, and K. J. Kitteridge, *Phil. Trans. R. Soc. London*, B311:545 (1985).

28. R. L. Willson, *in*: "Oxidative Stress", H. Sies, ed., Academic Press, London, p. 41-72 (1985).

29. K-D. Asmus, *in*: "Methods in Enzymology", 105:167 (1984).

30. G. E. Adams and P. Wardman, *in*: "Free Radicals in Biology", Academic Press, New York, Vol. 3, p. 53-95 (1977)

31. G. E. Adams, B. D. Michael, and R. L. Willson, "Advances in Chemistry Series" 81:289 (1968).

32. J. P. Keene, E. J. Land, and A. J. Swallow, *in*: "Pulse Radiolysis", M. Ebert, J. P. Keene, A. J. Swallow and J. H. Baxendale, eds.,Academic Press, London, p. 227-245 (1965).

33. R. L. Willson, *Chemical Commun.*, 1425 (1970).

34. K-D. Asmus, H. Möckel, and A. Henglein, *J. Phys. Chem.*, 77:1218 (1973).

35: G. E. Adams and R. L. Willson, *J. Chem. Soc. Faraday Trans. 1.*, 69:719 (1973).

36. R. L. Willson, *Trans. Farad. Soc.*, 67:3008 (1971).

37. K. B. Patel and R. L. Willson, *J. Chem. Soc. Faraday Trans. 1*, 69:814 (1973).

38. L. G. Forni and R. L. Willson, *Methods in Enzymology*, 105:179 (1984).

39. J. A. Farrington, E. J. Land, and A. J. Swallow, *Biochim. Biophys. Acta*, 590:273 (1980).

40. R. F. Anderson, *Biochim. Biophys. Acta*, 590:277 (1980).

41. P. M. Wood, *FEBS Letts.*, 44:22 (1974).

42. P. Wardman, *J. Phys. Chem. Reference Data*, 18:1637 (1989).

43. J. E. Packer, T. F. Slater, and R. L. Willson, *Life Sciences*, 23:2617 (1978).

44. J. E. Packer and R. L. Willson, D. Bahnemann, and K-D. Asmus, *J. Chem. Soc., Perkin Trans. 2*, 296 (1980).

45. J. Mönig, D. Bahnemann, and K-D. Asmus, *Chem. Biol. Interact.*, 47:15 (1983).

46. D. Bahnemann, K-D. Asmus, and R. L. Willson, *J. Chem. Soc. Perkin Trans. 2*, 1661 (1983).

47. D. Bahnemann, K-D. Asmus, and R. L. Willson, *J. Chem. Soc. Perkin Trans. 2*, 1669 (1983).

48. D. Bahnemann, K-D. Asmus, and R. L. Willson, *J. Chem. Soc. Perkin Trans. 2*, 890 (1981).

49. I. Wilson, P. Wardman, G. M. Cohen, and M, D'Arcy Doherty, *Biochem. Pharmacol.*, 35:21 (1986).

50. L. G. Forni, V. O. Mora-Arellano, J. E. Packer, and R. L. Willson, *J. Chem. Soc. Perkin Trans. 2*, 1579 (1988).

51. R. H. Bisby, S. Ahmed, R. B. Cundall, and E. W. Thomas, *Free Radical Res. Commun.*, 1:257 (1986).

52. M. J. Davies, L. G. Forni, and R. L. Willson, *Biochem J.*, 255:513 (1988).

53. M. J. Thomas and B. H. J. Bielski, *J. Amer. Chem. Soc.*, 111:3315 (1989).

54. S. Steenken, *J. Phys. Chem.*, 83:595 (1979).

55. B. S. Wolfenden and R. L. Willson, *J. Chem. Soc. Perkin Trans. 2*, 805 (1982).

56. L. G. Forni and R. L. Willson, *Biochem. J.*, 240:905 (1986).

57. L. G. Forni and R. L. Willson, *Biochem. J.*, 240:897 (1986).

58. J. L. Redpath and R. L. Willson, *Int. J. Radiat. Biol.*, 23:51 (1973).

59. L. G. Forni and R. L. Willson, *in*: "Protective Agents in Cancer", D. C. H. McBrien and T. F. Slater, eds., Academic Press, London, p. 159-173 (1983).

60. L. G. Forni, J. Mönig, V. O. Mora-Arellano, and R. L. Willson, *J. Chem. Soc. Perkin Trans. 2*, 961 (1983).

61. C. A. Dunster and R. L. Willson, *Int. J. Radiat. Biol.*, 55:873 (1989).

62. M. D'Aquino, C. A. Dunster, and R. L. Willson, *Biochem. Biophys. Res. Commun.*, 161:1199 (1989).

63. N. V. Raghavan, P. K. Das, and K. Bobrowski, *J. Amer. Chem. Soc.*, 103:4569 (1984).

64. C. A. Dunster and R. L. Willson, to be published.

65. R. L. Willson, *in:* "Iron Metabolism", R. Porter and D. W. Fitzsimons, eds., Ciba Foundation Symposium, Vol. 51, Elsevier/Excerpta Medica, pp. 331-354 (1977) and references cited.

66. G. Saez, P. J. Thornalley, H. A. O. Hill, R. Hems, and J. V. Bannister, *Biochim. Biophys. Acta*, 719:24 (1982).

67. A. J. F. Searle and R. L. Willson, *Biochem. J.*, 212:549 (1983).

68. C. Malvy, C. Paoletti, A. J. F. Searle, and R. L. Willson, *Biochem. Biophys. Res. Commun.*, 95:734 (1980).

69. L. S. Harman, D. K. Carver, J. Schreiber, and R. P. Mason, *J. Biol. Chem.*, 261:1642 (1986).

70. T. E. Eling, J. F. Curtis, L. S. Harman, and R. P. Mason, *J. Biol. Chem.*, 261:5023 (1986).

71. R. L. Willson, K-D. Asmus, and P. Wardman, *Nature*, 252:323 (1974).

72. P. O'Neill, *Radiat. Res.*, 96:198 (1983).

73. R. L. Willson, *Int. J. Radiat. Biol.*, 17:349 (1970).

74. G. E. Adams and R. L. Willson, *Trans. Faraday Soc.*, 65:2981 (1969).

75. M. Tamha, G. Simone, and M. Quintiliani, *Int. J. Radiat. Biol.*, 50:495 (1986).

76. J. Mönig, K-D. Asmus, L. G. Forni, and R. L. Willson, *Int. J. Radiat. Biol.*, 52:589 (1987).

77. A. J. Searle and A. Tomasi, *J. Inorg. Biochem.*, 17:161 (1982).

78. M. J. Davies, L. G. Forni and S. L. Shuter, *Chem.-Biol. Interact.*, 61:177 (1987).

79. F. Haber and R. Willstätter, *Chem. Berichte*, 64:2844 (1931).

FREE RADICAL TRANSFER INVOLVING SULPHUR PEPTIDE FUNCTIONS

Walter A. Prütz

Institut für Biophysik und Strahlenbiologie
Universität Freiburg
7800 Freiburg, F. R. Germany

Many biological processes involve series of electron transfer reactions through protein assemblies. Examples of such controlled biological electron transfer chains are the oxidative phosphorylation in mitochondria, and the photosynthesis in chloroplasts. Sulphur peptide functions commonly serve to stabilize, by disulphide bonds, the three-dimensional protein structure, and to hold certain reaction centers in position; for instance, in ferricytochrome c the active heme group is covalently bound by two cysteines and additional coordinative binding of the central iron is provided by a methionine and a histidine group, and in plastocyanine (an electron transfer protein in photosynthesis) the active Cu(II) center involves one cysteine, one methionine and two histidines as ligands. The specific structures of electron transfer proteins, and particularly the environment around the active sites, are thought to play a pertinent role in directing the electron transfer process, and various possible mechanisms of electron transfer through the protein matrix have been discussed.[1,2]

More random electron migration through proteins can be induced by ionizing radiation or UV-light, and already half a century ago free radical migration within proteins was predicted to extend the target size for radiation-induced genetic damage of cells. Sulphur centered radicals were, in fact, among the first intermediates detected by ESR spectroscopy in γ-irradiated dry proteins. Pulse radiolysis investigations have later abundantly shown that electrons become localized with high yield at cystine disulphide groups of proteins when irradiated in neutral aqueous solution. The present contribution mainly deals with radiation-induced free radical transfer between aromatic and sulphur peptide functions. More general accounts of radiation effects in peptides and proteins can be found elsewhere.[3-6]

SOME PROPERTIES OF PEPTIDES AND PROTEINS PERTINENT TO FREE RADICAL TRANSFER

The primary structure of a peptide or protein consists in a specific sequence of amino acids polymerized into a chain *via* amide bonds (–CO–NH–), commonly referred to as "peptide bonds":

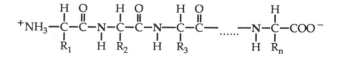

Sulfur-Centered Reactive Intermediates in Chemistry and Biology, Edited by
C. Chatgilialoglu and K.-D. Asmus, Plenum Press, New York, 1990

389

The chain length of proteins is usually in the order of n = 100 to 1000 (peptides are smaller), and 20 different kinds of amino acids are found in most proteins.

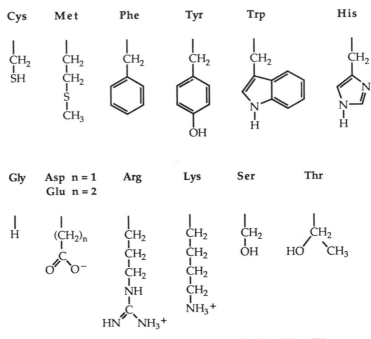

Scheme 1. Some specific amino acid side chains (R)

Glutathione (GSH) is an example of a short peptide which is thought to play an important role in the cellular defense against free radical action (see article by M. Quintiliani in this book):

Glutathione :

$$^+NH_3-\overset{\overset{\displaystyle COO^-}{|}}{CH}-CH_2-CH_2-CO-NH-\underset{\underset{\displaystyle CH_2-SH}{|}}{CH}-CO-NH-CH_2-COO^-$$

(γ-Glu–Cys–Gly)

The folding of polypeptide chains leads to specific structures (depending on amino acid sequence), such as α-helices and β-pleated sheets with parallel or antiparallel strands associated by a network of hydrogen bonds between the main chain peptide groups. Figure 1 shows as an example, the hydrogen bonding in an α-helix. The three-dimensional protein structure is additionally stabilized by covalent cystine disulphide bridges, noncovalent salt bridges and hydrophobic forces (Figure 2). Disulphide bridges are abundant structural elements particularly in extra-cellular proteins (e.g. serum albumin with 17 –SS– bridges), and as a rule, the disulphides of intra-cellular proteins often play a functional role as well (e.g. in glutathione reductase).

Theoretically both the polypeptide backbone and hydrogen bond networks (Figure 1) have been considered as possible channels for long-range electron transfer, but also other mechanisms such as direct outer-sphere collisional transfer, hydrophobic channel transfer etc. have been envisaged.[1,2] Of the side chains particularly the sulphur-containing and aromatic groups (Scheme 1) are of importance in the mediation of free radical transfer.

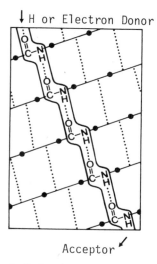

↓H or Electron Donor

Acceptor ✔

Fig. 1.   Cylindrical plot of an α-helical polypeptide (opened cylinder surface), showing the hydrogen bonds between peptide groups, which provide possible channels for long-range electron transfer. The filled circles represent the α-carbons along the peptide backbone.

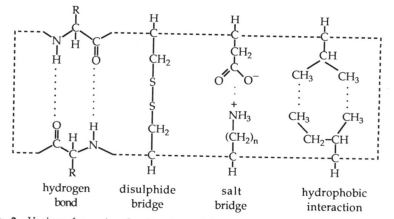

| hydrogen bond | disulphide bridge | salt bridge | hydrophobic interaction |

Fig. 2.   Various forces involved in the stabilization of tertiary protein structures

RADIATION-INDUCED FREE RADICAL REACTIONS

A great variety of free radical reactions can be initiated radiolytically in aqueous solutions through the water radiolysis products: (1 Gy = 1 J kg^{-1})

$$H_2O \xrightarrow{\text{radiation}} \quad ^\bullet OH \quad + \quad e^-_{aq} \quad + \quad H^\bullet \tag{1}$$
$$\text{yield (μM/Gy):} \quad\quad 0.29 \quad\quad 0.28 \quad\quad 0.06$$

Investigations of one-electron reduction of protein components is commonly performed in deaerated solution containing $^\bullet OH$-scavengers such as t-butanol or formate, i.e. with e^-_{aq}, $H^\bullet$ and $CO_2^{\bullet-}$ (in the case of formate) as reductants,

$$\bullet OH + HCOO^- \rightarrow H_2 + CO_2^{\bullet-} \tag{2}$$

Oxidative free radical reactions are commonly studied in N_2O-saturated solution, where e^-_{aq} is converted into $\bullet OH$,

$$e^-_{aq} + N_2O + H_2O \rightarrow N_2 + OH^- + \bullet OH \tag{3}$$

$\bullet OH$-radicals react rather unselectively with protein components; more selective one-electron oxidation of specific side chains can be achieved with halide and pseudo-halide radical anions derived from $\bullet OH$,

$$\bullet OH + X^- \rightarrow OH^- + X^\bullet \tag{4}$$

$$X^\bullet + X^- \rightleftharpoons X_2^{\bullet-} \tag{5}$$

Application of pulse radiolysis (see article by D.A. Armstrong in this book) has enabled estimates for the reaction kinetics of these radicals and of subsequent radical transformations. Reactivities of $\bullet OH$, $X^\bullet$ and $X_2^{\bullet-}$ radicals towards some amino acids are shown in Table 1.

Table 1. Bimolecular Rate Constants (in Units of 10^7 M^{-1} s^{-1}) for Reactions of Oxidizing Radicals with some Amino Acids (pH 7) (E^0: vs. NHE)

Radical	(E^0/V)	Trp	Tyr	Phe	His	Cys	Met	Val
$\bullet OH$	(1.91)	1000	1000	700	700	2000	700	70
$Br_2^{\bullet-}$	(1.63)	77	2.0	0.1	1.5	18	200	0.1
$N_3^\bullet$	(1.32)	410	10	0.1	0.1	1.4	0.1	0.1
$I_2^{\bullet-}$	(1.03)	0.1	0.1	0.1	0.1	11	0.1	0.1

Pulse radiolysis investigations have indeed revealed that the oxidative power of sulphur-centered free radicals decreases in the series $R_2S^{\bullet+} > (R_2S\bullet\bullet SR_2)^+ > (RSSR)^{\bullet+} > RS^\bullet > (RS\bullet\bullet SR)^-$.[7] ("$\bullet\bullet$" denotes weak $2\sigma/1\sigma^*$ bonds; for details see article by K.-D. Asmus in this book.) Radical transformation between sulphur peptide units can hence be expected to proceed according this series. Of particular importance in biological systems is probably the equilibrium between cysteine thiyl and disulphide radical anions,

$$Cys/S^\bullet + Cys \rightleftharpoons Cys_2/S\bullet\bullet S^- + H^+ \tag{6}$$

as it represents a cross-over between an oxidizing ($Cys/S^\bullet$) and a reducing ($Cys_2/S\bullet\bullet S^-$) entity, depending on pH and Cys-concentration ($K_6 = 6000$ M^{-1}). The redox behaviour of these intermediates can be expressed by the one-electron reduction potential, e.g. at pH 7, (vs. NHE): $E^0(Cys/S^\bullet; Cys/S^-) = 0.9$ V, $E^0(Cys_2/S\bullet\bullet S^-; 2\ Cys/S^-) = 0.65$ V and $E^0(Cys_2; Cys_2/S\bullet\bullet S^-) = -1.6$ V (see article by D.A. Armstrong in this book). (In some of the relevant literature the notation $.../SS^{\bullet-}$ is used to denote the disulfide radical anion, $.../S^\bullet$ denotes thiyl radical.)

Another feature of thiols, which might play a role in the free radical "repair" in biological systems, seems to be the coupling of hydrogen-atom with electron-transfer (see article by R.L. Willson in this book).

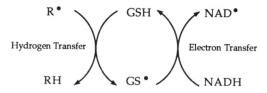

Free radical migration in proteins

In this section we will briefly consider radiation-induced free radical transfer in proteins (without specific reaction centers such as metallo-porphyrins).

Hydrated electrons (e^-_{aq}) tend to add rather randomly to the surface of proteins, primarily at peptide-bond carbonyls, and at solvent-exposed cysteine and cystine groups. The disulphide anion can easily be detected owing to its broad 410 nm absorption band. In proteins the electron becomes localized at disulphides at high yield (e.g. 20% in RNase A and 65% in lysozyme) immediately with the disappearance of e^-_{aq}, even in proteins with inaccessible $-SS-$ groups such as RNase A. These observations have prompted the conclusion that initial electron attachment at the protein surface is followed by rapid electron migration to disulphides, probably via hydrogen bond path.[8] Due to the primary random reaction of e^-_{aq} it appears ambiguous, however, to draw conclusion on the mechanism and pathway of electron migration in proteins by using this species as initiator.

The carbon dioxide radical anion $CO_2^{\bullet-}$ [reaction (2)] reacts more selectively, and thus is capable of electron transfer to disulphide but not to peptide-bond carbonyl groups, despite being a powerful one-electron reductant [$E^0(CO_2; CO_2^{\bullet-}) = -2$ V]. A question worth discussion is whether electron addition to peptide-bond carbonyl is a prerequisite to electron transfer via hydrogen bonded networks (see Figure 1); one may expect e^-_{aq} to be able, and $CO_2^{\bullet-}$ to be unable to initiate such transfer. Surprisingly, $CO_2^{\bullet-}$ was found to generate disulphide radical anions in RNase A via slow (100 μs) secondary processes, but the intermediates involved have not been identified.[8]

$^{\bullet}OH$ radicals also induce slow secondary radical transformations in proteins, but due to the low selectivity of primary reactions of $^{\bullet}OH$ (Table 1) it appears again difficult to characterize these processes. More selectively reacting one-electron oxidants $X^{\bullet}$ and $X_2^{\bullet-}$ [reactions (4) and (5)] have been applied extensively, particularly in attempts to probe functional sites in enzymes (see e.g. ref. 4). Such studies have, on the other hand, enabled the demonstration of radical transfer between specific sites in proteins.

The $Br_2^{\bullet-}$ species reacts selectively with methionine (Table 1) to form three-electron bonded sulphur-bromine adducts, $Met/S.^{\bullet}.Br$, absorbing around 390 nm. In RNase A evidence was obtained for a $Br_2^{\bullet-}$-induced radical transfer chain involving Met_{29}, Tyr_{25} and the adjacent 26-84 disulphide (see ref. 6 for references). (.../$SS^{\bullet+}$ denotes the disulphide radical cation, its electronic structure is a sulphur-sulphur $2\sigma/2\pi/1\sigma^*$ bond; see article by K.-D. Asmus in this book; .../$O^{\bullet}$ denotes oxyl radical).

$$Br_2^{\bullet-} \quad \overset{(7)}{\rightarrow} \quad Met_{29}/S.^{\bullet}.Br \quad \overset{(8)}{\rightarrow} \quad Tyr_{25}/O^{\bullet} \quad \overset{(9)}{\rightarrow} \quad (Cys_{26}-Cys_{84})/SS^{\bullet+}$$

$$\downarrow (10)$$

$$\text{dimers (o,o'-biphenol-coupling)}$$

The fast reaction (8) involving the side chains 29 and 25 ($k_8 = 4 \times 10^5$ s^{-1}) was indicated since Met_{29} is the only solvent-exposed (of four) methionine units, and Tyr_{25} is located in close proximity within an α-helical turn, enabling collisional contact of the reacting groups. Reaction

(8) is, in fact, much faster than the corresponding intramolecular transfer in the dipeptide Met–Tyr,

$$\text{Met/S} \cdot^{\bullet} \cdot \text{Br–Tyr} \quad \rightarrow \quad \text{Met–Tyr/O}^{\bullet} + \text{Br}^- + \text{H}^+ \qquad (k_{11} = 4 \times 10^4 \text{ s}^{-1}) \qquad (11)$$

Evidence for reaction (9) was only derived indirectly from the relatively low yield of dimerization *via* reaction (10), coupled with the observation of such interactions in the cystinyl peptide $(\text{Cys–Tyr})_2$.

The azide radical $(\text{N}_3{}^{\bullet})$ selectively oxidizes tryptophan (Table 1) *via* generation of indolyl radicals Trp/N$^{\bullet}$, which absorb around 520 nm. In peptides of the type Trp–x–Tyr efficient intramolecular radical transfers between the indol and phenol functions have been demonstrated by monitoring the first-order disappearance of the Trp/N$^{\bullet}$ absorption band and simultaneous appearance of the 405 nm band of Tyr/O$^{\bullet}$ (Table 2), (.../N$^{\bullet}$ denotes N-centered radicals)

$$\text{Trp/N}^{\bullet}\text{–x–Tyr} \quad \rightarrow \quad \text{Trp–x–Tyr/O}^{\bullet} \qquad (12)$$

Efficient intramolecular Trp/N$^{\bullet}$ → Tyr/O$^{\bullet}$ transfers, with half-lives in the order of milliseconds, have also been observed in numerous proteins such as alcohol dehydrogenase, apo-cytochrome c, concanavalin, β-lactoglobulin, lysozyme, trypsin etc.[10] It is not possible, however, to characterize the pathway of these transfers since the proteins studied contain several Trp and Tyr units. Erabutoxin b, a neurotoxin with 62 amino acids, contains just one Trp and one Tyr unit, Trp$_{25}$ and Tyr$_{29}$. However, despite the proximity of these functions there was no evidence for a Trp/N$^{\bullet}$ → Tyr/O$^{\bullet}$ transfer, unless the toxin was denatured by reduction of the four disulphide groups which stabilize the rigid structure. It can be concluded that the rigid conformation of native erabutoxin b does not enable a direct contact of the reactive groups, and that the peptide backbone does not provide a channel for the Trp/N$^{\bullet}$ → Tyr/O$^{\bullet}$ transfer.

Table 2. Rates of intramolecular transfer reaction (12) in various peptides at around pH 7 (in s^{-1})

n	Trp–(Gly)$_n$–Tyr	Trp–(Pro)$_n$–Tyr[a]	Trp–εAH–Tyr[b]	Gly–Trp–Tyr–Gly
0	7.3×10^4	7.7×10^4		2.0×10^4
1	5.1×10^4	2.6×10^4	1.3×10^5	
2	2.4×10^4	4.9×10^3		
3	3.3×10^4	1.5×10^3		

[a] Bobrowski et al.[9] (further references therein).
[b] –εAH– = –NH–(CH$_2$)$_5$–CO–.

Due to the problems encountered in identification of chemical reaction pathways induced by free radicals in proteins, several studies have been undertaken using simpler peptide model systems.

Free radical cascades in sulphur peptide model systems

An $^{\bullet}$OH-induced free radical transfer cascade has, for instance, been demonstrated by pulse radiolysis of a N$_2$O-saturated, aqueous solution containing methionine (25 mM),

(Cys–Gly)$_2$ (2 mM) and Gly–Tyr (1 mM) at pH 8.1, by detection of the delayed build-up of the Tyr/O$^\bullet$ absorption band:[11]

$$\overset{-H_2O}{^\bullet OH \longrightarrow} \underset{(13)}{Met/S \bullet^\bullet \bullet N^+} \overset{-CO_2}{\longrightarrow} \underset{(14)}{Met/C_\alpha{}^\bullet} \longrightarrow \underset{(15)}{Cys_2/S \bullet^\bullet \bullet S^-} \overset{-/+Cys}{\rightleftarrows} \underset{(6)}{Cys/S^\bullet} \rightleftarrows \underset{(16)}{Tyr/O^\bullet}$$

Rate constants for reactions of methionyl radicals, and structures of these, are shown in Table 3. The ring-closed Met/S$\bullet^\bullet \bullet$N$^+$ species, formed subsequent to $^\bullet$OH addition to the sulphur function, is a strong oxidant capable of oxidation of tyrosine. For unbound methionine, however, a rapid decarboxylation occurs [reaction (14)] with formation of the Met/C$_\alpha{}^\bullet$ species which is a powerful one-electron reductant, able to reduce (Cys–Gly)$_2$ [reaction (15)]. That tyrosine finally is oxidized by Cys/S$^\bullet$ [reaction (16)] and not by Met/S$\bullet^\bullet \bullet$N$^+$ was shown by the failure to detect Tyr/O$^\bullet$ formation when (Cys–Gly)$_2$ was absent in the above system. Reaction (16) is, in fact, an equilibrium (K~1 at pH 7), but since no Cys was present in the system the radical transfer is driven to the right. In the presence of Cys, however, the unpaired electron can be expected to be stabilized by formation of Cys/S$\bullet^\bullet \bullet$S$^-$ *via* the reverse reactions (16) and (6).

Decarboxylation *via* reaction (14) is prevented by going to peptides of the type Met–Gly, and the corresponding (Met–Gly)/S$\bullet^\bullet \bullet$N$^+$ species was found to interact also with Trp, Cys and N$_3^-$ (Table 3). The equilibrium between (Met–Gly)/S$\bullet^\bullet \bullet$N$^+$ and azide has indeed enabled an estimate of the reduction potential of this methionyl, which is shown in Figure 3 together with data for oxidizing Cys, Trp and Tyr species. The (Met–Gly)/S$\bullet^\bullet \bullet$N$^+$ radical can also be generated from dimeric methionyl radical cations Met$_2$/S$\bullet^\bullet \bullet$S$^+$, which are formed by $^\bullet$OH in solutions containing high concentrations of peptides like Gly–Met or Gly–Met–Gly. According to these data a cascading transfer of electron deficiency is feasible, for certain peptides and at neutral pH, through the following series:

$$\underset{(17)}{Met_2/S \bullet^\bullet \bullet S^+} \longrightarrow \underset{(18)}{Met/S \bullet^\bullet \bullet N^+} \longrightarrow \underset{(12)}{Trp/N^\bullet} \longrightarrow \underset{(16)}{Tyr/O^\bullet} \rightleftarrows Cys/S^\bullet$$

Rates for individual reaction steps in this series have been documented (Tables 2 and 3), but as yet we have not been able to demonstrate the oxidation of Cys by Trp/N$^\bullet$, which thermodynamically should be feasible; the question why this redox pair behaves quasi inert deserves to be discussed.

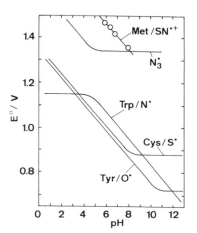

Fig. 3.   Reduction potentials vs. pH for various oxidizing peptide radicals.[11]

Table 3. Rate Constants of Radical Transitions and Structures of Sulphur-Centered Peptide Radicals [11]

Transition	pH	k (mol^{-1} dm^3 s^{-1})
Met/C$_\alpha{}^\bullet$ → Cys/S$^\bullet$	8.2	2.7 x 10^8
Met/C$_\alpha{}^\bullet$ → (Cys–Gly)/S•$^\bullet$•S$^-$	8.1	≈ 7 x 10^6
(Met–Gly)/S•$^\bullet$•N$^+$ → N$_3{}^\bullet$	6.8	8.0 x 10^7
←	6.8	≈ 4 x 10^6
(Met–Gly)/S•$^\bullet$•N$^+$ → Cys/S$^\bullet$	8.0	≈ 8 x 10^7
(Met–Gly)/S•$^\bullet$•N$^+$ → Trp/N$^\bullet$	6.8	4.0 x 10^7
Met$_2$/S•$^\bullet$•S$^+$ → N$_3{}^\bullet$	4.3	2.8 x 10^8
(Gly–Met)$_2$/S•$^\bullet$•S$^+$ → Cys/S$^\bullet$	8.3	3.5 x 10^8
(Gly–Met)$_2$/S•$^\bullet$•S$^+$ → (Met–Gly)/S•$^\bullet$•N$^+$	8.2	6.0 x 10^7
(Gly–Met–Gly)$_2$/S•$^\bullet$•S$^+$ → Cys/S$^\bullet$	8.2	1.7 x 10^8
(Gly–Met)$_2$/S•$^\bullet$•S$^+$ → Gly–Tyr/O$^\bullet$	8.9	9.0 x 10^7
(Gly–Met)$_2$/S•$^\bullet$•S$^+$ → Gly–Tyr/O$^\bullet$	7.5	2.3 x 10^7
(Gly–Met)$_2$/S•$^\bullet$•S$^+$ → Trp/N$^\bullet$–Gly	7.0	4.8 x 10^8
(Cys–Gly)/S$^\bullet$ → Gly–Tyr/O$^\bullet$	6.8	6.3 x 10^6
G/S$^\bullet$ → Gly–Tyr/O$^\bullet$	8.1	5.8 x 10^6
Ala/C$_\beta{}^\bullet$ → Cys/S$^\bullet$	8.2	5.0 x 10^6
G$^\bullet$ → G/S$^\bullet$	8.0	7.2 x 10^6

Structures:

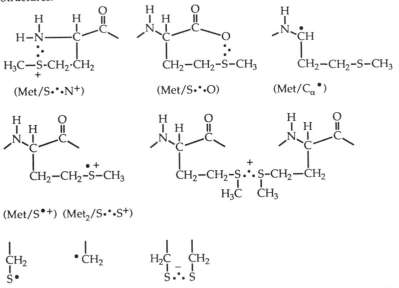

(Met/S•$^\bullet$•N$^+$)   (Met/S•$^\bullet$•O)   (Met/C$_\alpha{}^\bullet$)

(Met/S$^{\bullet+}$) (Met$_2$/S•$^\bullet$•S$^+$)

(Cys/S$^\bullet$)   (Ala/C$_\beta{}^\bullet$)   (Cys$_2$/S•$^\bullet$•S$^-$)   (For the Cys-derived radicals only the side chains are shown)

As with $^\bullet$OH also e^-_{aq}-induced radical transfer cascades can be demonstrated, e.g. in solutions containing the cyclic peptide c-Ala$_2$ (16 mM), (Cys–Gly)$_2$ (0.6 mM) and Gly–Tyr (4 mM), with t-butanol (1 M) as $^\bullet$OH-scavenger:[11]

$$e^-_{aq} \longrightarrow c\text{-Ala}_2/\overset{\bullet}{C}O^- \longrightarrow Cys_2/S\bullet^\bullet\bullet S^- \overset{-/+Cys}{\rightleftarrows} Cys/S^\bullet \rightleftarrows Tyr/O^\bullet$$
$$\quad\quad\quad\quad (19)\quad\quad\quad\quad (20)\quad\quad\quad\quad\quad (6)\quad\quad\quad (16)$$

These transitions again enter into the equilibria (6) and (16), which would be driven towards the $Cys_2/S\bullet^\bullet\bullet S^-$ intermediate in presence of Cys. The cyclic diketopiperazin c-Ala$_2$ serves here as model for a protein-main-chain carbonyl group (short linear peptides would rapidly deaminate upon e^-_{aq} addition). Reaction (20) actually suggests that electron transfer to di-sulphides *via* an intermediate is feasible in proteins (see section "Free radical migration in proteins"); another example of electron transfer *via* an intermediate is reaction (15). Addition of e^-_{aq} to thiols leads to rapid HS$^-$ elimination with formation of β-carbon-centered radicals (e.g. Ala/C$_\beta{}^\bullet$ and G$^\bullet$), which are capable of oxidizing the parent thiol (Table 3).

Glutathione

Some recent results obtained by pulse radiolysis of N$_2$-saturated solutions of GSH are presented in Figure 4.[11] The diagram shows pH-dependent yields of $G_2/S\bullet^\bullet\bullet S^-$, estimated from the 420 nm absorbance with $\varepsilon_{420} = 8000\ M^{-1}\ s^{-1}$, in comparison with the total yield of water radicals $G(e_{aq}^-) + G(H^\bullet) + G(^\bullet OH) = 0.65\ \mu M/Gy$ (dotted line). In the absence of formate (open circles) the data can be partly explained by a reaction sequence (Scheme 2) in which all water radicals engage to generate $G_2/S\bullet^\bullet\bullet S^-$. The drop in yield below the $pK_a = 9.2$ of GSH is probably mainly due to the reverse reaction (27), but the yield never reaches the value of $0.65\ \mu M/Gy$ at high pH. By scavenging $^\bullet$OH with formate we can virtually reach full conversion of all water radicals into $G_2/S\bullet^\bullet\bullet S^-$ at pH > 8 (filled circles). This indicates that $CO_2{}^{\bullet-}$ [reaction (28)] is more efficient in generating G/S$^\bullet$ than $^\bullet$OH [reaction (24)]. The time profile **a** at pH 9 shows the fast formation of $G_2/S\bullet^\bullet\bullet S^-$ initiated *via* the reactions (23) and (24), and the additional slow formation initiated *via* the reactions (21) and (22), with reaction (25) as rate-determining step. At pH 7 (time profile **b**) the oxidative and reductive pathways of $G_2/S\bullet^\bullet\bullet S^-$ formation cannot be distinguished kinetically, probably because reaction (27) is slow.

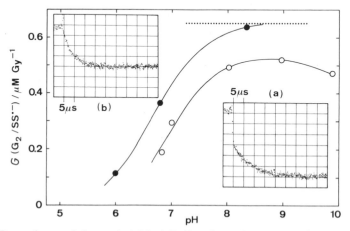

Fig. 4.   pH-Dependence of the total yield of $G_2/S\bullet^\bullet\bullet S^-$ at about 50 μs after pulse radiolysis of N$_2$-saturated GSH solutions: (o) 30 mM GSH, (●) 20 mM GSH + 1 M HCOONa. Inserted are time profiles of transmission changes at pH 9 (a) and pH 7 (b) for 30 mM GSH.

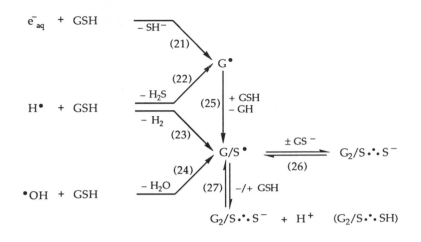

Scheme 2. Generation of $G_2/S \cdot \cdot \cdot S^-$ by radiolysis of GSH (Figure 4).

These results show that relevant amounts of $G_2/S \cdot \cdot \cdot S^-$ are formed by radiolysis even in a physiological pH range, and the question arises whether this species, as a powerful electron donor, may be of relevance in radiobiology, for instance in the "chemical repair" of radiation damage (see article by C. von Sonntag in this book).

CONCLUSIONS AND QUESTIONS

The direction and nature of free radical transfers which can be initiated radiolytically in peptide model systems can be expressed by the scheme:

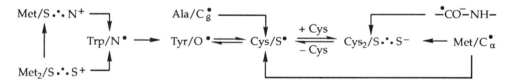

Transfers of both electrons and electron deficiencies can lead to formation of cystinyl and tyrosyl intermediates, i.e. $Cys/S \cdot \cdot \cdot S^-$, $Cys/S^\bullet$ and $Tyr/O^\bullet$, which hence may be of importance in terminal reactions such as crosslinking, disproportionation and "chemical repair". These observations give rise to a number of questions related to the problems of electron transfer in biological systems:

1. What are the mechanisms of free radical transfer through a protein matrix?

2. How can applications of radiolysis techniques help to obtain further insight into the mechanisms of transfer?

3. Is a transfer of radiation damage from proteins to DNA feasible and what are the intermediates involved?

4. Is "chemical repair" of radiation damage by interaction of $G_2/S \cdot \cdot \cdot S^-$ with DNA intermediates relevant,

$$DNA^{\bullet +} \quad + \quad G_2/S\bullet^{\bullet}\bullet S^- \quad \rightarrow \quad \text{"chemical repair"}$$

$$\downarrow O_2$$

inhibition of "chemical repair"

and can this mechanism be included in explaining oxygen effects?

5. Are the sulphur and aromatic peptide intermediates described important in physiological protein functions?

REFERENCES

1. G.R. Moore and R.J.P. Williams, *Coord. Chem. Rev.* 18:125 (1976).
2. S.S. Isied, *Prog. Inorg. Chem.* 32:443 (1984).
3. M.H. Klapper and M. Faraggi, *Quart. Rev. Biophys.* 12:465 (1979).
4. R.V. Bensasson, E.J. Land, and T.G. Truscott, "Flash Photolysis and Pulse Radiolysis", Pergamon Press, Oxford (1983).
5. W. Garrison, *Chem. Rev.* 87:381 (1987).
6. W.A. Prütz, *in*: "Radiation Research", Vol. 2 (8th International Congress of Radiation Research, Edinburgh), E.M. Fielden, J.F. Fowler, J.H. Hendry, and D. Scott, eds., Taylor & Francis, London, p. 134 ff. (1987).
7. M. Bonifacic and K.-D. Asmus, *Int. J. Radiat. Biol.* 46:35 (1984).
8. M. Faraggi, J.P. Steiner, and M.H. Klapper, *Biochem.* 24:3273 (1985).
9. K. Bobrowski, K.L. Wierzchowski, J. Holcman, and M. Ciurak, *Studia biophys.* 122:23 (1987).
10. J. Butler, E.J. Land, W.A. Prütz, and A.J. Swallow, *Biochim. Biphys. Acta* 705:150 (1982).
11. W.A. Prütz, J. Butler, E.J. Land, and A.J. Swallow, *Int. J. Radiat. Biol.* 55:539 (1989).

ELECTRON SPIN RESONANCE INVESTIGATION OF THE THIYL FREE RADICAL

METABOLITES OF CYSTEINE, GLUTATHIONE, AND DRUGS

Ronald P. Mason and D.N. Ramakrishna Rao

Laboratory of Molecular Biophysics
National Institute of Environmental Health Sciences
Research Triangle Park, NC 27709, USA

Most biochemicals, as opposed to aromatic drugs and industrial chemicals, are not easily metabolized through free radical intermediates. Cysteine and glutathione (GSH) are among the rather rare exceptions to this rule. The sulfhydryl group of L-cysteine plays many important roles in both the structure and function of proteins. These roles are modulated by the oxidation of L-cysteine. The ease of L-cysteine oxidation is also responsible for the radio-protection of intracellular GSH,[1] the bactericidal effect of cysteine,[2] and the general toxicity of cysteine. For this reason, N-acetylcysteine, and not cysteine, is used to treat the cases of acetaminophen (paracetamol) overdose.[3] The oxidation of thiol compounds, including L-cysteine, by ionizing radiation has been studied for a number of years. Sulhydryl compounds have been used to try to prevent radiation damage to normal tissue during radiation therapy. In fact, the high concentration of GSH in tissues provides significant radiation protection naturally. Among the stable products of oxidation of L-cysteine are L-cystine, L-cysteine sulfinic acid and L-cysteine sulfonic acid. These diamagnetic products of oxidation by metal ions or irradiation form via free radical intermediates, with the L-cysteine thiyl radical either dimerizing to form L-cystine,

$$RSH \xrightarrow{-H^\bullet} RS^\bullet \longrightarrow 1/2\ RSSR$$

or reacting with molecular oxygen or hydrogen peroxide to form the oxygen-containing products.

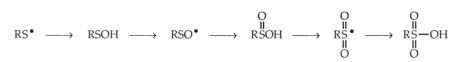

Although the importance of free radicals in the radiolytic and metal ion oxidation of L-cysteine is clear, no direct evidence of a role for cysteine-derived free radicals in an enzymatic reaction had been reported until recently. Among the most interesting possibilities for free radical formation are the thiol groups of sulfhydryl drugs and GSH.

We have used the spin-trapping ESR technique to study free radical metabolites formed via the oxidation of sulfhydryl compounds by the peroxidase prototype, horseradish peroxidase.[4] Spin trapping is a technique where a diamagnetic molecule (or spin trap) reacts

Sulfur-Centered Reactive Intermediates in Chemistry and Biology, Edited by
C. Chatgilialoglu and K.-D. Asmus, Plenum Press, New York, 1990

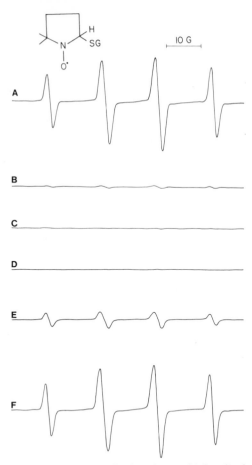

Fig. 1. The ESR spectrum of the DMPO-glutathione thiyl radical adduct produced in a system of horseradish peroxidase and hydrogen peroxide under aerobic conditions.

A. Incubation containing 10 mM glutathione, 50 µM H_2O_2, 90 mM DMPO, and 0.1 mg/ml horseradish peroxidase in Tris/HCl buffer.

B. Same as in A, but without addition of H_2O_2.

C. Same as in A, but without addition of horseradish peroxidase.

D. Same as in A, but heat-denatured enzyme.

E. Same as in A, but with 1000 units/ml catalase.

F. Same as in A, but with 40 µg/ml superoxide dismutase. (From ref. 6, with permission).

with a free radical to produce a more stable radical (or spin adduct) which is readily detectable by ESR.[5] Spin adducts are usually substituted nitroxide free radicals, which, for free radicals, are relatively stable, and in some cases have been identified by mass spectrometry.

The ESR spectrum of the DMPO-glutathione thiyl radical adduct (Figure 1A) is obtained in an incubation of 10 mM GSH, 50 µM H_2O_2 and 100 µg/ml horeradish peroxidase.[6] DMPO (5,5-dimethyl-1-pyrroline N-oxide) is a nitrone spin trap. Only a weak signal is obtained in the absence of hydrogen peroxide (Figure 1B). As expected, omission of horseradish peroxidase (Figure 1C) or use of heat-denatured horseradish peroxidase (Figure 1D) results in negligible signal. In accord with the dependence on hydrogen peroxide, catalase strongly inhibited formation of the GS-radical adduct (Figure 1E). Superoxide dismutase had no effect even at 40 μg/ml (Fig. 1F), ruling out a role for superoxide in this oxidation.

A wide variety of primarily aromatic compounds such as phenols are oxidized by horseradish peroxidase (HRP) to free radical metabolites in a reaction that requires hydrogen peroxide. The previously known evidence that thiyl free radicals can thus be generated were the reports by other workers of compound II formation in incubations of L-cysteine.[7]

$$HRP + H_2O_2 \rightarrow HRP\text{-compound I (green)} + H_2$$

$$HRP\text{-compound I} + GSH\ (GS^-) \rightarrow HRP\text{-compound II (red)} + GS^\bullet$$

$$HRP\text{-compound II} + GSH\ (GS^-) \rightarrow HRP + GS^\bullet$$

The formation of compound II, the second peroxidase intermediate, from compound I is evidence for a one-electron transfer and is thus a good inferential indication that the thiyl free radical is being generated, but our ESR investigations are the first to provide direct evidence of enzymatic thiyl free radical formation. The spin-trapping technique was used because thiyl free radicals themselves cannot be detected in solution with ESR for theoretical reasons.[8]

Since horesradish peroxidase is of plant origin, some may question the pharmacological relevance of work done with this enzyme. On the other hand, saliva, tears, and milk contain lactoperoxidase, which is thought to have an antimicrobial function *in vivo*. This mammalian peroxidase has a soret optical spectrum which is characteristic of the heme that is similar to that of thyroid peroxidase, intestine peroxidase, uterus peroxidase, eosiniphil peroxidase, and prostaglandin H synthase. As such, lactoperoxidase appears to be a useful prototype for most mammalian hemoprotein peroxidases. The ESR spectrum of the DMPO-cysteine thiyl radical adduct is obtained with 0.1 mg/ml lactoperoxidase and 10 mM cysteine in air (Figure 2A). The structure of this radical adduct has been confirmed with mass spectrometry.[9] The signal is totally inhibited by catalase (Figure 2B) and requires native lactoperoxidase (Figure 2C and 2D), but does not require added hydrogen peroxide. Under nitrogen, the thiyl radical formation requires 100 μM hydrogen peroxide and residual activity is inhibited by catalase.

D-penicillamine is a slow acting drug with an unknown mechanism of action used in the treatment of rheumatoid arthritis. Its structure differs from cysteine in that the methylene hydrogens have been replaced by methyl groups. As with cysteine, hydrogen peroxide is required in the absence of oxygen. The formation of this radical is also inhibited by catalase.

Glutathione thiyl free radical formation is hydrogen peroxide-dependent in air as it is in the case with the horseradish peroxidase. The omission of hydrogen peroxide gives a much weaker signal. N-acetylcysteine is similar to GSH in that hydrogen peroxide is required. The toxicity of cysteine relative to N-acetylcysteine may reflect the fact that the peroxidase-catalyzed oxidation of cysteine generates its own hydrogen peroxide, whereas the oxidation of N-acetylcysteine requires an additional source of hydrogen peroxide.

We have investigated the consumption of oxygen because free radicals often react with oxygen. In fact, oxygen is mother nature's spin trap. In order to implicate the role of free radicals in the oxygen uptake, we investigated the effect of the spin trap DMPO. Since DMPO is a radical trap, it is expected to be an antioxidant.

Studies with a Clark oxygen electrode show that oxygen is consumed in incubations of cysteine and lactoperoxidase. The rate of oxygen consumption decreased with the addition of DMPO either during or before the reaction.[9] These results are consistent with the cysteine thiyl free radical being formed and then trapped by DMPO before further reactions of the

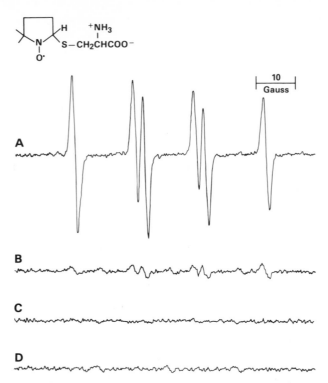

Fig. 2. The ESR spectrum of the DMPO-cysteine thiyl radical adduct produced in a system of lactoperoxidase under aerobic conditions.

A. Incubation containing 10 mM cysteine, 135 mM DMPO, and 0.1 mg/ml lactoperoxidase in 100 mM phosphate buffer plus 1 mM diethylenetri-aminepentaacetic acid.

B. Same as in A, but 5000 units/ml catalase.

C. Same as in A, but enzyme boiled 10 minutes before use.

D. Same as A, but no enzyme.

cysteinyl radical can lead to oxygen consumption. The reaction of radicals with DMPO is fast enough that the inhibition is complete. In agreement with the findings with ESR, the addition of catalase also inhibited oxygen uptake.

The generation of hydrogen peroxide and the consumption of oxygen by the lactoperoxi-dase-catalyzed oxidation of cysteine can be explained on the basis of known reactions of the cysteine thiyl radical.

$$RS^{\bullet} + RSH\ (RS^-) \rightleftarrows [RS\cdot^{\bullet}\cdot SR]^- + H^+$$

$$[RS\cdot^{\bullet}\cdot SR]^- + O_2 \rightarrow RSSR + O_2^{\bullet -}$$

$$2\ O_2^{\bullet -} + 2H^+ \rightarrow H_2O_2 + O_2$$

The thiyl free radical reacts with thiol anion to form the L-cystine disulphide anion free radical, which is air oxidized to L-cystine, forming superoxide (for general description of three-electron bonded species such as the disulfide radical anion see article by K.-D. Asmus in this book). Even in the absence of superoxide dismutase, superoxide disproportionates rapidly to form hydrogen peroxide, which will drive the horseradish peroxidase-catalyzed reaction.

The disproportionation of superoxide forms hydrogen peroxide that oxidizes the peroxidase, which forms thiyl radicals and, ultimately, more superoxide. What results is an enzymatic free-radical chain reaction, a most effective combination. The peroxidase-catalyzed oxidation of cysteine may be of physiological significance. It is particularly noteworthy that cysteine oxidation by lactoperoxidase is independent of exogenously added hydrogen peroxide because cellular hydrogen peroxide concentrations are generally very low, less than 0.1 μM.

Aromatic compounds such as phenols are better known substrates for peroxidases. In this regard, acetaminophen or paracetamol is a typical phenol. Using the fast-flow technique, we were able to observe the ESR spectrum of the acetaminophen phenoxyl free radical (Figure 3) in a system consisting of lactoperoxidase, hydrogen peroxide and acetaminophen.[10] No signal could be found if either hydrogen peroxide, lactoperoxidase, or acetaminophen was omitted from the reaction mixture.

An improved, highly resolved spectrum of the acetaminophen phenoxyl free radical was obtained from the reaction of acetaminophen with the inexpensive horesradish peroxidase and hydrogen peroxide. The simulated spectrum contains hyperfine splitting constants from all nuclei with spin in acetaminophen including the methyl of the acetyl group.[10] This constitutes an unambiguous proof of radical structure.

The reaction of free radical metabolites of drugs with biochemicals is undoubtedly important in their toxicity. We were clearly able to exclude a reaction with oxygen which would form superoxide.[11,12] Superoxide was not detectable using different approaches, and we were able to demonstrate that reports in the literature are dubious. There is some discussion in the literature as to whether GSH will reduce the acetaminophen phenoxyl free radicals (Scheme 1, eq. 1), regenerating acetaminophen and forming a thiyl radical.[13] Ross et al. reported the spin trapping of GSH-derived thiyl radical in incubations of acetaminophen,

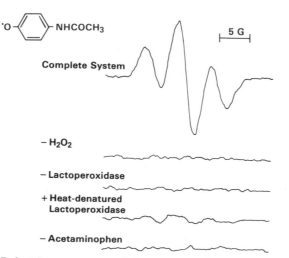

Fig. 3.   The ESR fast-flow spectrum of the N-acetyl-4-aminophenoxyl free radical produced in a system of acetaminophen, lactoperoxidase and hydrogen peroxide. The concentrations of acetaminophen, H_2O_2, and lactoperoxidase in the flat cell were 5.0 mM, 12.5 mM and 3.0 units/ml, respectively. Equal volumes of acetaminophen/H_2O_2 and lactoperoxidase in deoxygenated pH 7.5 phosphate buffer were mixed milliseconds prior to entering the flat cell at a total flow rate of 100 ml/min.

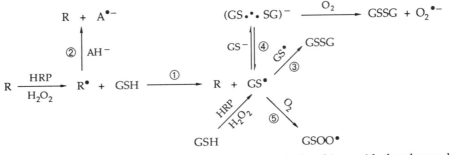

Scheme 1. A scheme for the reactions of ascorbate and glutathione with the phenoxyl radical of acetaminophen ($R^\bullet$).

horseradish peroxidase, and H_2O_2.[14] Since this reaction is thermodynamically uphill, there has been interest in how this type of reaction can occur.[15] We decided to elucidate this reaction using direct ESR spectroscopy with the fast-flow technique. This approach is more direct than the spin-trapping experiment. As stated previously, on theoretical grounds the thiyl free radical cannot be detected directly with ESR. Nevertheless, the existence of thiyl radicals can be demonstrated because thiyl radicals react with glutathione thiol anion to form a detectable glutathione disulfide anion radical (Scheme 1, equ. 4). This free radical is air-oxidized to glutathione disulfide (Scheme 1), so experiments must be done under nitrogen. The ESR literature further indicates $RSOO^\bullet$ radicals (Scheme 1, eq. 5) can also form as a result of an addition reaction of thiyl radicals with oxygen.[16,17] Indeed, we could demonstrate glutathione disulfide anion radical formation, $(GS \cdot^\bullet \cdot SG)^-$, (Figure 4, upper) by flowing acetaminophen and H_2O_2 against GSH and horseradish peroxidase.[18] We observed two ESR spectra with different g values. The triplet with some hyperfine structure at g = 2.0043 is due to the acetaminophen phenoxyl free radical, while the quintet, with its fourth and fifth line hidden under the triplet, has a g-value of 2.013, which is typical for the disulfide anion radical. The quintet results from a coupling of the free electron with four equivalent hydrogens. These are the four methylene hydrogens next to the sulfur atoms. A control experiment without acetaminophen demonstrated that the glutathione disulfide anion radical formation is not due to direct oxidation of GSH by the enzymatic system. This experiment demonstrated that acetaminophen catalyzed the oxidation of GSH at rates much faster than would occur in the presence of hydrogen peroxide and peroxidase alone. Omission of hydrogen peroxide or horseradish peroxidase prevents any radical formation, confirming the dependence on the enzymatic system. Omission of GSH yields an increase in the acetaminophen free radical signal, showing that GSH does reduce the acetaminophen phenoxyl free radical.

Qualitatively the same results can be obtained when GSH is substituted with cysteine. Again, the disulfide anion radical is detected when cysteine is added to a flow experiment where acetaminophen is oxidized in a peroxidase system. The chiral carbons of the cystine disulfide anion radical cause the inequivalency of the four methylene protons in pairs, leading to more hyperfine structure in the spectrum. Radical formation is also dependent on acetaminophen, hydrogen peroxide and horseradish peroxidase. If we add cystine instead of cysteine, no signal is detected, demonstrating that the sulfur-centered radical is not formed by the oxidation of cystine. Without the reducing sulfhydryl, the acetaminophen free radical concentration is increased to its original amplitude.

Ascorbate is a better reducing agent than GSH. It totally inhibits the acetaminophen free radical formation in the same enzymatic system we used for the GSH or cysteine incubations (Scheme 1, eq. 2). Omission of acetaminophen shows that ascorbate is also a substrate for the enzymatic system, but the rate of oxidation is much slower.[18]

406

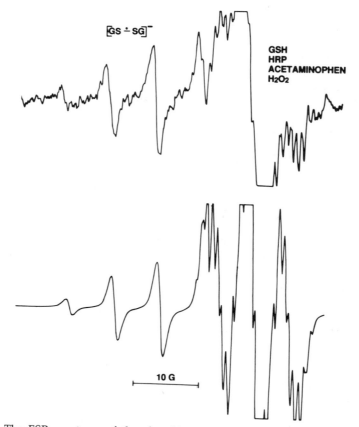

Fig. 4. The ESR spectrum of the glutathione disulfide radical anion and the phenoxyl radical of acetaminophen generated in a system of glutathione (100 mM), horseradish peroxidase type I (10 units/ml), hydrogen peroxide (25 mM), and acetaminophen (10 mM) in phosphate (0.1 M) buffer, pH 7.5. Below is the computer simulation of the glutathione disulfide radical anion and acetaminophen phenoxyl free radical (18).

If the experiments are performed in the presence of a 50% oxygen atmosphere, the cysteine disulfide anion radical disappears. Addition of both GSH and ascorbate leads only to the ascorbate free radical, implying ascorbate, and not GSH will be oxidized by the acetaminophen free radical in those cells that contain both reducing agents. Having a complete system, but lower pH of 6.5, does not lead to a stronger acetaminophen signal, implying the cysteine thiyl radical is formed in spite of the fact that the cystine disulfide anion radical cannot be detected. This pH effect indicates that the thiol anion is indeed the species trapping the thiyl radical, resulting in the formation of the disulfide anion radical. In these systems, oxidized glutathione comes primarly from its anion radical, and not the dimerization of thiyl radicals (Scheme 1, eq. 3) as is often stated.

Although much interest in the oxidation of glutathione by free radical metabolites has been shown *in vitro*, the oxidation of other biochemical reducing agents such as ascorbate may be more important *in vivo*. In summary: 1) the oxidation of cysteine, GSH and sulfhydryl drugs by lactoperoxidase forms RS$^\bullet$; 2) RS$^\bullet$ or RS$^\bullet$-derived radicals react with oxygen resulting in an oxygen-dependent free radical chain reaction. Only in the case of GSH and N–acetylcysteine is exogenous hydrogen peroxide necessary; 3) free radical metabolites

407

of drugs such as acetaminophen oxidize GSH to GS•, which combines with GS⁻ to form (GS•·SG)⁻; 4) ascorbate should outcompete GSH for the electrophilic acetaminophen phenoxyl free radical under most circumstances.

REFERENCES

1. M.Z. Baker, R. Badiello, M. Tamba, M. Quintiliani, and G. Gorin, *Int. J. Radiat. Biol.* 41:595 (1982).
2. G.K. Nyberg, G.P.D. Granberg, and J. Carlsson, *Appl. Environ. Microbiol.* 38:29 (1979).
3. L.F. Prescott, J. Park, A. Ballantyne, P. Adriaenssens, and A.T. Proudfoot, *The Lancet II* 432 (1977).
4. L.S. Harman, C. Mottley, and R.P. Mason, *J. Biol. Chem.* 259:5606 (1984).
5. C. Mottley and R.P. Mason, *Biol. Mag. Res.* 8:489 (1989).
6. L.S. Harman, D.K. Carver, J. Schreiber, and R.P. Mason, *J. Biol. Chem.* 261:1642 (1986).
7. J. Olsen and L. Davis, *Biochim. Biophys. Acta* 445:324 (1976).
8. B.C. Gilbert, H.A.H. Laue, R.O.C. Norman, and R.C. Sealy, *J. Chem. Soc. Perkin Trans.* 2 892 (1975)
9. C. Mottley, K. Toy, and R.P. Mason, *Mol. Pharm.* 31:417 (1987).
10. V. Fischer, L.S. Harman, P.R. West, and R.P. Mason, *Chem.-Biol. Interact.* 60:115 (1986).
11. V. Fischer, P.R. West, S.D. Nelson, P.J. Harvison, and R.P. Mason, *J. Biol. Chem.* 260:11446 (1985).
12. V. Fischer, P.R. West, L.S. Harman, and R.P. Mason, *Environ. Health Perspect.* 64:127 (1985).
13. R.H. Bisby and N. Tabassum, *Biochem. Pharm.* 37:2731 (1988).
14. D. Ross, E. Abano, U. Nilsson, and P. Moldéus, *Biochem. Biophys. Res. Commun.* 125:109 (1984).
15. I. Wilson, P. Wardman, G.M. Cohen, and M. D'Arcy Doherty, *Biochem. Pharmacol.* 35:21 (1986).
16. M.D. Sevilla, M. Yan, and D. Becker, *Biochem. Biophys. Res. Commun.* 155:405 (1988).
17. M.D. Sevilla, M. Yan, D. Becker, and S. Gillich, *Free Rad. Res. Commun.* 6:99 (1989).
18. D.N.R. Rao, V. Fischer, and R.P. Mason, *J. Biol. Chem.* 265:844 (1990).

SULPHUR COMPOUNDS AND "CHEMICAL REPAIR" IN RADIATION BIOLOGY

Clemens von Sonntag and Heinz-Peter Schuchmann

Max-Planck-Institut für Strahlenchemie
4330 Mülheim a.d. Ruhr, F.R. Germany

THE EFFECTS OF IONIZING RADIATION ON CELLS

When living cells are subjected to ionizing radiation, several functions may be impaired. Although the metabolic functions may continue, the cell may no longer be capable of propagation. This effect is called *reproductive cell death*. It is the most commonly-determined biological endpoint. On the other hand, the cell may still be able to divide, but some of the information of its genome is altered, a mutation has occurred. At much higher doses than required for these two effects to occur, the functioning of the cell may come to a complete stand-still (*metabolic cell death*). It is generally accepted that for the two first-mentioned events the essential target is the DNA.[1] In fact, it is observed that even *one* (unrepaired) DNA break means already reproductive cell death. The membrane has occasionally been discussed as an alternative important target, and it appears to be certain that it contributes to the metabolic cell death.[1]

It is remarkable how quickly the deleterious effects of ionizing radiation on cells have been recognized, and attempts made to put it to use for the benefit of mankind. Barely one month after the discovery of the X-rays by Röntgen, radiotherapy was attempted on a patient suffering from cancer.[1] In western society, about one-third of the population is expected to develop cancer, and for about one-fifth of the population this disease will be fatal. At present, radiotherapy, i.e. the attempt to inflict reproductive cell death on the cancerous cells, is still one of the most widely used tools to fight this disease.

On the other hand, the increased risk of cancer and also the second effect, the induction of mutations, is of increasing concern when the dangers of the nuclear age are considered. The induction of cancer by ionizing radiation is also thought to have DNA lesions as precursors.

THE SENSITIZING EFFECT OF OXYGEN

Taking reproductive cell death as an example, there is an enhanced effect when cells are irradiated in the presence of oxygen, compared to anoxic conditions. It has been found that the radiation sensitivity of living cells depends on the oxygen concentration. Although the maximum oxygen enhancement ratio (OER) is only in the order of three, this may have an important bearing on success or failure in the radiotherapy of solid tumors which are known to be poorly oxygenated. Since radiotherapy relies on the reduction of the number of

Sulfur-Centered Reactive Intermediates in Chemistry and Biology, Edited by
C. Chatgilialoglu and K.-D. Asmus, Plenum Press, New York, 1990

cancerous cells by many orders of magnitude, a sufficient number of hypoxic cells may withstand the treatment at otherwise just tolerable doses to be the cause of tumor regrowth.

In-vitro as well as *in-vivo* studies suggest that the oxygen effect is closely linked to the chemical action of thiols, which is now discussed.

THE PROTECTIVE EFFECT OF THIOLS; THE ALPER FORMULA

It has been noted for some time that an oxygen effect in the vicinity of three is never observed in *in-vitro* systems, *i.e.* when DNA is irradiated in aqueous solution, and strand breakage is measured as the relevant endpoint. The oxygen effect for both *in-vitro* and *in-vivo* systems, in terms of the oxygen enhancement ratio, can be represented by the Alper formula[2] (1) (m = maximum response at very high $[O_2]$, K = $[O_2]$ when the response is half its maximum value). A simple mechanistic scheme[2,3] [reactions (2) - (6)] has proven useful for the purpose of examining the scope of expression (1). The importance of reaction (4) in this scheme has been emphasized (Schulte-Frohlinde and Bothe, *Int. J. Radiat. Biol.*, in press).

$$OER = (m [O_2] + K) / ([O_2] + K) \qquad (1)$$

$$DNA \xrightarrow{\gamma} \text{damaged DNA} \qquad (2)$$

$$DNA \xrightarrow{\gamma} DNA^\bullet \text{ (various DNA radicals)} \qquad (3)$$

$$DNA^\bullet \rightarrow \text{damaged DNA} \qquad (4)$$

$$DNA^\bullet + RSH \rightarrow DNA \text{ (chemical repair)} \qquad (5)$$

$$DNA^\bullet + O_2 \rightarrow DNA\text{-}O_2^\bullet \text{ (oxygen fixation)} \qquad (6)$$

In reaction (2), DNA is damaged by the ionizing radiation in a way that is not chemically repairable by RSH, or modifiable by oxygen. In reaction (3), DNA radicals are formed and set the stage for the transformation reactions (4) - (6). Here we do not distinguish between different chemical causes bringing about the lesions through reactions (2) and (3), the *direct effect* where the energy of the ionizing radiation is absorbed by the DNA itself, or the *indirect effect* where reactive radicals generated in the environment of the DNA (notably the $^\bullet$OH radical from the radiolysis of water) attack the DNA.

The DNA radicals may be "repaired" by thiols [reaction (5)] yielding back either intact DNA directly, or a product which in subsequent enzymatic reactions may or may not be readily reconverted into functional DNA.[4]

In competition with this "chemical repair" reaction, oxygen may add to the DNA radical, yielding a DNA peroxyl radical (reaction (6); *oxygen fixation*).

Expression (1) holds exactly for the sub-scheme (3) - (6), for which it was originally developed by Alper. It also holds exactly for more elaborate schemes based on reactions (2) - (6) in the *in-vivo* context that take into account the ulterior enzymatic repair processes, which are of paramount importance. All reactions must produce some products which can be repaired by the repair enzymes and some products which cannot. However, some of the products from reaction (6) must be more difficult to repair. This is precisely the phenomenon of oxygen enhancement in the *biological* context.

There is a wealth of experimental results which are in apparent general agreement with this concept. Using fast-mixing devices, B.D. Michael and his colleagues have shown that the oxygen or the thiol are indeed required to be present at the very time of irradiation to bring about the oxygen (damage enhancement) effect or the thiol protective effect.[1] Thus, the action of the two is not linked to the subsequent stage of enzymatic repair. Also, drugs which reduce the thiol content of a cell increase its radiation sensitivity.[1] At the same time, the maximum OER goes down. These drugs act mainly on the level of glutathione (GSH) which is the most abundant thiol in cells (up to about 10^{-2} mol dm^{-3}). They usually do not reduce drastically the protein-bound thiols. Thus it is believed that the repair reaction is mainly due to the freely-diffusing low-molecular-weight thiol, glutathione.

Although the scheme as outlined above has some heuristic value and represents reasonably well the overall situation, details are not yet understood. For example, the oxygen effect is often biphasic, i.e. when plotted as a function of the oxygen concentration, sometimes there is not a monotonous increase in OER but an "intermediate plateau" region. In addition, in GSH-deficient cells exogenous thiols (including GSH) cannot bring the OER to the same level as found for GSH-proficient cells.[5] Thus the situation must be more complicated than described by this scheme. For further details see the article by P. Wardman in this book.

MODEL STUDIES ON THE EFFECT OF OXYGEN OR THIOLS ON RADICALS DERIVED FROM NUCLEOTIDES, POLYNUCLEOTIDES AND NUCLEIC ACIDS

In order to understand the chemistry behind these reactions, many model studies have been undertaken. These studies have yielded some information which point to the complexity of the system. Although it is generally true that oxygen reacts rapidly with organic radicals at rates which are close to diffusion-controlled [$k(R^\bullet + O_2) \approx 2 \times 10^9$ dm^3 mol^{-1} s^{-1}], this is not the case with the radicals that are formed by $^\bullet$OH attack on purines. This has been shown by pulse radiolysis,[6] but even under conditions where these radicals have a lifetime of about 0.1 s, their reaction with oxygen is far from quantitative.[7] This means that not all of the purine-derived radicals are accessible to oxygen fixation.

Pulse radiolysis data on single-stranded polynucleotides composed of uracil or cytosine units [poly(U) and poly(C)] indicate that the polynucleotide radicals react about five times more slowly with oxygen ($k = 4 \times 10^8$ dm^3 mol^{-1} s^{-1}) than with the radicals derived from the corresponding mononucleotides.[1] Taking into account that the only readily diffusing component in this system is the molecular oxygen, a factor of about two would be expected when the high ionic strength around the polynucleotide is neglected. Since oxygen is less soluble in salt solutions, one might expect a lower concentration of oxygen in the neighbourhood of the polynucleotide. Together these effects may well account for the observed factor of five. The effect of double-strandedness is especially surprising: the radicals formed in poly(A + U) (A refers to adenine) react with a rate constant of only 5×10^7 dm^3 mol^{-1} s^{-1}. In the case of double-stranded DNA, no decay of the base radicals is observed in the presence of oxygen (sugar radicals are not observable in such a pulse radiolysis experiment but could still contribute to the oxygen enhancement phenomenon). This can only mean that the diffusive access of oxygen into the interior of the helix is severely restricted, in contrast to that of $^\bullet$OH radicals. Thus we have no information on the rate constant of *oxygen fixation* in functional DNA, reaction (6).

A similar situation holds for the repair reaction. Although thiols react with many radicals with rate constants near 10^8 dm^3 mol^{-1} s^{-1} (*cf.* first article by C. v. Sonntag in this book), the uracil C(5)–OH adduct radical reacts with a rate constant of only about 10^6 dm^3 mol^{-1} s^{-1},[8,9] while its isomer, the uracil C(6)–OH adduct, is reported not to react with thiols at all.[10] In addition, only a fraction of the OH adduct radicals derived from purines react with thiols on the pulse radiolysis timescale (≈ 100 ns - 10 ms).[11] The reaction of the thiols with the purine radicals has been attributed to the reaction of the *oxidising* fraction of these radicals. There is however a strong pH effect which indicates that the thiolate ion reacts considerably faster than the thiol itself. Thus at pH 7.4 where rate constants in the order of

10^7 dm^3 mol^{-1} s^{-1} were determined for the dGMP-OH-adduct radicals, this reaction may well be an electron transfer reaction, undergone by the thiolate ions in equilibrium, rather than an H-transfer reaction from the thiol. The only radicals in the polynucleotides and in DNA that are expected to react quickly with thiols (and *do* react by H-transfer) are the sugar radicals, if model systems such as 2-deoxyribose ($k = 1.8 \times 10^8$ dm^3 mol^{-1} s^{-1}) and ribose-5-phosphate ($k \approx 5 \times 10^7$ dm^3 mol^{-1} s^{-1}) are a good guide.[12]

Two processes *must* have sugar radicals as precursors: strand breakage and base release. It has been shown that in the case of poly(U) the direct formation of sugar radicals [reaction (8)] is a comparatively minor process,[13] and that base radicals (mainly the uracil C(5)–OH adduct radicals) are the precursors of these sugar radicals [reactions (10) - (12)].[14] In poly(U), strand breakage is a relatively slow process ($k = 0.7$ s^{-1} at pH 7, increasing to ≈ 200 s^{-1} at pH 3)[13] and according to the present interpretation of the available data, the rate-determining step is the transfer of an H-atom from the sugar moiety to the base radical [reaction (11); *cf.* also Ref. 15]. Due to the slowness of this reaction, thiols quench the base radicals rather than the sugar radicals, despite the fact that the base radical reacts so slowly with thiols.[16,17]

$$H_2O \xrightarrow{\gamma} {}^\bullet OH, e_{aq}^-, {}^\bullet H, H^+ \qquad (7)$$

$$ {}^\bullet OH + poly(U) \rightarrow sugar\ radical \rightarrow strand\ break \qquad (8/9)$$

$$ {}^\bullet OH + poly(U) \rightarrow base\ radical \rightarrow sugar\ radical \rightarrow strand\ break \qquad (10\text{-}12)$$

Poly(U) is a poly-anion and will attract positively charged thiols while negatively charged thiols will be repelled. Fahey, Vojnovic and Michael (private communication) have studied the effectiveness of various thiols to prevent strand breakage. They found that negatively charged thiols are more than a hundred times less reactive than neutral thiols, which in turn are more than a hundred times less reactive than thiols with a unit positive charge. It is noted that the cellularly predominating free thiol, glutathione, is singly-negatively charged, i.e. it is expected to fall among the unreactive thiols.

Since base release is linked to strand breakage (both events have base and sugar radicals as precursors), the line of argument concerning strand breakage should similarly hold for base release. Considerable information on base release in poly(U) is available,[18-20] and use could be made of this information to corroborate the above finding.

In single-stranded DNA,[21] strand breakage occurs about a hundred times faster than in poly(U).[13] However, in DNA strand breakage is a much less important process (for polynucleotides see Ref. 22, 23), and it is not yet established whether base radicals contribute significantly to it. Quenching by thiols can occur nevertheless since sugar radicals also react about a hundred times faster with the thiol than base radicals. Like in poly(U), base release is a rather complex process; it does not occur on the level of free radicals but also on the ulterior level of damaged sugars (in poly(U) this is the major component[19]).

In DNA, base release has been used to measure the effect of the charge of the thiol on its ability to repair the precursor radicals.[24] Like poly(U), DNA attracts thiols which are positively charged, and repels negatively charged ones. As expected, glutathione which is negatively charged is less effective than the radiation protection agent WR 1065 whose net charge is +2. In the case of DNA and with base release as the measured endpoint, the effect of the net charge of the thiol is not as drastic as that reported for strand breakage in poly(U).

Since the oxygen (damage enhancement) effect and thiol protection are related to each other, the question arises as to what we know of the reaction of DNA model systems, and of DNA itself, in the presence of oxygen. In the low-molecular-weight models such as nucleobases or mononucleotides derived from pyrimidines, peroxyl radicals are formed which show the typical unimolecular and bimolecular reactions of peroxyl radials.[1] The situation is different in polynucleotides, because here the lifetime of the peroxyl radicals is drastically increased due to the macromolecular nature of the polynucleotide peroxyl radical (slow bimolecular decay). This enables some (surprisingly not all) polynucleotide base peroxyl radicals to attack a sugar moiety somewhere along the strand [reaction (13)]. This reaction leads to a sugar radical which is subsequently converted by oxygen into a sugar peroxyl radical [reaction (14)]. As a consequence of this sugar damage, strand breakage and base release occurs [reaction (15)]. Again, the polynucleotide investigated most extensively is poly(U).[20,25,26]

$$RO_2^\bullet \text{ (base peroxyl radical)} \to RO_2H \text{ (base hydroperoxide)} + {}^\bullet R' \text{ (sugar radical)} \quad (13)$$

$${}^\bullet R' + O_2 \to R'O_2^\bullet \text{ (sugar peroxyl radical)} \to \text{strand breakage and base release} \quad (14/15)$$

$$RO_2^\bullet + RSH \ (RS^- + H^+) \to RO_2H + RS^\bullet \quad (16)$$

Due to their rather long lifetime, the polynucleotide radicals can react with added thiols, most likely by forming hydroperoxides [reaction (16)]. This reaction has been investigated with the help of ESR, and it has been concluded that the rate constant for the reaction of such peroxyl radicals is in the order of 10^4 dm^3 mol^{-1} s^{-1}.[25] We have carried out some experiments with low-molecular-weight peroxyl radicals (the one derived from t-butanol) to substantiate the order of magnitude of the rate constant of peroxyl radicals with thiols. Although a reaction of the peroxyl radical with the thiol was observed, it appears to be so complex that it is not possible at present to arrive at a detailed description of this system.

Reaction (16) is interesting in so far as we may consider, as a reaction in competition with it, the transformation of peroxyl radicals into some other radicals. Such a reaction has been found with poly(U) as described above. Thus, depending on thiol concentration, different products will be formed. Considering this system now in the context of DNA damage and repair, one might speculate that in cells these different products will show different repair efficiencies. This would then lead to a change in the OER with changing thiol concentration.

OTHER MODELS REGARDING SENSITIZATION AND REPAIR

The usual chemical repair model employs thiol as the reducing agent. The model can, of course, be adapted with little change by including the thiolate ion as well. The major low-molecular-weight thiol in cells is glutathione. Its thiol function has a pK_a of 9.2, i.e. at pH 7, 0.6% of the total glutathione is present in the anionic form. It has been mentioned above that some base radicals are barely repaired by thiols but react readily with thiolate ions. Hence even at the low steady-state concentration of the GS$^-$ ions they may be responsible of the repair reaction.

Taking the thiolate ions into account does not drastically alter the concept. However, further concepts have been presented which deviate from the current view. It has been suggested by Prütz that the repair could be exerted by disulphide radical anions $(RS\bullet^\bullet SR)^-$.[27,28] This requires that in the first step some thiyl radicals are formed. There are a variety of sources which could generate such radicals in the neighbourhood of DNA. These radicals have oxidizing properties (see article by D.A. Armstrong in this book) and would not be able to reduce DNA damage. However, they readily complex with thiolate ions yielding disulphide radical anions which have reducing properties (see first article by C. v. Sonntag in this book). It is now thought that the disulphide radical anions are the reducing species rather than the thiol/thiolate couple. That this concept is at least feasible in the model system (DNA in aqueous solution) has been demonstrated.[28]

On the basis of the observation that thiyl radicals react rapidly (but also reversibly) with oxygen (see first article by C. v. Sonntag in this book), Quintiliani has proposed that in the oxygen effect thiylperoxyl radicals may be involved[29] as a damaging agent but that, on the other hand, thiyl radicals also act as oxygen scavengers and hence deplete the oxygen locally available for oxygen fixation. These are interesting suggestions, but at present we are far from knowing enough details about thiylperoxyl radical reactions and oxygen depletion parameters within a cluster of DNA radicals formed in a spur (whenever the ionizing energy happens to be deposited in very close vicinity of the DNA strand), to be able to judge how much promise these concepts hold.

REFERENCES

1. C. von Sonntag, "The Chemical Basis of Radiation Biology", Taylor and Francis, London (1987).
2. T. Alper, "Cellular Radiobiology", Cambridge University Press, Cambridge (1979).
3. M. Edgren, T. Nishidai, O.C.A. Scott, and R. Revesz, *Int. J. Radiat. Biol.* 47:463 (1985).
4. D. Schulte-Frohlinde, *Free Radical Res. Commun.* 6:181 (1989).
5. L. Revesz, *Int. J. Radiat. Biol.* 47:361 (1985).
6. R.L. Willson, *Int. J. Radiat. Biol.* 17:349 (1970).
7. M. Isildar, M.N. Schuchmann, D. Schulte-Frohlinde, and C. von Sonntag, *Int. J. Radiat. Biol.* 41:525 (1982).
8. G.E. Adams, G.S. McNaughton, and B.D. Michael, *Trans. Faraday Soc.* 64:902 (1968).
9. G. Nucifora, B. Smaller, R. Remko, and E.C. Avery, *Radiat. Res.* 49:96 (1972).
10. S.A. Grachev, E.V. Kropachev, and G.I. Litvyakova, *Izv. Akad. Nauk SSSR, Ser. Khim.* 2746 (1988).
11. P. O'Neill, *Radiat. Res.* 96:198 (1983).
12. M.S. Akhlaq, S. Al-Baghdadi, and C. von Sonntag, *Carbohydr. Res.* 164:71 (1987).
13. E. Bothe and D. Schulte-Frohlinde, *Z. Naturforsch.* 37c:1191 (1982).
14. D.G.E. Lemaire, E. Bothe, and D. Schulte-Frohlinde, *Int. J. Radiat. Biol.* 45:351 (1984).
15. K. Hildenbrand and D. Schulte-Frohlinde, *Free Rad. Res. Commun.* 6:137 (1989).
16. M.S. Akhlaq, H.-P. Schuchmann, and C. von Sonntag, *Int. J. Radiat. Biol.* 51:91 (1987).
17. D.G.E. Lemaire, E. Bothe, and D. Schulte-Frohlinde, *Int. J. Radiat. Biol.* 51:319 (1987).
18. D.J. Deeble and C. von Sonntag, *Int. J. Radiat. Biol.* 46:247 (1984).
19 D.J. Deeble, D. Schulz, and C. von Sonntag, *Int J. Radiat. Biol.* 49:915 (1986).
20. D.J. Deeble and C. von Sonntag, *Int. J. Radiat. Biol.* 49:927 (1986).
21. E. Bothe, G.A. Qureshi, and D. Schulte-Frohlinde, *Z.Naturforsch.* 38c:1030 (1983).
22. M. Adinarayana, E. Bothe, and D. Schulte-Frohlinde, *Int J. Radiat. Biol.* 54:723 (1988).
23. A.M. Önal, D.G.E. Lemaire, E. Bothe, and D. Schulte-Frohlinde, *Int. J. Radiat. Biol.* 53:787 (1988).
24. S. Zheng, G.L. Newton, G. Gonick, R.C. Fahey, and J.F. Ward, *Radiat. Res.* 114:11 (1988).
25. D. Schulte-Frohlinde, G. Behrens, and A. Önal, *Int. J. Radiat. Biol.* 50:103 (1986).
26. E. Bothe, G. Behrens, E. Böhm, B. Sethuram, and D. Schulte-Frohlinde, *Int. J. Radiat. Biol.* 49:57 (1986).
27. W.A. Prütz and H. Mönig, *Int. J. Radiat. Biol.* 52:677 (1987).
28. W.A. Prütz, *Int. J. Radiat. Biol.* 56:21 (1989).
29. M. Quintiliani, *Int. J. Radiat. Biol.* 50:573 (1986).

THIOL REACTIVITY TOWARDS DRUGS AND RADICALS: SOME IMPLICATIONS

IN THE RADIOTHERAPY AND CHEMOTHERAPY OF CANCER

Peter Wardman

Cancer Research Campaign
Gray Laboratory, Mount Vernon Hospital
Northwood, Middlesex, HA6 2JR, U.K.

The papers in these Proceedings attest to the special characteristics imparted by incorporation of sulphur in molecules: a reduction in ionization potential leading to increased ease of oxidation or to increased propensity to function as an electron donor, and a reduction in bond strengths (e.g. S-H compared to C-H) leading to a capacity for radical "repair" of species where C-H bonds have been broken, by hydrogen donation from a thiol. The protective or antioxidant role of thiols, resulting from these two properties, is currently attracting widespread attention. Major interest centres upon the importance of thiols in cancer therapy. Thiols have been recognised as radioprotectors for at least 40 years, and their activity characterised in innumerable model systems both *in vitro* and *in vivo*, ranging from pulse radiolysis of dilute aqueous solutions to clinical trials of thiols or pro-drug derivatives.

Another important property of thiols (RSH) is the reactivity of the thiolate anion (RS^-) as a nucleophile. Such reactivity is fundamental to the critical role of thiols in detoxification of even quite ordinary drugs (e.g. acetaminophen/paracetamol). Conjugation of xenobiotics with hydrophilic thiols (principally glutathione, GSH), is a major pathway by which the body eliminates drugs, the GSH conjugates usually being involved in further transformation to mercapturic acids before excretion. Intracellular thiol levels are now known to be an important determinant in both the therapeutic efficacy and the dose-limiting toxicity of several chemotherapeutic agents in widespread use (e.g. the alkylating agent, melphalan).

Science suffers from over-compartmentalization as it becomes increasingly specialized, and experience from chemotherapy is not always transferred to radiotherapy (and *vice versa*). Recent years have seen intense interest in the potential for modifying the radiotherapeutic response by both modulating tissue thiol levels and selective sensitization of hypoxic tumour cells to radiation. Apparently anomalous activity of some radiosensitizers is now recognised to arise from the nucleophilic reactivity of intracellular glutathione towards such exceptional radiosensitizers. However, a better appreciation of this aspect of thiol chemistry would have undoubtedly led to an earlier awareness of the consequences of susceptibility to nucleophilic attack of potential radiosensitizers and altered the emphasis and utilization of research resources. Another important feature of this non-radical reactivity of thiols is the catalysis of most such displacement or conjugation reactions by a major family of enzymes, glutathione-S-transferases (GSTases). These demonstrate, on the one hand, specificity towards GSH over other thiols, and on the other, a very wide spectrum of activity

Sulfur-Centered Reactive Intermediates in Chemistry and Biology, Edited by
C. Chatgilialoglu and K.-D. Asmus, Plenum Press, New York, 1990

towards xenobiotic substrates. An important substrate for GSTases is hydroperoxymethyl-uracil, so that the enzyme-catalysed non-radical reactions of GSH may be important in oxidative damage to DNA mediated by free radicals.

This paper outlines some aspects of these four major features of thiols and sulphur-centred intermediates: electron and hydrogen-donation to radical centres, nucleophilic reactivity, and catalysis by GSTases. Other papers in these Proceedings offer more detailed discussion of some of these aspects, and the aim here is to present an overview with emphasis on the implications in cancer therapy by both radiation and drugs. References are largely restricted to major compilations of recent research papers, reviews and books for reasons of brevity and ease of reading.

GLUTATHIONE IN BIOLOGY

Glutathione is a naturally-occurring thiol, common in mammals where it is by far the most abundant non-protein thiol, an adult human having about 30 g glutathione widely distributed in most tissues. Measurements in mice are typical of mammals, with concentrations in brain, heart, and lung often in the range 1-2 mmol kg^{-1}, whilst levels in kidney or especially liver are higher, typically ≈ 4 or 8 mmol kg^{-1} respectively. Plasma concentrations are lower, e.g. in mice ≈ 30 µmol dm^{-3}.

A tripeptide, glutathione is synthesised via glutamic acid and cysteine to form γ-glutamylcysteine, and then conjugated with glycine. The thiol function is thus a cysteinyl residue. (In plants, other thiols are also important such as homoglutathione, whilst in some bacteria simpler thiols such as H_2S are sometimes found in substantial amounts.) Enzymes are involved not only in the biosynthesis of glutathione (a feature which has been utilized to modulate intracellular GSH levels, see below), but also in its degradation and inter- and intra-organ transport.

In addition to some heterogeneity between glutathione levels in different organs, there is a marked cellular heterogeneity with intracellular compartmentalization of GSH. Thus in hepatocytes (liver cells), there is a mitochondrial pool of GSH which is a physiologically distinct reservoir separate from cytoplasmic GSH. Reactive drugs may deplete the latter much faster than the former and so the effects of drugs which either react directly with intracellular GSH or with an action dependent on intracellular GSH levels may be complicated by the lack of rapid equilibration of GSH across the mitochondrial membrane. With many agents, including radiation, damage in the cell nucleus is especially important. Although the nucleus is known to have substantial "pores" in the nuclear membrane permitting the selective passage of low molecular weight material, and cytoplasmic and nuclear equilibration of GSH might be expected, the confirmation of this point is not without technical difficulties and it remains a subject of current debate (see below).

RADIATION SENSITIVITY - A BALANCE BETWEEN OXIDATION AND REDUCTION

The specific lesions responsible for radiation lethality in mammalian cells have not been fully characterised at the molecular level, although the importance of damage to nuclear material, especially doubled-stranded breaks in DNA, is undisputed. Detailed pathways for the molecular events which could lead to DNA strand breaks, either indirectly from water radicals (e.g. $^{\bullet}OH$) in the immediate vicinity of DNA, or by direct energy absorption, have been evaluated and are referred to in the paper by von Sonntag in these Proceedings, and in the book by the same author (see Bibliography). Much of the initial damage is repaired; whilst "repair" can be discussed both at the molecular level (e.g. hydrogen donation to a carbon-centred radical from a thiol), and at the enzymatic level (e.g. where enzymes join or cut out damaged DNA), we can construct a simple model which describes, at

416

least algebraically, some of the main features of radiosensitivity. We recognise the competition between a criticial species, $B^\bullet$ leading either to cellular lethality or to non-lethal damage involving some form(s) of "repair":

$$B^\bullet \rightarrow \text{lethal damage} \tag{1}$$

$$B^\bullet \rightarrow \text{non-lethal damage} \tag{2}$$

The removal of oxygen is radioprotective in virtually all biological organisms, up to about a factor of 3 in terms of radiation dose required for constant biological response. Alper and Howard-Flanders recognised many years ago that the action of oxygen could be added to this simple model of chemical competition:

$$B^\bullet + O_2 \rightarrow \text{lethal damage} \tag{3}$$

Denoting the response in O_2 relative to that in its absence by r (a radiation dose-modifying factor), it was shown that the dependence of r upon O_2 concentration in cells could be fitted to the hyperbola:

$$r = (m[O_2] + K)/([O_2] + K) \tag{4}$$

This function can be easily derived from the simple model described by eqs. (1) - (3), m being the maximum response obtained as $[O_2] \rightarrow \infty$ and K equates practically to $[O_2]$ when $r = (m - 1)/2$.

If we wish to give this model a more chemical "flavour" we can denote k_1 and k_2 as the sum of the first-order rate constants (in s^{-1}) for *all* the processes which equations (1) and (2), respectively, can describe, and k_3 the second-order rate constant for reaction (3). Pathways (1) and (2) will include unimolecular processes as well as bimolecular reactions in which $[B^\bullet]$ is very small compared to the concentration of other reactants. Since the steady-state concentration of $B^\bullet$ is likely to be exceedingly low, this latter type of reaction may be common. Alexander and Charlesby recognised the radioprotection of thiols could be ascribed to a hydrogen-donation reaction, e.g. of the type:

$$B^\bullet + RSH \rightarrow BH + RS^\bullet \tag{5}$$

In the simple model above, when $[B^\bullet] << [RSH]$, modulation of protective thiol concentration is equivalent to varying the rate constant sum k_2 which will include contributions from bimolecular reactions such as (5) as well as unimolecular pathways, i.e. all first-order terms. It is easily shown that $m = (1 + k_2/k_1)$ and $K = (k_1 + k_2)/k_3$. Hence $K/m = k_1/k_3$ and so if the oxygen concentration is fixed and only k_2 is varied (e.g. by depleting or adding thiols), K/m should be invariant.

Innumerable studies have been made of the effects of adding or depleting thiols in cells and tissues. Whilst the initial success of the model in accounting for the variation of r with $[O_2]$ was an important stimulus, such a simple model is now, not surprisingly, recognised as inadequate. The ratio K/m is *not* invariant as pathway (2) is modulated, K being more susceptible to changes in thiol concentrations than m. This is particularly true when the sensitizing effect of oxygen is replaced by that of other oxidants, such as nitroaryl radiosensitizers. The efficacy in hypoxia (the concentration required for constant response) of such compounds can be markedly altered by adding or depleting thiols, as illustrated in Figure 1. Misonidazole is a simple 2-nitroimidazole and in these experiments either additional GSH was present (added extracellularly, with the problem of the detailed intracellular consequences unanswered) or endogenous levels were depleted by inhibiting biosynthesis (see below).

417

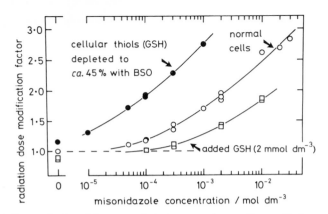

Fig. 1. Effect of glutathione modulation on the radiosensitizing effect of misonidazole in hypoxic V79 hamster fibroblasts *in vitro*. Data from R.J. Hodgkiss and R.W. Middleton, *Int. J. Radiat. Biol.* 43:179 (1983).

Whatever the mechanisms involved in the chemical modification of radiosensitivity (discussed in greater detail by the author elsewhere, and particularly by von Sonntag), it is clear that radiation lethality reflects *in part* a competition between "fixation" by oxygen or other oxidants and "repair" by thiols. "Fixation" could involve electron transfer from a bioradical to the sensitizer, but more likely involves radical-addition as an initial step followed by (possibly very rapid) heterolysis to yield the products of electron transfer. "Repair" is a crude term lacking specificity: its limitations or meaning varies with the endpoint being measured. To most radiobiologists discussing molecular models of the protective effects of thiols, "repair" is synonymous with "restitution" as described in eqn. (5). However, as von Sonntag *et al.*, and Raleigh *et al.* have illustrated, the production of radical centres from sp^3 hybridised centres in a biomolecule is *inevitably* accompanied by loss of stereochemical identity, and isomers of the original biomolecule may be generated even though the radical centre is restituted. This is molecular "mis-repair", a term also used in connection with enzymatic repair and hence again a possible source of confusion.

Another factor which von Sonntag *et al.* and Asmus *et al.* have drawn attention to is the reactivity of the thiyl radical, $RS^{\bullet}$ produced in (5) -- or in many other situations in medicine and biology -- reactivity which has been ignored by most radiobiologists. Such reactivity is illustrated in Figure 2, which was obtained some years ago by the author but which eluded interpretation until recent studies by Schöneich *et al.* The initial aim of the experiments was to use benzyl viologen radical-cation ($BV^{\bullet+}$) as a convenient chromophore to characterise the competition between carbon-centred radicals reducing BV^{2+} and being repaired by GSH, initially in the simplest of model systems:

$$(CH_3)_2\overset{\bullet}{C}OH + BV^{2+} \rightarrow (CH_3)_2CO + H^+ + BV^{\bullet+} \tag{6}$$

$$(CH_3)_2\overset{\bullet}{C}OH + GSH \rightarrow (CH_3)_2CHOH + GS^{\bullet} \tag{7}$$

The intial yield of $BV^{\bullet+}$ a short time after reactions (6) and (7) were completed gave, as expected, an excellent fit to simple competition kinetics and confirmed the use of this system to measure rate constants for radical repair such as that in (7) (a rate constant already well characterized). However, the subsequent yield of further $BV^{\bullet+}$ and the kinetics of its appearance on the sub-millisecond timescale was unexpected. The author now recognizes, from not dissimilar experiments by Schöneich *et al.* with penicillamine and 4-nitroacetophenone,

418

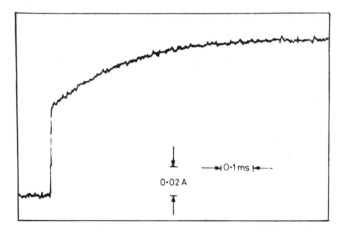

Fig. 2. Prompt and delayed formation of benzyl viologen radical cation after generating $(CH_3)_2C^\bullet OH$ by pulse radiolysis of a solution containing 2-propanol (0.5 mol dm^{-3}), GSH (1 mmol dm^{-3}), benzyl viologen (50 µmol dm^{-3}), N$_2$O (saturated), pH 2.8. Pulse 0.2 µs/ $\approx$ 1.6 Gy; absorption change at 600 nm (2 cm cell). Unpublished work by the author.

that the explanation lies in the hydrogen-*abstracting* properties of thiyl radicals, i.e. the reverse of (7):

$$GS^\bullet \;+\; (CH_3)_2CHOH \;\;\rightarrow\;\; GSH \;+\; (CH_3)_2\overset{\bullet}{C}OH \tag{8}$$

A full kinetic analysis is complex; in the experiment illustrated in Figure 2 the slow, delayed formation of BV$^{\bullet+}$ ($k \approx 4.5 \times 10^3$ s^{-1}) in the presence of 0.5 mol dm^{-3} 2-propanol yields only a crude estimate of $k_8 \approx 10^4$ dm^3 mol^{-1} s^{-1} since over the $\approx$1 ms timescale other reactions occur under the reaction conditions used. Thus some loss of GS$^\bullet$ by reaction with itself, and probably some reduction of GS$^\bullet$ by BV$^{\bullet+}$ (reactions known or expected to be close to diffusion-controlled) occurs in < 1 ms even with total radical concentrations < 1 µmol dm^{-3}. These experiments are now being repeated, benefiting from the hindsight provided by Schöneich *et al.*, who minimised unwanted reactions by using higher alcohol concentrations.

The concentration of GSH and of O$_2$ will determine the potential importance of subsequent reactions of GS$^\bullet$ in biology because of competing equilibria and electron-transfer "sinks":

$$GS^\bullet \;+\; GS^- \;\rightleftarrows\; (GS\!\cdot\!^\bullet\!\cdot\!SG)^- \tag{9}$$

$$GS^\bullet \;+\; O_2 \;\rightleftarrows\; GSOO^\bullet \tag{10}$$

$$(GS\!\cdot\!^\bullet\!\cdot\!SG)^- \;+\; O_2 \;\rightarrow\; GSSG \;+\; O_2^{\bullet-} \tag{11}$$

Some of these aspects are considered further below. (The disulfide radical anion is a $2\sigma/1\sigma^*$ three-electron bonded species; for this kind of electronic structure and its implications see also the article by K.-D. Asmus in this book).

Glutathione can thus be involved in radiation damage in a wide variety of types of reaction. Eqn. (5) respresents that most quoted, but hydrogen donation to a carbon-centred radical is only one of many possible scenarios, particularly when oxygen or other radiosensitizers are involved. These questions are considered in more detail in other papers in this volume. Electron transfer to short-lived cations (e.g. guanine) may only be a factor with thiol functions in the immediate vicinity of DNA, but electron transfer from GS^- to the radical product(s) of oxidation of guanine by $^\bullet OH$ is well documented. Oxidants such as O_2 or nitroaryl compounds add to the C(6)-yl radicals of pyrimidines, whereas thiols generally seem to "repair" the C(5)-yl radical. However, a possible reaction between the nucleophilic thiol and the intermediate peroxyl or nitroxyl radical-adduct of the radiosensitizer has received little attention. Nitroxyls, in particular, exhibit known reactivity towards thiols and such a reaction could conceivably compete with the base-catalysed heterolysis of the C(6)-yl nitroxyl adduct of pyrimidines to reduce oxidative damage to the base. More work in this area is clearly needed, but it is apparent that the *kinetic* consequences of these and other more complex possibilities (such as base/sugar intramolecular radical transfer) have not been seriously considered in terms of the implications to algebraic or kinetic models of the radiobiological response.

RADIOPROTECTION BY XENOBIOTIC COMPOUNDS

A huge research effort has been devoted to assessing the potential, and optimal molecular requirements, of thiols as radioprotectors. Unfortunately, military interests dominated the research and the opportunity to make advances at the fundamental level was lost. However, clinical trials of selected thiols as radioprotectors of normal tissues have provided some intriguing new leads.

The most widely-investigated radioprotector in recent years has been not a thiol, but a phosphorothioate ester, WR-2721, which is a pro-drug broken down hydrolytically to a free thiol, WR-1065 in a reaction catalysed by alkaline phosphatase:

$$H_2N(CH_2)_3NH(CH_2)_2SPO_3H_2 \rightarrow H_2N(CH_2)_3NH(CH_2)_2SH \tag{12}$$

Evidently, there is selective uptake or more probably hydrolysis (or at least selective protection) of some normal tissues compared to tumours, resulting in part from differences in the distribution of the hydrolytic enzyme or perhaps from differences in vascularity. (The two are not unrelated.) Therapeutic gain would be achieved if effective radiation doses to tumours could be increased whilst using the thiol to selectively protect normal tissues. However, most animal experiments with WR-2721 demonstrating potential therapeutic gain involved administered doses of about 200-600 mg/kg, whereas it turned out that the tolerated single dose in humans is about 18 mg/kg. This order-of-magnitude difference between mouse and man seems likely to be a most difficult obstacle to further progress, at least in its use as a simple radioprotector.

However, one aspect of the chemistry of WR-2721 or at least its thiol derivative is attracting attention. The dibasic nature of the molecule (largely doubly protonated at physiological pH) leads to "ion condensation" with the negatively-charged phosphate backbone of DNA. This concentrating effect of DNA will vary markedly with net charge on the thiol. It is of particular importance in interpreting experiments with different thiols in model systems involving polynucleotides, although the effects will be very much less at high physiological ionic strengths (typically ≈ 0.16).

Whilst several papers have explored these "ion condensation" effects, little attention has been given to another effect of the protonated amino functions in WR-1065. These reduce the thiol pK_a (pK_{13}) to around 7.3, and will also have a considerable but presently unquantified effect on the important thiyl/disulphide radical anion equilibrium (14) compared to the values for other simple thiols. It could be argued that equilibrium (13) is the most important factor controlling the relative reactivity of different thiols in biology and medicine, since Brønsted coefficients for thiolate reactions are typically less than unity.

$$RSH \rightleftarrows RS^- + H^+ \tag{13}$$

$$RS^\bullet + RS^- \rightleftarrows (RS\bullet\bullet SR)^- \tag{14}$$

Hence in the author's view, the low pK_a of WR-1065 is *at least as important* as the poly-cationic nature (at pH 7.4 it does not even have a +2 net formal charge).

Further justification for more attention to studying the importance of equilibria (13) and (14) is provided by the evidence of Prütz *et al.* that the radical anion $RS\bullet\bullet SR^-$ can serve as an electron donor to "repair" DNA damaged by radical attack. Radical transfer from protein components can also contribute to $(RS\bullet\bullet SR)^-$ formation. Whilst much more work is needed to evaluate these possibilities, the critical role of equilibria (13) and (14) in controlling the balance between an oxidizing radical $(E_{m7}[RS^\bullet,H^+/RSH] \approx +1 \text{ V})$ and a reducing radical $(E_{m7}[RSSR/(RS\bullet\bullet SR)^-] \approx -1.6 \text{ V})$ is obviously central to the biological chemistry of thiyl radicals.

Whilst the radioprotective role of thiols is generally ascribed to fast, free-radical reactions, recent rapid-mix experiments have shown some protective effect of e.g. dithio-threitol to persist even if the protector is added up to ≈ 1 s after irradiation. It seems that thiol-reactive species are persisting for times rather longer than first thought although the problems of intracellular diffusion complicate interpretation, as does the special charac-teristics of dithiothreitol, e.g. the stability of the disulphide radical anion.

MODULATING GLUTATHIONE LEVELS FOR THERAPEUTIC GAIN

Methods which have been used to *increase* glutathione levels in cells or tissue include the administering of cysteine precursors such as N-acetylcysteine, or other prodrugs including the oxothiazolidine **1** or the methylthiazolidine **2**.

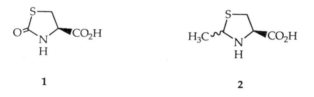

1   **2**

The former is enzymatically converted to cysteine whilst the latter undergoes spontaneous hydrolysis.

More numerous methods are available for *decreasing* glutathione concentrations. Inhibiting GSH biosynthesis using buthionine sulphoximine (BSO) is especially useful, developed in the studies of Meister *et. al.* on inhibitors of γ-glutamylcysteine synthetase, a key enzyme in the biosynthetic pathway. Clinical use of BSO is fraught with difficulty because of the ubiquitous protective role of GSH, but it is possible that differences in turnover and inter-organ transport of GSH can be exploited to deplete GSH in target organs whilst maintaining sufficient GSH, e.g. in the liver, for it to fulfil essential roles. Certainly in the mouse model, studies at the Gray Laboratory have used BSO to deplete tumour GSH to 10 - 20% of initial levels whilst more rapid re-synthesis or transport to the liver maintained, on average, high levels in the latter.

Inhibition of other enzymes, such as glutathione reductase to prevent the reduction of GSSG to GSH, has proven less useful because of the efficiency of the enzyme. Intentional reduction of GSH by administering chemicals reacting with 1 : 1 stoichiometry is bounded by restrictions such as the total body burden of GSH being ≈ 0.1 mol in humans.

The startling effects of BSO on misonidazole radiosensitization *in vitro* shown in Figure 1 have, to date, not been translated to results *in vivo*. Aside from possible problems of impeded diffusion of a hydrophilic drug from the capillaries to distant hypoxic cells, the problem of subcellular compartmentalisation noted earlier merits further attention. Other problems are the relative rates of GSH biosynthesis in oxic and hypoxic cells and the role of other reductants as protectors. Thus ascorbate has been demonstrated to compete with misonidazole in thiol-depleted cells, and ascorbate is absent in most *in vitro* test systems but is present in tumours in substantial amounts.

So far as radiosensitivity goes, it is very probably GSH accessibility to nuclear DNA that is important. Recent work by the author and his colleagues has shown some promise in the use of acridine orange (AO$^+$) as a probe intercalated with DNA which exhibits delayed fluorescence. The triplet state precursor of the fluorescing singlet is a powerful oxidant and is quenched by thiols, presumably in an electron-transfer reaction:

$$(AO^+)^* + RS^- \rightarrow AO^\bullet + RS^\bullet \tag{15}$$

Such quenching studies can be performed using mammalian cells stained with AO$^+$ and a commercial luminescence spectrometer. They offer potential in answering key questions such as the nature of the reductive environment in the immediate vicinity of DNA after treatment of cells with BSO or different thiols. They also illustrate the "ion condensing" properties of charged thiols with DNA in solution at low ionic strengths.

REACTIVITY OF GLUTATHIONE AS A NUCLEOPHILE

The thiolate anion, GS$^-$ is the active nucleophilic form and hence pH is of paramount importance in controlling reactivity ($pK_{16} \approx 9$ but there is the problem of multiple ionization constants):

$$GSH \rightleftarrows GS^- + H^+ \tag{16}$$

Glutathione is unexceptional as a nucleophile towards most substrates, typical Brønsted β coefficients being in the range 0.2 - 0.5 (most reactions of thiolate anions with electrophiles increase in rate with thiol pK_a). A classical reaction, used in the assay of GSTases, is displacement of halogen from chlorodinitrobenzene. Figure 3 shows the likely mechanism, which involves a Meisenheimer intermediate.

Whilst the Michael addition of GSH across activated double bonds (also involving GS$^-$ as the reactive form) was recognised at an early date to be an important feature of the activity of N-ethylmaleimide as a radiosensitizer, the widely documented reactivity of GSH as a nucleophile was initially overlooked by radiobiologists who found anomalous radiosensitization by a series of nitroimidazoles (typically 1-methyl-4-nitroimidazoles) substituted with adjacent good leaving groups (e.g. in the 5-position) such as halogen or methylsulphonyl. (The author, too, was slow to grasp the simple chemical explanation!)

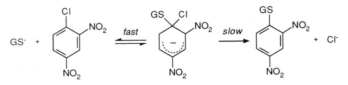

Fig. 3.   Probable general mechanism for the nucleophilic displacement of activated halogen in nitroaryl compounds, illustrated by 2,4-dinitro-chlorobenzene.

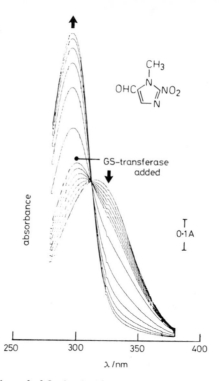

Fig. 4. Reaction of 1-methyl-2-nitroimidazole-5-carboxaldehyde (60 µmol dm^{-3}) with GSH (2 mmol dm^{-3}) at pH 7.66, 25 °C, followed by absorption spectrophotometry. Successive scans at 3 min intervals. After seven scans, GSTase (Sigma G8366, 0.3 U/ml) was added.

When glutathione reactivity of such compounds was appreciated, more extensive susceptibility towards nucleophilic attack of glutathione with nitroaryl compounds which had been evaluated as radiosensitizers became apparent.

A typical experiment from these studies is shown in Figure 4, which shows the absorbance changes as GSH reacts with a 2-nitroimidazole. In this case, quantitative release of nitrite (NO$_2^-$) was measured, illustrating that NO$_2$ was a better leaving group than CHO, the reverse of the experience with halonitroimidazoles. The experiment shown also demonstrated catalysis by GSTase.

Reactivity of GSH towards many other types of electrophile has been demonstrated, e.g. epoxides, alkyl halides, quinones and quinoneimines, arylhalides, carbonium ions, diverse α,β-unsaturated systems and iso(thio)cyanates. Another important aspect is the attack at disulphides leading to thiol/disulphide exchange:

$$GS^- + R^1SSR^2 \rightleftarrows GSSR^2 + R^1S^- \tag{17}$$

since this is one route to perturbing intracellular or membrane-bound protein thiols on adding extracellular GSH. Recent work by Dennis *et al.* shows that thiol/disulphide exchange leads to accumulation of intracellular cysteine when cells are treated with dithiothreitol, if the medium contains cystine.

CATALYSIS OF DISPLACEMENT AND CONJUGATION REACTIONS OF GLUTATHIONE

Glutathione transferases are a family of enzymes with high specificity towards GSH compared to other thiols but with a wide spectrum of activity towards electrophilic substrates for displacement or conjugation reactions. In addition, some GSTase isoenzymes have activity as peroxidases active towards organic peroxides. The soluble forms are dimeric and they are found in plants, insects, protozoa and molluscs as well as man, but possibly not in bacteria. Some of the reactions exhibit very low Michaelis-Menten K_m values. Reaction of GS$^-$ with the quinoneimine metabolite of paracetamol (acetaminophen) occurs at a high non-enzymic rate ($k \approx 3 \times 10^6$ dm^3 mol^{-1} s^{-1}) yet it is catalysed by GSTase isoenzymes with K_m in the micromolar range. GSTase activity is particularly high in the liver and kidney.

The study of GSTases is expanding rapidly with the characterisation of different isoenzymes in different species and the application of molecular-biological techniques to study gene expression. Whilst detailed discussion is outside the scope of this article, as a broad generalisation it may be concluded that most reactions of GSH with electrophiles, particularly hydrophobic molecules, will occur faster in biology than chemical rate data might predict because of the ubiquitous nature and high activity of GSTases.

IMPORTANCE OF GLUTATHIONE IN CANCER THERAPY

There are an increasing number of reports of enhanced toxicity of chemotherapeutic agents when intracellular GSH concentrations are reduced (or protection when they are increased). The compounds susceptible to modulation of activity in this way include alkylating agents such as nitrogen mustard and melphalan; anthracyclines such as adriamycin; bleomycin; mitomycin-c, *cis*-diamminedichloro-platinum(II); cyclophosphamides. Whilst detailed discussion is impractical here, a simple chemical explanation is apparent in some cases. Thus the haloalkyl substituents in melphalan (Figure 5) would be expected to be susceptible to nucleophilic attack - they are to hydrolysis - and indeed the glutathione conjugates shown have been reported. Further, the reactions are catalysed by GSTases. This latter factor explains a recent report of a melphalan-resistant human plasma cell line which, unlike in some earlier reports, did not have elevated intracellular GSH levels but did have a 1.5-fold induction of a GSTase. Melphalan is one chemotherapeutic agent which is a candidate for an approach using combination therapy with BSO to help overcome drug resistance.

Recent work by Ketterer *et al.* characterising the GSTase isoenzyme specificity for catalysing the reduction by GSH of a thymine hydroperoxide (hydroperoxymethyluracil) is being extended to study the detoxification of peroxidized DNA, illustrating the differing requirements for catalysis of reactions at the surface of a macromolecule compared to reactions with small molecules. Such studies are relevant not only to the repair or detoxification of radiation-induced damage in DNA, but have much wider implications in oxidative stress from drugs and in ordinary metabolic processes such as those associated with ageing.

REACTIONS OF GLUTATHIONE WITH INTERMEDIATES IN DRUG METABOLISM

Other papers in these Proceedings describe the generation of thiyl radicals during oxidative drug metabolism. Figure 6 illustrates the principle reactions involved in the catalytic

Fig. 5. Melphalan and the glutathione conjugates formed by nucleophilic attack.

424

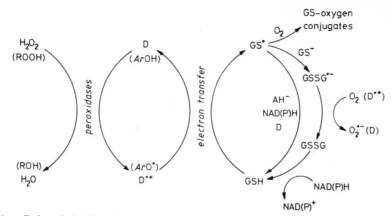

Fig. 6. Role of thiyl radicals in the redox cycling of oxidizable drugs (electron donors, D, e.g. phenols *ArOH*) via peroxidases. From Wardman (1988) (see Bibliography).

role of readily-oxidizable substrates of peroxidases in the conjugation of glutathione thiyl radicals with GSH and O_2. Whilst peroxidases can oxidize GSH directly, amines or phenols can act catalytically as "mediators" to accelerate kinetically sluggish reactions.

Reaction of a radical electron donor such as a phenoxyl radical, PhO$^\bullet$, with GSH may be thermodynamically unfavourable ($K_{18} \ll 1$):

$$PhO^\bullet + GS^- \rightleftarrows PhO^- + GS^\bullet \tag{18}$$

but kinetically driven by efficient removal of GS$^\bullet$ from the equilibrium by reactions (9) - (11) noted earlier. The potential oxidizing role of the thiylperoxyl radical, GSOO$^\bullet$ has been discussed, but experiments by the author illustrated in Figure 7 suggest that the reactivity of GSOO$^\bullet$ towards the oxidizable substrate, chlorpromazine (CPZ$^+$) is at least an order of magnitude lower than that of GS$^\bullet$. The data reflect competition between (19) and (9):

$$GS^\bullet + CPZ \rightarrow GS^- + CPZ^{\bullet+} \tag{19}$$

$$GS^\bullet + O_2 \rightleftarrows GSOO^\bullet \tag{9}$$

the latter being suggested to be an equilibrium by Tamba *et al.* The fit to the data uses the expression:

$$k_{obs} = k_{19}/(1 + K_9[O_2]) \tag{20}$$

which is that appropriate if $k_{21} \ll k_{19}$:

$$GSOO^\bullet + CPZ \rightarrow CPZ^{\bullet+} + products \tag{21}$$

An independent estimate of $K_9 = 2770 \pm 370 \ dm^3 \ mol^{-1}$ is derived making these assumptions, in good agreement with the earlier estimate of $3200 \ dm^3 \ mol^{-1}$ and lending some support to the equilibrium (9) proposed. (The conditions used ensured the relaxation time τ_9 was much shorter than the half-life of reaction (19), as required for this analysis.)

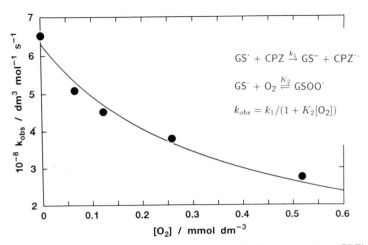

Fig. 7. Effect of oxygen on the rate of formation of chlorpromazine (CPZ) radical-cations after pulse radiolysis (0.2 μs, ≈ 1.5 Gy) of solutions containing GSH (1 mmol dm⁻³), CPZ (80 - 160 μmol dm⁻³) and O_2/N_2O (saturated with mixtures of 0 - 40% v/v O_2 in N_2O), at pH 4.8. Rate constants = $(k_{obs}/s^{-1})/$ [CPZ] obtained from increase in absorbance at 525 nm. Unpublished work (1988) by the author.

CONCLUSIONS

Although this outline of some aspects of the chemical properties of the most biologically-important thiol suffers from its breadth of coverage at the expense of depth or detail, it illustrates how widespread the implications of glutathione reactivity and concentrations are, and how important it is to further characterize both. Whilst it was prepared to complement the more comprehensive discussions of individual types of reaction to be found elsewhere in the Proceedings, emphasising a biological overview rather than chemical detail, it is hoped that this summary of some simple chemical reactions in a biological context will help provide a stimulus for further discussion. Other important aspects, such as the role of thiols in anticarcinogenesis, are the subject of much current work and the interested reader is directed to the bibliography, where much more comprehensive surveys are to be found.

ACKNOWLEDGEMENTS

This work is supported by the Cancer Research Campaign.

BIBLIOGRAPHY: Suggestions for Further Reading

Brown, J. M., ed., "Chemical Modifiers of Cancer Treatment", *Int. J. Radiat. Oncol. Biol. Phys.* 12:1019-1545 (1986).
Cerutti, P. A., Nygaard, O. F., and Simic, M. G., eds., *in* "Anticarcinogenesis and Radiation Protection" Plenum Press, New York, 1987.
Dennis, M.F., Stratford, M. R. L., Wardman, P., and Watfa, R. R., "Increase in Cellular Cysteine after Exposure to Dithiothreitol: Implications in Radiobiology", *Int. J. Radiat. Biol.* 56:877 (1989).

Dolphin, D., Avramovic, O., and Poulson, R., eds., "Glutathione. Chemical, Biochemical and Medical Aspects", Parts A and B, John Wiley, New York, 1989.

Fielden, E. M., Fowler, J. F., Hendry, J. H., and Scott, D., eds., *in* "Radiation Research. Proceedings of the 8th International Congress of Radiation Research, Edinburgh" Vol. 2, Taylor & Francis, London, 1987.

Jocelyn, P. C., *in* "Biochemistry of the SH Group" Academic Press, London, 1972.

Larsson, A., Orrenius, S., Holmgren, A., and Mannervik, B., eds., *in* "Functions of Glutathione" Raven Press, New York, 1983.

Malaise, E. P.,. Guichard, M., and Siemann, D. W., eds. "Chemical Modifiers of Cancer Treatment", *Int. J. Radiat. Oncol. Biol. Phys.* 16:885-1345 (1989).

Mantle, T. J., Pickett, C. B., and Hayes, J. D., eds., "Glutathione S-Transferases and Carcinogenesis" Taylor & Francis, London, 1987.

Meister, A. and Anderson, M. E., "Glutathione", *Ann. Rev. Biochem.* 52:711 (1983).

Nygaard, O. F. and Simic, M. G., eds., "Radioprotectors and Anticarcinogens" Academic Press, New York, 1983.

O'Brien, P. J., "Radical Formation during the Peroxidase Catalyzed Metabolism of Carcinogens and Xenobiotics: The Reactivity of these Radicals with GSH, DNA, and Unsaturated Lipid", *Free Rad. Biol. Med.* 4:169 (1988).

Patai, S., "The Chemistry of the Thiol Group", Parts I and II, John Wiley, London, 1974.

Prütz, W. A., "Chemical Repair in Irradiated DNA Solutions Containing Thiols and/or Disulphides. Further Evidence for Disulphide Radical Anions Acting as Electron Donors", *Int. J. Radiat. Biol.* 56:21 (1989).

Ross, D., "Glutathione, Free Radicals and Chemotherapeutic Agents. Mechanisms of Free-radical Induced Toxicity and Glutathione-dependent Protection", *Pharmac. Ther.* 37:231 (1988).

Sies, H. and Ketterer, B., eds., "Glutathione Conjugation: Mechanisms and Biological Significance", Academic Press, London, 1988.

Schöneich, Ch., Bonifacic, M., and Asmus, K.-D., "Reversible H-atom Abstraction from Alcohols by Thiyl Radicals: Determination of Absolute Rate Constants by Pulse Radiolysis", *Free Rad. Res. Communs.* 6:393 (1989).

Sonntag, C. v., "The Chemical Basis of Radiation Biology" Taylor & Francis, London, 1987.

Tamba, M., Simone, G., and Quintiliani, M., "Interaction of Thiyl Free Radicals with Oxygen: A Pulse Radiolysis Study", *Int. J. Radiat. Biol.* 50:595 (1986).

Wardman, P., "Radiation Chemistry in the Clinic: Hypoxic Cell Radiosensitizers for Radiotherapy", *Radiat. Phys. Chem.* 24:293 (1984).

Wardman, P., "The Mechanism of Radiosensitization by Electron-affinic Compounds", *Radiat. Phys. Chem.* 30:423 (1987).

Wardman, P., "Conjugation and Oxidation of Glutathione via Thiyl Free Radicals", *in*: "Glutathione Conjugation: Mechanisms and Biological Significance", H. Sies and B. Ketterer, eds., Academic Press, London, 1988.

Wardman, P., Dennis, M.F., and White, J., 1989, "A Probe for Intracellular Concentrations of Drugs: Delayed Fluorescence from Acridine Orange", *Int. J. Radiat. Oncol. Biol. Phys.* 16:935 (1989).

Weiss, J. F. and Simic, M. G., eds., "Perspective in Radioprotection", *Pharmac. Ther.* 39:1-414 (1988).

Wilson, I., Wardman, P., Lin, T.-S., and Sartorelli, A. C., "Reactivity of Thiols towards Derivatives of 2- and 6-Methyl-1,4-naphthoquinone Bioreductive Alkylating Agents", *Chem.-Biol. Interactions* 61:229 (1987).

IN VIVO HEMOGLOBIN THIYL RADICAL FORMATION AS A CONSEQUENCE OF

HYDRAZINE-BASED DRUG METABOLISM

Ronald P. Mason and Kirk R. Maples

Laboratory of Molecular Biophysics
National Institute of Environmental Health Sciences
Research Triangle Park, NC 27709, USA

For a variety of reasons, the possibility of free radical metabolism has not received much attention in the past, although Michaelis of the Michaelis-Menten equation was interested in free radical metabolites and their importance in biochemistry in the 1930's. One reason for the late development of this area is that most biochemicals, as opposed to aromatic drugs and industrial chemicals, are not easily metabolized through free radical intermediates. Another reason is that unless something can be demonstrated with a whole animal, there will always be some question of its actual existence in biology. To detect something as ephemeral as a free radical formed inside a whole animal is certainly not easy. In such investigations, the toxic effects of free radicals are presumably the result of purely chemical reactions; therefore, a knowledge of the radical's chemistry under physiological conditions is necessary.

It is important to show that the characteristics of the toxicity are consistent with the known enzymatic and nonenzymatic free radical reactions. In some cases the manipulation of an animal model provides evidence that a toxicity is indeed free radical-mediated. *In vivo* detection of free radical metabolites is a very challenging task. Production rates of free radicals in animals are inherently slow; therefore, the highest possible sensitivity is of paramount importance. Water, with its high dielectric constant, is the worst solvent for ESR studies in that only very small samples can be studied, which decreases the molar sensitivity just when sensitivity is needed most.

Several approaches to this problem have been tried over the last thirty years. Freezing water lowers its dielectric constant so larger samples can be studied. The freeze-quench technique is useful for enzymes where solutions can be frozen in milliseconds, but tissues must be ground to fit in ESR sample tubes, and this leads to mechanically induced radicals or artifacts. In addition, the resulting powder spectra are poorly resolved, and their interpretation in complex biological systems is very difficult, if not impossible. Lyophilized, or freeze-dried tissue is plagued by the same problems of artifacts and poor resolution. Low-frequency ESR enables the study of larger samples, and perhaps a small animal could be studied directly, that is, via *in vivo* spectroscopy. Unfortunately, ESR sensitivity, like NMR sensitivity, is strongly dependent on frequency, and low-frequency instruments are unlikely to achieve the molar sensitivity of the standard X-band instruments. Spin trapping, in that it ideally integrates free radicals formed over time, is the most attrative approach to the detection of free radicals *in vivo*. Spin trapping is a technique where a diamagnetic molecule (or spin trap) reacts with a free radical to produce a more stable radical (or spin adduct)

Sulfur-Centered Reactive Intermediates in Chemistry and Biology, Edited by
C. Chatgilialoglu and K.-D. Asmus, Plenum Press, New York, 1990

which is readily detectable by ESR.[1] Spin adducts are usually substituted nitroxide free radicals, which, for free radicals, are relatively stable. Some radical adducts are so stable that their structure has been determined by mass spectrometry and NMR.

Our approach to *in vivo* spin trapping has been to examine biological fluids directly for spin adducts using the TM_{110} ESR cavity and 17 mm flat cell, which gives the largest possible aqueous sample size in the active region of the cavity, about 100 μl.[2] This gives the highest molar sensitivity and high resolution. No background signals other than the ascorbyl semidione doublet have been detected. The detection of free radical metabolites in urine, blood or bile is little different from the detection of the products of drug metabolism by HPLC as practiced in pharmacology departments or by the pharmaceutical industry.

For over 100 years the reaction of phenylhydrazine with red blood cells has been receiving attention.[3] This reaction has been shown to induce the oxidative denaturation of hemoglobin, and, in a series of processes, to give rise to hemolytic anemia. Although free radical formation due to the reaction of phenylhydrazine with the oxyhemoglobin in red blood cells has been established *in vitro*, analogous ESR spectroscopic evidence for *in vivo* free radical formation has not been reported, therefore we decided to study the *in vivo* reaction of phenylhydrazine with blood.

Seven years ago the phenyl radical was detected using the spin trap DMPO (5,5-dimethyl-1-pyrroline N-oxide) with *in vitro* incubations of human red blood cells by Hill and Thornalley[4] and the following year by Augusto *et al.*[5] This phenyl radical may be responsible for the well known hemolytic anemia caused by phenylhydrazine. At the very least it is closely related to the species responsible for red blood cell destruction.

We found an unexpected result in rat blood from an animal given an LD_{50} dose of phenylhydrazine in the stomach and DMPO in the peritoneal cavity (Figure 1A). The LD_{50} is the toxic dose that will kill half of the animals. This radical adduct is bound to a macromolecule as indicated by its broad, asymmetric lineshape. We believe it to be a sulfhydryl radical adduct of hemoglobin. The signal can be detected at a dose of only 1 mg of phenylhydrazine/kg (Figure 1B). The formation of the radical adduct requires the spin trap DMPO. The β-H splitting of the DMPO radical adduct appears as a doubling of each line of the typical immobilized nitroxide spectrum, giving six lines in all.

We are convinced that the rest of the radical adduct is due to hemoglobin bound through one of its sulfhydryl groups. *If this is the case, then this is the first report of target macromolecule free radical formation as a consequence of the metabolism of a toxic chemical.*

First, the radical adduct, as measured by ESR, co-chromatographs with oxyhemoglobin as measured by its optical spectrum.[6] Second, we employed purified oxyhemoglobin *in vitro* with phenylhydrazine and DMPO. We were able to detect the identical immobilized adduct found *in vivo*. When other workers examined the phenylhydrazine/oxyhemoglobin reaction, only freely rotating six-line DMPO/phenyl radical adducts were detected. In their studies, Hill and Thornalley[4] employed 1% suspensions of human red blood cells in buffer. When we repeated their experiment employing 1% rat red blood cells, we obtained the same freely rotating phenyl radical adduct of DMPO as they did (Figure 2F). However, when we used undiluted rat red blood cells, an immobilized spectrum identical to the *in vivo* results was obtained (Figure 2A). We found that the concentration of the immobilized DMPO radical adduct was dependent on the extent of dilution of the rat red blood cells prior to addition of phenylhydrazine and DMPO (Figures 2A - 2F). Once formed, dilution of the immobilized adduct samples with buffer did not affect the degree of immobilization. Therefore, the ratio of DMPO to rat red blood cells must determine the ratio of the phenyl radical adduct to the immobilized radical adduct of hemoglobin.

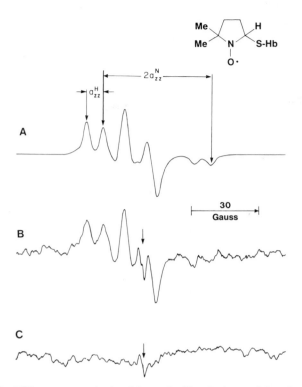

Fig. 1.   A.   The ESR spectrum obtained from the blood of a rat 2 hr after being given
                188 mg/kg phenylhydrazine per os (orally) and 500 μl/kg DMPO intra
                peritoneally.
          B.   Same as in A, but the phenylhydrazine dosage was 1 mg/kg.
          C.   Same as in A, but the rat was given only DMPO.
          In both B and C, the *arrow* marks the location of the small doublet due to
          the presence of the ascorbate radical. (From ref. 6, with permission).

Apparantly, at *in vivo* oxyhemoglobin concentrations represented by 50% red blood cells, DMPO cannot successfully compete with hemoglobin for the phenyl radical. Thus, when we employed 50% red blood cells *in vitro*, we detected the DMPO/thiyl free radical adduct and not the DMPO/phenyl radical adduct obtained using 1% red blood cells.

When we used ^{13}C-labelled phenylhydrazine, the hemoglobin-derived radical adduct was unchanged, implying it does not contain the phenyl ring; however, the phenyl radical adduct gave a 12-line spectrum due to the ^{13}C-hyperfine coupling of 7.45 G.[7] Hill and Thornalley had already proven the structure of the phenyl radical adduct with mass spectrometry,[4] so ^{13}C-coupling was certainly expected. That ^{13}C-substitution had no effect on the hemoglobin-derived adduct is more important.

We have proven that the immobilized adduct contains hemoglobin and does not contain phenyl near the radical center, but proof that the linkage is a thioether is difficult to obtain. Of the amino acids which compose hemoglobin, cysteine is the most easily oxidized. It is known that with the use of sulfhydryl-blocking agents, e.g., iodoacetamide, maleimide, or N-ethylmaleimide, the number of free sulfhydryl groups in a protein can be markedly diminished. Pretreatment of rat red blood cells with either iodoacetamide, maleimide, or N-ethylmaleimide markedly lowered the adduct concentration. This is really our only evidence for a thioether linkage in the radical adduct.

431

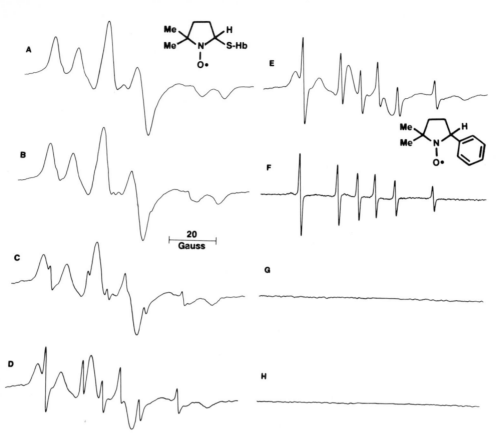

Fig. 2. The effect of diluting rat RBCs (red blood cells) with isotonic (pH 7.4) phosphate buffer prior to addition of DMPO (100 mM) and phenylhydrazine (1 mM). (From ref. 6, with permission).

A. Undiluted RBCs.   B. RBCs (50 %) in buffer.
C. RBCs (25 %) in buffer.   D. RBCs (10 %) in buffer.
E. RBCs (5 %) in buffer.   F. RBCs (1 %) in buffer.
G. Only buffer, no RBCs.   H. Undiluted RBCs, DMPO, and no phenylhydrazine.

We have also performed the *in vivo* experiment with phenylhydrazine using the spin trap PBN (α-phenyl-N-*t*-butyl nitrone) in place of DMPO (Figure 3A). An immobilized PBN adduct which was dependent on the administration of phenylhydrazine was detected (Figure 3B). This PBN radical adduct spectrum is characterized by a $2a_{zz}^{N}$ of 61.6 G. The PBN/hemoglobin thiyl radical adduct spectrum is different from the DMPO/hemoglobin thiyl radical adduct spectrum because the a_{β}^{H} doublet splitting for PBN radical adducts is typically much smaller than that for the DMPO radical adducts and it is lost in the linewidth. Therefore, a three-line spectrum instead of a six-line spectrum is obtained. Thus, the PBN radical adduct spectrum resembles that reported for human hemoglobin spin-labelled at the β-93 cysteine residue with a maleimide spin label.[8]

We have extended our *in vivo* investigation to the antihypertensive agent hydralazine, the antidepressant drugs phenelzine and iproniazid, and the antitubercular agent isoniazid.[7] When rats were adminstered DMPO and either iproniazid or phenelzine, the immobilized DMPO/hemoglobin thiyl radical adduct was readily detected in the blood of

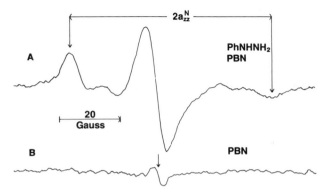

Fig. 3. A. The ESR spectrum obtained from the blood of a rat 2 hr after being given 188 mg/kg phenylhydrazine ($PhNHNH_2$) orally and 500 mg/kg PBN intraperitoneally.

B. The ESR spectrum obtained from the blood of a rat that received only PBN (500 mg/kg ip). The *arrow* marks the position of the overmodulated signal due to the ascorbate radical.

(From ref. 7, with permission).

the rats. We were still able to detect this radical adduct in rats which had received doses of iproniazid and phenelzine as low as 3 mg/kg and 15 mg/kg, respectively. However, when we administered either isoniazid or hydralazine at their respective LD_{50} levels, no ESR signal could be detected. When these drugs were tested *in vitro* with red blood cells and DMPO, only phenelzine and hydralazine yielded the immobilized DMPO/hemoglobin thiyl radical adduct signal, so the *in vivo* results differ from the *in vitro*. Our inability to detect the hemoglobin thiyl radical adduct with iproniazid *in vitro* was consistent with iproniazid being metabolized *in vivo* into a more reactive species. This finding was in accord with the work of Nelson et al.[9] who demonstrated that the hepatic arylamidase hydrolyzes iproniazid to isopropylhydrazine, a more reactive hydrazine. In order to test this, we pretreated rats with the arylamidase enzyme inhibitor bis-*para*-nitrophenylphosphate prior to administration of DMPO and iproniazid. We found inhibition indicating that iproniazid does not react in the bloodstream of rats to yield the immobilized DMPO/hemoglobin thiyl radical adduct, but is first metabolized to a more reactive species, isopropylhydrazine. Based on the aforementioned results of Nelson et al.,[9] the arylamidase reaction most likely occurs in the liver, which releases isopropylhydrazine into the blooodstream, where it reacts with oxyhemoglobin.

Hydralazine, on the other hand, did react *in vitro* with red blood cells to yield radicals which could oxidize hemoglobin to yield the DMPO/hemoglobin thiyl radical adduct. The absence of a detectable signal *in vivo* from hydralazine-treated rats probably reflects a rapid biotransformation of hydralazine into a less reactive metabolite, probably by acetylation. In conclusion, a DMPO/hemoglobin thiyl radical adduct has been detected *in vivo*, and this species is formed by the reaction of phenylhydrazine and even some hydrazine-based drugs with oxyhemoglobin.

REFERENCES

1. C. Mottley and R.P. Mason, *Biol. Mag. Res.* 8:489 (1989).
2. R.P. Mason, K.R. Maples, and K.T. Knecht, *Elec. Spin Res.* 11B:1 (1989).

3. G. Hoppe-Seyler, *Z. Physiol. Chem.* 9:34 (1885).
4. H.A.O. Hill and P.J. Thornalley, *FEBS Lett.* 125:235 (1981).
5. O. Augusto, K.L. Kunze, and P.R. Ortiz de Montellano, *J. Biol. Chem.* 257:6231 (1982).
6. K.R. Maples, S.J. Jordan, and R.P. Mason, *Mol. Pharmacol.* 33:344 (1988).
7. K.R. Maples, S.J. Jordan, and R.P. Mason, *Drug Met. Disp.* 16:799 (1988).
8. J.S. Hyde and D.D. Thomas, *Ann. N.Y. Acad. Sci.* 222:680 (1973).
9. S.D. Nelson, J.R. Mitchell, J.A. Timbrell, W.R. Snodgrass, and G.B. Corcoran III, *Science (Wash. D.C.)* 193:901 (1976).

THE BIOLOGICAL ACTIONS OF THE GLUTATHIONE/DISULFIDE SYSTEM:

AN OVERVIEW

Marcello Quintiliani

Institute of Biomedical Technologies (CNR)
Via G. B. Morgagni 30/E
00161 Roma, Italy

Glutathione (GSH) and its disulfide (GSSG) have been the object of considerable interest, during the last ten or fifteen years, on the part of investigators acting in many different areas of biological research. Such interest is justified by the multifunctional properties of this tripeptide, involving, for example, metabolism of intermediates, enzyme mechanisms, biosynthesis of macromolecules, drug metabolism, radiation, cancer, oxygen toxicity, transport, immune phenomena, endocrinology, environmental toxins, and aging.

Reviewing the biological actions of the glutathione/disulfide system (GSH/GSSG), in the context of a discussion on "Sulfur-centered reactive intermediates in chemistry and biology", seems to be quite appropriate. However, due to the existence of many recent and excellent reviews covering various aspects of the subject, the present contribution will try to give a comprehensive, but concise, overview of the relevant aspects, providing the appropriate references for a more detailed coverage. The number of references has been kept as low as possible, giving the preference to reviews and books.

MAIN PHYSICO-CHEMICAL PROPERTIES OF GLUTATHIONE

GSH is the most abundant non-protein thiol in the great majority of cells and organisms, where its concentration is usually in the millimolar range (in mammalian liver and kidney it ranges between 5 and 10 mM).

GSH is a tripeptide containing L-glutamic acid, L-cysteine and glycine. The unusual peptide bond between glutamic acid and cysteine is at the γ-carboxyl of the former rather than at its α-carboxyl group, as usually observed. The pK_a values of the acidic and basic groups in GSH are as follows: Glu-α-NH_3^+, 9.5; -SH, 9.2; Gly-COOH, 3.7; and Glu-α-COOH, 2.5.[1] The dissociation of the -SH group to give the thiolate ion (-S$^-$) is important for the GSH functions. In general, the reactivity of thiols is mainly due RS$^-$ ions which are strongly nucleophilic. Moreover, RS$^-$ ions can lose an electron yielding the corresponding thiyl radical RS$^\bullet$.

GSH, like many other simple thiols, can easily undergo oxidation of its SH group. The complete oxidation of the sulfur atom yields the glutathione sulfonic acid (GSO_3H). There are, of course, intermediate oxidation states, some of them very unstable like sulfenic acid (GSOH). A very stable product, however, is obtained in the oxidation to the disulfide and this is, by far, the most frequent process.

Sulfur-Centered Reactive Intermediates in Chemistry and Biology, Edited by
C. Chatgilialoglu and K.-D. Asmus, Plenum Press, New York, 1990

The oxidation-reduction reactions of GSH that play major roles in its biological functions are: one electron reactions; sulfydryl-disulfide interchanges; and two electron oxidations via the formation of an intermediate.

Type 1 reactions are among those radical reactions that constitute the main object of interest of the present meeting, and they will be discussed later. Many aspects are also presented and discussed in other articles of this book, so that they do not require more extensive consideration at this point.

GSH-involving sulfydryl-disulfide interchanges are the most common reaction of this type occurring intracellularly due both to the high intracellular concentration of GSH and to the fact that low molecular weight intracellular thiols consist almost entirely of GSH. Such reactions play a role in the refolding of proteins containing multiple disulfide links, and in the formation of GSS-protein mixed disulfides. Disulfide links are important in many proteins in order to insure the stability of their folded conformation, while the reversible formation of mixed disulfides with low molecular weight disulfides has been postulated as a possible mechanism for the regulation of metabolic pathways in mammalian organisms.[2]

The two-electron oxidation, via the formation of an intermediate, occurs, for instance, in the oxidation of thiols by iodine or by some properly substituted azoesters. The pertinent reactions are as follows:

$$GS^- + I_2 \quad \rightarrow \quad GSI + I^- \tag{1}$$

$$GS^- + GSI \quad \rightarrow \quad GSSG + I^- \tag{2}$$

$$GS^- + YCON=NCOY + H^+ \quad \rightarrow \quad YCON(SG)NHCOY \tag{3}$$

$$GS^- + YCON(SG)NHCOY + H^+ \rightarrow \quad GSSG + YCONHNHCOY \tag{4}$$

The reaction with azo-compounds is particularly interesting because the ability to oxidize GSH to GSSG, displayed by some compounds with a general formula as shown in reaction 3, can also occur within living cells. The most convenient GSH-oxidizing agent of this type is diamide [diazenedicarboxylic acid bis-(N,N-dimethylamide)], in which Y in the general formula corresponds to $(CH_3)_2N$. Diamide is easily soluble in aqueous media and quite stable in solution at neutral pH. It rapidly penetrates cells and oxidizes intracellular GSH in a matter of seconds without being toxic at the concentrations required for the complete conversion of GSH to its disulfide. Cells can then recover their GSH content through the action of GSH-reductase. In practice, diamide has been the first compound allowing to modulate cellular GSH content with no major limitations due to toxicity. Its uses in studies on the GSH role in cellular funtions and radioresistance have been reviewed by J.W. Harris in 1983 ([3], pp. 255-274).

GLUTATHIONE MECHANISM AND CHEMICAL FUNCTIONS

In multicellular organisms, glutathione is mainly located inside the cells where it is almost completely maintained in its reduced form. As stated above, in mammalian cells its concentration usually ranges between 2 and 5 mM.

GSH as such cannot penetrate cell membranes, but is synthesized intracellularly and eventually transported as such out of the cells. In mammalian organisms the glutathione concentration in extracellular fluids and in plasma is much lower than intracellularly, and the prevalent form is the oxidized one. This is not due to a low rate of GSH export, but rather to an active interorgan transportation. Transported glutathione will then be subject to partial degradation and intracellular resynthesis as described below.[9]

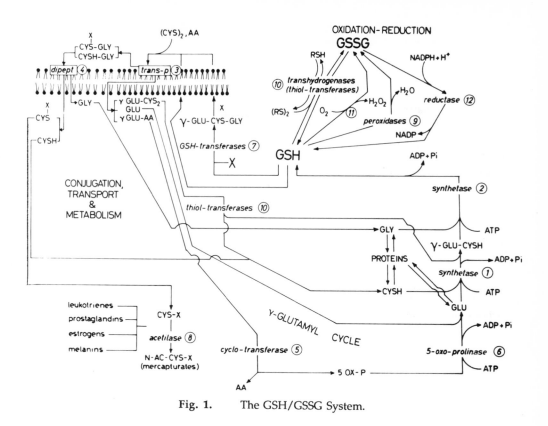

Fig. 1.   The GSH/GSSG System.

The GSH/GSSG system is involved in a complicated network of metabolic and functional interactions which are sinoptically summarized in Figure 1, at least with regard to the most important of them. Such interactions are mostly connected with the so-called γ-glutamyl cycle, whose definiton is mainly due to the work carried out by Alton Meister and his associates during the last 15 years.[5] The cycle consists of the reactions involved in the synthesis and degradation of GSH. GSH is synthesized intracellularly by γ-glutamylcysteine synthetase (No. 1 in Figure 1) and by glutathione synthetase (No. 2) and degraded by γ-glutamyltranspeptidase (No. 3), dipeptidase (No. 4), γ-glutamylcyclotransferase (No. 5), and 5-oxoprolinase (No. 6).

γ-Glutamyltranspeptidase initiates GSH degradation by removing the γ-glutamyl moiety. Such moiety can be simply transferred to water, but most frequently it is transferred to a suitable amino-acid acceptor, e. g. methionine, glutamine, and cystine. Transpeptidation also occurs with S-substituted GSH conjugates, becoming in this way an essential step in the synthesis of mercapturic acid and of other metabolic products indicated in Figure 1. Any transpeptidation process also produces cysteinylglycine, whose cleavage is catalyzed by dipeptidase. It is worth noting at this point that transpeptidase and dipetidase are located on the outer side of cell membranes, therefore GSH must be transported outside the cells (either through a channel or some carrier system) in order to account for the extracellular GSH/GSSG and to be accessible to enzymes located on the outer membrane. The products formed by transpeptidase and dipetidase (free amino acids, γ-glutamyl amino acids and S-substituted cysteines) are transported back into the cells. Here, γ-glutamyl cystine is reduced to γ-glutamyl cysteine and free cysteine, and the former can be used directly as substrate for glutathione synthetase. Other γ-glutamyl amino acids are substrates of γ-glutamylcyclotransferase, which cyclizes the γ-glutamyl moiety to 5-oxoproline, releasing the free amino

437

acids. 5-oxoproline is finally decyclized by 5-oxoprolinase to glutamate in a reaction coupled to cleavage of adenosine triphosphate to adenosine diphosphate and phosphate.[4,5,6]

In the previous paragraph, mention was made of S-substituted GSH conjugates. These are the products of the reactions, catalyzed by a family of enzymes named glutathione S-transferases (No. 7),[7] between GSH and a wide variety of electrophilic compounds, either of exogenous or of endogenous origin. Glutathione S-transferases are present in relatively abundant concentrations in the cytosol of cells of most aerobic organisms. These enzymes are among the biological catalysts participating in the processes of intracellular detoxification of mutagens, carcinogens, and other noxious chemical agents. GSH transferases promote the reaction of the nucleophile GSH, as the thiolate anion, with a variety of electrophilic compounds. Conjugates produced by the action of glutathione S-transferases end up as S-substituted cysteines. These are finally acetylated (No. 8) to form mercapturic acids, which are transported and ultimately excreted in the urine or feces. The same reactions can be utilized in part for the synthesis of biologically important metabolites, such as those listed in Figure 1.

An interesting feature of the conjugation reactions catalyzed by GSH S-transferases is that they can produce cellular depletion of free GSH when suitable compounds are involved in the process. Suitable compounds are those which can be introduced into the *in vitro* or *in vivo* systems in large enough concentration as to form stable adducts with a large fraction of cellular GSH, without inducing toxic effects, and without affecting the protein thiol content to a significant extent. Diethylmaleate (DEM) is the most widely used compound of this type ([3]; pp. 297-323). *In vitro*, the treatment of cultured mammalian cells with DEM can reduce their intracellular content of free GSH to about zero; and a similar result, even if less extensive, can be achieved with the *in vivo* treatment. The depletion of GSH content is complete both *in vitro* and *in vivo* in about an hour, while its recovery requires a few hours.

A further possibility of depleting the cellular content of GSH is offered by the selective inhibition of its synthesis. The most convenient compound presently available is buthionine sulfoximine (BSO), an irreversible inhibitor of γ-glutamylcysteine synthetase.[5,6] The drug is very effective in reducing GSH content and can be safely used, *in vitro* and *in vivo*, without inactivating other enzymes or perturbing other aspects of metabolism. With BSO, GSH depletion occurs because the tripeptide is normally used up by cells, but not replaced by the *novo* synthesis. High levels of depletion can be achieved, after several hours of BSO action, either in single cell systems or in organisms. The recovery takes longer times, because the inhibited enzyme molecules must be replaced.

The availability of manageable GSH-depleting drugs, such as DAM, DEM, and BSO, has been very useful in studies aimed to elucidation of GSH functions. An important result of these studies has been the development of selective enzyme inhibitors allowing to modulate cellular GSH content and also to selectively inhibit reactions involved in GSH metabolism.

The possibility to increase the GSH content can be realized by substances interacting at some step of the appropriate metabolic pathway. The best known example for a compound increasing the concentration of cell-GSH is: L-2-oxothiazolidine-4-carboxylate (OTZ). OTZ is structurally related to 5-oxoproline and is, therefore, a suitable substrate for 5-oxoprolinase to act on. The enzyme cleaves the heterocyclic ring of OTZ producing S-carboxy-cysteine which spontaneously decarboxylates to yield cysteine. In this way, OTZ acts as an intracellular cysteine delivery system promoting GSH synthesis. OTZ is active in animals as well as in cultured mammalian cells. In this last system the GSH content can be triplicated in two hours on addition of 10 mM OTZ to the culture medium. *In vivo*, the results are less dramatic but still clear cut and validly supported by the finding that OTZ protects experimental animals against the toxic effects of xenobiotics, such as acetaminophen, the toxicity of which is known to be counteracted by GSH ([5]; pp. 571-585).

Other compounds exist which can, more or less effectively, increase the intracellular concentration of GSH. Amongst them, GSH esters appear to be of particular interest. As stated before, GSH as such cannot penetrate the membrane of most of the cells, while its esters, being more hydrophobic, can do so and, once the ester group is cleaved by cellular esterases, can produce high intracellular GSH levels.

In general terms, it can be said that studies on the modulation of cellular GSH levels resulted in a better understanding of the mechanisms how cells defend themselves from oxidative damage, toxic compounds and radiation. Moreover, these studies led to the development of selective enzyme inhibitors which also are potential therapeutic agents.[8]

In addition to transferase and transpeptidase processes, glutathione is also involved in enzyme catalyzed redox processes with GSH-peroxidases (No. 9), GSH-transhydrogenases (No. 10), and GSSG-reductase (No. 12). GSH-peroxidases reduce hydrogen peroxide and organic hydroperoxides thereby oxidizing GSH to GSSG. In this way, they play an important role, together with the other biological antioxidants, in the processes which operate the detoxification of active oxygen species produced by metabolism under various normal and abnormal conditions. The most common GSH-peroxidase has four subunits, each one contaning one atom of selenium, probably in the form of Se-cysteine.[9] Evidence has also been produced that several animal tissues contain a non-selenium GSH-peroxidase activity, acting on artificial organic hydroperoxides but not on hydrogen peroxide. The nature of the Se-peroxidase and the mechanism involved are not yet defined and still under discussion. However, it remains the fact that perfused livers from Se-deficient rats do not release GSSG when hydrogen peroxide is infused, but they do so on the infusion of *tert*-butylhydroperoxide.

GSH-transhydrogenases is a collective name for several GSSG generating-enzymes. Since nonenzymatic oxidation of thiols (No. 11) is probably insignificant within intact cells, the disulfide required for the synthesis of protein disulfide bonds, whether intramolecular or in mixed disulfides, must be generated enzymatically. Only a limited number of enzymes able to perform this process intracellularly have been identified and a few of them have been well characterized.[2]

GSSG-reductase is a flavoprotein enzyme, widely distributed in bacteria, plant and animal cells, which reduces GSSG to GSH in a reaction coupled with the oxidation of NADPH (see Figure 1). NADP produced in the reaction is, on its turn, reduced back to NADPH through the oxidative pentose phosphate pathway. In this way cellular glutathione is maintained almost totally in its reduced form.

Finally, GSH functions as coenzyme for several enzymes acting in widely different metabolic patways, such as formaldehyde dehydrogenase, maleylacetoacetate isomerase, glyoxalase, prostaglandin endoperoxidase isomerases, DDT-dehydrochlorinase, and others.

THE PHYSIOLOGY OF GLUTATHIONE

An overview of glutathione metabolism and of the main biochemical reactions in which the tripeptide is involved and is given in the literature.[9] It provides the basis, at the molecular biological level, for the understanding of the multiform and complex role of glutathione in cells and organisms. These aspects will not be reviewed, however, in this paper because they are not directly related to the subject of the meeting.

RADICAL REACTIONS IN GLUTATHIONE FUNCTIONS

Many other papers in this volume deal with radical reactions of GSH/GSSG system that it would be fully redundant to repeat again what has been already extensively discussed. Most of what is known on the subject is based on the experimental work carried out using ionizing radiation in order to produce the radicals to be investigated. Obviously, concepts and reactions derived from studies on irradiated systems do generally apply also under conditions where radicals are generated in processes not involving radiation.

Aerobic life on our planet is characterized by the reduction of molecular oxygen to water in processes in which appropriate substrates are oxidized providing the energy necessary for the dynamic of living matter. However, a small fraction of the total oxygen consumed is not completely reduced and yields species such as superoxide anions, hydroxyl free radicals, and hydrogen peroxide all of which can be potentially damaging to respiring cells. Because of their toxicity, they are referred to as active oxygen species, being radicals or radical precursors. Other active oxygen species are: singlet oxygen, and organic peroxyl radicals and hydroperoxides. They are responsible of radical processes occurring in living cells and organisms, in the absence of radiation, during the course of many physiological and pathological phenomena including inflammation, ageing, carcinogenesis, drug action and drug toxicity, and others.

Being very reactive, free radicals will obviously react with all cell components, producing chemical modifications and damages of proteins lipids, carbohydrates, and nucleic acids. These reactions will generate a variety of metabolic and cellular disturbances, leading to various end points such as cell killing, cell mutation, and alterations of cell membrane functions, namely permeability, transport processes, antigenic characters, and so on.[4,9,10] Radical reactions with lipids are particularly important, because polyunsaturated lipids, by reaction with radicals, can originate a chain peroxidative process. Such process will mainly occur in membranes where polyunsaturated lipids are abundant. Any radical promoted hydrogen abstraction from a methylene group will form a carbon-centered radical, which then easily reacts with molecular oxygen to give a peroxyl radical which again can abstract a hydrogen atom from another lipid molecule to give a lipid hydroperoxide and another carbon-centered radical. Once the process is initiated, it tends to continue. Lipid hydroperoxides are fairly stable at physiological temperature, but rapidly decompose in the presence of transition-metal complexes. It should be noted at this point that many metal complexes capable of doing this are present *in vivo*.

The fate of lipid peroxyl radicals does not only involve hydroperoxides. Alternative pathways do exist, even if not all of them have been as yet completely identified, leading to different end products. Malondialdehyde is the main defined reaction product measurable in biological systems which undergo lipid peroxidation. In any case, whatever pathway is followed, radicals are produced which maintain the radical chain peroxidation reaction.

The defense against oxidizing radicals takes place in living matter in various different ways. GSH, of course, plays an important role among antioxidant systems. It can act by itself reacting with radicals via electron or hydrogen transfer, or by simply intercepting radicals able to initiate biologically relevant reactions; or else through enzyme catalyzed reactions. Glutathione peroxidases, in fact, can either remove hydrogen peroxide, which could react with ferrous ions to give hydroxyl radicals, or react with a variety of hydroperoxides reducing them to alcohols. Together with glutathione, other important antioxidant compounds are vitamin E and C, and the enzymes superoxide dismutase and catalase.

Obviously, the reactions of glutathione with radicals generate glutathione radicals. The most common of these is the thiyl radical (GS$^\bullet$) which is originated in all electron or hydrogen transfer reactions, as well as in the reaction of GSSG with the hydrated electron (via the disulfide radical anion):

$$GSSG + e^-_{aq} \rightarrow (GS\cdot^\bullet\cdot SG)^- \tag{5}$$

$$(GS\cdot^\bullet\cdot SG)^- \rightleftarrows GS^\bullet + GS^- \tag{6}$$

It seems reasonable to postulate, therefore, that significant amounts of thiyl radicals are produced in living matter, in all circumstances where the GSH/GSSG system is involved in radical processes. As a consquence, the fate of thiyl radicals in living systems appears to be a relevant point to discuss.

Thiyl radicals can, of course, react with each other to give GSSG, however, in living matter, radical-radical reactions are not very probable, so that other, more likely reactions, should be considered. In the absence of oxygen, the back reaction of (6) has a high probability, because of the favourable equilibrium constant, however, at physiological pH the dissociation of GSH is only a few percent (equilibrium 6 requires the thiolate anion).

Thiyl radicals may also accept electrons, so that, for example, reaction (7), between GS$^{\bullet}$ and ascorbate, present in animal tissues in the millimolar range, is likely to occur. The redox potential of the former is 0.7-1 V and that of the latter 0.3 V.[11,12]

$$GS^{\bullet} + AH^- \rightarrow GS^- + A^{\bullet -} + H^+ \qquad (7)$$

GS$^{\bullet}$ and other thiyl radicals are generally considered very unreactive with respect to abstraction of hydrogen atoms from other molecules and from biologically important molecules in particular. While exceptions can occur under particular experimental conditions,[11] the above statement finds substantial support in model experiments in which the inactivation by ionizing radiation of simple biological systems (such as biologically active DNA and enzymes in solution; or virus particles in suspension) has been investigated in the presence of GSH and of other thiol compounds. A constant protective effect by thiols has been found in most cases with respect to the effect observed in their absence, and little if any evidence for a secondary interaction of thiyl radicals with target molecules.[13] The protection can be explained by the scavenging of inactivating radicals, and, possibly, by target radical repair through hydrogen donation from thiols.

The fate of thiyl radicals in the presence of oxygen is still the object of discussion. What is known for sure is that oxygen strongly affects the radiolysis of thiols, increasing the yield of their consumption in irradiated solutions, with G(–RSH), up to values suggesting short chain reactions.[14,15] At physiological pH, G(–GSH) is around 20, and about 90% of the GSH consumed is converted to GSSG, the other products being mainly glutathione sulfinic acid and γ-glutamylserylglycine. At low pH, G(–GSH) amounts to about half of that measured at pH 7, while that for the production of sulphinic acid is triplicated and that for hydrogen peroxide production rises from 0 to about 1.[12] This is evidence for oxygen being consumed during the radiolysis of thiols, even though direct measurements of such consumption are not available. However, pulse radiolysis experiments on thiols in solution show that oxygen inhibits, linearly with its concentration, the formation of the strong absorption due to the disulfide radical anion formed in the back reaction (6). The formation of thiyl peroxyl radicals according to reaction 8 has therefore been postulated, implying competition between reaction (8) and the reverse of reaction (6), as shown below with reference to GSH:[16]

$$GS^{\bullet} + O_2 \rightleftarrows GSOO^{\bullet} \qquad (8)$$

$$GS^{\bullet} + GS^- \rightleftarrows (GS\cdot^{\bullet}\cdot SG)^- \qquad (9)$$

It is interesting to note that there are experimental indications suggesting that reaction (8) is also an equilibrium reaction. However, this suggestion is somewhat controversial.[17,18] In fact, the indications mentioned above are based on the assumption that the weak transient absorption with maximum at 540 nm, which appears when GSH is pulse irradiated at neutral or acid pH is due to GSOO$^{\bullet}$. Other thiols, such as mercaptoethanol and cysteine show similar spectra with the respective maxima at 560 and 530 nm.[14] Now, while various authors agree on the existence of reaction (8), even though only towards the right hand direction, the assignment of the absorption in the visible region to thiol peroxyl radicals has been questioned on the basis of quantum machanical calculations ([11] and Guerra and Chatgilialoglu in this book). If this is true, kinetic measurements carried out using this absorption might not directly reflect the properties of thiol peroxyl radicals. A valid contribution to the elucidation of this point can probably be offered by the work reported by Sevilla and coworkers[19] who have recently been able to dectect and characterize, by ESR investigation of frozen aqueous glasses, the cysteine thiyl peroxyl radical. The radical, showing an ESR spectrum with g-values typical for a peroxyl radical, appears in oxygen saturated samples irradiated at 77 K and is annealed at 155-160 K. Samples showing the ESR spectrum of cysteine thiyl peroxyl radicals have also an intense violet colour and and absorb visible light with a maximum at 540 nm just like the transient species shown by pulse radiolysis experiments. Evidence is also shown that on further annealing the peroxyl radical converts to the corresponding sulfinyl radical ($RSO^\bullet$). These findings probably do not solve all the questions, but certainly provide further support to the postulate that oxygen reacts with thiyl radicals whichever the way by which they are produced.

Much less controversial is the reaction with oxygen of disulfide radical anions, including that of glutathione:

$$(GS{\cdot}^{\bullet}{\cdot}SG)^- + O_2 \quad \rightarrow \quad GSSG + O_2^{\bullet-} \tag{10}$$

There is little doubt that the reactions of thiyl radicals with oxygen can have important biological consequences. This can be generically suggested by various different indications in biological experimental systems, or by the observations from cases of inherited diseases altering the glutathione metabolism.[3,4,11] However, detailed molecular mechanisms are still far from being identified even in relatively simple systems like, for instance, DNA or enzymes in solution.[13] Much more research will be needed to achieve such a goal.

ACKNOWLEDGEMENT

Personal research work reported in this paper was partially supported by a grant from the Italian National Research Council (CNR), Special Project "Oncology".

REFERENCES

1. N.S. Kosower, and E.M. Kosower, *Int. Rev. Cytol.* 54:109 (1978).
2. D.M. Ziegler, *Ann. Rev. Biochem.* 54:305 (1985).
3. O.F. Nygaard and M.G. Simic, eds., "Radioprotectors and Anticarcinogens" Academic Press, New York and London (1983).
4. A. Larsson, S. Orrenius, A. Holmgren, and B. Mannervik, eds., "Functions of Glutathione: Biochemical, Physiological, Toxicological, and Clinical Aspects." Raven Press, New York (1983).
5. A. Meister, A., ed., Glutamate, Glutamine, Glutathione and Related Compounds *in*: "Methods in Enzymology", S.P. Colowick and N.O. Kaplan, eds., Vol. 113, Academic Press, Inc., New York and London (1985)
6. A. Meister and M.E. Anderson, *Ann. Rev. Biochem.* 52:711 (1983).
7. B. Mannervick and U.H. Danielson *CRC Crit. Rev. Biochem.* 23: 283 (1988).
8. J.B. Mitchell, J.A. Cook, W. DeGraff, E. Glatstein, and A. Russo, *Int. J. Radiat. Oncol. Biol. Phys.* 16: 1289 (1989).
9. B. Halliwell and J.M.C. Gutteridge, "Free Radicals in Biology and Medicine." Clarendon Press, Oxford (1985).
10. H. Sies, "Oxidative Stress", Academic Press, London (1985).

11. C. von Sonntag, "The Chemical Basis of Radiation Biology", Taylor & Francis, London (1987).
12. M. Lal, *Can. J. Chem.* 54: 1092 (1976).
13. M. Quintiliani, *Int. J. Radiat. Biol.* 50: 573 (1986).
14. J.E. Packer, The radiation chemistry of thiols, *in*: "The Chemistry of the Thiol Group", S. Patai, ed., John Wiley & Sons, London (1974).
15. M. Quintiliani, R. Badiello, M. Tamba, and G. Gorin, Radiation chemical basis for the role of glutathione in cellular radiation sensitivity, *in*: "Modification of Radiosensitivity in Biological Systems", IAEA, Vienna, pp. 29 (1976).
16. M.Z. Baker, R. Badiello, M. Tamba, M. Quintiliani, and G. Gorin, *Int. J. Radiat. Biol.* 41: 595 (1982).
17. M. Tamba, G. Simone, and M. Quintiliani, *Int. J. Radiat. Biol.* 50: 595 (1986).
18. J. Mönig, K.-D. Asmus, L.G. Forni, and R.L. Willson, *Int. J. Radiat. Biol.* 52:589 (1987).
19. M.D. Sevilla, M. Yan, and D. Becker, D., *Biochem. Biophys. Res. Commun.* 155: 405 (1988).

INDEX

Iproniazid, 432
Iron complexes, 355
Isodesmic equation, 13
Isomerization
  cis-trans, 377
Isoniazid, 432
Isotopomers, 243

Kinetics, 327
  time-resolved measurements, 110, 111,
    112, 158, 369, 397, 419
Kornblum's rule, 29

Lactoperoxidase, 403
Laser flash photolysis, 109, 118, 164
Linoleic acid, 371, 382
Linolenic acid, 371, 382
Lipid peroxidation, 367, 440
Lipoamide, 365
Lipoic acid, 129, 148, 378
Lipoxygenase
  inhibitory activity, 288
2-Lithiobenzenethiol synthons, 272
α-Lithiomethanethiol synthons, 272
Lone pair-lone pair interaction, 231
"Long distance" electron transfer, 356

Malondialdehyde, 440
Marcus Theory, 74, 90
Markovnikov, 202
anti-Markovnikov, 202
Mass spectrometry, 185
Mercaptoethanol, 372, 442
2-Mercaptophenylphosphine, 277
Metabolites
  free radical, 401
Metal chalcogenide, 104, 105
Metal complexes of TTCN, 355
Metal oxide, 104, 107, 108
Methanesulfenyl chloride, 198
Methionine
  decarboxylation, 147
  oxidation, 127, 389
Methyl acrylate, 377
6-endo-Methylthiobicyclo-
    [2.2.1]heptanes (2-endo-
    substituted), 227
Methylviologen, 110
Metiazinic acid, 379
MINDO/3 calculations, 1, 3
Misonidazole, 417
MNDO calculations, 1, 3
Møller-Plesset, 5, 13
Molecular dynamics, 1
Molecular mechanics, 1, 2
Molecular orbital calculations, 1, 3

Molybdenum-thiolate complexes, 237, 269,
    278, 280
Molybdoenzymes, 269
Muconic acid, 381
Multiple Scattering Xα (MSXα), 7, 32

NADH, 380, 393
Naphthalene
  1,8-"peri", 235
Natural bond orbital (NBO), 13
Neighboring group participation, 89, 213
"Nernstian" behavior, 108
Neutralization/reionization mass
    spectrometry (NRMS), 185
Nickel complexes, 272
p-Nitroacetophenone (PNAP), 368
Nitrogenase enyzmes, 257
Nitrogen fixation, 269
Nitroxyls, 420
Norbornane derivatives, 89, 170, 224, 227
Nucleic acids, 411
Nucleophiles
  reactivity of sulfur-centered nucl., 17
  n-, π-, 200
Nucleophilic attack, 83
Nucleotides, 411

OH$^\bullet$ radicals, 121, 379, 391
Oleic acid, 371, 382
Onion, 283
  lachrymatory factor, 284
d-Orbital participation, 14
Organometallic reagents, 291
Organosulfur carbanions, 257
"Overpotential", 88
5-Oxoprolinase, 437
5-Oxoproline, 437
L-2-Oxothiazolidine,-4-carboxylate, 438
Oxygen effect, 410
Oxygen fixation, 411
Oxyl radicals, 360

Papain, 348
Participation
  neighboring group, 89. 213
  neighboring thioether, 216
  d-orbital, 14
Peak potentials, 355
Penicillamine, 348, 369, 403
Pentadienyl-radicals, 371
  absorption spectrum, 373
Peptides, 389
  sulfur-containing, 389
  structure of radicals, 396
Perester decomposition, 164, 215, 217
Peroxidase, 403